ELECTRICAL MACHINES
DIRECT AND ALTERNATING CURRENT

A 62,500-kw outdoor steam turbine generator unit of the Florida Power and Light Company. (*Allis-Chalmers Mfg. Co.*)

Other books by Charles S. Siskind:

Electrical Circuits: Direct and Alternating Current, 2d ed
Electricity: Direct and Alternating Current, 2d ed
Electrical Control Systems in Industry

ELECTRICAL MACHINES

DIRECT & ALTERNATING CURRENT

CHARLES S. SISKIND
*Late, Associate Professor of Electrical Engineering
Purdue University*

SECOND EDITION

McGRAW-HILL BOOK COMPANY
New York   St. Louis   San Francisco   London
Toronto   Sydney   Mexico   Panama

ELECTRICAL MACHINES
DIRECT AND ALTERNATING CURRENT

Copyright © 1959 by the McGraw-Hill Book Company, Inc.

Copyright 1950 by the McGraw-Hill Book Company, Inc. Printed in the United States of America. All rights reserved. This book, or parts thereof, may not be reproduced in any form without permission of the publishers.

Library of Congress Catalog Card Number: 58-10008

19 20 21 22 23 - MAMM - 7

ISBN 07-057728-5

PREFACE

Although the scope and character of this volume are essentially similar to those of the first edition, it does include numerous changes and additions of subject matter. Moreover, since modifications were made, in part, at the suggestion of practicing engineers, technicians, and educators, it is both significant and gratifying to note that the text has been widely used in industry as well as in formal courses of instruction in technical schools, colleges, and universities.

Because of the great importance of *control methods* in the operation of modern direct- and alternating-current motors, they are given special treatment in the second edition. Such related topics as acceleration, reversing, dynamic braking and plugging, regeneration, and speed control are discussed in considerable detail, particularly in connection with the many kinds of manual and automatic controllers. Also added is a completely new section, in Chapter 12, devoted to the subject of *rectifiers;* these include the various metallic and arc-discharge types of equipment such as copper-oxide, selenium, and silicon rectifiers, gas tubes, and mercury-arc tubes and tanks. Furthermore, all theoretical and practical discussions are supplemented by helpful wiring diagrams and sketches as well as by carefully selected photographs.

An extremely valuable aspect of the book is the solution of a large number of illustrative examples; these generally follow derived equations and emphasize the usefulness of mathematical relationships to analyze and predict machine performance. Many new examples are solved in this edition. Moreover, each chapter is concluded by a set of questions and problems that follow the text material closely for assignment and home-study purposes; these portions of the book have also been considerably enlarged.

Finally, many parts of the book have been reworded, modified, and rewritten where, it was felt, such changes helped to clarify the principles and practices of energy conversion equipment.

CHARLES S. SISKIND

CONTENTS

PREFACE v

INTRODUCTION

CHAPTER 1. ELECTRICAL MACHINERY GENERALIZATIONS . . . 1
Rotating electrical machines—Armature windings—Field poles—Types of direct-current generator—Voltage characteristics of direct-current generators—Speed of direct-current generators—Alternating-current generators—Types and characteristics of direct-current motors—Starting direct-current motors—Commutating poles for direct-current machines—Types and characteristics of alternating-current motors—Starting alternating-current motors—Transformers

PART I. ELECTRICAL MACHINES—DIRECT CURRENT

CHAPTER 2. DIRECT-CURRENT GENERATOR AND MOTOR PRINCIPLES 23
Principle of generator action—General voltage equation for direct-current generator—Direction of generated voltage—The elementary alternating-current generator—The commutation process—Principle of motor action—Force and torque developed by direct-current motors—Commutation in direct-current motors—Main fields in direct-current machines

CHAPTER 3. DIRECT-CURRENT DYNAMOS, CONSTRUCTION AND ARMATURE WINDINGS 45
Generator and motor construction—Types of armature winding—Coil span for all types of winding—Commutator pitch for lap windings—Parallel paths in simplex- and multiplex-lap windings—Simplex-wave windings—Number of parallel paths in simplex-wave windings—Multiplex-wave windings—Armatures with more commutator segments than slots—Dead, or dummy, elements in armature windings—Equalizer connections for lap windings—Frog-leg windings

CHAPTER 4. DIRECT-CURRENT GENERATOR CHARACTERISTICS . . 94
Types of direct-current generator—No-load characteristics of generators—Building up the voltage of a self-excited shunt generator—Behavior of a shunt generator under load—Controlling the terminal voltage of shunt generators—Compound generator operation under load—Degree of compound-

ing adjustment—Series generator behavior under load—Armature reaction in direct-current generators—Interpoles for direct-current generators—Compensating windings for direct-current generators—Commutation and reactance voltage—Need for operation of generators in parallel—Operation of shunt generators in parallel—Operation of compound generators in parallel

CHAPTER 5. DIRECT-CURRENT MOTOR CHARACTERISTICS . . . 134

Operating differences between motors and generators—Classification of direct-current motors—Counter electromotive force (counter emf); voltage generated by a motor—Starting a direct-current motor—Starters for shunt and compound motors—Controllers for series motors—Controllers for shunt and compound motors—The automatic starter for shunt and compound motors—Loading a motor; effect upon speed and armature current—Torque characteristics of direct-current motors—Speed characteristics of direct-current motors—Speed regulation of direct-current motors—Differential-compound motors—Speed control of direct-current motors—Armature reaction in direct-current motors—Reversing the direction of rotation of direct-current motors

CHAPTER 6. EFFICIENCY, RATING, AND APPLICATIONS OF DYNAMOS 183

Power losses in dynamos—Efficiency of direct-current generators—Efficiency of direct-current motors—Importance of efficiency—Rating of generators and motors—Selection of generators and motors—Special dynamos and applications—Dynamotors—Booster systems—Electrical braking of direct-current motors—Series-parallel control of railway motors—Direct-current dynamo applications

PART II. ELECTRICAL MACHINES— ALTERNATING CURRENT

CHAPTER 7. ALTERNATING-CURRENT GENERATORS 225

Alternator construction—Frequency of alternating-current generators—The revolving field—The stator—Generated voltage in an alternator—Armature windings for alternators—Coil pitch and pitch factor—Distribution factor—Corrected voltage of an alternator—Alternator regulation—Voltage drops in alternator armatures—Alternator phasor diagram—Synchronous reactance and synchronous impedance—Alternator efficiency—Operation of alternators in parallel

CHAPTER 8. TRANSFORMERS 268

Transformer action—Transformer construction—Transformer voltages and the general transformer equation—Voltage and current ratios in transformers—Ratio of transformation—Tapping a transformer—Loading a transformer—Regulation calculations using voltage values—Leakage reactance—Equivalent resistance, reactance, and impedance—Equivalent circuit of a transformer—The short-circuit test—The open-circuit test—Regulation calculations using short-circuit data—Efficiency calculations using short-circuit and open-circuit data—Maximum efficiency—All-day efficiency—Autotransformers—Instrument transformers—Transformer polarity—Parallel operation of transformers—Three-phase transformer connections—Three-phase transformers—The constant-current transformer—The induction voltage regulator

CONTENTS

CHAPTER 9. POLYPHASE INDUCTION MOTORS 346

Induction motor principle—The stator—The rotor—The stator winding—The revolving field—Slip and rotor speed—Generated voltage and frequency in a rotor—Rotor current and power—Rotor torque—Starting torque—Induction-motor efficiency—The blocked-rotor test—Starting induction motors—Induction motor starting methods—Operating characteristics of squirrel-cage motors—Operating characteristics of wound-rotor motors—Speed control of induction motors—Electric braking of alternating-current motors

CHAPTER 10. SINGLE-PHASE MOTORS. 401

Types of single-phase motor—The universal (series) motor—The shaded-pole motor—The reluctance-start motor—The split-phase motor—The repulsion-start induction motor—The repulsion motor—The repulsion-induction motor—The reluctance motor—The hysteresis motor—Automatic starters for split-phase motors

CHAPTER 11. SYNCHRONOUS MOTORS. 449

General facts concerning synchronous motors—Synchronous motor construction—Exciters—Starting synchronous motors—High-starting-torque synchronous motors—Principles of operation—Loading a synchronous motor—Power-factor adjustment of synchronous motors—The synchronous condenser—The dual-purpose synchronous motor—The synchronous-induction motor—Hunting and damping of synchronous motors—The supersynchronous motor—Synchronous motor applications

CHAPTER 12. CONVERTERS AND RECTIFIERS 481

Converters—types and uses—Synchronous converter construction—The single-phase converter—The polyphase converter—Heating and ratings of synchronous converters—Starting synchronous converters—Transformers and transformer connections for converters—Armature reaction and commutating poles in converters—Controlling the direct-current voltage of a converter—Parallel operation of converters—The inverted converter—Frequency converters—Phase-converters—Types of rectifiers—Half-wave and full-wave rectification—The copper-oxide rectifier—The selenium rectifier—The hot-cathode gaseous rectifier—The mercury-arc rectifier—The two-anode single-phase mercury-arc rectifier—Multianode three-phase mercury-arc rectifiers—Ignitron rectifiers

APPENDIX 1. NATURAL SINES, COSINES, TANGENTS, AND COTANGENTS 544

APPENDIX 2. LOGARITHMS OF NUMBERS 548

INDEX 551

ANSWERS 561

CHAPTER 1

ELECTRICAL MACHINERY GENERALIZATIONS

Rotating Electrical Machines. Rotating electrical machines are widely used for the purpose of converting energy from one form to another. The two most frequently used types of such machines are *generators* and *motors*. In the first of these, the generator, mechanical energy is converted into electrical energy. In the motor, electrical energy is converted into mechanical energy. Two other types of rotating electrical machines, not used so often as generators and motors, are *rotary converters* and *frequency converters*.

When an electric generator is in operation, it is driven (rotated) by a mechanical machine usually called a *prime mover*. The latter may be a steam turbine, a gasoline engine, an electric motor, or even a hand-operated crank. As will be pointed out later, *generator action can take place when, and only when, there is relative motion between conducting wires* (usually copper) *and magnetic lines of force*.

When an electric motor is in operation, it is supplied with electrical energy and develops torque, that is, a tendency to produce rotation. And if the rotating element of the motor is free to turn, it will do so and thereby cause mechanical rotation of itself and its application.

All rotating electric generators consist essentially of two important parts: (1) an even set of electromagnets or permanent magnets and (2) the laminated steel core containing current-carrying copper wires, the latter being called the *armature winding*. In the d-c generator, the armature winding is mechanically rotated through the stationary magnetic fields created by the electromagnets or permanent magnets; in the a-c generator, the electromagnets or the permanent magnets and their accompanying magnetic fields are rotated with respect to the stationary armature winding. In the d-c motor, current is sent *into* the armature winding, the latter being placed inside a set of radially supported magnet poles.

Figure 1 shows the *field* structures of a d-c generator and an a-c machine. The large multipolar frame in the rear is for a 3,000-kw (kilowatt) 600-volt 250-rpm (revolutions per minute) generator and clearly shows the many main and commutating poles (discussed later) bolted to the outside yoke; also visible is the brush rigging on the far side. In the foreground is an

18-pole field for a 3,500-kw 6,600-volt three-phase 60-cycle 400-rpm a-c machine. Note particularly the two rings mounted on the shaft; the two ends of the entire field winding are connected to these rings so that stationary brushes riding on the latter can "feed" direct current into the rotating structure. Figure 2 shows a laminated stationary stator core and its armature winding for a 1,330-kw 480-volt three-phase 60-cycle

FIG. 1. Field structures for a d-c generator (*rear*) and an a-c machine (*foreground*). (*Allis-Chalmers Mfg. Co.*)

225-rpm a-c generator. A field similar to that depicted in the foreground of Fig. 1 would be rotated *inside* such a frame as this. A completely assembled d-c motor rated at 400 hp, 230 volts when operating at 250 rpm is illustrated in Fig. 3. The armature commutator and the carbon brushes are plainly visible on the near side.

In the a-c motor, current is sent into the armature winding, which is usually placed in a stationary laminated iron core; the rotating element may or may not be a set of magnet poles. Since there are many kinds of a-c motor construction, no general statement can be made concerning them as can be done in the case of d-c motors. The machine shown in

Fig. 2 Wound stationary armature for a low-speed a-c generator. Note the split stator frame. (*General Electric Co.*)

Fig. 3. Completely assembled d-c motor rated at 400 hp, 230 volts, 250 rpm. (*Allis-Chalmers Mfg. Co.*)

Fig. 4 is of the so-called *synchronous* type, in which a small d-c generator is placed on the shaft extension to supply d-c excitation to the rotating field.

In the rotary converter, electrical energy of one form is changed into electrical energy of another; the usual arrangement is to change a-c energy into d-c energy, although the reverse is sometimes done. To accomplish this remarkable change in a *single* rotating machine, the input (a-c energy, for example) is first converted into mechanical energy, so that the rotating

Fig. 4. A-c synchronous-type motor rated at 700 hp, to be operated at 500 rpm when connected at a 6,000-volt 50-cycle three-phase source. The small generator on the shaft extension is of 5-kw capacity. (*Allis-Chalmers Mfg. Co.*)

part functions as an electric motor; the resulting rotation then causes the machine to become a generator, thereby converting mechanical energy into d-c electrical energy. It should be understood that the a-c energy input first produces motor action, that is, mechanical energy; the mechanical motion of the revolving element then develops generator action, causing it to deliver d-c energy. Figure 5 shows a 300-kw 275-volt rotary converter, the operating speed of which is 1,200 rpm. In the center portion are the yoke frame and its six poles, inside of which rotate the armature core and its winding; to the right may be seen the six rings into which the six-phase alternating current is "fed," while to the left are the commutator and the six sets of brushes *from which* direct current is "delivered" to the external load.

Fig. 5. A 300-kw 1,200-rpm 275-volt rotary converter for automatic operation. (*General Electric Co.*)

The frequency converter has very limited application; its function is to change a-c electrical energy at one frequency into a-c electrical energy at another frequency. In the usual arrangement for such a change of frequency, *two* rotating machines are directly coupled together; one of them operates as an a-c motor when connected to an a-c source having a given frequency, while the other, *driven*, machine functions as an a-c generator to deliver electrical energy at some other frequency. Figure 6

Fig. 6. A small frequency-converter set. (*The Louis Allis Co.*)

illustrates a frequency-converter set mounted on a base. The machine on the left operates as an a-c motor and drives the one on the right. The stationary part of the latter is also connected to an a-c source, usually the same supply to which the motor is connected, but alternating current at a different frequency may be "taken" from the rings and brushes (visible in the photograph).

It should be clear, therefore, that in every type of rotating electrical machine there is always an actual conversion of energy from one form, electrical (or mechanical), to another form, mechanical (or electrical).

Armature Windings. The armature windings of all types of motors and generators, whether of direct or alternating current, are always wound on *laminated* steel cores of good magnetic permeability. And, strange as this may seem to the student, it is nevertheless true that *the current in the armature windings of all motors and generators*, whether of direct or alternating current, *is always alternating.* Alternating voltages are always generated in the *windings* of a-c and d-c generators; in the a-c generator, the generated alternating electromotive force (emf) is transmitted directly to the load; in the d-c generator, the generated alternating emf is first rectified by a commutator and its brushes, that is, changed to direct current, *before* it is transmitted to its load. The a-c motor receives its energy directly from an a-c source and, without any change whatever in form, uses it as alternating current in its winding to develop torque. In the d-c motor, however, direct current is delivered to the brushes but flows as alternating current in the armature winding *after* passing through the brushes and commutator. This extremely important commutation process will be discussed fully in Chap. 2.

Field Poles. The electromagnets (called *field poles*) used in all d-c generators and motors, in a-c generators, and in one type of a-c motor are very simple in construction. There are always an even number of them in a given machine, and each one consists of a laminated steel core, of rectangular cross section, surrounded by one or more copper coils. One face of the steel core is concentric with the laminated armature core and has a larger cross-sectional area than the portion around which the copper coil of wire is placed. The spread-out portion of the pole core, or *shoe*, permits the magnetic flux to enter the armature core over a wider area than would be possible with a core having straight sides. When the field structure is assembled for a stationary-field type of machine, the electromagnets are bolted to a yoke ring, as in Fig. 7, so that they project radially *inward* toward the rotating armature. In the rotating-field type of machine driven by a slow-speed prime mover, the electromagnets are bolted to a hub fastened to the shaft, as in Fig. 8, so that they project radially *outward* toward the stationary armature core; this construction is called a *salient-pole field.* When the alternator is driven by a high-speed turbine, the field

Fig. 7. Assembled field structure for a d-c motor. Note the four large (main) pole cores and their windings and the two narrow pole cores and their windings. (*The Louis Allis Co.*)

winding is placed in a slotted core; this construction is called a *non-salient-pole field*.

Types of Direct-current Generator. Practically speaking, there are only two general types of d-c generator. They are distinguished by the way in which the flux is produced by the electromagnets (assuming the usual electromagnetic excitation); the type of generator is, however, absolutely independent of the manner in which the armature winding is placed on the armature core and connected to the commutator.

Shunt Generators. If the excitation is produced by a single winding connected to its own positive and negative brushes, the machine is called a *self-excited shunt generator*, or, briefly, a *shunt generator*. In some special cases in which the excitation is produced by a single winding connected to the positive and negative bus bars fed by *another* d-c generator, the machine is called a *separately excited shunt generator*. Although their operating characteristics differ slightly, self- and separately excited shunt generators are usually classified under the same heading, "shunt generator," because their field windings are constructed and excited similarly. The latter point will become clearer later, when it will be shown that the shunt-field winding contains many turns of fine wire, has a comparatively high resistance, and, when connected directly to the full-voltage source of a generator will take a relatively low current, entirely independent of the load or armature current.

Compound Generators. The second type of generator has two complete sets of field windings for excitation purposes: (1) the shunt field and (2) the series field. Both field windings are placed over one set of pole cores and act together to create a common set of magnetic fluxes. Such a

FIG. 8. Assembled field structure (14 poles) for an a-c generator. Note the two slip rings connected to the field winding. Direct current is fed to the winding through brushes and to the rings from an outside source. (*Westinghouse Electric Corp.*)

machine is called a *compound generator*. Its shunt field is, in every respect, exactly like that used in the shunt generator. Its series-field winding, however, differs greatly from the shunt-field winding because it contains very few turns of heavy wire, has an extremely low resistance, and is excited by the comparatively large load current. Since the two types of generator differ only by the addition or omission of a series field, it is quite possible to operate a compound generator as a shunt generator by completely disconnecting the series field from the machine or to change a shunt generator into a compound generator by properly winding a series field directly over the shunt field and connecting it to operate with the latter.

In some special installations in which it is used for purposes of control, a generator contains a series field only; it is then called a *series generator*. Such machines are not often used, but when they are, it is usually in conjunction with other equipment such as shunt or compound generators and batteries. They are discussed in Chap. 6 (see Figs. 130 and 131).

Voltage Characteristics of Direct-current Generators. The most important characteristic of a generator is its voltage behavior with respect to loading. When the armature of a machine is delivering no current to a consuming device, it is said to be operating at *no load;* it will, of course, be sending a comparatively low value of current into its own shunt field if it is a self-excited shunt generator. If the armature is delivering rated (name-plate) current to one or more electrical consuming devices, the machine is said to be operating at *full load.* Now then, if a shunt generator is operating at full load at a given voltage (rated voltage, for example) and the load is suddenly removed so that the machine is performing at no load, the voltage will always rise above the full-load value. That is, there will be a *change* in voltage between full load and no load. A similar removal of the load on a compound generator may have almost any effect upon the voltage, the change or lack of change depending upon several factors, the most important of which are the number of series-field turns and the way in which the series field is connected with respect to the shunt field. When full load is removed from a compound generator, the voltage may drop, remain constant, or rise. Just why compound generators behave in the manner indicated will be considered in detail in Chap. 4.

Speed of Direct-current Generators. It is customary for a d-c generator, whether shunt or compound, to operate at a speed that remains substantially constant at all times. There are, however, a few exceptions to this practice, as, for example, in automobile and train generators, the speeds of which vary over an extremely wide range; in such applications, generators must be specially designed and supplemented by other equipment, such as batteries and control devices, if they are to function properly. Obviously, the speed of a generator is determined and controlled by the machine that drives it, that is, its prime mover. Therefore, if the prime mover is a constant-speed machine such as a turbine, a steam engine, a gas engine, a water wheel, or a synchronous motor, the generator speed will be constant. And since this is the most desirable and efficient method of operating a d-c generator, it is customary to maintain its speed at a definite predetermined value at all times. Strictly speaking, slight speed changes are generally permissible and even expected, and such departure from absolute constancy in speed is not particularly objectionable.

If, for some reason, a d-c generator is operated at a speed much **higher** or lower than that intended by the manufacturer, it will usually not perform with complete satisfaction unless corrective measures are taken to forestall faulty operation. For example, a lower speed will cause a generator to overheat, while a higher speed will give rise to poorer commutation, that is, sparking at the commutator, and wider voltage changes with variations in load. Furthermore, the fields of d-c generators operating at speeds other than those recommended by the manufacturer would have to be modified to accommodate such changes; for example, if the speed of a generator were to be *increased without a change in voltage*, it would be necessary to *weaken* the field.

Alternating-current Generators. A-c generators, generally called *alternators*, are nearly always constructed so that the armature core and its winding are stationary (Fig. 2) while the field poles rotate (Fig. 8). Note that this is just the reverse of the practice with d-c generators. Moreover, the speed of rotation of the field *must be kept absolutely constant* for the very good reason that the frequency of the generated voltage in cycles per second, which is directly proportional to the speed, must be maintained at a constant value at all times. This requirement, therefore, imposes a very important obligation on the prime mover, the speed of which must not change, even slightly, if the frequency is to be maintained at a constant value.

The voltage developed by an a-c generator varies greatly with changes in load; as the load increases, the voltage tends to fall. This tendency for the voltage to fluctuate is an inherent one and cannot be corrected by the addition of another field winding, a series field, as is done in d-c generators. To maintain a constant voltage *at the load*, it is usually customary to equip an a-c generator with a regulator, a device that tends to maintain the terminal voltage constant regardless of the load. The latter functions in such a manner that a change in the a-c voltage (a decrease, for example) is immediately accompanied by a corresponding increase in the d-c flux, thereby compensating for those factors that are responsible for the downward voltage trend.

A-c generators are usually constructed to generate much higher voltages than are generators built for d-c service. Voltages as high as 2,300, 4,600, 6,600, and 13,200 are quite common in a-c machines, whereas 115, 230, and 600 volts have been standardized for d-c systems, with 1,200- and 1,500-volt service used occasionally in special installations such as interurban railways. Furthermore, the maximum kilowatt capacity of a-c generators is considerably greater than that of generators constructed for d-c operation; a-c generators have been built in sizes exceeding 200,000 kw. whereas d-c machines having outputs of 5,000 to 10,000 kw would be considered very large. There are two reasons why a-c machines can be built in large sizes and made to develop high voltages:

1. No commutator is required, a commutator being a very definite limiting factor in the construction and operation of a d-c generator.

2. The armature winding can be placed in a stationary part of the machine, the stator, where it is possible to provide good insulation strength for the high-voltage winding. And since it is more economical, on the basis of cost per kilowatt capacity, to construct large machines, it should be clear that a-c generating equipment is generally more desirable than d-c machines.

One of the most important reasons for the use of a-c generators in comparatively large power systems is that alternating current can be transformed efficiently from one voltage to another by the use of transformers. At the beginning of a long transmission line, the voltage can be raised to a high value, say, 110,000, 167,000, or 220,000 volts. At such high voltages, electrical energy can be transmitted much more efficiently and over greater distances than would be possible at low d-c voltages, which cannot be transformed economically at present. And at the receiving end of the system, the high voltage is reduced to a safe value, again by the use of transformers, so that the energy may be used in the factory, the farm, the office, and the home. On the other hand, it should be stated that d-c service is sometimes more economical and satisfactory where there is a large concentration of load requiring the operation of d-c motors. This is particularly true in the business districts of large cities, where loads such as elevators are best served by d-c motors and controls.

Types and Characteristics of Direct-current Motors. There are three general types of d-c motor, and these, like d-c generators, are classified on the basis of the kind of excitation used.

Shunt Motor. If a comparatively *high-resistance* field winding of many turns of fine wire is employed for this function, it is connected in parallel with (*in shunt with*) the armature; such an arrangement of an armature and field winding in shunt with each other is called a *shunt motor*. Figure 7 shows the field of such a motor.

Series Motor. When an extremely *low-resistance* field winding of very few turns of heavy wire is used, it is connected *in series* with the armature; this arrangement of an armature and field winding in series with each other is called a *series motor*.

Compound Motor. A machine that is excited by a combination of a shunt field (many turns of fine wire) connected in shunt with the armature and a series field (few turns of heavy wire) in series with the armature is referred to as a *compound motor*.

Note particularly that it is the kind of excitation provided by the field, and nothing more, that differentiates one type of motor from another; the construction of the armature has nothing whatever to do with the type of motor.

Each type of motor has very definite operating characteristics in so far as starting torque, overload capacity, and speed variation with changes in load are concerned. All these points will be discussed fully in connection with other subjects, but it should be mentioned here that the following general operating characteristics distinguish the three types of motor from one another:

1. On the basis of the same horsepower and speed rating, a series motor develops the highest starting torque and the shunt motor the least, while the compound motor falls somewhere between the first two.

2. The overload capacities follow the same general order as for starting torques.

3. The speed variation for changes in load are the least for the shunt motor, the greatest for the series motor, and somewhat larger than the shunt motor for the compound machine.

4. Both the shunt and compound motors operate at a very definite stable speed when all mechanical load is removed, while the series motor is quite unstable and tends to operate at an extremely high, often runaway, speed when the mechanical load is removed.

The speed of all three types of motor can be controlled quite readily and in essentially the same way. As will be explained in Chap. 5, there are three methods that may be employed to alter the speed of a d-c motor: (1) by changing the flux through resistance control, (2) by changing the voltage across the armature through resistance control, and (3) by changing the voltage across the armature when the latter is supplied with power from a separate voltage-controlled generator. The very fact that the speed of all types of d-c motor can be controlled efficiently and simply is one of the most important reasons for the widespread use of these machines in preference to a-c motors, particularly when speed control is an essential requirement.

Starting Direct-current Motors. An extremely difficult period in the operation of d-c motors is the *starting* period. It is during this comparatively short time that the machine must accelerate, often under heavy load conditions, and come up to operating speed. Since the *armature current* (not the shunt-field current) is usually much higher than normal during this starting period, arcing at the commutator is likely to be very severe. Unless proper precautions are taken to prevent the destructive effects of such commutator burning as well as other serious electrical stresses, the motor will usually fail to operate satisfactorily and will often break down after a number of starts.

During the starting period, an external resistance must be added to limit the current in the armature circuit, where the current must pass between brush and commutator and where the serious effects of poor commutation are likely to result. This inserted armature resistance is in the form of a rheostat so that it may be cut out gradually as the motor comes up to speed. Its value in ohms is chosen in order to limit the armature current from about 1¼ times to twice the rated amperes at the instant the machine is started. As the motor accelerates, the armature generates an opposing voltage, called a *counter electromotive force* (counter emf), and in doing so causes the current to fall; this counter emf (to be discussed fully in Chap. 5) has the same effect in lowering the armature current as would an impressed voltage across the armature terminals *reduced* by this same counter emf. It is therefore permissible to cut out the added armature resistance gradually as the motor comes up to speed because the counter emf gradually increases in value as the armature accelerates. The counter emf cannot, however, be equal to the impressed voltage; it will usually be a maximum of about 80 to 95 per cent of the latter.

In practice, starting resistors for d-c motors are frequently placed inside manually operated starters and are properly tapped for connection to a series of contact points mounted on a slate panel. A spring-loaded arm is then arranged to be moved over these contact points as the motor starts and gradually comes up to speed. In the final position, that is, the running position, when all the resistance has been cut out, an electromagnetic holding coil prevents the movable spring-loaded arm from returning to the OFF position unless the power is suddenly disconnected. In automatic starters, relay contacts are used to short-circuit steps of resistance as a motor accelerates.

Note particularly that a starter, of which there are many types of construction, functions primarily *to limit the current in the armature circuit during the starting period only*. Such a device is a necessary part of nearly every d-c motor installation. In the case of small machines, up to perhaps ¾ hp, starters need not be employed, because the inrush current is generally low.

Commutating Poles for Direct-current Machines. When current passes between brushes and commutator in a d-c machine, generator or motor, the ends of those coils joined to commutator segments that are bridged by the brushes are short-circuited for an instant. Since each coil so short-circuited has an extremely low value of resistance, it is essential that its generated voltage be zero, or very nearly so, at this particular instant. If the generated voltage in the commutated coil (the one undergoing short circuit) is not zero, the current passing through the local circuit, consisting of a low-resistance coil, a brush, and two small air gaps between brush and commutator, will be extremely high. Obviously, a large value of current confined to a small brush area will cause excessive arcing and commutator burning, a condition which would soon lead to motor failure. For each value of load current there is a correct brush position that will make the short-circuit current zero, or nearly so. Since it is physically impractical to shift the brushes to a new ideal position for every load change, however, it was necessary to develop an electrical method that would automatically cause the generated voltage to be zero, regardless of the magnitude of the load. The ingenious scheme finally developed to perform this remarkable function—that is, causing the generated voltage in the short-circuited coil to be zero for all values of load—involves the use of special, narrow poles located between the large main poles. These poles are actually carefully designed electromagnets excited by the very current that changes with the load, i.e., the armature current. Its influence upon the commutating zone is therefore in direct proportion to the value of the load current. Note particularly that the corrective measure is provided electromagnetically, a method that is accurate, automatic, and simple. Figure 7 clearly shows the two narrow commutating poles mounted between the wide main poles.

Practically all d-c machines, except very small ones or those supplying unvarying loads, the brushes of which may be carefully set for the best commutation conditions, have commutating poles. When properly designed, they provide virtually sparkless commutation under the most severe conditions of variable loading. The theory of these poles, often called *interpoles*, will be discussed in detail in Chap. 4.

Types and Characteristics of Alternating-current Motors. Whereas only three general types of d-c motor are found in practice, a comparatively large number of different constructions are available for use in a-c systems. The reasons for this situation will become clear in later chapters, when it will be learned that each type of a-c motor is confined to narrower operating characteristics, especially with regard to such important matters as torque, overload capacity, speed variation, speed control, and starting procedures. Furthermore, a-c motors must be constructed for operation on single-phase service or polyphase (either two- or three-phase) service; in one type of construction they must perform satisfactorily on d-c service as well as on single-phase alternating current.

Classification of Single-phase Motors. Single-phase motors generally have low horsepower ratings and are used to operate mechanical devices and machines requiring a comparatively small amount of power. Their greatest fields of application are in the fractional-horsepower range, that is, below 1 hp. Motors larger than the latter, up to perhaps 10 hp, are sometimes used on farms and in small shops and factories where polyphase power is not available. As will be pointed out in Chap. 9, polyphase motors generally have better operating characteristics than single-phase machines and cost less per horsepower, so that it is usually true that single-phase motors are used in the larger sizes only because two- or three-phase service is not available.

In the single-phase classification may be listed the following types of motor: *shaded-pole, reluctance, split-phase* (with or without capacitor starting), *repulsion, repulsion-start, repulsion-induction, series* (a-c only or universal), and *synchronous.*

Shaded-pole and *reluctance motors* are built in very small sizes from about $\frac{1}{500}$ to $\frac{1}{6}$ hp; they are cheap to construct, have low starting torque, little overload capacity, and low efficiency and may be speed-controlled.

Standard *split-phase motors* are manufactured in sizes up to about $\frac{3}{4}$ hp; they are comparatively low in cost, have fair starting torque, not much overload capacity, and fair efficiency, and operate at nearly constant speed. Split-phase motors equipped with capacitors have high starting torque and may or may not be arranged to continue to run with a capacitor; when their capacitor is used only during the starting period, they are called *capacitor-start split-phase motors;* when two values of capacitor are provided, one for starting and another for running, they are called *two-value capacitor motors.* However, whether or not these motors are provided with capacitors, they are all, nevertheless, split-phase motors.

Series motors are usually constructed for service on direct or alternating current up to 60 cycles, in which case they are called *universal motors.* When properly designed, they will operate with complete satisfaction on direct or alternating current, developing high starting torque, having excellent overload capacity and good efficiency, and permitting the speed to be controlled over very wide limits. Such motors are not as trouble-free as those described above (shaded-pole, reluctance, and split-phase types), because they have the usual commutator and brushes and their accompanying commutation problems.

Synchronous motors, as the name implies, operate at synchronous speed, that is, a definite, constant speed determined only by the frequency of the supply and the number of poles on the machine. They have very little starting torque, practically no overload capacity, and are quite inefficient; they have, however, the one important characteristic possessed by none of the motors previously discussed, that is, absolute constancy of speed, a requirement that is very important for timing devices.

Classification of Polyphase Motors. Polyphase motors, that is, machines served with two- or three-phase power, may be classified as follows: *induction* (squirrel-cage or wound-rotor types), *commutator* (Schrage or Fynn-Weichsel), or *synchronous.*

Squirrel-cage induction motors are widely used because they have, generally speaking, desirable all-purpose characteristics. They are comparatively low in cost per horsepower, have good starting torque and overload capacity, are highly efficient, and are particularly rugged and trouble-free. These motors will operate in an atmosphere containing dirt, moisture, or corrosive or explosive fumes and can even be constructed to perform submerged in oil or water. They are, practically speaking, constant-speed motors in the sense that change in load does not affect the speed by more than about 5 per cent. Such motors are, however, at a disadvantage when it becomes necessary to control the speed, because it is usually difficult or expensive, from the standpoint of additional equipment, to do so. When speed control becomes a necessary requirement of an application, the squirrel-cage rotor is often replaced by a *wound rotor*, the winding ends of which are connected to slip rings. Speed control is then accomplished by connecting a resistor controller to the brushes riding on the slip rings; the greater the resistance inserted, the lower the speed, and vice versa. Wound-rotor motors, therefore, differ from squirrel-cage motors only by the construction of the rotor, the stator of both types being exactly similar. In addition to its speed-control feature, the wound-rotor induction motor also develops considerably more starting torque. It does, however, have a lower full-load efficiency and a greater speed variation with load changes than does the squirrel-cage type of motor.

Synchronous motors for polyphase service are generally constructed with a stator core and winding similar to those used on induction motors (squirrel-cage or wound-rotor), but with a rotor consisting of a set of salient poles. The latter must be excited with direct current from a small exciter, that is, a self-excited shunt generator, mounted on an extension of the motor shaft or coming from a separate d-c source. Direct current is fed to the rotor field through brushes and slip rings (see Fig. 4). Since synchronous motors, as such, have no starting torque, it is always necessary to provide the rotor poles with a complete squirrel cage built into the pole faces. The motor can then be started in much the same way as are squirrel-cage induction machines; when *nearly* synchronous speed is reached, the d-c rotor field is excited, after which the motor continues to run at *exactly* synchronous speed. The outstanding advantages of this type of motor are (1) absolutely constant speed, determined only by the frequency of the supply and the number of rotor poles, and (2) the possibility of adjusting the motor power factor to any desirable value. Synchronous motors, when properly designed, have good starting torque, overload capacity, and efficiency. They are more expensive than induction machines in the smaller sizes, but cost about as much as squirrel-cage or wound-rotor motors in ratings of more than 100 hp. As a rule, synchronous motors are used in applications requiring infrequent starting, where the load is substantially constant, and where high power factor or power-factor correction is desirable and profitable.

Although the speed of a wound-rotor motor can be changed over a wide range by the insertion of resistance in the rotor circuit, the efficiency of operation is very low at reduced speed. To offset this disadvantage, particularly in large motors where energy cost is important, special types of machines have been developed. A special motor construction that has wide speed-control possibilities is the *Schrage* motor. In the Schrage motor, the stator is of the usual construction found in induction machines, but the rotor differs greatly from any of those already described. The latter has two windings, one on top of the other, placed in deep slots. The *primary* winding, in the bottom of the slots, is connected to slip rings and is fed, through brushes, with polyphase alternating current. The other winding, on top of the primary and next to the rotor surface, is connected to a commutator. Finally, the stator winding, called the *secondary*, is connected to brushes riding on the commutator. Speed control is accomplished by shifting the brushes over the commutator, the method used being an ingenious mechanical lever construction permitting the motion of all brushes simultaneously by the manual or motor-controlled operation of a handle. Such motors are high in cost per horsepower but have good efficiency, starting torque, and overload capacity. They are used only when it is extremely important that the speed be varied over a wide range. An additional advantage of this motor is that power-factor adjustment is also possible.

Starting Alternating-current Motors. Like small d-c motors, single-phase machines are generally started by simply closing the line switch; no auxiliary starting equipment is necessary to limit the initial rush of current. The same practice applies to the starting of small polyphase motors and to some large ones that have been specially designed so that the starting current is limited to a reasonable value even though full voltage is applied to the stator at the instant of starting. In the majority of cases, however, it is necessary to limit the initial rush of current to reasonable values, about $1\frac{1}{4}$ to 5 times rated value, by one of several methods:

1. Inserting resistances in the line wires and then cutting out these resistances gradually as the motor comes up to speed.

2. Inserting reactances in the line wires and then cutting out these reactances gradually as the motor comes up to speed.

3. Applying reduced voltage to the motor at the instant of starting, using autotransformers for this purpose, and connecting the motor directly to the power source after the motor has reached full speed; such autotransformers, called *compensators*, are completely disconnected from the line when the running position is attained.

4. Using a wound-rotor type of motor, which employs a resistor controller for the starting function and which may also, as was previously pointed out, serve as a speed-control device.

5. Using the Y-Δ (star-delta) method, in which the stator is connected

in *star* at the instant of starting and in *delta* after the motor has reached normal speed; this method requires that the stator winding be designed so that it is connected *delta* when it is operating normally. As in the case of d-c motor starting, it is essential that the initial current taken by a-c motors be limited to reasonably low values in order that the voltage drop in the line wires be held down to accepted standards; high starting currents will cause the line voltage to dip, so that lights will flicker and motors already in operation may slow down momentarily.

FIG. 9. Power transformer of 20,000-kva capacity used to step down the voltage at a substation from 132,000 to 34,500 volts. (*Allis-Chalmers Mfg. Co.*)

Transformers. It is undoubtedly true that the principle of transformer action and the practical application of this principle in connection with the construction of transformers and induction-type motors are responsible for the widespread use of alternating current as a primary source of electrical energy. The transformer is a simple, efficient, and comparatively inexpensive device used primarily in a-c circuits for the purpose of changing the voltage from one value to another. There are no moving parts in the transformer, which means that mechanical losses, always present and responsible for much of the heating of rotating and reciprocating machines, are entirely absent. Actually, a transformer is a device that transfers electrical energy from one electric circuit to another without a change of frequency. This energy transfer usually takes place with a voltage change, although the latter is not always necessary or even desirable. When the electric circuits are insulated from each other, as they are in most transformers, they are conventional and are generally referred to as *transformers*. In some special cases, the electric circuits are joined together, in which case the device is referred to as an *autotransformer*. The electric transformer winding that is connected to the source of supply is called the *primary*, while the winding that feeds the load is known as the *secondary*. Some transformers are designed to raise the primary voltage to a higher value, in which case they are known as *step-up* transformers; others are constructed to reduce the primary voltage to a lower value, in which case they are called *step-down* transformers. In step-up transformers the current on the secondary side is lowered in the same ratio as the voltage is raised, while in step-down transformers the current on the secondary side is raised in the same ratio as the voltage is lowered. Transformers have many applications in a-c circuits that require both the raising and lowering of the primary voltage as well as the lowering and raising of the primary current. When used in groups in polyphase circuits, they are especially valuable in performing many important functions, one of which, apart from its voltage-changing use, is to change the number of phases from two to three, three to two, three to six, or several other combinations. The subject of transformers will be dealt with in Chap. 8. Figure 9 shows a large power transformer of 20,000-kva (kilovolt-ampere) capacity, used for stepping down the voltage at a substation from 132,000 to 34,500 volts.

Part I

ELECTRICAL MACHINES
DIRECT CURRENT

CHAPTER 2

DIRECT-CURRENT GENERATOR AND MOTOR PRINCIPLES

Principle of Generator Action. It was previously stated that an electric generator is a machine that converts mechanical energy into electrical energy. This implies, of course, that mechanical motion is imparted to one part of a machine that is made to move with respect to another. In the electric generator, this is done by placing a large number of properly connected copper wires on a cylindrical, laminated steel core and mechanically rotating this structure inside a set of carefully shaped electromagnets. The rotating element is called an *armature*, while the stationary set of electromagnets is called the *field*. If it is assumed that the magnetic lines of force leave the cylindrically shaped pole-core faces and pass across the air space (called the *air gap*) and thence into the rotating armature core, it is clear that *the moving copper conductors cut the lines of force* as they are rotated mechanically. This flux-cutting action on the part of the copper conductors is responsible for the generated voltage in the latter.

Figure 10 illustrates simply the general arrangement of the various parts with respect to one another and shows how this flux-cutting action is accomplished in a four-pole generator. Understand that the mechanical rotation is provided by another machine such as a steam or gas engine or a turbine. And note particularly that in this four-pole generator, one half of the wires cut lines of force under two *north* poles and one half of the wires cut lines of force under two *south* poles.

Briefly summarized, the foregoing *principle of generator action* requires (1) *the presence of magnetic lines of force*, and (2) *motion of conductors cutting the flux*, before (3) *voltage is generated*.

Faraday's Law. If the conductors are mounted on a constant-speed rotating armature, no voltage will be generated in some of them while they are moving parallel to the flux lines or passing through a region where there is no flux. On the other hand, other conductors will be moving perpendicularly with respect to the flux lines, so that maximum cutting action will result; in such cases, the voltage will also be a maximum. It should be clear, then, that *the magnitude of the generated voltage is directly proportional to the rate at which a conductor cuts magnetic lines of force*. This law, called *Faraday's law*, and *Ohm's law* are generally regarded as the two most

24   ELECTRICAL MACHINES—DIRECT CURRENT

important laws in the realm of electrical science. Faraday's law implies simply that higher voltages may be generated by moving conductors more rapidly *across* lines of flux, by increasing the number of flux lines *across* which the conductors move, or by increasing both the speed of the conductors and the flux *across* which they move.

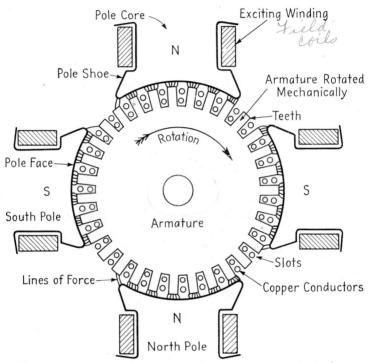

FIG. 10. Sketch illustrating the arrangement of the various elements of a d-c generator.

It was found that when a conductor moves at a constant speed across a uniformly dense magnetic field, that is, a field in which the flux density in lines of force per square inch is constant, 1 *volt is generated for every* 100,000,000 (10^8) *lines cut per second*. If the flux density is not constant, the generated voltage will be an *average* value determined by the total number of lines of force that are cut and the length of time it takes to do so. This experimentally verified fact leads to the formula

$$E_{av} = \frac{\phi}{t \times 10^8} \quad \text{volts} \tag{1}$$

where E_{av} = average generated voltage in a conductor
 ϕ = total flux cut
 t = time, seconds, during which cutting takes place

DIRECT-CURRENT GENERATOR AND MOTOR PRINCIPLES 25

In the actual generator, a great many conductors are placed on a uniformly rotating armature as the former cut the flux under two or more poles (always an even number of them). Since modern armature windings (discussed in Chap. 3) always have the conductors divided into two or more parallel paths (always an even number of them), in each of which there are an approximately equal number of conductors in series, the generated voltage is determined only by the "string" of conductors joined in series and not by the number of parallel paths through which the current may pass. The number of parallel paths, as will be learned later, determines the current rating of the generator, whereas the number of series conductors per path is a measure of the terminal voltage of the machine; both current and voltage ratings determine the power rating in watts. The situation existing in a generator with regard to voltage and current is analogous to dry-cell connections. If voltage and current ratings of 1.5 volts and 5 amp are assured per cell, the following tabulation shows the relative ratings for 120 cells connected in various numbers of parallel paths:

No. of parallel paths	E, volts	I, amp	P, watts
2	90	10	900
4	45	20	900
6	30	30	900
8	22.5	40	900
10	18	50	900
12	15	60	900

The same set of values would result if each cell were replaced by a generating conductor whose voltage and current ratings are 1.5 volts and 5 amp. Note particularly that as the voltage decreases, the current increases by exactly the same proportion as the increase in the number of parallel paths. *The power rating is, however, independent of the manner in which the cells or conductors are connected.*

EXAMPLE 1. A four-pole d-c generator has an armature winding containing a total of 648 conductors connected in two parallel paths. If the flux per pole is 0.321×10^6 maxwells and the speed of rotation of the armature is 1,800 rpm, calculate the average generated voltage.

Solution

Number of conductors in series per path = $648/2$ = 324
Flux cut per revolution = $4 \times 0.321 \times 10^6 = 1.284 \times 10^6$ maxwells
Revolutions per second of armature = $\dfrac{1,800}{60}$ = 30 rps
Seconds per revolution of armature = $1/30$ = 0.0333 sec

E_{av} (per conductor) $= \dfrac{1.284 \times 10^6}{0.0333} \times 10^{-8} = 0.386$ volt

E_g (total generated voltage) $= 0.386 \times 324 = 125$ volts

EXAMPLE 2. In Example 1, calculate the rated current in each conductor (per path) if the power delivered by the armature is 5 kw.

Solution

Total armature current $= \dfrac{\text{watts}}{\text{volts}} = \dfrac{5,000}{125} = 40$ amp

Current per armature circuit (per conductor) $= {}^{40}\!/_2 = 20$ amp

General Voltage Equation for Direct-current Generator. The foregoing discussion of voltage generation should make it clear that in the modern d-c generator, with many poles, with a large number of armature conductors, with armature winding connections that may result in two or more (an even number always) paths in parallel, and with comparatively high armature rotating speeds, the resulting voltage can be calculated only by considering all those factors responsible for the measured value. Remembering that the generated voltage depends upon the *rate* at which flux is cut and that 1 volt results from the cutting of 10^8 lines of force per second, the following analysis will lead to a very useful fundamental equation. Each one of the Z conductors cuts $\phi \times P$ lines of force per revolution, where ϕ is the flux supplied by *each* of the poles P. If the speed of the armature is represented by rpm, the speed in revolutions per second is rpm/60. Assuming a parallel armature paths, the number of conductors in series per path will therefore be Z/a.

Now then, if ($\phi \times P$) is multiplied by rpm/60, the product would represent the flux cut by each conductor per second. This product ($\phi \times P \times$ rpm/60) next multiplied by 10^{-8} would give the voltage generated in each conductor, because 1 volt is generated for every 10^8 lines cut per second. Finally, multiplying ($\phi \times P \times$ rpm/60 $\times 10^{-8}$) by (Z/a) would yield the total generated voltage E_g. Thus the fundamental voltage equation, one of the most important in this book, will become

$$E_g = \dfrac{\phi \times P \times \text{rpm} \times Z}{a \times 60} \times 10^{-8} \quad \text{volts} \tag{2}$$

where $E_g =$ total generated voltage
$\phi =$ flux per pole, maxwells
$P =$ number of poles, an even number
rpm $=$ speed of armature, revolutions per minute
$Z =$ total number of armature conductors effectively used to add to resulting voltage
$a =$ number of armature paths connected in parallel (determined by type of armature winding)

DIRECT-CURRENT GENERATOR AND MOTOR PRINCIPLES 27

EXAMPLE 3. An 85-kw six-pole generator has an armature containing 66 slots, in each of which are 12 conductors. The armature winding is connected so that there are six parallel paths. If each pole produces 2.18×10^6 maxwells and the armature speed is 870 rpm, determine the generated voltage.

Solution

Total number of armature conductors $Z = 66 \times 12 = 792$

$$E_g = \frac{2.18 \times 10^6 \times 6 \times 870 \times 792}{6 \times 60} \times 10^{-8} = 250 \text{ volts}$$

EXAMPLE 4. How many armature conductors are there in a generator, given the following information: $\phi = 2.73 \times 10^6$ maxwells; $P = 4$; rpm $= 1{,}200$; $a = 2$; $E_g = 240$?

Solution

Rewriting Eq. (2) in terms of Z gives

$$Z = \frac{E_g \times a \times 60 \times 10^8}{\phi \times P \times \text{rpm}} = \frac{240 \times 2 \times 60 \times 10^8}{2.73 \times 10^6 \times 4 \times 1{,}200} = 220$$

Direction of a Generated Voltage. The direction of the generated voltage in a conductor, or more correctly in a coil of wire, as it is rotated to

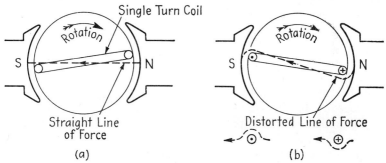

FIG. 11. Sketch illustrating how the direction of the generated voltage in a coil may be determined by the stretching rubber-band analysis.

cut the lines of force produced by the electromagnets in a generator, will depend upon two factors only: (1) the direction of the flux, which is, of course, determined by the magnet polarity, and (2) the direction of motion of the conductor or coil. Two analyses will be given, either of which may be used to predict the direction of the generated voltage.

Consider Fig. 11a, which represents an elementary two-pole generator

28   ELECTRICAL MACHINES—DIRECT CURRENT

with a single-turn coil, shown in section. The student should imagine the machine lying with its shaft axis perpendicular to the plane of the paper, with the plane of the paper cutting the poles and with the armature core and the two conductors forming the sides of the coil so that the outline of these shapes is indicated. Note that the two sides of this coil are diametrically opposite so that while one side is under a *north* pole the other side is under a *south* pole. In a four-pole machine, the distance between the two sides of the coil would be one-fourth the circumference because adjacent poles always have opposite polarities; in a six-pole machine, the coil span would be one-sixth of the circumference. With our attention centered on a single line of force stretching from the center of the *north* pole to the center of the *south* pole, let it be *assumed* that this line of force has its ends fastened to both poles at the centers and that it is furthermore given the property of elasticity, like a rubber band. With the coil in position a, the line of force is straight across from pole to pole. With the coil turned into position b in a clockwise direction, the elastic line of force (rubber band) is distorted out of shape, so that the *left* conductor is shown with the flux line partially wrapped around in a *counterclockwise* direction, and the *right* conductor is shown with the same flux line partially wrapped around in a *clockwise* direction.

Now then, if the familiar "right-hand rule"* of current direction in a conductor and the resulting direction of the flux around the conductor is applied here, it will be concluded that the direction of the generated voltage in the right conductor will be into the paper (shown with a cross inside the circle) and the direction of the generated voltage in the left conductor will be toward the observer (shown with a dot inside the circle). Since the directions in the two sides of the coil are such as to aid one another in sending current in a continuous path, it should be clear why the coil span must always be the distance between poles of opposite polarity, that is, adjacent poles.

The procedure for determining the direction of the generated voltage in a coil of wire depends upon properties *arbitrarily assigned to the fictitious lines of force,* namely, that the ends are somehow fastened to the pole faces and that they stretch like rubber bands. A much more satisfactory method, based on sound electrical principles, will now be given. When lines of force pass through a coil of wire and the latter is moved, *Lenz's law* states, *the direction of the generated voltage in the coil is such that it tends to produce a current flow opposing a change of flux through the coil.* Referring to Fig. 12a assume two poles that create a magnetic field from right (*north*) to left (*south*) through a coil of wire $mnpq$ in a vertical plane. In

---

* Grasp the conductor with the right hand so that the thumb points in the direction of the current; the encircling fingers will then indicate the direction of the lines of force around the wire.

DIRECT-CURRENT GENERATOR AND MOTOR PRINCIPLES

this position, maximum flux passes through the coil. If the coil is now rotated in a clockwise direction so that it occupies the oblique position shown in Fig. 12b, *less* flux will pass through it. By Lenz's law, therefore, the voltage generated in this coil will tend to establish a current whose direction will oppose a *reduction* of flux through it. This tendency can only mean that a current will attempt to flow in the coil from m to n to p to q, so that, by the right-hand rule, flux will be created by the coil from right to left. As the coil continues to move to the position represented by Fig. 12c, the *opposition* to flux change through the coil *increases* because less and less main flux passes through it. Thus it is seen that for clockwise rotation, the side of the coil under the *north* pole will always have a voltage direction *away* from the observer, while the side of the coil under the *south* pole will always have a voltage direction *toward* the observer. This corresponds exactly with the "stretching rubber band" analysis.

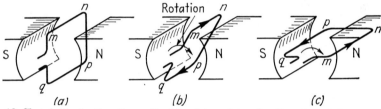

Fig. 12. Sketch illustrating the method for determining the direction of the generated voltage by Lenz's law.

The Elementary Alternating-current Generator. The armature of a generator contains a large number of coils of wire, all connected together in such a manner as to contribute to the desired terminal voltage. Each one of these coils will actually generate an *alternating* voltage as it is rotated on the laminated slotted steel core inside the several *north* and *south* poles. If clockwise rotation is assumed again, the direction of the voltage will be from m to n to p to q when conductor mn is under a *north* pole and pq is under a *south* pole (see Fig. 12). Then, when conductor mn moves under a *south* pole, the direction of its voltage will reverse, so that current will tend to flow from n to m; at the same time, when conductor pq moves under a *north* pole, the direction of its voltage will also reverse, so that current will tend to flow from q to p. Thus it is seen that, for one half of a revolution (in a two-pole generator), the voltage is directed around the loop from m to n to p to q; for the second half of a revolution, the voltage is directed around the loop from q to p to n to m. Therefore, if the ends of this coil were connected to a resistor of some sort, say, a lamp, the current would flow through the latter in one direction for the time required to make one half of a revolution and in the opposite direction during the period required for the next half of a revolution. Such a current is called an *alternating* current because its magnitude and direction change periodically.

Figure 13 represents such an elementary a-c generator, with two poles and a single coil. In order that the current be allowed to flow through the stationary external resistor AB indefinitely, the rotating ends of the coil are shown connected to a pair of rings x and y, upon which ride brushes joined to the load. Note that with mn moving under a *north* pole and with pq moving under a *south* pole, the current through the load resistance is from B to A; when mn moves under a *south* pole and pq cuts lines of force

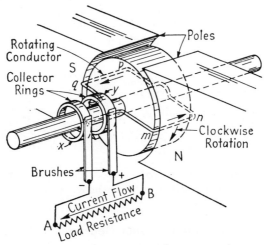

FIG. 13. Elementary two-pole a-c generator.

under a *north* pole, the current will be from A to B. Assuming one direction, B to A, as positive and the opposite direction, A to B, as negative, one cycle of the alternating current is completed in one revolution for this two-pole machine. Had there been four poles instead of two, two cycles would have been generated per revolution; for six, eight, or P poles, there would have been three, four, or $P/2$ cycles per revolution, respectively.

In general terms, therefore, the *frequency* of the alternating current in cycles per second is $P/2 \times$ revolutions per second or, more conveniently, where rpm/60 = rps,

$$f = \frac{P}{2} \times \frac{\text{rpm}}{60} = \frac{P \times \text{rpm}}{120} \qquad (3)$$

EXAMPLE 5. An a-c generator has six poles and operates at 1,200 rpm. (a) What frequency does it generate? (b) At what speed must the generator operate to develop 25 cycles? 50 cycles?

DIRECT-CURRENT GENERATOR AND MOTOR PRINCIPLES

Solution

(a) $$f = \frac{6 \times 1{,}200}{120} = 60 \text{ cycles}$$

(b) $$\text{rpm}_{25} = \frac{120 \times 25}{6} = 500$$

$$\text{rpm}_{50} = \frac{120 \times 50}{6} = 1{,}000$$

EXAMPLE 6. How many poles are there in a generator that operates at a speed of 240 rpm and develops a frequency of 60 cycles?

Solution

$$P = \frac{120 \times f}{\text{rpm}} = \frac{120 \times 60}{240} = 30 \text{ poles}$$

The diagrams in Fig. 14 represent graphically the manner in which the

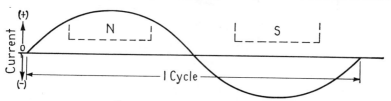

(a) Two poles—one cycle per revolution.

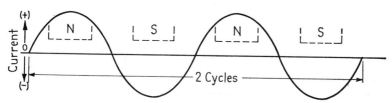

(b) Four poles—two cycles per revolution.

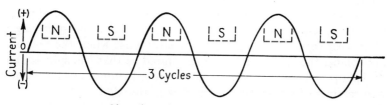

(c) Six poles—three cycles per revolution.

FIG. 14. Sketch illustrating the relation between the number of poles and the generated a-c frequency in cycles per revolution.

number of cycles per revolution increases with an increase in the number of poles.

The Commutation Process. The d-c generator is *fundamentally* an a-c generator because, internally, in the armature conductors, the current reverses periodically as the wires move to cut lines of force successively under the *north* and *south* poles. Absolutely nothing can be done to change this action in the modern type of generator. What can be changed, however, is the way in which the current may be made to flow in the *external* circuit, that is, through the load resistance AB of Fig. 13. If the two slip rings in Fig. 13 are replaced with a *split ring* like that shown in Fig. 15, the current will always flow from A to B. The current will change in magnitude, but *there will be no reversal of current flow.*

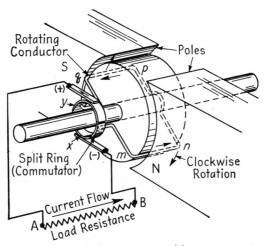

Fig. 15. Elementary two-pole d-c generator with two-segment commutator.

Consider Fig. 15, in which a two-part commutator (split ring) replaces the two slip rings. Brushes are located so that they touch two segments exactly on top and bottom. Note particularly that conductor mn is permanently connected to semiring x and that conductor pq is permanently connected to semiring y; also, that the split ring rotates with the rotating coil $mnpq$. Brushes and poles are, of course, stationary. Observe that, when the coil is in a horizontal plane, the split ring is in position with the split along the horizontal axis; when the coil is in a vertical plane, the split ring has been rotated so that the split is along a vertical plane. These adjustments are extremely important if this mechanism, a rectifier of alternating current, is to function properly.

When conductor mn is moving downward (clockwise rotation) and cutting flux under a *north* pole, semiring x will be negative; at the same time, conductor pq will be moving upward and cutting flux under a *south*

DIRECT-CURRENT GENERATOR AND MOTOR PRINCIPLES 33

pole, thus making semiring y positive. The upper brush will therefore be positive, while the lower brush will be negative; the current through the load will be from A to B. During the next half of the revolution, conductor mn will change places with conductor pq under the poles, and this exchange will cause the generated voltages in the two conductors to reverse their direction. But when this happens, semiring x will move to a position on top, while semiring y will be located on the bottom. Thus, *as the emfs in the conductors mn and pq reverse direction, the semirings, to which they are connected, automatically change places under the stationary brushes.* It follows, therefore, that *the polarity of the brushes does not change.* Hence, the current through the load resistance will always be from A to B. It is true, of course, that the magnitude of the current will change as the conductors mn and pq occupy different positions under the poles, but *there will be no reversal of current through the load.*

The mechanism described in the preceding paragraphs is called a *commutator*—a simple two-part commutator—and the process performed by it in rectifying the alternating current (that is, changing the *internal* alternating current to an *external* direct current) is called the *commutation process.*

In modern d-c generators—d-c generators only in the sense that the current in the external circuit is unidirectional—there are usually many coils of wire in the armature winding and, more often than not, four or more poles. (Two poles are used only when the machine is small.) In such cases, the ring is divided into many segments, and there will usually be as many brushes as there are poles. The brushes are always interconnected, so that one set of alternate ones is joined together to form the positive terminal, while the other set of alternate brushes is joined together to form the negative terminal.

It is interesting to note that the rectified current supplied by *each coil* of a multipolar machine pulsates as many times per revolution as there are number of poles. Thus, for a two-pole generator there will be two pulsations per revolution for each coil; for a four-pole generator, four pulsations per revolution; for a six-pole generator, six pulsations per revolution. Figure 16 depicts graphically the pulsations of the current in the load circuit supplied by a single-turn armature coil when there are two, four, and six poles.

When several coils are joined together properly so that their combined effect acts additively, the result is not only increased voltage, but also voltage pulsations that are not so violent; in other words, the voltage wave becomes smoother as the number of coils are increased. Figure 17 shows the effect upon the resultant voltage wave with two coils in series and also with three coils in series. Obviously, when there are a great many armature coils, the external voltage wave between brushes approaches a smooth

34   ELECTRICAL MACHINES—DIRECT CURRENT

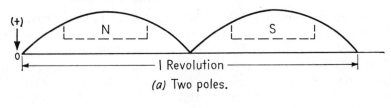

(a) Two poles.

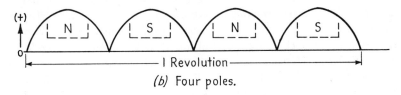

(b) Four poles.

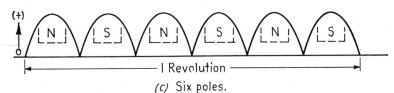

(c) Six poles.

FIG. 16. Sketch illustrating how the rectified current pulsates in two-, four-, and six-pole generators in making one revolution.

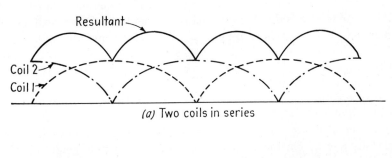

(a) Two coils in series

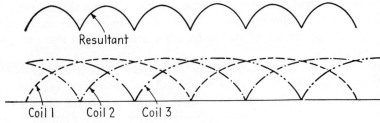

(b) Three coils in series

FIG. 17. Sketch illustrating the combined additive effects of two and three coils. Note how the voltage is increased and how the wave form tends to become smoother.

DIRECT-CURRENT GENERATOR AND MOTOR PRINCIPLES   35

unvarying line that approximates the pure d-c voltage supplied by a battery. Strictly speaking, however, a d-c generator does not deliver a pure direct current, as does a storage battery, for example, but approaches such a current very closely as the number of coils and commutator segments are increased. Incidentally, this is why a commutator-type machine causes interference with the reception of a radio set, unless proper corrective measures are taken.

Principle of Motor Action. When an electric *generator* is in operation, it is driven mechanically and develops a *voltage*, which in turn can send a current through a load resistance. When an electric *motor* is in operation, it develops *torque*, which, in turn, can produce mechanical rotation. Thus the electric motor converts electric energy into mechanical energy.

Before a motor can develop torque, which is a tendency to produce rotation, it is first necessary that a force—or, more correctly, forces—be

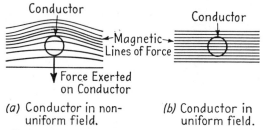

(a) Conductor in non-uniform field. (b) Conductor in uniform field.

FIG. 18. Sketches illustrating that a force action is produced upon a current-bearing conductor placed in a nonuniform field, and no force action results when the field is uniform and the conductor carries no current.

created. This is done simply by placing conductors in magnetic fields and then sending electric currents in the proper directions through the conductors. In the actual motor, many coils of copper wire are placed in the slots of a cylindrical laminated steel core. The ends of the coils are then connected to a commutator, previously described, upon which rest brushes. When current is fed to the coils through the brushes and commutator, forces will be experienced by the conductors only if the armature is in a magnetic field.

Briefly summarized, the foregoing *principle of motor action* requires (1) *the presence of magnetic lines of force* and (2) *current through conductors lying in the magnetic field* before (3) *force, and therefore torque, is produced.*

It was learned in the study of the generator action that the direction of the magnetic field, the direction of the motion of the conductors, and the direction of the resulting generated voltage are all mutually perpendicular with respect to one another. And now, in the study of motor action, we learn that the direction of the magnetic field, the direction of the current through the conductors, and the resulting force produced by the conductors are also all mutually perpendicular with respect to one another.

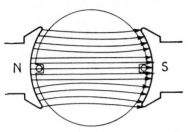

(a) Field produced by magnet poles

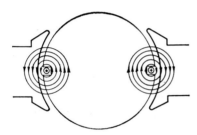

(b) Field produced by current-carrying conductors

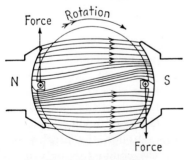

(c) Resultant field and force produced by magnet poles and current-carrying conductors

FIG. 19. Sketches illustrating how motor action is produced by the interaction of the magnetic fields created by the main poles and the current-carrying conductors.

The first important point to be made in connection with the study of motor action is this: if a current-bearing wire is in a *nonuniform* magnetic field so that the flux density on one side of the conductor is greater than that on the other side, the conductor will experience a force action in a direction away from the higher density to the lower density; if, however, the wire carries no current and is in a *uniform* magnetic field, absolutely nothing will happen. In Fig. 18a, a current-carrying conductor, perpendicular to the plane of the paper, is represented in a nonuniform magnetic field with a higher flux density above the wire than below it; in this case, the force action exerted upon the conductor will be down. In Fig. 18b, no force action is exerted upon the conductor, because the latter is supposedly in a uniform magnetic field and carries no current. In the actual d-c motor, of course, the nonuniform flux distribution results from the interaction of two magnetic fields, one being the field produced by the stationary main poles and the other the field created by a large number of current-carrying conductors on the armature core.

Consider Fig. 19, in which a pair of conductors is shown in the two diametrically located slots on the armature core of a two-pole machine. If the field is energized, the flux distribution is quite uniform and symmetrical, as shown in Fig. 19a. If the conductors carry currents as indicated in Fig. 19b (no field excitation), the magnetic fluxes surrounding both conductors are approximately circular paths. However, when the field is excited and the currents flow in the two conductors, the resulting flux distribution is far from uniform in the region of the conductors, as represented by Fig. 19c. Notice particularly that the flux densities are greatest on the bottom of the left conductor (under the north pole) and on the top of the right conductor (under the south pole) and are least on the top of the left conductor and the bottom of the right conductor. As a result of this nonuniform flux distribution, the left conductor will tend to move upward, while the right conductor will tend to move downward. And if the armature is free to turn on a shaft, the resulting torque will produce rotation in a clockwise direction. Of special interest is the fact that the direction of the main field, the position of the current-carrying conductors, and the forces exerted upon the conductors are mutually perpendicular.

If the directions of the currents in the two conductors are reversed, so that the left conductor carries current away from the observer and the right conductor carries current toward the observer, the armature will tend to rotate in a counterclockwise direction. If the polarity of the field is reversed, so that the left pole is *south* and the right pole is *north*, the armature will again tend to turn counterclockwise. Thus it is seen that the direction of rotation of a d-c motor can be reversed by changing *either* the direction of current flow through the armature conductors or the polarity of the field. A motor will not reverse its direction of rotation if *both* the field polarity and the direction of the current flow through the armature are changed. Interchanging the two line wires connected to a d-c motor effects both these changes at once and does not, therefore, cause the motor to reverse.

Force and Torque Developed by Direct-current Motors. The foregoing discussion should make it obvious that the force action exerted by a current-carrying conductor placed in a magnetic field depends, among other things, upon (1) the strength of the main field and (2) the value of the current through the conductor, because the resultant nonuniform magnetic field is determined by both the main field and the flux set up by the current-carrying conductor. Since every unit of length of wire is active in producing flux, it follows that the total force exerted by the conductor will also depend upon how long the wire happens to be. Experiment has shown that a force of 1 dyne will be exerted upon a conductor 1 cm long carrying a current of 10 amp when placed under a pole the area of which is 1 sq cm and producing one line of force (flux density = one line per square centimeter). If the pole creates more than one line of force, the *flux density* will be increased in the same proportion; furthermore, if the pole area is increased with corresponding increases in the wire length, the force will be still greater. This analysis, therefore, leads to the equation

ELECTRICAL MACHINES—DIRECT CURRENT

$$F' = \frac{B' \times I \times l'}{10} \quad \text{dynes} \qquad (4)$$

where B' = flux density, lines per square centimeter
I = current in conductor, amp
l' = length of conductor, cm

If the units of F, B, and l are specified in more practical terms, that is, pounds, lines per square inch, and inches,* respectively, Eq. (4) becomes

$$F = \frac{(B/6.45) \times I(l \times 2.54)}{10 \times 980 \times 453.6} = \frac{B \times I \times l}{11{,}300{,}000} \quad \text{lb} \qquad (5)$$

where B = flux density, lines per square inch
I = current in conductor, amp
l = length of conductor, in.

EXAMPLE 7. A conductor is 8 in. long and carries a current of 140 amp when placed perpendicularly to a magnetic field the intensity of which is 58,000 lines per square inch. Calculate the force exerted by the conductor.

Solution

$$F = \frac{58{,}000 \times 140 \times 8}{11{,}300{,}000} = 5.75 \text{ lb}$$

EXAMPLE 8. The armature of a d-c motor has 648 conductors, 65 per cent of which are directly under the poles where the flux density is 48,000 lines per square inch. If the core diameter is 7 in. and its length 4 in. and the current in each conductor is 20 amp, calculate: (*a*) the total force tending to rotate the armature; (*b*) the torque exerted by the armature in pound-feet.

Solution

(*a*) $$F = \frac{(648 \times 0.65) \times 48{,}000 \times 20 \times 4}{11{,}300{,}000} = 143.5 \text{ lb}$$

(*b*) $$T = 143.5 \times \frac{7}{2} \times \frac{1}{12} = 41.8 \text{ lb-ft}$$

As the foregoing example illustrates, the turning moment, or torque, of a motor is created by the combined action of many current-carrying conductors in a strong magnetic field. Moreover, only those conductors that are directly influenced by the magnetic field (65 per cent in the example given) contribute to the tendency of the armature to rotate; those conductors that lie between the pole tips, in the interpolar spaces, are in weak fields and are therefore more or less inactive. That is why only 65 per

* 1 in. = 2.54 cm; 1 sq in. = 6.45 sq cm; 1 gm = 980 dynes; 1 lb = 453.6 gm.

DIRECT-CURRENT GENERATOR AND MOTOR PRINCIPLES   39

cent of the total number of conductors were used in the calculation of Example 8.

EXAMPLE 9. A d-c motor has an armature containing 192 conductors, 70 per cent of which lie directly under the pole faces at any given instant. If the flux density under the poles is 52,000 lines per square inch and the armature diameter and length are 12 in. and 4.5 in., respectively, calculate the current in each armature conductor for a torque of 120 lb-ft.

Solution

$$\text{Torque} = 120 \text{ lb-ft} = 120 \times 12 \text{ in.-lb} = 1{,}440 \text{ lb-in.}$$

$$\text{Force} = \frac{\text{torque}}{\text{armature radius}} = \frac{1{,}440}{6} = 240 \text{ lb}$$

$$240 = \frac{(192 \times 0.7) \times 52{,}000 \times I \times 4.5}{11{,}300{,}000}$$

$$I = \frac{240 \times 11{,}300{,}000}{(192 \times 0.7) \times 52{,}000 \times 4.5} = 86.3 \text{ amp}$$

Commutation in Direct-current Motors. When a d-c motor is in operation, a unidirectional current is fed to the armature conductors through brushes and the commutator. For any given position of the armature, the current directions adjust themselves so that in those conductors under the *north* poles, the flow is in one direction, while in those conductors under the *south* poles, the flow is in the opposite direction. The dividing axes between the ingoing and outgoing currents are along lines exactly midway between *north* and *south* poles, assuming uniform flux densities under the latter. A four-pole machine is represented by Fig. 20, in which, it will be noticed, all conductors under the two *south* poles carry currents away from the observer; all the conductors under the two *north* poles carry currents toward the observer. For such a combination of poles and current-carrying conductors, the armature will turn in a clockwise direction. As the armature rotates, those conductors under the *south* poles gradually exchange places with those under the *north* poles; that is, as the conductors in the slots pass from one side to the other side of the axes indicated by *a*, *b*, *c*, and *d*, the currents change direction from *in* to *out* or from *out* to *in*, as the case may be. For example, the conductors between *a* and *b* under the upper south pole and the conductors between *c* and *d* under the lower *south* pole will all occupy positions under the two *north* poles, left and right, in a quarter of a revolution. Furthermore, the conductors between *b* and *c* under the right *north* pole and the conductors between *d* and *a* under the left *north* pole will all occupy positions under the two *south* poles, bottom and top, in the same quarter of a revolution. It should therefore be clear that the function of the commutator and the brushes in a d-c motor is to act as an *inverter*, that is, to change direct current to alternating current, because *the current in the armature conductors must be alternating if rotation in the same direction is to continue.*

It should be noted particularly that (1) in the d-c generator the commutator and brushes function to change the internally generated alternating

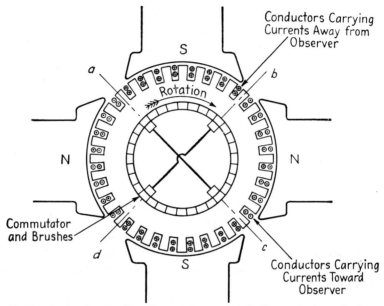

Fig. 20. Sketch showing the directions of the currents in the conductors of a four-pole motor for clockwise rotation. Note the dividing axes midway between adjacent *north* and *south* poles.

current to a load-applied direct current, and (2) in the d-c motor the commutator and brushes perform an inverse function by changing the externally applied direct current to alternating current flowing in the armature conductors. In both cases, whether generator or motor, the current in the armature winding is always alternating, while the current in the external circuit is always direct.

Main Fields in Direct-current Machines. The polarities of the main field poles in d-c generators and motors remain unchanged while such machines are in operation. This implies, of course, that the poles are either permanent magnets or electromagnets excited with direct current. Most modern machines are of the latter type, although on some small generators and motors it has sometimes been found desirable to use extremely powerful *alnico* permanent magnets for the field poles. Where

DIRECT-CURRENT GENERATOR AND MOTOR PRINCIPLES

permanent magnets are employed, the main field flux is constant in magnitude; no flux control is possible. When the main poles are electromagnets, the excitation current can be varied by using a rheostat in the field winding circuit; the flux can therefore be varied over wide limits. Since the voltage of a generator [see Eq. (2)] and the speed of a motor may be varied easily and efficiently by controlling the flux, the electromagnetic type of field is the preferred construction; permanent-magnet types of generators and motors have been used to a limited extent in applications in which the loading is substantially constant and where a constant field flux is desirable.

The electromagnet construction permits excitation to be produced in several ways, among which are the following:

1. The excitation of each pole may be produced by a single coil of many turns of comparatively fine wire.
2. Excitation may be produced by a single coil of few turns of comparatively heavy wire.
3. Excitation may be produced by two coils of wire, one of them having many turns of comparatively fine wire and the other having a few turns of rather heavy wire.

Since the flux produced by an electromagnet depends upon the product of the number of turns of wire and the current, that is, *amperes* $\times$ *turns*, it should be clear that the winding with the fine wire and many turns will carry a relatively low current, while the winding with the heavy wire and few turns will carry a rather high current. In the first case, the winding resistance is always high enough so that full line voltage may be applied to it without exceeding the normal low value of current; in the second case, the resistance of the winding is always low enough so that, when the winding is connected in series in the armature circuit or directly in series in one of the line wires, the low voltage drop across this winding does not subtract appreciably from the source voltage.

When a generator or motor has a main field winding with many turns of fine wire, the winding is connected directly across the brush terminals where full voltage exists; it is then called a *shunt* machine. When the generator or motor has a main field winding with few turns of heavy wire, the winding is connected in series in one of the line wires where the full armature current flows; it is then called a *series* machine. When the generator or motor has a main field consisting of two windings, one of them having many turns of fine wire and the other having few turns of heavy wire, the shunt winding is connected to the line terminals *or* across the brush terminals, while the series winding is connected in series in one of the line wires *or* in series in the armature circuit; it is then called a *compound* machine.

ELECTRICAL MACHINES—DIRECT CURRENT

Questions

1. Distinguish between a *generator* and a *motor*.
2. What is an armature? a field?
3. What two important requirements are necessary before generator action is possible?
4. What two important requirements are necessary before motor action is possible?
5. State Faraday's law.
6. In what two ways is it possible to generate a higher voltage in a moving conductor?
7. How many lines of force must be cut per second if 1 volt is to be generated in a moving conductor?
8. In an actual generator, what effect has the number of parallel paths in the armature winding upon the terminal voltage?
9. What effect has the number of armature paths upon the current-carrying ability of a generator?
10. Is the power output of a generator affected by the number of parallel paths in the armature winding? Give reasons for your answer.
11. Derive the fundamental voltage equation for the d-c generator.
12. Write the fundamental voltage equation as a function of the total number of armature conductors and the number of armature-winding circuits.
13. What two factors determine the *direction* of the generated voltage in a conductor moving through a magnetic field?
14. Considering a d-c generator, in what two ways can the polarity (plus and minus) of the brushes be changed?
15. Using the fictitious "rubber-band" comparison, explain how the direction of the generated voltage may be determined.
16. State Lenz's law.
17. Using Lenz's law, explain how the direction of the generated voltage may be determined.
18. What kind of current flows in the armature conductors of a d-c generator?
19. What is an alternating current? What is meant by the *frequency* of an alternating current?
20. What is a commutator? Describe its construction.
21. What factors determine the frequency of a generated alternating voltage?
22. Carefully describe the commutation process, using appropriate sketches to illustrate your answer.
23. Why is it desirable to have many coils of wire and commutator segments on the armature of a d-c generator?
24. What kind of current is theoretically delivered to a load by a d-c generator?
25. What is meant by *torque*?
26. What factors determine the force exerted by a conductor on the armature of a d-c motor?
27. Will a force be exerted by a conductor carrying a current when it is placed parallel to a magnetic field? Explain carefully.
28. What happens to the existing uniform field if a conductor carrying a current is placed in this field?
29. Using a sketch showing a two-pole motor with a single coil placed with its two sides under the pole centers, describe how torque is developed by the coil.
30. Why is no torque developed by those conductors occupying positions in the interpolar spaces of a motor?

DIRECT-CURRENT GENERATOR AND MOTOR PRINCIPLES 43

31. What is the essential difference between commutation in a d-c generator and commutation in a d-c motor?
32. Describe the commutation process in a d-c motor.
33. What two types of field winding are used in d-c machines? How do they differ from each other with regard to the number of turns of wire? the size of wire? the manner in which they are connected?
34. When is it desirable and permissible to use permanent magnets in d-c machines? What kind of material is generally used for permanent magnets?
35. Make sketches showing two ways in which the two field windings, shunt and series, may be connected for compound generator or motor operation.

Problems

1. Thirty-six dry cells are connected in four parallel groups of nine cells in series per group. If the voltage and current rating of each cell is 1.45 volts and 4 amp, respectively, (a) calculate the voltage, current, and power rating of the entire combination. (b) Recalculate the problem for nine parallel groups of four cells in series per group.
2. Calculate the average voltage generated in a moving conductor if it cuts 2.5×10^6 maxwells in $\frac{1}{40}$ sec; in $\frac{1}{80}$ sec.
3. A six-pole d-c generator has an armature winding with 504 conductors connected in six parallel paths. Calculate the generated voltage in this machine if each pole produces 1.65×10^6 maxwells and the armature speed is 1,800 rpm.
4. The armature winding of the generator of Prob. 3 is modified so that it has two parallel paths instead of six. At what speed should the machine be driven if it is to develop the same voltage as before the change, assuming all other conditions to remain unchanged?
5. Calculate the voltage generated by a four-pole d-c machine given the following particulars: number of slots in armature = 55; number of conductors per slot = 4; flux per pole = 2.62×10^6 maxwells; speed = 1,200 rpm; number of parallel paths in armature = 2.
6. If the armature winding of Prob. 5 had had four parallel paths, what would have been the generated voltage?
7. A four-pole machine generates 250 volts when operated at 1,500 rpm. If the flux per pole is 1.85×10^6 maxwells, the number of armature slots is 45, and the armature winding has two parallel paths, calculate (a) the total number of armature conductors; (b) the number of conductors in each slot.
8. How many total conductors and conductors per slot would be necessary in the armature of Prob. 7 if the winding had had four parallel paths?
9. The speed of the generator of Prob. 7 is decreased to 1,350 rpm. (a) What will be the generated voltage if the flux per pole is maintained at the same value, i.e., 1.85×10^6 maxwells? (b) To what value of flux per pole should the excitation be adjusted if the generated voltage is to remain the same, i.e., 250 volts?
10. What is the frequency of the alternating voltage generated in the armature conductors of (a) a six-pole 900-rpm machine? (b) an eight-pole 750-rpm machine? (c) a 10-pole 500-rpm machine?
11. How many poles are there in a generator in which the armature frequency is 30 cps when operating at a speed of 600 rpm?
12. At what speed is an armature rotating in a 12-pole machine if the frequency in the armature conductors is 50 cps?

44 ELECTRICAL MACHINES—DIRECT CURRENT

13. Calculate the force exerted by each conductor, 6 in. long, on the armature of a d-c motor when it carries a current of 90 amp and lies in a field the density of which is 52,000 lines per square inch.
14. What torque will the conductor of Prob. 13 develop if it lies on an armature the diameter of which is 9 in.?
15. The armature of a d-c motor has 31 slots with 16 conductors in each slot. Only 68 per cent of the conductors lie directly under the pole faces, where the flux density is uniform at 46,000 lines per square inch. If the armature-core diameter is 7.25 in., its length 4.25 in., and the current in each conductor is 25 amp, calculate: (a) the force exerted by the armature tending to produce rotation; (b) the torque in pound-feet.
16. Using the data of Prob. 15, calculate the torque that will be developed if the flux density is reduced by 5 per cent while the current is increased to 40 amp.
17. What total current must the armature of a d-c motor carry, given the following information: armature slots = 72; conductors per slot = 6; pole arcs cover 70.5 per cent of circumference; flux density = 58,000 lines per square inch; armature-core length = 8 in.; armature-core diameter = 21 in.; number of armature paths in parallel = 6; torque = 1,050 lb-ft.
18. What must be the total armature-winding current in Prob. 17 if the torque increases to 1,200 lb-ft while the flux density drops by 4 per cent?

CHAPTER 3

DIRECT-CURRENT DYNAMOS, CONSTRUCTION AND ARMATURE WINDINGS

Generator and Motor Construction. The term *dynamo* is a symbol of power. It is often used figuratively to indicate great activity and accomplishment on the part of human beings or animals. Electrically, a dynamo is defined as a rotating electrical machine that converts mechanical energy into electrical energy *or* electrical energy into mechanical energy. The two kinds of dynamo are, therefore, electric generators and electric motors, terms more appropriate and definitive to modern electrical machines.

A dynamo may be divided, for purposes of description, into two sections, namely, that portion which is stationary and that portion which rotates; the former may be called the *stator*, while the latter may be called the *rotor*. The most important part of the stator is the field, with its customary laminated steel core and windings. The field poles are usually bolted to a field *yoke* or frame, to which are also fastened the end bells with their *bearings* and the *brush rigging*. The yoke may have a base with feet or a supporting bracket upon which the whole structure rests. The rotor is built up of a laminated steel core, slotted to receive the insulated copper armature winding. A shaft through the core center supports both the armature winding and a securely bolted commutator. The armature-coil ends are fastened and soldered to the commutator, the latter being so located that carbon brushes in the brush rigging line up and rest upon it. Spring tension is applied to the brushes so that good uniform contact is made between them and the commutator.

Each one of the *field-pole cores* is built up of a stack of steel laminations, about 0.025 in. thick per lamination, having good magnetic qualities; rivets are driven through holes in the sheets to fasten together a stack of such laminations equal to the axial length of the armature core. The shape of the assembled pole core is such that a smaller cross section is provided for the field winding or windings, while a spread-out portion called the *pole shoe* permits the flux to spread out over a wider area where the flux enters the armature core. The upper projecting face of the shoe also provides a ledge upon which the field winding can have mechanical support. This construction is very desirable for several reasons:

1. The reduced cross section permits the use of less copper wire for the field coils.
2. The increased area of the pole shoe reduces the reluctance of the air gap between the pole face and the armature core.
3. It permits the entire pole core and windings to be assembled before the former is bolted to the yoke frame.

Figure 21 is a photograph of a stack of steel laminations riveted together; it clearly shows the pole shoe, the pole core, the ledge, and the threaded bolt holes where the core is bolted to the yoke frame.

FIG. 21. Main laminated pole-core assembly for a d-c motor or generator. (*General Electric Co.*)

As was previously pointed out, there are several *field-winding* constructions: (1) a shunt field in which there are many turns of fine wire, (2) a series field in which there are comparatively few turns of heavy wire, and (3) a compound field in which both a shunt and series winding are used. Figure 22 depicts three steps in the manufacture of one of the sets of field coils for a compound motor or generator. The photograph on the left shows the fine-wire shunt coil, while the picture on the right shows the shunt coil taped and surrounded by a few turns of very heavy wire representing the series coil. After the field coils are properly wound, they are dipped in an insulating varnish and baked in an oven; this operation adds stiffness, mechanical strength, and good insulating properties to the winding. Figure 23 shows a field assembly, after the dipping and baking operations, ready to be placed over a pole core such as that illustrated in Fig. 21. A complete pole with the field coil placed over the core may be seen in Fig. 24. This photograph shows a shunt-field coil surrounding the laminated pole core ready to be bolted to the yoke ring.

Fig. 22. Three steps in the manufacture of shunt and series field coils for a typical main field of a compound motor or generator. (*General Electric Co.*)

Fig. 23. Complete assembly of shunt and series coils for main field of d-c motor or generator after being dipped and baked. Note that the shunt coil is on the inside and the series coil is on the outside. (*General Electric Co.*)

The outside frame of the machine, or yoke, is usually a circular iron or steel ring of rectangular section, sometimes rounded on the surface for added strength. It is to this yoke ring that the field-pole assemblies are bolted. Figure 25 clearly shows the construction for a d-c motor of small size in which there are four main poles and two interpoles. It should also be noted that feet for motor support and an outlet box have been welded

Fig. 24. Shunt field-coil and pole-core assembly for d-c motor or generator. (*Westinghouse Electric Corp.*)

Fig. 25. Wound magnet frame for a small d-c motor showing four main poles and two interpoles bolted to yoke ring. (*General Electric Co.*)

to the yoke. In very small machines, in which weight is not particularly important, cast iron is used for the yoke frame; in the larger dynamos, however, cast steel is used because this material makes it possible to reduce the weight by as much as 60 per cent without any increase in the reluctance of the magnetic circuit. In some recent designs, the yoke has been rolled to the proper diameter from a rectangular slab of steel; the ends are welded where they butt together.

DIRECT-CURRENT DYNAMOS 49

Like the field core, the *armature core* is a stack of steel laminations, but it is circular in section. The circumferential edge is slotted to a convenient depth to receive the copper armature winding, the number of slots being very carefully selected in conjunction with the number of commutator segments, on the basis of good design. If the armature is small in diameter, circular holes are punched in the center of the laminations for the shaft. Such an arrangement is represented by Fig. 26, which shows

FIG. 26. Assembled armature core on the shaft of a small two-pole motor. (*General Electric Co.*)

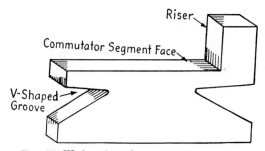

FIG. 27. Wedge-shaped commutator segment.

an 18-slot laminated armature core pressed on its shaft. On large machines, a *spider*, consisting of a hub and projecting arms, is provided, so that the annular laminations may be rigidly fastened to the shaft. The latter construction permits air to flow freely between the radial arms to keep the armature ventilated and cooled.

The *commutator* is a built-up group of hard-drawn copper bars, wedge-shaped in section when viewed on end, and having V-shaped grooves at each end. Figure 27 is a sketch of one such commutator segment with identifying notations. An exploded view of a typical commutator for a small d-c machine is shown in Fig. 28, with the completely assembled view in the lower right. Exactly how the various parts are put together may

Fig. 28. Typical commutator, showing an exploded view and the complete assembly for a small d-c motor or generator. (*General Electric Co.*)

Fig. 29. Partially assembled commutator for a small d-c motor or generator. (*General Electric Co.*)

be observed in the partially assembled commutator of Fig. 29. Note particularly how the mica V-rings fit snugly into the V-shaped grooves of the commutator segments to insulate the latter from the steel shaft, how mica insulators separate the individual segments, and how the threaded steel ring tightens the various components together. The completely assembled commutator, like that shown in Fig. 28, is later forced on the shaft to a proper distance from the laminated armature core. On small machines, this latter operation is usually performed after the armature winding is in place. A view of a core and commutator without the armature winding is shown in Fig. 30 and should indicate the general constructional details already discussed.

FIG. 30. Unwound armature, showing laminated, slotted core and commutator for a d-c motor or generator. (*General Electric Co.*)

FIG. 31. D-c armature completely wound, dipped in insulating varnish, and baked, ready for operation. (*The Louis Allis Co.*)

The *armature winding* is virtually the heart of the dynamo; it is where the voltage is generated in the generator or where torque is developed in the motor. The armature-coil ends are soldered to the commutator, after which the latter is turned down on a lathe and undercut. Undercutting is performed on most commutators by a special undercutter tool that grooves the mica between segments to a depth of about 0.03 in. The subject of armature windings will be treated in some detail subsequently, but at this point it will be sufficient to state that a winding consists of a group of insulated copper coils, all alike, placed in the slots of the armature core with the coil ends properly connected to the commutator. A completed armature ready for operation is shown in Fig. 31.

The *brush rigging* consists of groups of brush holders properly fastened together and bolted to the yoke. It is customary to have as many brush arms as poles, although in some special cases only two sets of brush arms are used, regardless of the number of poles. (This will be discussed later under Number of Parallel Paths in Simplex-wave Windings, p. 65.) Figure 32 shows a simple four-pole brush rigging before it is properly fastened to the yoke frame. Note how each carbon brush fits snugly into a brush holder and how spring tension is applied so that good electrical

Fig. 32. Standard brush rigging for a d-c motor or generator before being fastened to the yoke frame. (*General Electric Co.*)

contact can be made between brush and commutator. In large machines, a brush arm may have several brush holders, into each of which is placed a carbon or copper-graphite brush so that the brush rides freely, without chattering, on the commutator. Figure 33 clearly illustrates the eight-pole brush rigging fastened to the yoke frame for a 1,000-hp motor. Each brush arm may be seen to have seven brushes, and also visible are the flexible copper pigtails for good electrical connection between the insulated brush arms and the respective brushes. In practice, each brush is held down firmly on the commutator by a spring that exerts a pressure of about 1 to 2 lb per sq in. (psi). Obviously, the surfaces of the brushes in contact with the commutator must be ground or sandpapered smooth so that perfect contact exists between them. When more than two brush arms are used, in multipolar machines, alternate sets are electrically connected together; one of the junctions becomes a positive armature terminal, while the other junction becomes the negative armature terminal. (In a generator the positive terminal "feeds" current to the load, while in the motor it is the terminal that is connected to the positive bus of the source.)

FIG. 33. Brush rigging for an eight-pole motor shown bolted to the frame. Note that each brush arm has seven brush holders and brushes. (*Allis-Chalmers Mfg. Co.*)

Types of Armature Winding. There are only two general types of armature winding, *lap* and *wave*. They are distinguished from each other in several ways, but from the standpoint of construction they differ only by the manner in which the coil ends are connected to the commutator bars. In *simplex-lap* windings the coil ends are connected to adjacent commutator segments, while in *simplex-wave* windings the coil ends are connected to commutator segments very nearly, but never exactly, equal to the distance between poles of the same polarity, that is, alternate poles. Since one cycle always occurs in a distance covered by a *pair* of poles, this distance is arbitrarily called 360 *electrical degrees*. This means that in simplex-wave windings the coil ends are connected to commutator segments very nearly, *but never exactly*, equal to 360 electrical degrees. Figure 34 illustrates how the two types of coil are connected to commutator segments. In this regard it should be emphasized that the distance between the two coil *sides* is exactly the same for both types of winding if

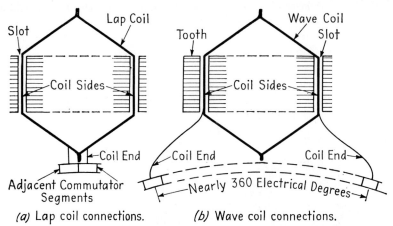

(a) Lap coil connections.   (b) Wave coil connections.

Fig. 34. Sketches showing how the coil ends are connected to the commutator in lap- and wave-wound armature windings.

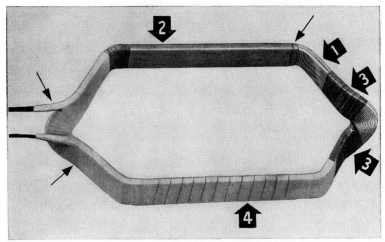

Fig. 35. Lap coil showing details of construction: (1) rectangular insulated conductors; (2) insulating cell; (3) treated cloth; (4) final layer of linen tape. (*Westinghouse Electric Corp.*)

the machines have the same number of poles. All two-pole machines have lap windings, while most four-pole machines, up to about 75 kw or 100 hp for 115- or more volt service, have wave windings.

Figure 35 is a photograph of a *lap* coil, showing how the coil ends are brought out midway between the sides so that connections to the commutator may be readily made. This coil was specially prepared to show the insulated rectangular wire (1), the insulating cell (2), the treated cloth (3), and the final layer of linen tape (4).

Figure 36 is a photograph of a *wave* coil, showing how the coil ends are brought out at the sides so that they may easily be bent outward for connection to commutator bars about 360 electrical degrees apart.

Another construction, which combines the advantages of both lap and wave types and which is used on machines manufactured by the Allis-Chalmers Manufacturing Company, is called a *frog-leg* winding. As will be shown later, it is actually a combination of a lap and a wave winding, so that it should not be classified as a fundamentally different type.

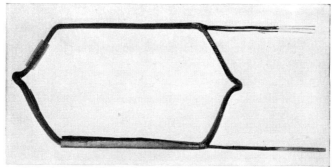

FIG. 36. Wave coil shown partly insulated. Note how the coil ends are brought out at the coil sides for connection to segments about 360 electrical degrees apart. (*General Electric Co.*)

Figure 37 depicts one complete coil for a frog-leg winding and illustrates very clearly that it consists of a lap coil (see Fig. 35) and a wave coil (see Fig. 36) constructed as a single unit. The term *frog-leg* is used to indicate the similarity between this type of coil and the legs of a frog.

Coil Span for All Types of Winding. Perhaps the most important aspect of all types of winding, and this applies not only to all d-c windings but also to most polyphase and some single-phase a-c windings, is that *the distance between the two sides of a coil must be equal (or very nearly so) to the distance between two adjacent poles.* This always means that *the coil span must be* 180 *electrical degrees, or approximately so.* The reason for this important requirement follows from these facts:

1. In a generator, the voltage generated in one side of a coil must be *away* from the observer at the same instant that the voltage generated in the other side of the same coil is *toward* the observer if the voltages generated in the two coil sides are to *aid* one another. This rule can only mean that if one coil side is under the center of a *north* pole the other side must

be very close to the center of a *south* pole, 180 electrical degrees away (see Fig. 38a).

In a motor, the force exerted by one side of a coil must be clockwise, for example, at the same instant that the force exerted by the other side of

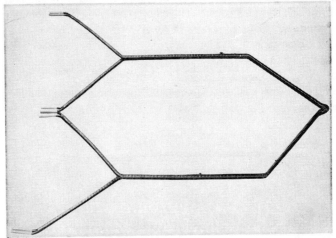

Fig. 37. Frog-leg coil showing the distinct lap and wave elements. (*Allis-Chalmers Mfg. Co.*)

the same coil is also clockwise if both forces are to develop torque for rotation in the *same* direction. This can only mean that if one coil side is under the center of a *north* pole and carries current *toward* the observer, the other side of the same coil must be very close to the center of a *south* pole because it carries current *away* from the observer (see Fig. 38b).

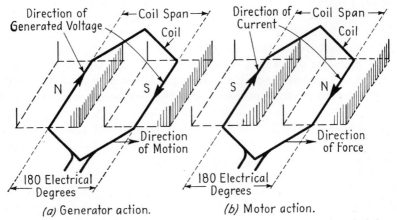

Fig. 38. Sketches illustrating why the coil span must be about 180 electrical degrees in a generator or a motor.

DIRECT-CURRENT DYNAMOS

Figure 38 represents a coil, shown under a pair of adjacent poles, in which the coil span is 180 electrical degrees. Actually, the coil span is made a few electrical degrees less than 180 in most windings, because commutation is improved somewhat when this is done.

The foregoing discussion should make it clear that the two slots into which the sides of the coil should be placed may be determined by calculating how many slots there are in 180 electrical degrees. Since the distance between adjacent poles is this number of degrees, it is merely necessary to divide the total number of slots by the number of poles to determine the coil span *in slots*. Thus, if there are 36 slots and four poles, a 180-electrical-degree coil span is equivalent to nine slots; this means that coils will occupy slots 1 and 10, 2 and 11, 3 and 12, etc., particularly when the machine has interpoles (to be discussed later). If the number of slots is not exactly divisible by the number of poles, it is customary to drop the fraction after dividing slots by poles. Thus, if there are 35 slots and four poles, the coil span is eight; this means that coils will occupy slots 1 and 9, 2 and 10, 3 and 11, etc.

This leads to the general equation for the coil span, more often called *coil pitch:*

$$Y_s = \frac{S}{P} - k \qquad (6)$$

where Y_s = coil pitch, in slots
S = total number of armature slots
P = number of poles
k = any part of S/P that is subtracted to make Y_s an integer

EXAMPLE 1. Calculate the coil pitches (coil spans) and indicate the slots into which the first coils should be placed for the following armature windings: (*a*) 28 slots, four poles; (*b*) 39 slots, four poles; (*c*) 78 slots, six poles; (*d*) 121 slots, eight poles; (*e*) 258 slots, 14 poles.

Solution

(*a*) $\quad Y_s = \dfrac{28}{4} - 0 = 7 \qquad$ slots 1 and 8

(*b*) $\quad Y_s = \dfrac{39}{4} - \dfrac{3}{4} = 9 \qquad$ slots 1 and 10

(*c*) $\quad Y_s = \dfrac{78}{6} - 0 = 16 \qquad$ slots 1 and 17

(*d*) $\quad Y_s = \dfrac{121}{8} - \dfrac{1}{8} = 15 \qquad$ slots 1 and 16

(*e*) $\quad Y_s = \dfrac{258}{14} - \dfrac{6}{14} = 18 \qquad$ slots 1 and 19

Commutator Pitch for Lap Windings. As previously pointed out, the ends of the coils of simplex-lap windings are connected to adjacent commutator segments. Since this applies to every one of the simplex-lap coils, they are all connected together in succession at successive commutator segments. This means that coil 1 is connected to segments 1 and 2, coil 2 is connected to segments 2 and 3, coil 3 is connected to segments 3 and 4, etc. Segment 2 is therefore the place where coils 1 and 2 are joined together, segment 3 is the place where coils 2 and 3 are joined together, etc. As the connections continue in this regular order, the final coil will have its last end joined to the starting segment, that is, segment 1. The winding is then said to close upon itself, or reenter; the term *reentrancy* is used to designate this important fact. Should any of the wires be broken or disconnected at the commutator while the machine is in service, the winding would have an "open," that is, it would not be reentrant; under this condition the machine would operate unsatisfactorily or not at all.

If, on the other hand, the ends of the first lap coil are joined to segments 1 and 3 instead of 1 and 2, the succeeding coils will be joined to segments 2 and 4, 3 and 5, 4 and 6, etc. When this is done, the winding is called *duplex lap.* Obviously, if the commutator has an *even* number of segments, say, 36, the odd-numbered segments will join together one half of all the coils, while the even-numbered segments will join together the other half of all of the coils. Such a winding is *doubly reentrant.* On the other hand, a duplex-lap winding on an armature having an odd number of segments will be *singly reentrant* because the first half of the winding will end on the last segment, which is odd-numbered; the second half of the winding will then start on the last numbered segment, continue through all the even-numbered segments, and finally reenter on segment 1, the starting point. Figure 39 illustrates the connections for simplex-lap and duplex-lap windings; only two coils are shown for the former and three coils for the latter. These sketches should be studied carefully, particularly with regard to the commutator connections.

If this analysis of multiplex-lap windings is continued, *triplex-lap* windings would have the first coil ends connected to segments 1 and 4, *quadruplex-lap* windings would have the first coil ends connected to segments 1 and 5, etc. And the degree of reentrancy, as before, would depend only upon the relation between the number of commutator segments and the "plex" (sim*plex*, du*plex*, tri*plex*, quadru*plex*, etc.) of the winding.

In general:

1. The *commutator pitch*, symbolized by Y_c, which designates the coil end connections to the commutator, is equal to the "plex" of lap-wound armature windings. Thus, Y_c is equal to 1, 2, 3, 4, etc., respectively, for simplex-, duplex-, triplex-, quadruplex-, etc., lap windings.

2. The degree of reentrancy of lap windings is equal to the *highest common factor* between the number of commutator segments and the "plex" of the winding. Thus, with a 36-segment commutator, the reentrancy is

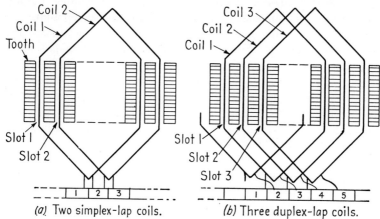

(a) Two simplex-lap coils. (b) Three duplex-lap coils.
Fig. 39. Coil end connections to the commutator for simplex- and duplex-lap windings.

2, 3, and 4, respectively, for duplex-, triplex-, and quadruplex-lap windings. With 38 segments the reentrancies would be 2, 1, and 2, respectively, for duplex-, triplex-, and quadruplex-lap windings.

EXAMPLE 2. Determine the commutator pitches Y_c and the degrees of reentrancy for the following windings:

a. simplex-lap, 29 segments
b. duplex-lap, 29 segments
c. duplex-lap, 42 segments
d. triplex-lap, 47 segments
e. triplex-lap, 48 segments
f. quadruplex-lap, 394 segments

Solution

a. $Y_c = 1$, reentrancy = 1
b. $Y_c = 2$, reentrancy = 1
c. $Y_c = 2$, reentrancy = 2
d. $Y_c = 3$, reentrancy = 1
e. $Y_c = 3$, reentrancy = 3
f. $Y_c = 4$, reentrancy = 2

Parallel Paths in Simplex- and Multiplex-lap Windings. When the current passes through any armature winding, it always divides into an *even* number of parallel paths. In a simplex-lap winding, the current divides into P paths, where P is the number of poles; in a *duplex-lap* winding, the current divides into $2 \times P$ paths; in a *triplex-lap* winding, it divides into $3 \times P$ paths; in a *quadruplex-lap* winding, $4 \times P$ paths serve the total current. Thus, in general, there are $m \times P$ parallel armature winding paths in a lap winding, where m represents the multiplicity (plex) of the winding. For example, a six-pole duplex-lap winding has 12 parallel paths through which the total current passes; an eight-pole triplex-lap winding has 24 parallel paths.

Multiplex-lap windings are generally restricted to low-voltage high-current machines because, practically speaking, it is desirable to limit the current per path to values no greater than about 250 to 300 amp. When the current per path is in excess of these general practical limits, commutation becomes difficult. For example, a 50-kw 25-volt four-pole machine

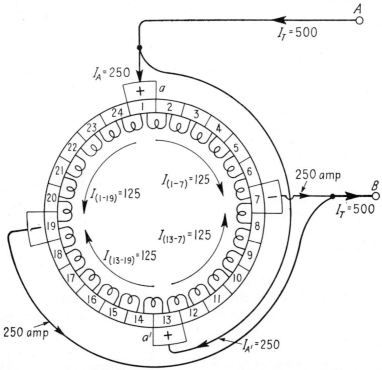

Fig. 40. Schematic winding diagram showing how the total current divides into four paths in the simplex-lap winding of a four-pole machine.

would have a total armature current of about 2,000 amp. If a simplex-lap winding were used, the current per path would be 500 amp, or about twice the upper limit; if a duplex-lap winding were used, the current per path would be about 250 amp. This explains the purpose of multiplex-lap windings, although it must be mentioned in passing that when it is found necessary to use them, they must be very carefully constructed and equalized if unbalanced currents are to be avoided.

Figure 40 represents a simplex-lap winding for a four-pole machine. The diagram has been greatly simplified by showing the individual coils connected to adjacent commutator segments; the student should visualize each coil with the two sides properly placed in armature core slots. If a current of 500 amp is assumed to enter the armature at A, 250 amp pass into brush a and 250 amp pass into brush a'. I_A then divides into two equal parts, $I_{1-7} = 125$ and $I_{1-19} = 125$, while $I_{A'}$ also divides into two equal parts, $I_{13-7} = 125$ and $I_{13-19} = 125$. Thus it is seen that there are four parallel paths, each of which carries one-fourth the total current. If a sketch were made showing a four-pole duplex-lap winding, the total current would be seen to divide itself into eight parallel paths with 62.5 amp per path.

EXAMPLE 3. The table following gives several armature winding problems with their solutions:

	PROBLEMS				SOLUTIONS			
Number	Slots	Segments	Poles	Winding*	Y_s	Y_c	Parallel paths	Reentrancy
1	27	27	4	SL	6	1	4	1
2	66	66	6	DL	11	2	12	2
3	63	63	6	DL	10	2	12	1
4	160	160	6	TL	26	3	18	1
5	147	147	8	TL	18	3	24	3
6	222	222	8	QL	27	4	32	2

* SL = simplex-lap; DL = duplex-lap; TL = triplex-lap; QL = quadruplex-lap.

After verifying the solutions given, the student is urged to draw simple sketches, such as Fig. 39, to illustrate each problem.

Figure 41 emphasizes the importance of tracing winding diagrams from positive to negative brushes. Figure 41a shows a complete simplex-lap diagram for a four-pole 20-commutator-segment armature. All commutator segments are properly numbered, and every coil has been given a letter. Coil a appears in heavy lines, with its slot pitch $Y_s = 5$ and its commutator pitch $Y_c = 1$. Note particularly how the brushes are located and connected. Figure 41b is a dissected sketch of the same diagram, showing how the individual coils are connected to the respective commutator segments and especially how the total current divides into four parallel paths. A careful study of the complete diagram and its schematic counterpart should prove extremely useful to the student in his understanding of the lap type of windings.

Simplex-wave Windings. From the standpoint of armature winding *construction*, there is only one difference between lap and wave windings; this is simply the manner in which the coil ends of the two types are connected to the commutator. In *lap* windings, the coil ends are con-

62 ELECTRICAL MACHINES—DIRECT CURRENT

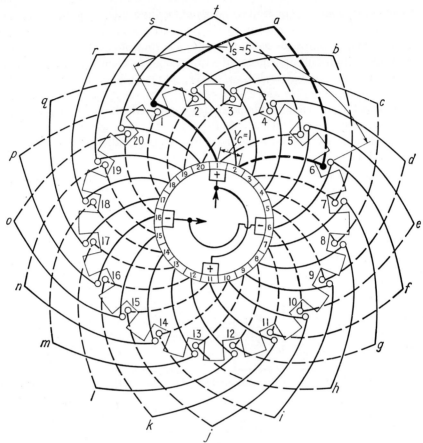

(a) Complete winding diagram of four-pole simplex-lap winding.

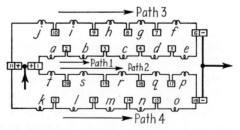

(b) Schematic diagram of four-pole simplex-lap winding showing four parallel paths.
Current directions indicated for motor action.

Fig. 41. Simplex-lap diagrams for a four-pole 20-commutator-segment armature.

nected to segments quite close together (Fig. 39); also, the commutator pitch Y_c is a small number, such as 1, 2, and 3, for simplex, duplex, triplex, etc., windings. In *wave* windings, however, the commutator pitch Y_c is a comparatively large number because the coil ends must be joined to segments approximately 360 electrical degrees apart (Fig. 34b). The word *approximately* must be taken literally because a wave winding is impossible if the coil ends are joined to segments exactly 360 electrical degrees apart.

To understand why the foregoing statements are true, consider the following facts about a *simplex-wave winding:*

1. Starting at any commutator segment, the *entire* winding must be traced from segment to segment and from coil side to coil side before closure occurs, that is, before the winding reenters.

2. If the coil ends are connected to segments exactly 360 electrical degrees apart, every group of $P/2$ coils would constitute a complete reentrancy because there are 360 electrical degrees in every pair of poles $(P/2)$.

An example should make this clear. Assume a 48-segment commutator. If $P = 4$, segments 1 and 25 are exactly 360 electrical degrees apart. One coil, therefore, the sides of which are in slots on one side of the armature, has its ends connected to these segments; another coil, the sides of which are in slots on the opposite side of the armature, has its ends joined to the *same* segments. Thus these two coils would constitute a complete closure, which is a violation of statement 1 above (Fig. 42a). If $P = 6$, then segments 1 and 17, 17 and 33, and 33 and 1 are all 360 electrical degrees apart; therefore, if these three coils were connected, one to each of the above pairs of segments, the winding would close, which would be a violation of statement 1 above (Fig. 42b). Again, if $P = 8$, then segments 1 and 13, 13 and 25, 25 and 37, and 37 and 1 are all 360 electrical degrees apart; therefore, if these four coils were connected, one to each of the above pairs of segments, the winding would close, which would again be a violation of statement 1 above (Fig. 42c).

For a simplex-wave winding to have only one degree of reentrancy, the number of commutator segments must be selected with relation to the number of poles, so that the commutator pitch Y_c can be made a little *more* or *less* than 360 electrical degrees. Also, after tracing the winding *once* around the commutator, the last coil end should arrive one segment behind or one segment ahead of the starting segment. In this way, further tracing of the winding will result in dropping back or going ahead one segment at a time, until the entire winding is traced and the first segment is reached; thus the winding will reenter only after every coil and segment is accounted for.

A consideration of the foregoing therefore leads to the following general

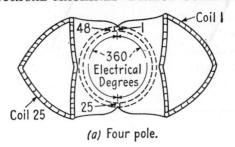

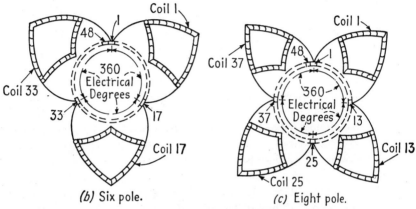

(b) Six pole.   (c) Eight pole.

Fig. 42. *Improper* commutator connections for simplex-wave windings.

rule: *a simplex-wave winding is* NOT *possible if* segments $\div P/2$ *is an integer*. However, *a simplex-wave winding* IS *possible if the following equation results in an integer*:

$$Y_c = \frac{C \pm 1}{P/2} \qquad (7)$$

where Y_c = commutator pitch
C = total number of commutator segments
P = number of poles

Note that in this equation it is necessary to add 1 to C or subtract 1 from C *before* dividing by the number of pairs of poles $(P/2)$. And if the result is *not* an integer, a simplex-wave winding is impossible; if the result *is* an integer when either $+1$ or -1 is used, a simplex-wave winding is possible.

EXAMPLE 4. Determine the commutator pitch Y_c for a four-pole simplex-wave-wound armature having 21 segments. Also list the commutator segments in the proper order as the coils are traced through the entire winding from segment 1 until it closes.

DIRECT-CURRENT DYNAMOS

Solution

$$Y_c = \frac{21 \pm 1}{2} = 10 \text{ or } 11$$

Note the two possibilities for this armature. Using $Y_c = 10$, the succession of commutator segments is as follows:

1–11–21–10–20–9–19–8–18–7–17–6–16–5–15–4–14–3–13–2–12,

then reentering segment 1

Using $Y_c = 11$, the succession of commutator segments is as follows:

1–12–2–13–3–14–4–15–5–16–6–17–7–18–8–19–9–20–10–21–11,

then reentering segment 1

EXAMPLE 5. Calculate the commutator pitches Y_c for the following pole and commutator segment combinations: (*a*) six poles, 34 segments; (*b*) eight poles, 63 segments; (*c*) 10 poles, 326 segments. In each case, trace the winding around the commutator once; start at segment 1 and show that after one trip around the commutator, a segment is reached that is one *behind* or one *ahead* of the starting segment.

Solution

(*a*) $$Y_c = \frac{34 - 1}{3} = \frac{33}{3} = 11$$

Tracing, 1–12–23–34 (one behind segment 1)

(*b*) $$Y_c = \frac{63 + 1}{4} = \frac{64}{4} = 16$$

Tracing, 1–17–33–49–2 (one ahead of segment 1)

(*c*) $$Y_c = \frac{326 - 1}{5} = \frac{325}{5} = 65$$

Tracing, 1–66–131–196–261–326 (one behind segment 1)

Number of Parallel Paths in Simplex-wave Windings. There are two extremely important electrical differences between simplex-lap and simplex-wave windings. These are:

1. In a simplex-lap winding there are as many parallel paths between positive and negative brushes as there are poles (see Fig. 41), while the number of parallel paths in a simplex-wave winding is two, regardless of the number of poles.

2. The conductors in each of the P paths of a lap winding are distributed under *two* poles, a *north* and a *south* pole, whereas the conductors

in each of the two paths of a wave winding are distributed under *all* the poles.

The fact that simplex-wave windings always have *two* parallel paths, each of which contains conductors more or less uniformly distributed around the entire armature-core surface, serves to provide machines wound in this manner with certain unique advantages. In the first place, wave-wound dynamos need have but two sets of brushes (although as many brush sets as poles are frequently employed), while satisfactory operation of lap-wound machines is possible only if there are as many brush sets as there are poles. The use of but two sets of brushes in multi-polar machines is often very convenient, particularly in installations in which brushes cannot be replaced without difficulty. Furthermore, in a wave-wound machine equipped with as many brush sets as poles, if one or more of the brush sets develop poor contact with the commutator, satisfactory operation is still possible; this is not true of lap-wound machines.

A second advantage possessed by wave-wound dynamos is that sparkless commutation is more likely to occur than in those having lap armature windings. The reason for this is that each of the two parallel paths contains conductors distributed completely around the entire circumference; in lap armatures, each of the P parallel paths contains conductors lying under two poles only. If the fluxes produced by all the poles are not exactly the same, the voltages generated in both of the paths of the wave-wound armature are still exactly equal because the two paths are affected similarly. This is not true in lap-wound armatures, because the conductors in one path may be cutting fluxes under stronger poles than may the conductors of some other path. Under this condition, current must flow from the path in which the generated voltage is higher to the one in which the generated voltage is lower. Since this circulating current must pass across the brush contacts, sparking is likely to result. To avoid this difficulty, lap-wound armatures must be specially constructed with equalizer connections, which will be discussed on page 78.

In order to understand why machines having simplex-wave armature windings need be provided with only two brushes, why they have only two parallel paths, and why each path consists of conductors distributed completely around the circumference, it will be desirable to study Fig. 43. This represents a complete winding diagram for a four-pole 21-slot 21-commutator-segment armature. Single-turn coils are shown for simplicity.

In Fig. 43a, it should be noted that coil a is connected to segments 1 and 11; coil b, to segments 2 and 12; coil c, to segments 3 and 13; and so on until the last coil u is connected to segments 21 and 10. Moreover, one positive brush covers segment 1 and "feeds" current to one end of coil l; the other positive brush, diametrically opposite, "feeds" current into the

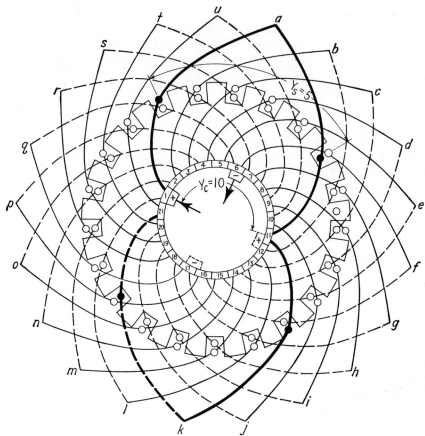

(a) Complete winding diagram of four-pole Simplex-wave winding.

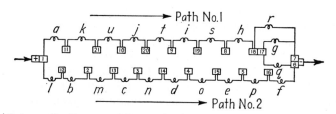

(b) Schematic diagram of four-pole simplex-wave winding showing two parallel paths. Current directions are indicated for motor action.

FIG. 43. Winding diagram for a four-pole 21-slot 21-commutator-segment armature.

same coil *l* through coil *a*. For this reason, the second positive brush, shown with broken lines, may be eliminated. Also, one negative brush "takes" current from one end of coil *f*, while the second negative brush "takes" current from coil *f* through coil *q*; the latter brush, therefore, can also be eliminated. Obviously, two brushes are necessary for a wave-wound armature having four poles. If the winding of a machine having six, eight, or more poles were drawn, it would also be found that only one positive and one negative brush would be necessary.

Figure 43*b* is a schematic diagram of the same winding, showing the various coil elements spread out and properly connected to the commutator segments. This dissected sketch is electrically accurate, although the segments and coils are not placed in their actual positions with respect to one another. It clearly shows, however, that only two parallel paths exist between positive and negative brushes and that each path contains coils uniformly distributed around the entire armature. At the instant shown, coils *g* and *q* are short-circuited at segments 6 and 7 by the negative brush; path 1 contains nine coils in series, while path 2 contains 10 coils in series. In armatures having a great many coils and segments, a difference of one or two coils between the two paths would be less significant than in the 21-slot armature chosen in this example.

Multiplex-wave Windings. It was shown in the foregoing section that there are only two parallel paths between positive and negative brushes in simplex-wave windings, regardless of the number of poles in the machine. For generators and motors of up to about 75-kw capacity with voltages that result in total armature currents of no more than 600 amp, the current per path will not exceed 300 amp. This value represents a sort of *practical* upper limit, above which serious commutation difficulties appear. For total current ratings larger than 600 amp, it becomes necessary to use armature windings having more than two parallel paths, that is, windings that are not simplex wave. Simplex-lap windings that have as many parallel paths as poles are customarily employed in such cases, but their use always involves a certain degree of unbalance in the generated voltages in the various paths. The reason for this situation is that, in lap windings, the conductors in each of the *P* paths are distributed under *two* poles, a *north* and a *south* pole. Since the voltage generated in any given path with respect to that in any other path is determined by the flux created under the corresponding two poles, it follows that equality of generated emfs cannot exist unless there is absolute symmetry and magnetic balance around the entire circumference. Practically speaking, this condition is rarely attainable, so that when lap windings are used, it is always necessary to employ corrective measures to prevent the flow of circulating currents that result from such unbalance. In contrast to the foregoing, it should be understood that magnetic unbalance *cannot* produce electrical unbalance in simplex-wave-wound machines, because the conductors in each of the two parallel paths are distributed completely around the entire circumference; any nonuniform flux distribution that may exist acts similarly upon the two paths.

In order to retain the advantages of the wave type of winding and still keep within the upper limit of current per path indicated above, designers have attempted to use multiplex-wave windings. Such windings have 2 × "plex" paths in parallel, regardless of the number of poles. Thus duplex-, triplex-, and quadruplex-wave windings have four, six, and eight paths in parallel, respectively; for total armature currents in the range of approximately 600 to 1,200 amp, duplex-wave windings can be used, and for currents of about 1,200 to 1,800 amp, it is possible to employ triplex-wave windings. Such windings, however, have had limited use because other operating difficulties have developed. In some cases, the arcing at the commutator that these windings were supposed to correct appeared in another way by causing sparking and blackening at every second, third, or fourth segment, a condition that resulted when slight differences existed in the electric circuits.

The greatest field of application for multiplex-wave windings, however, has been in connection with the so-called *frog-leg winding*, previously mentioned, for which a construction combining a multiplex-wave and a simplex-lap winding has been developed; thus the frog-leg winding has the unique feature of retaining the advantages of both types without their inherent disadvantages. Frog-leg windings will be discussed on page 80.

It will be recalled that a simplex-wave winding requires that the commutator pitch Y_c be so chosen with respect to the number of segments that after tracing the winding *once* around the commutator, the last coil end arrives one segment behind or one segment ahead of the starting segment (see Examples 4 and 5). By similar reasoning, it is possible to show that:

1. A *duplex-wave* winding requires Y_c to be so chosen that after the winding is traced once around the commutator, the last coil end arrives *two* segments behind or ahead of the starting point.

2. A *triplex-wave* winding requires Y_c to be so chosen that after the winding is traced once around the commutator, the last coil end arrives *three* segments behind or ahead of the starting point.

3. In general, a *multiplex-wave* winding requires Y_c to be so chosen that after the winding is traced once around the commutator, the last coil end arrives m segments behind or ahead of the starting point.

For the above condition to be fulfilled, it is necessary that Eq. (7), p. 64, be modified as follows:

$$Y_c = \frac{C \pm m}{P/2} \tag{8}$$

where m is the "plex" of the multiplex-wave winding, the other symbols are as defined on p. 64, and the result is an integer.

When this equation is used, it should be clear that Y_c must result in a whole number; if it does not, the particular multiplex-wave winding is not possible for the combination of segments, poles, and "plex." In order to emphasize the points made above and to test the validity of Eq. (8), a number of problems will now be solved.

EXAMPLE 6. Determine the commutator pitch Y_c for a four-pole duplex-wave-wound armature having 120 commutator bars. Trace the winding around the commutator once, starting at segment 1, and show that after one trip around the commutator, a segment is reached that is two segments *behind* or *ahead* of the starting segment.

Solution

$$Y_c = \frac{120 \pm 2}{2} = 59 \text{ or } 61$$

Tracing for $Y_c = 59$, segments 1–60–119 (two *behind* segment 1)
Tracing for $Y_c = 61$, segments 1–62–3 (two *ahead* of segment 1)

Note that in this example it is possible to use either of two values for the commutator pitch. In general, this is true for all four-pole duplex-wave windings, where the number of segments is an *even* number.

EXAMPLE 7. Using the procedure illustrated in Example 6, make calculations for a 126-segment six-pole triplex-wave winding.

Solution

$$Y_c = \frac{126 \pm 3}{3} = 41 \text{ or } 43$$

Tracing for $Y_c = 41$, segments 1–42–83–124 (three *behind* segment 1)
Tracing for $Y_c = 43$, segments 1–44–87–4 (three *ahead* of segment 1)

EXAMPLE 8. The following table gives several armature winding problems with their solutions:

	PROBLEMS				SOLUTIONS
Number	Segments	Poles	Winding*	Y_c	Tracing
1	65	6	DW	21	1–22–43–64
2	90	6	TW	29	1–30–59–88
3	124	8	QW	30	1–31–61–91–121
4	213	10	TW	42	1–43–85–127–169–211

* DW = duplex-wave; TW = triplex-wave; QW = quadruplex-wave.

Each of the solutions in the foregoing table will become clearer if the student illustrates them with sketches similar to Fig. 44 for Problem 2.

Armatures with More Commutator Segments than Slots. The discussion thus far has avoided the fact that, except for very small machines, armatures usually have more commutator segments than slots. In

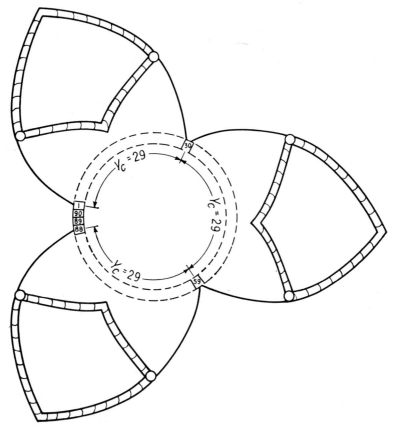

FIG. 44. Sketch illustrating Problem 2 of Example 8. This is for a 90-segment six-pole triplex-wave winding. Note that after tracing the winding once around the commutator, segment 88 is reached, three segments behind 1.

Example 3, a group of problems was chosen to illustrate how to calculate the coil pitch for lap windings and, for simplicity, it was assumed that there were as many segments as slots. In practice, however, it is more satisfactory to design armatures with fewer slots than segments for the following reasons:

1. As the number of commutator segments is increased, the voltage between those that are adjacent to each other decreases and the number

of turns of wire in the coil or coils connected to adjacent segments also decreases. The result is that there is less sparking at the commutator because of the improved commutation.

2. As the number of slots is reduced, the armature core teeth become mechanically stronger, so that, from the standpoint of handling in manufacture, there is less damage to laminations and coils.

3. Assuming that a comparatively large number of segments has been selected so that good commutation will result, the choice of an armature core with one-half, one-third, one-fourth, etc., as many slots means that fewer coils will have to be constructed; this reduces the cost of manufacture.

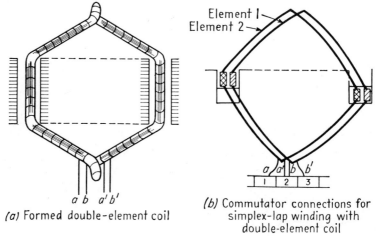

(a) Formed double-element coil

(b) Commutator connections for simplex-lap winding with double-element coil

FIG. 45. Sketches representing manner in which a lap coil is formed and connected to the commutator when there are twice as many segments as slots.

It should be thoroughly understood that the pitch calculations for Y_s and Y_c are made in exactly the same way, regardless of whether the number of slots and segments is the same or not. The important point that must be kept in mind, however, when the number of slots does not equal the number of segments, is that formed coils are no longer simple and double-ended; for example, an armature with twice as many segments as slots has double coils, each one of which connects to a pair of segments; an armature with three times as many segments as slots has triple coils, etc. Exactly how such coils are handled in the armature winding will now be explained.

When an armature has twice as many commutator segments as slots, each of the completely formed coils is a sort of *double-element coil*, in the sense that it serves in a double capacity. Such coils are necessary because each slot must hold as many wires as are ordinarily held by two slots when the number of slots equals the number of segments. When they are con-

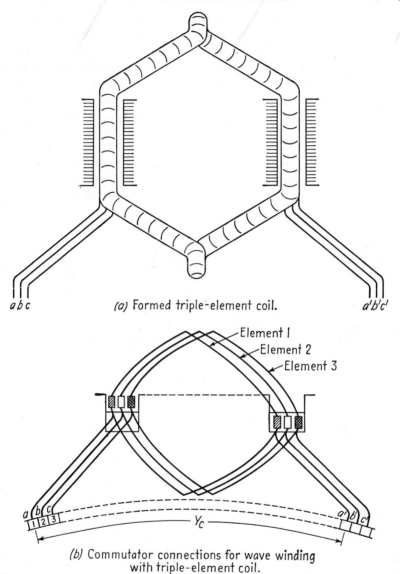

(a) Formed triple-element coil.

(b) Commutator connections for wave winding with triple-element coil.

FIG. 16. Sketches representing manner in which a wave coil is formed and connected to the commutator when there are three times as many segments as slots.

structed, the coils are wound with two wires at the same time, with as many turns used as necessary; when completed, they will have four ends, as indicated in Fig. 45. Note that coil ends a and a' correspond to one element, while coil ends b and b' correspond to the second element. Connections to the commutator for a simplex-lap winding are shown in Fig.

45b, but properly joining the coil ends to the correct segments makes it possible to provide any "plex" that may be desired. For example, if a and a' were joined to segments 1 and 3, while b and b' were joined to segments 2 and 4, the winding would be duplex lap. Note that the coil pitch Y_s remains unchanged. It should be understood, of course, that what is done for one coil is repeated for all the remaining coils, the total number of coils being equal to the number of armature slots.

Figure 46 represents a triple-element coil for a wave winding. Note that there are six ends, two for each element. Commutator connections must be made in the usual manner after the commutator pitch Y_c has been calculated for the winding desired. And here again it should be made clear that calculations for Y_s and Y_c are absolutely independent of the

FIG. 47. Armature before commutator was pressed on the shaft. (*General Electric Co.*)

ratio of segments to slots. The sketch shows that the ends of element 1 are a and a', the ends of element 2 are b and b', while the ends of element 3 are c and c'. Connections to the commutator are indicated in Fig. 46b, where the commutator pitch Y_c is represented by the distance a to a' (or b to b', or c to c').

A triple-element wave coil was shown in Fig. 36, which the student should compare with Fig. 46a.

Figure 47 depicts a lap type of armature winding in a core in which each coil has four elements. This means that the commutator to be used with this armature must have four times as many segments as slots. In small machines such as this, which is for a two-pole motor, the commutator is pressed on the shaft *after* the winding has been completed; this practice makes it possible to insert the coils in the slots more easily because there is ordinarily very little space between the core and the commutator. Figure 48 shows the same armature with the commutator pressed on the shaft and most of the coil ends connected to the proper segments. Note particu-

Fig. 48. Armature of Fig. 47 after the commutator was pressed on the shaft and most of the coil leads were connected. (*General Electric Co.*)

larly that the four coil ends from a four-element coil are carefully separated to show that the armature has four times as many segments as slots.

In larger machines, it is customary to press the commutator on the shaft before the formed coils of the armature winding are put in place. An unwound armature is shown in Fig. 30, and one in the process of being completed may be seen in Fig. 49. A careful study of this view should

Fig. 49. Partly wound armature for a four-pole motor or generator. There are 29 slots and 87 segments, so that triple-element coils are required. (*General Electric Co.*)

76 ELECTRICAL MACHINES—DIRECT CURRENT

indicate to the student that the winding is wave, with a coil pitch Y_s of 7, since the coil span is from slots 1 to 8. It should also be observed that the coils are of triple-element construction because three leads may be seen to emerge from the sides of the coils. Actually, this armature has 29 slots and 87 commutator segments, for which pitch calculations as shown in the two following equations for a four-pole simplex-wave winding yield:

$$Y_s = \frac{29}{4} - \frac{1}{4} = 7$$

$$Y_c = \frac{87 - 1}{2} = 43$$

A simple diagram of one coil of such a winding is illustrated in Fig. 46b, in which the slots might be numbered 1 and 8 and the three segments to which a', b', and c' are connected might be identified as 44, 45, and 46, respectively.

Dead, or Dummy, Elements in Armature Windings. In designing standard lines of motors and generators, manufacturers find it necessary, for economic reasons, to use a certain armature-core lamination for several specifications of voltage, speed, and number of poles. This practice requires that commutators with definite numbers of segments be used with the core-slot combination selected. When lap windings are used, the number of segments may be *exactly* two, three, four, or more times the number of slots, because the commutator pitch Y_c is independent of this number. When wave windings are necessary, however, the number of segments must be chosen so that Eq. (8) is fulfilled. For example, if this equation is applied to four- and eight-pole simplex-wave windings, it becomes apparent that the number of segments must be an odd number; for duplex-wave windings, the number of segments must be even when there are four and eight poles. In any case, it should be understood that the designer is free to apply a commutator of any selected number of segments to the shaft of an armature, regardless of the core-slot combination, since commutators are made or purchased independently of core laminations.

Further consideration of the four-pole simplex-wave winding, an extremely popular winding, discloses the fact that no additional problem is presented if the slot-segment combinations are, for example, 15–45, 17–51, or 31–155. In such cases, the number of segments is properly odd. But a combination like 26 slots and 52 segments is not permissible, nor is 35 slots and 70 segments, nor any combination of slots and segments having a ratio of 1 to 2. However, if an arrangement is used in which the number of segments is *one less than 2, 3, 4, etc., times the number of slots* such as 16 slots and 31 segments, then a four-pole simplex-wave

DIRECT-CURRENT DYNAMOS 77

winding *is* possible. Other permissible combinations are 16–47, 18–71, or 26–103; in each case, the number of segments is one less than exactly 2, 3, or 4 times the number of slots.

When the above practice is followed, and it frequently is, it will always be found that there is one complete element of a multielement coil that cannot be used electrically; there are not sufficient segments for exactly

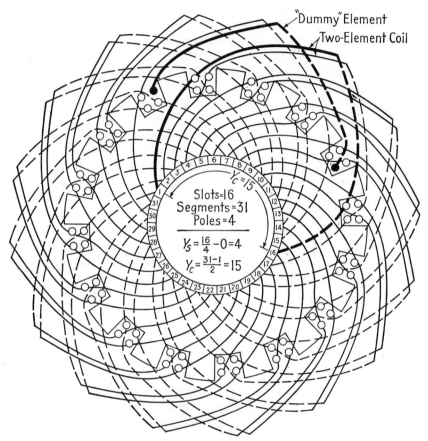

Fig. 50. Winding diagram for four-pole simplex-wave armature with dummy element.

two ends of one element, because the ratio of segments to slots is *not* a whole number. Thus, when the winding is connected, one element is left open in the armature and serves only to keep the revolving structure balanced mechanically. The unconnected element is called a *dead*, or *dummy*, element. A machine such as this will perform satisfactorily, assuming that other conditions are correct.

Figure 50 is a complete winding diagram for a four-pole simplex-wave winding and illustrates the idea of the dead, or dummy, element. It should be studied carefully because it represents an extremely common arrangement in practice. Note that this armature has 16 slots and 31 segments.

Equalizer Connections for Lap Windings. In large *lap-wound* machines, trouble is frequently encountered when the air gaps between stationary poles and rotating armatures are not all alike or when the reluctances of the several magnetic paths are not the same. These conditions cause the generated voltages in the various parallel paths to differ widely, and as a result, large circulating currents flow in the armature winding. These circulating currents not only tend to heat the armature above the temperature caused by the normal load current, but also cause an undue amount of sparking at the brushes, since such currents must circulate from one path to another through the contacts between brushes and commutator. Sparking should, of course, be avoided, because it causes undue burning and wear of the commutator and brushes and, if carried too far, may result in flashover from positive to negative brushes, a condition representing a *short circuit* across the supply lines.

It is true that very small differences in the air gaps under the various poles or lack of similarity in the flux distributions of the poles can produce the inequalities described, and even if extraordinary care is taken by machine assemblers to make careful adjustments, such air-gap differences appear in time as the bearings wear. As was pointed out previously, the circulating currents can flow in lap-wound armatures only because the conductors of each path are not distributed completely around the circumference, as in wave windings, but occupy positions under one pair of poles at a time. For example, in a six-pole lap type of winding, each of the six parallel paths could conceivably generate a different voltage, as a result of which circulating currents would flow from the higher voltage paths through the brush contacts to the lower voltage paths. The circulating currents actually flow to *equalize* the terminal voltage between positive and negative sets of brushes; that is, the paths where the generated voltages are too high suffer a voltage drop, while those where the generated voltages are too low incur a voltage rise. Moreover, it is a significant fact that a voltage difference of as little as 1 per cent may cause circulating currents to flow that may be as much as 50 per cent of the rated armature current.

To overcome the detrimental effects resulting from the circulating currents, it is customary to use *equalizer connections* in all lap-wound armatures. These are low-resistance copper wires that connect together points in the armature winding which should, under ideal conditions, be at exactly the same potential at all times but which, because of mechanical and electrical differences, are not. Such equalizer connections perform two important functions: (1) they relieve the brushes of the circulating current load by causing the latter to be bypassed, and (2) they create a magnetic effect that actually reduces the flux under those poles where there is too much magnetism and increases the flux under those poles where there is too little magnetism.

Consider Fig. 51, which schematically represents a simplex-lap winding for a six-pole machine. Since coils a, b, and c occupy identical positions under similar (*north*) poles, they are exactly 360 electrical degrees apart;

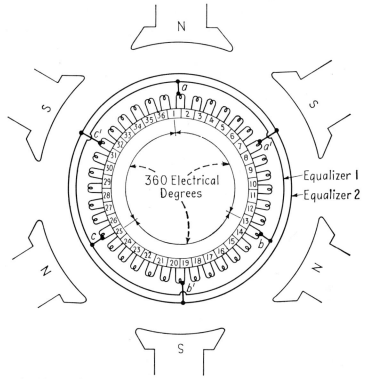

FIG. 51. Sketch showing two equalizer connections in a six-pole lap-type armature winding.

the same may be said of coils a', b', and c', which are under identical *south*-pole positions. Obviously, the voltages in coils a, b, and c should be equal and in the same direction, and the same should be true for coils a', b', and c'. If for any reason this is not true, circulating currents will flow to equalize any potential differences, and what is particularly significant is that such currents will pass across brush contacts (not shown in the figure). To bypass the brushes, equalizer connections 1 and 2 can be made in the manner indicated in the sketch. A difference in potential between points a and b, for example, will cause a circulating current to flow from a to b

through equalizer 1 and back through all the coils from b to a. This current will not only avoid the brushes but will actually *reduce the magnetic unbalance that causes the original potential difference.*

Note that in this sketch only two equalizer connections are shown. To achieve 100 per cent equalization, all coils must be properly connected, which, in this 36-coil example, would require 12 equalizers, since each one connects together three coils.

Since equalizers must connect points that are exactly 360 electrical degrees apart, it follows that *the total number of coils in an armature winding must be divisible by half the number of poles, or $P/2$* (in this case 36/3 = 12). In many lap-wound armatures, the equalizers are more easily made at the commutator instead of at the coils; in such cases, it is common to refer to the armature as having a *cross-connected commutator*. In this example, the 12 equalizers would join the following segments:

$$
\begin{array}{llll}
1\text{--}13\text{--}25 & 5\text{--}17\text{--}29 & 9\text{--}21\text{--}33 \\
2\text{--}14\text{--}26 & 6\text{--}18\text{--}30 & 10\text{--}22\text{--}34 \\
3\text{--}15\text{--}27 & 7\text{--}19\text{--}31 & 11\text{--}23\text{--}35 \\
4\text{--}16\text{--}28 & 8\text{--}20\text{--}32 & 12\text{--}24\text{--}36
\end{array}
$$

EXAMPLE 9. Determine the number of equalizer connections at the commutator of a 160-segment eight-pole armature for 100 per cent equalization and make a table showing those that are joined together at each one.

Solution

Number of equalizers for 100 per cent equalization $= \dfrac{160}{8/2} = 40$

Equalizer connections:

1–41–81–121	11–51–91–131	21–61–101–141	31–71–111–151
2–42–82–122	12–52–92–132	22–62–102–142	32–72–112–152
3–43–83–123	13–53–93–133	23–63–103–143	33–73–113–153
4–44–84–124	14–54–94–134	24–64–104–144	34–74–114–154
5–45–85–125	15–55–95–135	25–65–105–145	35–75–115–155
6–46–86–126	16–56–96–136	26–66–106–146	36–76–116–156
7–47–87–127	17–57–97–137	27–67–107–147	37–77–117–157
8–48–88–128	18–58–98–138	28–68–108–148	38–78–118–158
9–49–89–129	19–59–99–139	29–69–109–149	39–79–119–159
10–50–90–130	20–60–100–140	30–70–110–150	40–80–120–160

Frog-leg Windings. As explained in the previous article, lap windings operate unsatisfactorily unless equalizers are provided. These latter connections improve the commutating process materially, although they generally add to the expense of the machine and to the difficulties experienced in making repairs. If it were possible to obtain a multiplicity of armature circuits by the use of wave windings, the necessity for installing equalizers would be eliminated because wave windings are self-equalizing by virtue of the fact that each path is made up of a series of coils lying under all the poles at the same time. The designers of early machines therefore tried to use multiplex-wave windings, but it was soon discovered that, unless each of the simplex sections of the winding was identical with every other section, their use resulted in blackening and burning of those segments to which the most heavily loaded section happened to be connected. Such a condition, of course, introduced the original difficulty of sparking at the commutator, a difficulty it was supposed to correct.

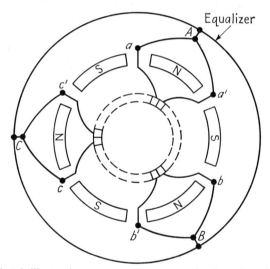

Fig. 52. Sketch illustrating one equalizer connection in a six-pole machine.

After a number of sad experiences, the simplex-lap windings were tried again, this time with the use of equalizer connections; it is this arrangement that is generally used on most machines today.

The equalizer connections, it should be remembered, are low-resistance conductors between points on the armature winding that are 360 electrical degrees apart; they are placed on the armature outside the influence of the magnetic field. *They are non-potential-generating wires and carry equalizing currents only.* Their usual location is at the rear end of the armature winding, each of the equalizers being tapped in at the rear bend of the coils. Sometimes, when this is possible, equalizer rings connect together commutator segments exactly two pole pitches apart. The diagram in Fig. 52 indicates one equalizer ring connected to points A, B, C for a six-pole machine.

Further inspection of Fig. 52 discloses the fact that, theoretically, points A, B, and C are at the same potential. Since this is so, it is at once evident that points a and a', b and b', and c and c' are also at the same potential because these points are all connected to the equalizer and are themselves *lying outside the influence of the magnetic field.* This means, of course, that, since no voltage is generated in the portion of the coils between a and A, a' and A, b and B, b' and B, c and C, and c' and C, it is quite possible, without affecting the winding in any way, to connect points a' and b by connection E; points b' and c by connection F; and points c' and a by connection D. Figure 53 indicates the change suggested here.

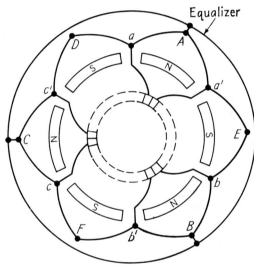

FIG. 53. A simple modification of Fig. 52 in which wires D, E, and F have been added without altering the winding electrically. Note that this sketch represents a sort of lap-wave winding.

If the student looks carefully at Fig. 53, he will observe that it really represents a sort of combination lap-wave winding. The lap-winding portion is the original one, whereas the wave-winding portion was introduced when the second set of connections was made, i.e., E, F, and D.

Proceeding a step further in our discussion of what is to lead to another winding construction, suppose that the wire representing each of the single-turn coils of Fig. 53 is slit in half lengthwise from each of the commutator segments up to points a, a', b, b', c and c'. Electrically, no change has taken place from such an imaginary slitting process. The connections are as given in Fig. 54.

It will be observed that in Fig. 54 the equalizer shown in Fig. 53 has been omitted for the reason that *any wave element, such as E, and the* succeeding lap element, such as B, connect two points on the commutator exactly two pole pitches apart. In addition, *the net voltage theoretically generated in elements E and B in series between segments m and n is zero.* Elements E and B together therefore have the two important characteristic properties that must be possessed by an equalizer connection and may serve in the place of the removed connection. No attempt has been made in our diagrams thus far to indicate how the various coils are interconnected to form complete windings. It has been our purpose simply to choose those elements in the winding that could best be represented as broken up into two distinct halves depicting lap and wave elements.

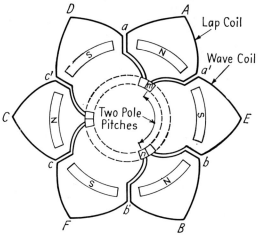

FIG. 54. A further modification of Figs. 52 and 53 in which the arrangement of coils appears like a combination of a lap and wave winding.

Referring again to Fig. 54, note that only three elements are shown surrounding the three *north* poles. Three additional elements can similarly be *imagined* to surround the three *south* poles. If the same reasoning is applied to these three elements and they are sketched in as was done for the original three elements, another diagram similar to Fig. 54 will result. Then, if this imaginary diagram and Fig. 54 are combined, the sketch will look like Fig. 55.

It should be noted in Fig. 55 that there are four coil sides in every slot instead of the customary two. (The number of commutator bars is assumed to be equal to the number of slots.) This fact results from the imaginary lengthwise slitting of each conductor. Similarly, each commutator segment will have four leads connected to it instead of two, for the same reason. (Only two connections appear in the figure.)

Examining next *one* of the new coils somewhat more carefully, notice that it contains a lap element and a wave element, i.e., that it is a sort of combination lap-wave element. Figure 56 illustrates such a coil.

For simplicity, all the elements have been drawn with but one turn each. Although single-turn elements are frequently employed on large

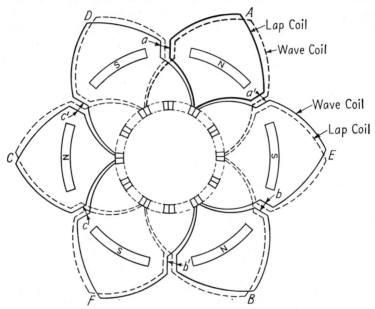

FIG. 55. Sketch illustrating groups of lap and wave coils to form part of a frog-leg winding.

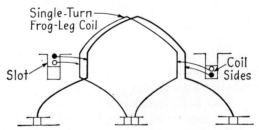

FIG. 56. Sketch illustrating how one frog-leg coil fits into the slots of an armature—one-turn coil.

armatures, it is sometimes necessary to have two turns per element. A coil with two turns per element is represented in Fig. 57.

Because of its general appearance, this new combination lap-wave coil has been appropriately named *frog-leg coil* by the engineers of the Allis-Chalmers Manufacturing Company who developed it, and the winding itself has come to be known as the *frog-leg winding*. Each of the coils of this frog-leg winding is formed in a manner quite similar to that of ordinary lap or wave coils. Single-element lap or wave coils are wound with *single* conductors (or their equivalent in the case of large cross-sectional areas), and their ends are brought out either at the lower bend, in the case of lap coils, or at the sides, in the case of wave coils. The frog-leg coil, however, being a combination "lap-wave" coil, is wound with *two* conductors, and two corresponding ends are brought out at the lower bend to be connected to adjacent commutator bars; the other corresponding ends are bent outwards from the coil sides to be connected to segments at some distance from each other. It should be emphasized that exactly the same result could be obtained electrically by first placing a lap winding in the armature and then proceeding to place a wave winding having the proper "plex" directly over the lap winding in the *same slots* and connecting it to the *same commutator segments*.

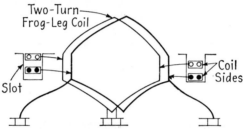

FIG. 57. Sketch illustrating how one frog-leg coil fits into the slots of an armature—two-turn coil.

The next problem is to determine precisely where to connect the ends of the two frog-leg coil elements. Consider first a machine having the same number of slots as commutator segments. Under such circumstances, each frog-leg coil will contain two ends belonging to the lap element and two ends emerging from the wave element. If all the separate lap elements are regarded as forming a *distinct lap winding* and all the wave elements as forming a *distinct wave winding*, it is easy to see that the number of parallel paths in the lap winding *must* equal the number of parallel paths that the wave winding should have because both windings are *in parallel* with each other. That is, since both windings are connected in multiple with one another and are wound with the same number of turns and size of conductor, the number of elements in each circuit of either the lap winding or the wave winding *must be the same* if the generated electromotive forces in *all* circuits are to be the same. Since a simplex-lap winding has as many paths as poles and since a wave winding having a "plex" equal to $P/2$ also has as many paths as poles, it follows that only such a combination will produce satisfactory operation.

In practice, the lap portion of the frog-leg winding is always simplex, so that it is necessary to give the wave portion a "plex" equal to $P/2$.

In making calculations for the coil pitch Y_s and the commutator pitch Y_c of the wave section of the frog-leg winding, it is necessary to consider the lap and wave portions separately. The coil pitches Y_s for lap and wave sections are frequently the same, but the value of Y_c for the wave section must be calculated on the basis of the proper "plex" that it must have. Table 1 indicates the proper lap and wave "plex" combinations that will produce satisfactory operation in machines having 4 to 16 poles.

TABLE 1. LAP-WAVE "PLEX" AND PATHS COMBINATIONS

Poles	Lap		Wave		Frog-leg
	Plex	Paths	Plex	Paths	Total paths
4	1	4	2	4	8
6	1	6	3	6	12
8	1	8	4	8	16
10	1	10	5	10	20
12	1	12	6	12	24
14	1	14	7	14	28
16	1	16	8	16	32

In this discussion, it is well to remember that the real purpose of this type of winding is to eliminate the equalizer connections and yet to retain their advantages. The wave portion of the frog-leg winding, acting together with the lap portion, serves to replace the equalizers, but acts, in addition, as *a current-carrying winding*. It is thus possible to obtain 100 per cent equalization of the winding and also to make the maximum use of all the copper placed on the armature. The engineers of the Allis-Chalmers Manufacturing Company have claimed unusually satisfactory results with this winding and have entirely eliminated the use of equalizers on all their designs.

Frog-leg windings have as many circuits in parallel as duplex-lap windings (see Table 1) because the simplex-lap portion supplies P circuits and the multiplex-wave section also provides P circuits, the total being $2 \times P$ paths in parallel. Since both lap and wave sections are identical electrically, it is obvious that each will carry one-half the total current.

Two problems involving frog-leg windings will now be solved.

EXAMPLE 10. Determine the coil and commutator pitches for a 28-slot 28-segment four-pole frog-leg armature winding. Make a simplified sketch showing several of the frog-leg coils and how they are arranged in the slots and connected to the commutator.

DIRECT-CURRENT DYNAMOS 87

Solution

Lap portion will be simplex.   Wave portion must be duplex.

$$Y_s = \frac{28}{4} + 0 = 7 \quad \text{for both lap and wave sections}$$

$$Y_c = \frac{28 - 2}{2} = 13 \quad \text{for the wave section}$$

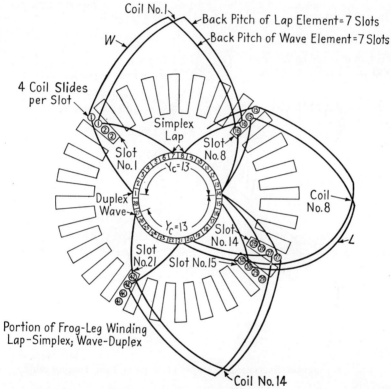

Fig. 58. Diagram illustrating Example 10, showing three frog-leg coils.

Figure 58 is a diagram for this problem showing only three frog-leg coils. All the other coils have been purposely omitted in order to simplify the drawing. A very important point to notice is that elements W and L of coils 1 and 8, respectively, are connected in series between commutator segments 1 and 15, exactly 360 electrical degrees apart. This winding will have eight parallel paths (see Table 1).

EXAMPLE 11. Determine the coil and commutator pitches for a 24-slot 48-segment six-pole frog-leg armature winding. Make a simplified

88 ELECTRICAL MACHINES—DIRECT CURRENT

sketch, as in Example 10, showing several frog-leg coils arranged in the proper slots and connected to the commutator.

Solution

Lap portion will be simplex. Wave portion must be triplex.

$$Y_s = \frac{24}{6} + 0 = 4 \quad \text{for both lap and wave sections}$$

$$Y_c = \frac{48 - 3}{3} = 15 \quad \text{for the wave section}$$

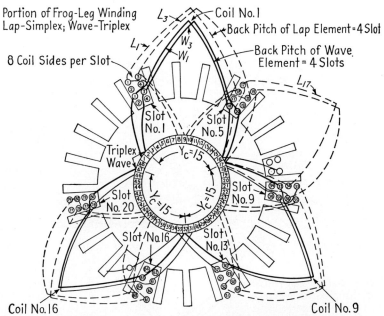

FIG. 59. Diagram illustrating Example 11, showing three frog-leg coils.

Figure 59 shows only those coils that are necessary to clarify the solution of the problem. Note particularly that elements W_1 and L_{17} are connected in series between commutator segments 1 and 17, exactly 360 electrical degrees apart. It should also be pointed out that this winding will have 12 parallel paths.

Instead of using coils that have a single turn of wire each (see Fig. 37), it is sometimes necessary to employ coils that require two or more turns. Such a coil is shown in Fig. 60, which indicates very clearly that the lap and wave sections have three elements each. This implies, therefore, that the armature has three times as many commutator segments as slots. Figure 61 shows a frog-leg winding in the process of being placed in an

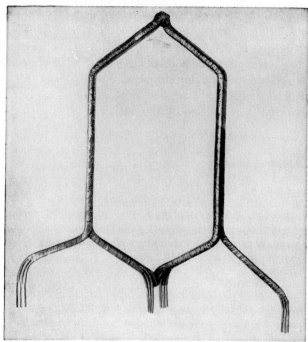

Fig. 60. Frog-leg coil, showing lap and wave elements having more than one turn of wire. (*Allis-Chalmers Mfg. Co.*)

Fig. 61. Frog-leg winding being placed into an armature of a 250-hp 250-volt motor. (*Allis-Chalmers Mfg. Co.*)

armature of a 250-hp 250-volt motor. Note the great care that is taken to keep the winding properly insulated.

Frog-leg windings in which the coil span of the wave section is one slot more than the lap section are sometimes used. This is necessitated by certain design considerations that are beyond the scope of this book. For the method of calculating the proper pitches of such constructions, the student is referred to a bulletin by the author.*

Questions

1. Define the term *dynamo*.
2. Name the various parts of the dynamo and indicate which parts are stationary and which rotate.
3. From what material is each part listed in Question 2 made?
4. What is meant by *pole core? pole shoe?*
5. Why are laminations used in constructing the field core?
6. Why are laminations used in constructing the armature core?
7. Why are pole cores constructed with a shoe?
8. What two kinds of field winding are used for generators and motors? Describe the construction of each.
9. What is the purpose of dipping and baking field coils?
10. What is the *yoke?* What purposes does it serve? How is it usually constructed?
11. Why is cast steel or rolled steel more desirable than cast iron for the yoke material of dynamos?
12. Describe a commutator construction.
13. What purpose does the commutator serve in a generator? in a motor?
14. What is the function of the armature winding in a generator? in a motor?
15. Under what conditions is it necessary to use a brush rigging in which each arm contains several brush holders and brushes?
16. What keeps the brushes in good contact with the commutator?
17. How is the brush surface made to conform with the commutator surface?
18. What purpose is served by the brush pigtails?
19. Which terminal is positive on a generator? on a motor?
20. Name the two general types of armature winding.
21. Distinguish between a simplex-lap and a simplex-wave winding with regard to construction; with regard to the number of parallel paths.
22. In general, what is a frog-leg winding?
23. What is the important requirement with regard to the coil span of all types of armature winding?
24. Explain what would happen if the coil span were 360 electrical degrees in a generator; in a motor.
25. State the formula to determine the coil span Y_s for an armature winding.
26. What is the commutator pitch in a simplex-lap winding? a duplex-lap winding? a multiplex-lap winding?
27. What is meant by the term *reentrancy?*
28. Under what condition is the reentrancy of a duplex-lap winding single? double?
29. Under what condition is the reentrancy of a triplex-lap winding single? triple?
30. What is the purpose of multiplex-lap windings?

*C. S. Siskind, Frog-leg Windings for Direct-current Generators and Motors, *Purdue University Research Engineering Bulletin*, **70**, 1940.

DIRECT-CURRENT DYNAMOS

31. State the general rule for determining the reentrancy of a multiplex-lap winding.
32. What general rule can be used to determine the number of parallel paths in a multiplex-lap winding?
33. To what values of machine ratings are multiplex-lap windings restricted?
34. Why must the ends of wave coils never be connected to the commutator exactly 360 electrical degrees apart?
35. State the general rule for determining the commutator pitch in a simplex-wave winding.
36. Explain why only two brush sets need be used in a machine in which the armature is wave wound.
37. How are the conductors in each path of a lap winding distributed around the armature?
38. How are the conductors in each path of a wave winding distributed around the armature?
39. Under what conditions is it desirable to use two brush sets in machines having wave-wound armatures?
40. Under what conditions is it desirable to use as many brush sets as poles in machines having wave-wound armatures?
41. What are multiplex-wave windings?
42. Under what circumstances would it be desirable to use multiplex-wave windings?
43. Why is it possible to have circulating currents in lap-wound armatures?
44. Why is it impossible to have circulating currents flowing in wave-wound armatures?
45. State the rule for determining the commutator pitch in a multiplex-wave winding.
46. In tracing a simplex-wave winding once around the commutator, at what segment should one arrive with respect to the starting point?
47. Answer Question 46 for a duplex-wave winding; a triplex-wave winding; a multiplex-wave winding.
48. Why is it usually more satisfactory to construct armatures that have more commutator segments than slots?
49. Are pitch calculations Y_s and Y_c affected when an armature has more segments than slots?
50. What is meant by a *double-element* coil? a *triple-element* coil? When are they necessary?
51. Under what condition is it desirable to wind an armature *before* the commutator is pressed on the shaft?
52. What is meant by a "dead" or "dummy" element?
53. When does an armature winding have a "dummy" element?
54. What purpose is served by a "dummy" element?
55. Under what conditions do circulating currents flow in lap-wound armatures?
56. Why are circulating currents detrimental to good operation?
57. What is an equalizer connection?
58. What two important functions are served by equalizer connections in lap-wound armatures?
59. State the general rule for determining the number of equalizer connections to be used in an armature that is to be equalized 100 per cent.
60. Exactly why does a frog-leg winding eliminate the necessity for using equalizer connections? Explain carefully.
61. Is it possible to place a lap winding on top of a properly designed wave winding and have this combination of windings perform as satisfactorily as a frog-leg winding? Explain.
62. What is the origin of the term *frog-leg winding?*

92 ELECTRICAL MACHINES—DIRECT CURRENT

63. If the lap section of a frog-leg winding is always simplex, what must be the multiplicity of the wave section?
64. How many parallel paths are there in a P-pole frog-leg winding?

Problems

1. Calculate the coil pitches and indicate the slots into which the first coils should be placed for the following armature windings: (a) 36 slots, four poles; (b) 57 slots, four poles; (c) 76 slots, six poles; (d) 132 slots, eight poles; (e) 270 slots, 10 poles; (f) 234 slots, 12 poles.
2. How many parallel paths are there in the armature windings of Prob. 1, if (a), (b), and (c) are simplex-wave wound and the others are simplex-lap wound?
3. The armature of a six-pole generator carries a total of 350 amp. What current flows in each path if the winding is (a) lap? (b) wave?
4. How many parallel paths are there in the windings of armatures that are wound (a) duplex-lap for four poles? (b) triplex-lap for six poles? (c) duplex-lap for eight poles? (d) triplex-lap for 10 poles? (e) quadruplex-lap for six poles?
5. Make simple sketches like Fig. 39 showing how lap coils are connected to the commutator for simplex-, duplex-, triplex-, and quadruplex-lap windings.
6. Determine the degrees of reentrancy for the following lap windings: (a) duplex, 36 segments; (b) duplex, 35 segments; (c) triplex, 117 segments; (d) triplex, 118 segments; (e) quadruplex, 286 segments.
7. The total armature current carried by a 50-hp four-pole 230-volt motor is 180 amp. If the winding is wave, how much current is handled by each path and by each brush arm if there are (a) four brush arms? (b) two brush arms?
8. Determine the commutator pitches Y_c for the following wave-wound armatures: (a) 75 segments, four poles; (b) 93 segments, four poles; (c) 229 segments, six poles; (d) 227 segments, eight poles.
9. In each of the combinations of Prob. 8, trace the winding around the commutator and show that the proper segment, ahead of or behind the first one, is reached.
10. How many parallel paths are there in the following armature windings: (a) six-pole duplex-wave? (b) six-pole triplex-wave? (c) eight-pole simplex-wave? (d) eight-pole duplex-wave? (e) eight-pole triplex-wave? (f) eight-pole quadruplex-wave?
11. An eight-pole triplex-wave-wound armature carries a total of 660 amp. If there are eight brush arms, calculate the current in each armature winding path and in the brush arm.
12. A commutator has 456 segments. Indicate whether or not the following windings are possible; (a) simplex-wave for six poles; (b) triplex-wave for six poles; (c) duplex-wave for eight poles; (d) simplex-wave for 10 poles.
13. In each of the possible combinations of Prob. 12, calculate the commutator pitch.
14. For each of the answers of Prob. 13, trace the winding around the commutator and show that the proper segment, ahead of or behind the first one, is reached.
15. A 17-slot 33-segment armature is wound simplex-wave for four poles. Calculate the coil and commutator pitches and specify whether or not there will be a "dead" element.
16. Will there be a "dead" element in each of the following combinations of slots and segments: (a) 33 slots, 99 segments? (b) 76 slots, 227 segments? (c) 39 slots, 77 segments? (d) 54 slots, 216 segments?
17. A 500-kw 600-volt 10-pole generator has an armature with 108 slots, 324 commutator segments, and a simplex-lap winding with a total of 648 conductors. Calculate: (a) the number of conductors per slot; (b) the number of winding ele-

ments per coil; (c) the number of turns per winding element; (d) the full-load current per conductor; (e) the current per brush arm.

18. The emf generated in each conductor of a six-pole simplex-wave armature winding is 0.48 volt. If the armature has 42 slots and 125 commutator segments, and there are four turns per armature-winding element, calculate the terminal voltage of the generator.
19. The following information is given for a d-c generator: terminal volts = 460; rpm = 900; number of slots = 130; number of commutator segments = 390; coil pitch = slot 1 to slot 22; type of winding = simplex lap; air gap flux density = 57,000 maxwells per sq in.; pole-face area = 5.75 × 6 in. Calculate the number of turns per armature-winding element.
20. A 54-slot 216-segment armature has a four-pole lap armature that is equalized 100 per cent at the commutator. (a) How many equalizer connections are there? (b) To what two segments is the first equalizer connected?
21. How many equalizers are there in a six-pole 50 per cent equalized armature in which there are 216 segments?
22. Make a table for Prob. 21 similar to that given for Example 9, p. 80.
23. How many parallel paths are there in (a) a 10-pole frog-leg winding? (b) an eight-pole frog-leg winding?
24. Determine the coil and commutator pitches for a 72-slot 288-segment six-pole frog-leg armature winding.

CHAPTER 4

DIRECT-CURRENT GENERATOR CHARACTERISTICS

Types of Direct-current Generator. There are three general types of d-c generator: shunt, series, and compound. Each type is distinguished by its field winding or windings (that is, by the manner in which the excitation is produced), which are absolutely independent of the kind of armature winding. If the excitation is produced by a field winding that is connected to full, or nearly full, *line voltage*, the machine is known as a *shunt generator*. If the excitation originates in a field winding connected in series with the armature, so that the flux depends upon the *current*

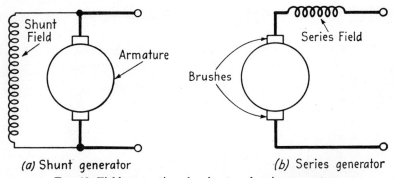

FIG. 62. Field connections for shunt and series generators.

delivered to the load, the machine is called a *series generator;* such generators are seldom found in use today. If the excitation is produced by two field windings, one connected to the full, or nearly full, line voltage and the other excited by the line or armature current, a comparatively large current, the machine is called a *compound generator*. Figure 62 shows the schematic wiring diagrams for the shunt and series types of generator, while Fig. 63 represents two ways of making connections for the compound generator.

The field coil for a shunt generator is illustrated in Fig. 24, while Fig. 23 shows the shunt- and series-field coils for one pole of a compound generator. Note, in Fig. 23, that the series coil is wound *over* the shunt coil; this is good general practice because the series field, carrying high values

94

DIRECT-CURRENT GENERATOR CHARACTERISTICS 95

of current, is kept cool more readily when placed on the outside. Another point to recognize is that the shunt field contains many turns of comparatively fine wire, while the series field is wound with few turns of heavy wire.

When a generator is in operation, whether or not it is delivering a load current, the shunt field is always excited; the series field is excited only when a load current is being supplied. In a series generator, therefore, the terminal voltage is very low at light loads because the excitation or

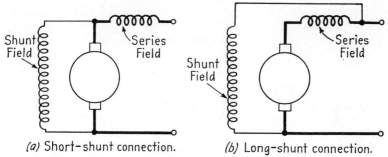

FIG. 63. Two arrangements for compound generator operation. Self-excitation.

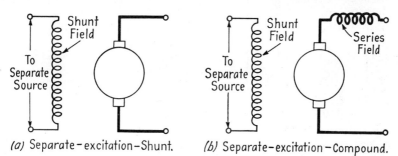

FIG. 64. Separately excited generators.

load current is low under this condition. [Remember that the voltage of a generator depends, among other things, upon the flux. See Eq. (2).] In a compound generator, the series field, acting together with the fairly constant shunt excitation, adds to or subtracts from the latter as the load changes.

The shunt field of a shunt or compound generator may be excited by current supplied to it by its own armature or may be connected to an outside, i.e., a separate, source of supply. In the first of these machines, the generator is said to be *self-excited* (Fig. 63); in the second arrangement the generator is said to be *separately excited* (Fig. 64).

No-load Characteristics of Generators. When a shunt or compound generator operates without load—that is, when it is driven by a prime mover, is properly excited, and has none of the load switches closed—a voltage will appear at the terminals that are normally connected to the electrical devices. This generated voltage will depend, for a given machine, upon two factors: (1) the speed of rotation and (2) the flux. If the flux is kept constant while the speed is increased or decreased, the voltage will rise or fall, respectively, in direct proportion to the changed speed. Similarly, if the speed is held constant while the flux (not the field current) is varied, the voltage will change in direct proportion to the change in magnetism. The first of these statements may be shown to be true experimentally by driving a separately excited generator (Fig. 64)

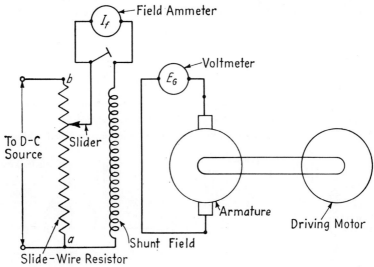

FIG. 65. Separately excited shunt generator connections to determine experimentally the no-load characteristics.

over as wide a range of speed as possible while the shunt-field current is kept absolutely constant. The driving machine may conveniently be a shunt type of d-c motor with a field rheostat for speed control. Figure 65 is a diagram of connections for such an experiment, with the motor wiring omitted for simplicity (the wiring will be included in Chap. 5). A slide-wire resistor is used because with it the field current can be adjusted readily and carefully to any desired value and also because, in a companion experiment described below, it will be necessary to bring the field current down to zero. To perform such an experiment, it will be desirable first to adjust the generator speed to its highest *permissible* value and at the same time to set the field current I_f so that a high voltmeter reading is recorded. Then, as the speed is gradually lowered without changing the field excitation, lower readings of E_G and revolutions per minute (rpm) are recorded. A plot of the dependent variable E_G vs. the independent variable rpm will yield a straight line, as indicated by Fig. 66.

To show that the generated voltage E_G is directly proportional to the *flux* is much more difficult because magnetism measurements are not made as readily as are those of amperes and volts. This determination is not particularly important from a practical point of view, however, because it is more desirable to know how the no-load generated voltage is affected by changes in *field current*. This relationship is not a direct one for *all* changes in excitation, because magnetic saturation sets in after the field current is increased beyond a certain value. Starting with the slider at point *a* (Fig. 65), where the field current is zero, there will be an initial residual generated voltage E_r, because of residual magnetism. As the field current is gradually increased by moving the slider toward *b*, E_G will be proportional to I_f for a limited range. Then, as the iron saturates with increasing values of flux, the generated voltage does not increase as rapidly as the field current. This is shown in the curve of Fig. 67, in which E_G is plotted against I_f for constant speed. In performing such an experiment, it is best to operate the generator at some *constant speed* approximating that given on the name plate. Starting with zero I_f, gradually and *progressively* increase its value until point *b* on the slider is reached,

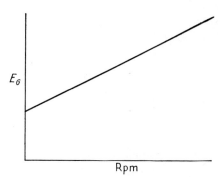

FIG. 66. Direct proportionality relationship between the no-load generated voltage E_G and the speed rpm with constant excitation.

recording data for a number of field-current values. A so-called *saturation curve* (sometimes called *magnetization curve*) can then be plotted; it should be similar in shape to the curve of Fig. 67. The place where the curve departs from the straight line is known as the *knee*, and the upper part, where the voltage is leveling off toward the *x* axis, is where magnetic saturation sets in. The slight curvature at the lower end is due to magnetic inertia.

Such a curve emphasizes the extremely important fact that *the generated voltage is directly proportional to the flux* [see Eq. (2)] *and not the field current*. Since it is also true, as previously shown, that the generated emf is directly proportional to the speed, it is easily possible to calculate the voltage at *any* speed after the magnetization curve (Fig. 67) is plotted. Thus, if E is 100 volts at 1,500 rpm, E_H will be 110 volts at 1,650 rpm [$(1,650/1,500) \times 100 = 110$] and E_L will be 90 volts at 1,350 rpm

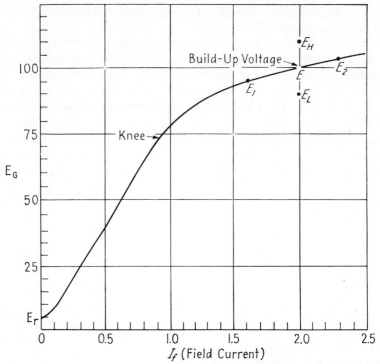

Fig. 67. Saturation curve for d-c shunt generator operating at constant speed.

$[(1{,}350/1{,}500) \times 100 = 90]$. In fact, if a saturation curve is determined by experiment for one speed, it is quite possible to calculate values for another such curve at any other speed by the simple method of proportionality. The student is urged to try this.

Building Up the Voltage of a Self-excited Shunt Generator. When a self-excited shunt generator (Fig. 68) is operated at a speed approximating the name-plate value with the field switch open, the armature will generate a residual voltage, i.e., an emf resulting from the cutting of a residual magnetic field. A voltmeter connected across the armature terminals will then register this voltage E_r, as indicated in Fig. 67. When the field switch is closed, the voltmeter will immediately indicate a much higher value than E_r *if the machine fulfills all the conditions for building up.* As a rule, no difficulty will be experienced in getting such a generator into service from a standing start if it has operated properly in the past; it is necessary merely to bring it up to speed, close the field switch, and make a minor field-rheostat adjustment to obtain the desired voltage. After that, the generator may be made to supply load current from the positive and negative terminals.

DIRECT-CURRENT GENERATOR CHARACTERISTICS

The fact that a self-excited shunt generator *will* build up, i.e., rise from its residual value E_r to its normal operating value, means that several important conditions have been fulfilled. Moreover, the *value of the voltage* to which a given generator will build up will be determined by other factors.

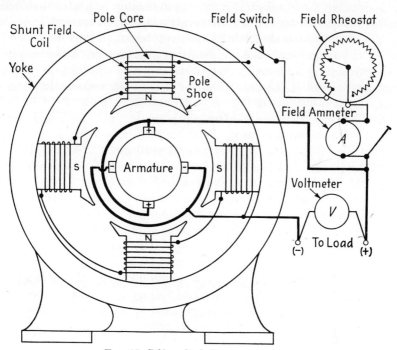

FIG. 68. Self-excited shunt generator.

Four Requirements for Build-up

1. The first requirement for the build-up process is *a small voltage resulting from residual magnetism*. A generator will *not* build up if the residual flux is insufficient; in 110- and 220-volt generators, this flux should be of such magnitude that about 4 to 10 volts is developed. The residual voltage represented by E_r causes a small current to flow through the exciting coils of the field winding. The magnetomotive force thus produced acts upon the residual field either to strengthen or to weaken it. If the excitation is in such a direction that the residual field is weakened, the generator will build *down*, not up. However, if the field is strengthened because the field winding *is* properly connected to the positive and negative brushes, as in Fig. 68, the generated voltage will rise; this rise, in turn, will cause an increase in field current, a further increase in flux, and a still larger generated voltage. This cumulative process will continue until a point

of equilibrium is reached; the interesting build-up may be seen by observing the field ammeter and the voltmeter as the pointers swing up scale. Equilibrium implies that the generated voltage causes a field current to flow that is just sufficient to develop the required excitation for the generated voltage. Thus, at a point on the saturation curve below the build-up value, such as E_1 in Fig. 67, the excitation is greater than necessary for this particular voltage; the result is that build-up continues. On the other hand, if the available excitation is less than that required for a particular voltage, such as E_2, build-up cannot proceed to this value.

An example should make this clear. Suppose the resistance of the shunt-field circuit is exactly 50 ohms. If $E = 100$ volts is the build-up point, the field current is $100/50 = 2.0$ amp. At point $E_1 = 95$ volts, the field current available is 1.9 amp (95/50), but only 1.6 amp is needed for this voltage; the generator, therefore, continues to build up further. At point $E_2 = 103$ volts, the field current available is 2.06 amp (103/50), but 2.3 amp is needed; the generator cannot proceed to build up to this point. It follows, then, that the point of equilibrium is that voltage which, when divided by the total field resistance, corresponds exactly to the field-current value on the saturation curve. For a *given* saturation curve there is only one such build-up point that corresponds to a definite value of field resistance. [For example, the *total* field resistance for build-up to $E_1 = 95$ volts (Fig. 67) should be 59.4 ohms.]

Since a self-excited shunt generator cannot build up unless it is given an initial "boost" or "push" by a residual voltage, it should be clear that a new machine or one that has lost its residual flux because of a long period of idleness must be separately excited to create the necessary magnetism. This is usually done while the armature is at rest by connecting the shunt field only to a separate d-c source, such as a battery, for a few seconds. This practice is generally referred to as *flashing the field*.

2. A second requirement for build-up is a field-circuit resistance that is less than the so-called "critical value" for the speed used in operating the generator. Since the field-circuit resistance consists of the field winding, the field rheostat, and all the connecting wires, it is important that there be no breaks or loose connections. Furthermore, the brushes must make good contact and ride freely on a clean commutator because the field current is fed through the brush contacts. The *critical resistance* mentioned above is merely the total field resistance below which the generator will build up and above which it will refuse to do so. By gradually cutting out resistance in the field rheostat from its "all-in" position, it is readily possible to pass through the critical value to a desired build-up voltage.

3. A third requirement is that the speed be high enough for the shunt-field resistance used. In this connection, it should be stated that a gen-

erator is usually operated at some definite speed originally fixed by the manufacturer. However, should it be desirable to operate at some higher or lower speed (if, for example, pulley diameters or gear ratios do not permit operation at the rated speed), the field rheostat must be adjusted for the new speed. In general, it will be found that for a given field resistance, a *critical speed* exists below which the generator will not build up and above which it will build up.

4. Finally, there must be a proper relation between the *direction of rotation* and the field connections to the armature. If a generator will not build up when operated clockwise, for example, it will build up when the direction of rotation is counterclockwise, assuming other conditions to be satisfied. Furthermore, the polarity of the brushes will be reversed when the direction of rotation is reversed. In this connection it should be stated that the polarity of the residual magnetism, *north* or *south*, has

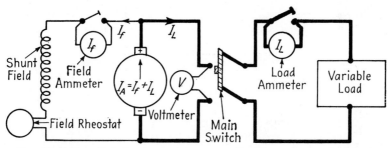

Fig. 69. Self-excited shunt generator connections to perform a load test.

absolutely no effect upon whether or not a generator builds up; it merely determines the polarity of the brushes, positive or negative.

Behavior of a Shunt Generator under Load. After a self-excited shunt generator builds up to a required voltage, *a no-load voltage*, it is ready to supply power to a number of electrical loads up to, and a little above, its rated capacity. In the case of a separately excited shunt generator, the armature always develops a voltage as soon as the field is connected to its separate source of supply; the points made concerning build-up do not, in general, apply to separately excited generators. Figure 69 is a simple wiring diagram of connections for a self-excited shunt generator supplying power to a variable load. Note that the armature is the origin of the comparatively low shunt-field current *and* the main line (or load) current. The shunt-field current, about 2 to 10 per cent of the total armature amperes, would be supplied by an outside source in a separately excited machine.

One of the most important characteristics of any generator is its behavior with regard to the terminal voltage when the load current is increased. In the shunt type of generator, the voltage always falls as more current is delivered to the load. There are three reasons for this:

1. As more current is delivered by the armature, the *voltage drop* in the armature $I_A R_A$ increases, thus making a lower emf available at the load terminals.

2. When the armature terminal voltage falls, the field winding suffers a corresponding reduction in current, which, in turn, reduces the flux; the latter further reduces the generated emf.

3. When the armature winding carries increasing values of load current, the armature core becomes an electromagnet, apart from the effect of the main poles; this electromagnetic action of the armature core reacts with the main field flux further to reduce the flux, the result being that the generated emf suffers an additional drop.

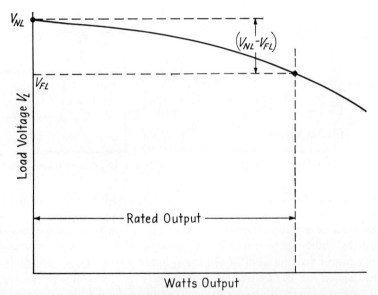

FIG. 70. Characteristic *load vs. output* curve of self-excited shunt generator.

All three actions occur simultaneously, since they are all interdependent upon one another. In this analysis, it should be clearly understood that the *generated voltage*, which depends upon the flux (other factors remaining unchanged), is always *greater* than the *terminal* or *load voltage* by exactly the amount of the voltage drop in the armature circuit. This is why it is important to keep the armature-circuit resistance as low as possible.

In order to determine how a self-excited shunt generator behaves under load, a load test must be performed. A standard procedure involves operating the generator at rated speed as the load is varied from zero current to somewhat above rated output. Proper adjustments should be made with the field rheostat, so that the full-load current is delivered at name-plate voltage. Measurements are recorded of speed, which should be kept constant, load current I_L, field current I_f, and terminal voltage V. For each load, the power output is calculated by multiplying V by I_L after which a graph similar to Fig. 70 is plotted.

It should be noted particularly that the load voltage falls from its no-load voltage V_{NL} to its full-load value V_{FL}, the latter representing the rated name-plate value. Or, to put it another way, if the generator is delivering its rated power at rated voltage V_{FL} and the load switch is quickly opened, the terminal voltage will rise to the no-load value V_{NL}. This rise in voltage (V_{FL} to V_{NL}) is especially important because it is an indication of how poorly or how well a generator performs. The ideal (never attained in shunt generators) would, of course, be zero change in voltage; under this condition, the voltage would remain essentially constant. A poor generator, on the other hand, would be one whose voltage changes considerably between full load and no load.

A convenient standard of reference used to measure the performance of a generator is *referring the change in voltage between full load and no load* (V_{FL} to V_{NL}) *to the full-load voltage* V_{FL}. This ratio is represented as a percentage and is called the *per cent voltage regulation*, or briefly, *per cent regulation*. In equation form, it becomes

$$\text{Per cent regulation} = \frac{V_{NL} - V_{FL}}{V_{FL}} \times 100 \qquad (9)$$

EXAMPLE 1. The voltage of a 100-kw 250-volt shunt generator rises to 260 volts when the load is removed. What full-load current does the machine deliver, and what is its per cent regulation?

Solution

$$I_{FL} = \frac{100{,}000}{250} = 400 \text{ amp}$$

$$\text{Per cent regulation} = \frac{260 - 250}{250} \times 100 = 4 \text{ per cent}$$

EXAMPLE 2. A 25-kw 230-volt shunt generator has a regulation of 8.7 per cent. (a) What will be the terminal voltage of the generator at no load? (b) If the change in voltage is assumed to be uniform between no-load and full-load kilowatts, calculate the kilowatt output of the generator when the terminal voltages are 240 and 235 volts.

(a) $$8.7 = \frac{V_{NL} - 230}{230} \times 100$$

$$V_{NL} = \frac{8.7 \times 230}{100} + 230 = 250 \text{ volts}$$

(b) $$P_{240} = \frac{250 - 240}{250 - 230} \times 25 = 12.5 \text{ kw}$$

$$P_{235} = \frac{250 - 235}{250 - 230} \times 25 = 18.75 \text{ kw}$$

Controlling the Terminal Voltage of Shunt Generators. It is possible to prevent the terminal voltage V from changing as the load changes by merely adjusting the field rheostat as the voltage changes in accordance with the load. For example, if the voltage falls with increase in load, the field-rheostat resistance must be decreased; this raises the flux and with it the generated emf. If the load changes gradually, or if the expense of automatic regulating equipment is not warranted, the field rheostat may be adjusted manually with each change in terminal voltage. Automatic control is, of course, much more desirable and satisfactory, not only because it eliminates the need for watching the generator as it operates, but also because it performs its regulating function quickly and accurately without attention.

Automatic regulators are designed and constructed in several ways, but all operate on the fundamental principle that a terminal-voltage change is accompanied by an inverse change in flux. In one of these, the *Tirrill regulator* (formerly used almost exclusively for such service but now superseded by other types that are quicker acting and more reliable), a pair of short-circuiting contacts are connected across a portion of the field rheostat and are actuated by the electromagnetic action of a relay. When the terminal voltage drops, the relay causes the short-circuiting contacts to close; this is accompanied by a decrease in the field resistance, an increase in the field current, and an increase in the field flux. The generated and terminal voltages then rise. On the other hand, if the terminal voltage rises, the relay causes the contacts to open; this decreases the field flux and with it the generated and terminal voltages. In practice, the field rheostat was usually adjusted so that the generator terminal voltage was about 25 to 35 per cent below rated value when the machine was delivering full-load current with the relay contacts held open.

In a type of automatic regulator now widely used, the so-called *Diactor* (direct-acting) unit, the same functions are accomplished without vibrating contacts but by special rheostatic elements that are acted upon directly by a unique design of torque element; the latter, energized by the varying line potential, is made to tilt stacks of plates at different angles so that proper values of resistance are connected across a manually set rheostat in series with the shunt field. More recently, the saturable reactor has been adapted to voltage regulators for generators. In its simplest form, this extremely interesting magnetic amplifier is made to act upon a special control field wound directly over the main field of the generator; in this way, the total flux may be adjusted to the proper

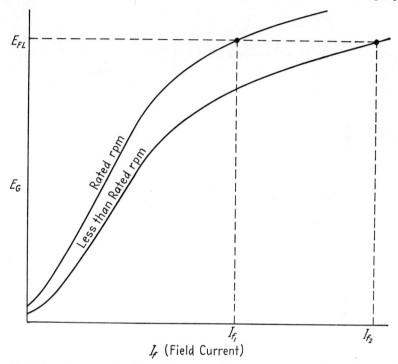

FIG. 71. Saturation curves for operation of a self-excited shunt generator at two different speeds.

value without the necessity for making changes directly in the main field circuit.

The tendency on the part of a shunt generator to lose voltage with increasing values of load may be minimized by operating the machine at a lower speed. When this is done, the field flux must be increased, which means that the iron portions of the magnetic circuit are more highly saturated. Referring to Fig. 71, it will be noticed that the field current for a machine operating at less than rated speed is much greater for a given generated emf than the field current required for normal speed. This fact indicates that the *effect* of armature reaction (to be discussed on p. 113) is reduced because a given number of armature ampere-turns has less effect upon a strongly saturated field than upon a weak field. The result is that the rise in voltage between full load and no load is not so great at the reduced speed (see Fig. 72). An objection to operation at lower speed, however, is that the machine is likely to overheat because (1) the copper loss in the field is increased and (2) ventilation, that is, cooling, is not as good because the fanning action of the armature is reduced.

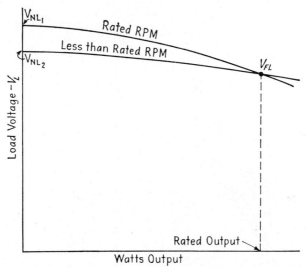

FIG. 72. Comparison of regulation curves of a self-excited shunt generator for operation at two different speeds.

Compound Generator Operation under Load. The addition of a second field winding, connected in series in the line circuit (Fig. 63a) or in the armature circuit (Fig. 63b), provides a generator with two sources of excitation. The shunt-field excitation is usually more or less steady and is affected only slightly as the terminal voltage fluctuates. The effect of the series field is quite variable, however, since its ampere-turns depend upon the load current; when the load current is zero, it produces no component of flux and when the load current is high, it creates an appreciable component of flux. Exactly how much flux it must develop depends upon the extent to which it must compensate for voltage drop. In the actual *compound generator*, that is, a machine having both shunt and series excitation, the series field is wound directly over the shunt field. Figure 23 represents a shunt- and series-field-coil assembly, while Fig. 73 shows a cutaway view of a small d-c generator in which the shunt- and series-field coils are clearly visible. This photograph is also an excellent view of the arrangement of the various parts of the machine and should be studied carefully.

DIRECT-CURRENT GENERATOR CHARACTERISTICS 107

Obviously, the shunt- and series-field coils around each of the main poles should be so connected that they create flux in the *same* direction if the tendency of the generator to lose voltage is to be counteracted. When this is done, the machine is said to be *cumulative-compounded*. If, for

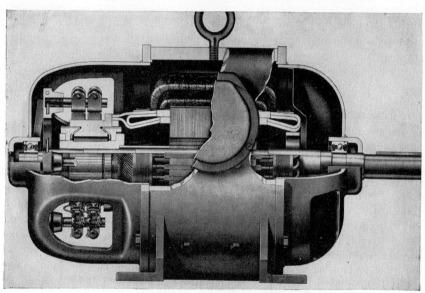

FIG. 73. Cutaway view of a compound generator showing the arrangements of the various parts including one shunt- and series-field coil. (*General Electric Co.*)

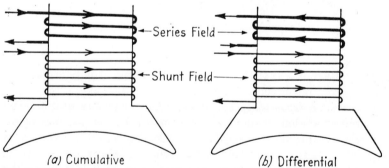

FIG. 74. Current directions in series- and shunt-field coils of cumulative- and differential-compound generators.

some special reason, the action of the series field must oppose, that is, "buck," the shunt field, the machine is referred to as *differential-compounded*. Figure 74 illustrates both the cumulative and the differential connections.

For testing a compound generator, the same procedure should be followed as was described for testing shunt machines, p. 102. In Fig. 75, the generator is first adjusted to deliver full load at rated voltage. After that, readings are taken as the load is reduced in steps to zero. A set of readings for some overload may also be taken if desired. When the data are plotted, it will be found that one of the three typical curves, represented by the group marked *cumulative-compound* (Fig. 76), will be obtained. (The shunt and differential-compound curves are drawn for comparative purposes.) If the terminal voltages at no load and full load are equal, the machine is said to be *flat-compound*. If the full-load voltage is higher than the no-load value, it is called an *overcompound* generator. If the

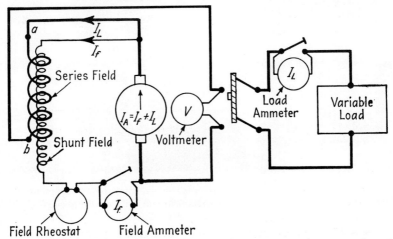

FIG. 75. Cumulative-compound generator (short-shunt) connections to perform load test.

full-load voltage is less than the zero-load voltage, it is referred to as an *undercompound* machine.

The *degree of compounding*, i.e., whether a given generator is flat-, over-, or undercompounded, is determined primarily by the number of series-field turns or, more particularly, by the full-load series-field ampere-turns with respect to the shunt-field ampere-turns. Many series-field turns will produce overcompounding, whereas few series-field turns will give the generator an undercompounded characteristic. In practice, it is customary to use flat-compound generators where the transmission distance between the generator and the load is short. Overcompound generators are employed where the load is a considerable distance from the generator because the machine not only must generate sufficient voltage to overcome its own internal drop but also must compensate for the transmission-line voltage drop. This would be true of a generator

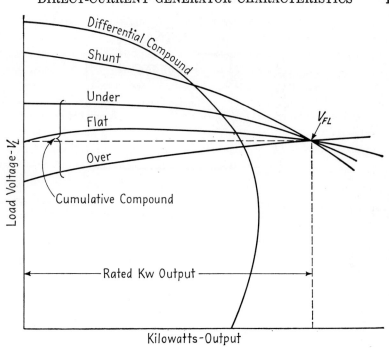

FIG. 76. Characteristic *voltage-load* curves for compound and shunt generators.

used for traction service. A flat-compound generator would, however, be installed for power service in a small building.

Degree of Compounding Adjustment. Standard compound generators, that is, those which are manufactured in comparatively large numbers, are usually constructed with sufficient series-field turns to operate overcompounded. Then, by connecting *a very low resistance shunt directly across the series field*, points a and b in Fig. 75, the no-load voltage may be brought up to almost any desired value to meet individual demands. Such a low-resistance shunt is called a *diverter* because it *diverts*, or bypasses, part of the load current through a section of wire that creates no flux. Thus the series field is less effective in creating flux to boost the generated emf to an extent determined by the diverted current. Figure 77 is a wiring diagram illustrating the connections of a compound generator with a *series-field diverter*; it indicates how the total line current divides into two parts I_{SE} and I_D, the series-field and diverter currents, respectively. When the resistance of such a diverter is extremely large, the diverted current will be small and the generator characteristic will be overcompounded. On the other hand, if the resistance of the diverter approaches that of a short circuit, practically all the load current will be diverted and the operating external characteristic will resemble that of a

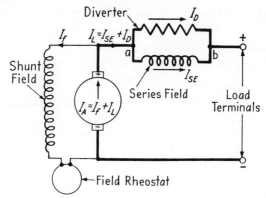

Fig. 77. Schematic wiring diagram of a compound generator with a diverter connected across series field.

shunt generator. Since the series-field resistance R_{SE} and the diverter resistance R_D are in parallel, the total line current I_L will divide so that I_{SE} and I_D are related to each other by an inverse ratio of the respective resistances.

Thus
$$\frac{I_{SE}}{I_D} = \frac{R_D}{R_{SE}} \qquad (10)$$

Since
$$I_L = I_{SE} + I_D$$

it follows that
$$I_{SE} = I_L \times \frac{R_D}{R_D + R_{SE}} \qquad (11)$$

where I_{SE} = series-field current
I_D = diverter current
R_{SE} = series-field resistance
R_D = diverter resistance

In practice, it is customary to adjust the resistance of the diverter R_D by experiment. The no-load voltage of the generator is first set for the desired value by manipulating the field rheostat, with the machine operating at the proper speed. Load is then applied so that rated current is delivered; the terminal voltage is observed under this condition. If the latter is higher than the required value, a diverter is connected across the series field and the full-load test is repeated. If the full-load voltage is still too high, the diverter resistance is reduced; if the full-load voltage is less than desired, the diverter resistance is increased. After several trials, the experimenter will be able to make the proper diverter-resistance adjustment for the required operating conditions. If should be remembered in performing this test that both the diverter and the series field have extremely low values of resistance and carry comparatively high values of current. Adjustments must therefore be made with care, after which good permanently soldered joints are prepared. In practice, the diverter material is manganin, german silver, or any other high-resistivity material with a low temperature-resistance coefficient.

EXAMPLE 3. The series field of a compound generator has a resistance of 0.018 ohm. If the full-load current is 120 amp and it is necessary to divert 36 amp so that the full-load voltage will be brought down to the desired value, calculate: (a) the value of the diverter resistance, (b) the length of a square manganin wire (resistivity = 265) whose cross-sectional area is 15,616 circular mils.

Solution

(a) $$\frac{120 - 36}{36} = \frac{R_D}{0.018}$$

$$R_D = \frac{84}{36} \times 0.018 = 0.042 \text{ ohm}$$

(b) $$0.042 = \frac{265 \times l(\text{feet})}{15,616}$$

$$l = \frac{0.042 \times 15,616}{265} = 2.48 \text{ ft, or 30 in.}$$

EXAMPLE 4. Each of the series-field coils of a four-pole 50-kw 250-volt compound generator has 6½ turns of wire. The resistance of the entire series field is 0.012 ohm, and the diverter resistance is 0.036 ohm. Calculate the number of ampere-turns of each series-field coil at full load.

Solution

$$I_L = \frac{50,000}{250} = 200 \text{ amp}$$

$$I_{SE} = 200 \times \frac{0.036}{0.036 + 0.012} = 150 \text{ amp}$$

$$NI = 6\tfrac{1}{2} \times 150 = 975 \text{ amp turns per series coil}$$

Series-generator Behavior under Load. When a generator has a single field that is connected *in series* with the armature, the load current is simultaneously the excitation current; called a *series generator*, its voltage will depend upon the current delivered to the load. On open circuit, when the load is zero, the series-field ampere-turns is likewise zero and the generated voltage is that due to the cutting of the residual flux, i.e., the residual value E_r, Fig. 78. However, if the generator terminals are closed through a load resistance, a current I will flow, in which event the series field will create additional flux and cause the machine to generate a higher voltage; at the same time the armature will develop a demagnetizing action, and a voltage drop will occur in the armature and series-field resistances R_A and R_{SE}. Therefore, the voltage V_t that will appear at the series-generator terminals will be stabilized at a value that is some function of the *net* generated voltage (due to the combined actions of the series-field and armature mmfs) and the $I(R_A + R_{SE})$ voltage drop. The terminal voltage V_t will, obviously, rise as the load amperes are increased and continue to do so as long as the resultant generated voltage rises *more* rapidly than those factors already noted tend to reduce it. However, at loads that are considerably above normal values, the iron portions of the magnetic circuits become highly saturated, under which

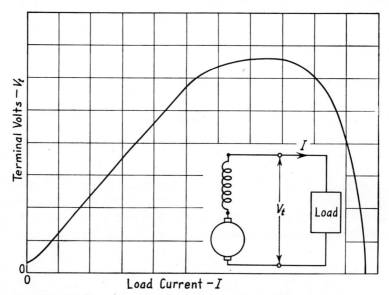

FIG. 78. Characteristic *voltage-load* curve for a series generator.

condition the subtractive effects exceed the slowly rising generated emf; the terminal voltage then begins to drop. Thus, as Fig. 78 shows, the external *volts-vs.-load current* characteristic curve rises rapidly from its initial E_r value during the light-load stages, then tapers off to a maximum, and finally drops to zero.

Because of the varying nature of the terminal voltage with respect to load, the series generator has few practical applications, and then only when the V_t-vs.-I curve is advantageous to the installation. In this country it is sometimes used in a d-c system for voltage-boosting purposes or to minimize leakage currents in grounded d-c systems so that electrolytic action in underground structures may be reduced; in Europe it is sometimes employed in the *Thury* high-voltage d-c systems for the transmission of electrical energy.

Armature Reaction in Direct-current Generators. *When the armature of a d-c generator carries a load current, it becomes an electromagnet,* apart from the magnetic effect induced in it by the main poles. Assuming a simple *bipolar* machine, with the *north* and *south* poles placed horizontally, the magnetic field resulting from these main poles will be directed from left to right, as shown in Fig. 79a; the magnetic neutral will then be a vertical plane perpendicular to the plane of the paper and at right angles to the direction of the field. If it is further assumed that the armature is rotating clockwise, the generated emfs under the *north* pole will be away from the observer, while the generated emfs under the *south* pole will be toward the observer; these directions are indicated by crosses (+) and dots (·) in Fig. 79b. Now then, if a load is delivered by the armature, the current directions in the armature winding will be exactly the same as

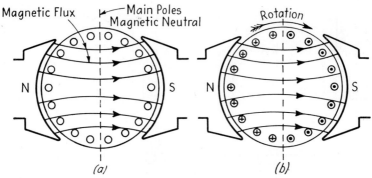

Fig. 79. Sketches showing main-pole flux distribution and directions of emfs and currents in armature windings.

those indicated for the emfs if the brushes are positioned on the commutator so that they are on the exact neutral plane *with respect to the armature conductors being short-circuited.* This latter point is important because it does not necessarily mean that the brushes line up exactly on the magnetic neutral of the main poles. Figure 80 should make this clear. Remembering that an armature coil must not be short-circuited by a brush unless its *coil sides* are in a neutral magnetic zone, it should be clear that there are two possible extreme locations for the brushes *with respect to the poles.* In Fig. 80a, in which the ends of an armature coil connect to commutator segments midway between the coil sides, *the brush neutral lines up with the center of the pole.* In Fig. 80b, in which the ends of an armature coil connect to commutator segments in line with one side of a coil, *the brush neutral lines up with the magnetic neutral of the main poles.*

Continuing further with Fig. 79, the next step in this analysis of the reaction of the armature with respect to the main pole flux is to recognize the fact that the current in the armature winding creates a field of its own and that *this field is superimposed on the main field.* This is shown in Fig. 81a and is represented by the broken lines as distinguished from the continuous main field flux. Note particularly that this armature flux is directed downward through the armature, whereas the main field is from left to right; that is, the two fields are directed 90° with respect to each other. The two fields then react with each other to create a resultant field, represented by Fig. 81b.

The resultant field has a new direction obliquely downward. Since a neutral plane must always be at right angles to the resultant field, the new magnetic neutral is shifted in a clockwise direction, which, it should be noted particularly, is the same as the direction of rotation. Obviously, if

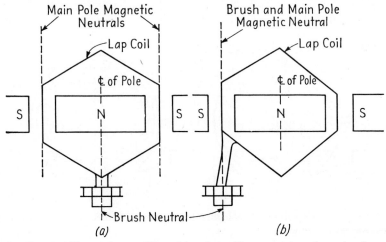

FIG. 80. Sketches illustrating position of brushes with respect to pole center for symmetrical and unsymmetrical coils.

sparkless commutation is to result, the brushes must be shifted by an angle of $A°$.

Another effect of this *armature reaction*, not quite so apparent, is that the generated voltage is actually *reduced.* To understand why this is so it is necessary to recognize the fact that the armature flux weakens the field on one half of each pole and strengthens the field on the other half (see Fig. 81a). If the decrease were equal to the increase, the magnitude of the resultant flux would remain unchanged. This is not the case, however, because the decrease is usually greater than the increase, because of magnetic saturation. In most practical cases, the reduction in flux may be from about 1 to 4 per cent between the no-load and full-load values.

The shift of the magnetic neutral in the direction of rotation seriously affects commutation because sparking will occur at the brushes unless they are shifted to the new magnetic neutral. Furthermore, the brushes

DIRECT-CURRENT GENERATOR CHARACTERISTICS

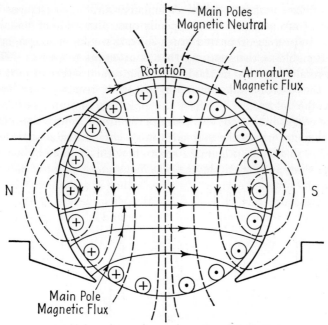

(a) Main poles and armature flux distributions.

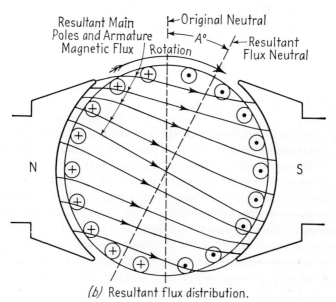

(b) Resultant flux distribution.

FIG. 81. Sketches illustrating flux distributions of loaded two-pole generator.

must be shifted back and forth continually as the load changes because the effect of armature reaction depends upon the value of the armature current. In practice, repeated brush shifting would, of course, be almost as objectionable as the sparking that it attempts to correct. This has led to several corrective methods that counteract, either in part or completely, the detrimental effects of armature reaction. In two such schemes, the reluctance between the pole tips and the surface of the armature core is increased; this reduces the armature reaction flux in the interpolar zone, where the coil sides must cut no flux if sparkless commutation is to result. Figure 82 shows how this is done by using *chamfered pole shoes* and by employing *pole laminations with one pole tip*. In Fig. 82a, the rounded surface of the pole shoe is not concentric with the circular armature core; this reduces the effectiveness of the armature magnetomotive force (mmf) at the pole tips and in the commutation zone.

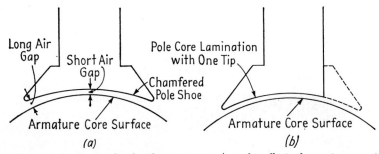

FIG. 82. Special pole-core laminations to counteract the effect of armature reaction.

In Fig. 82b, the same result is accomplished by cutting off one pole tip; in assembling the laminations, the pole tips are alternated from one side to the other, so that the cross-sectional area of the iron is one-half as much under the pole tips as under the center section.

To summarize, it should be understood that armature reaction (1) is produced by the load current in the armature conductors that results in a magnetic field whose direction is displaced 90 electrical degrees with respect to the main field, (2) depends upon and is directly proportional to the load current, (3) shifts the neutral plane in the same direction as that of the rotating armature, (4) reduces the total flux, and (5) can be counteracted to some extent by the use of specially shaped pole laminations.

Interpoles for Direct-current Generators. One of the most important developments in the design of d-c machines was the use of *interpoles* to correct the objectionable commutation effects of armature reaction. These are narrow poles placed exactly halfway between the main poles and directly in line with the no-load magnetic neutral, usually called the *mechanical* neutral. *The exciting windings for these poles are always*

DIRECT-CURRENT GENERATOR CHARACTERISTICS

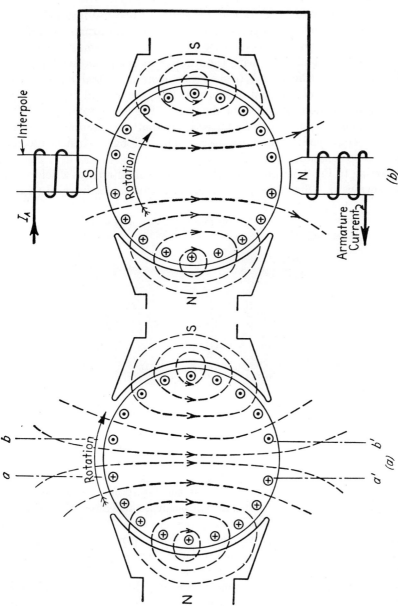

Fig. 83. Sketches showing how interpoles counteract effect of armature reaction in interpolar zones of a d-c generator.

permanently connected in series with the armature because interpoles must produce fluxes that are directly proportional to the armature current. The armature and interpole mmfs are thereby affected simultaneously by the *same* armature current, with the result that the armature flux *in the commutating zone*, which tends to shift the magnetic neutral, is neutralized by an appropriate component of interpole flux; the neutral plane is therefore fixed in position regardless of the load. The use of interpoles (often called *commutating poles*) is so widespread that noninterpole machines are rarely employed; if they are used at all, it is only when the load is known to be constant at all times.

Referring to Fig. 83a, note that the armature mmf creates a field vertically downward. Since the part of this field that is detrimental to good

FIG. 84. Completely assembled interpole for d-c machine. (*General Electric Co.*)

commutation is in a narrow zone between lines aa' and bb', it should be clear that poles located in this area and creating an mmf of the proper value will counteract the effect of armature reaction. A pair of such interpoles is shown in Fig. 83b. Note that the interpolar zones are free of magnetism. This latter condition will occur at all values of load if the number of turns on each interpole is properly chosen because the current in the interpole winding is the same armature current that is originally responsible for the sparking at the brushes. Figure 84 is a photograph of one such completely assembled interpole ready for bolting to the yoke frame, while Fig. 7 shows a machine in which the interpoles are clearly visible in their relative positions with respect to the main poles. In order to emphasize further the importance of this modern construction, a sketch showing the arrangement of the various parts of a generator (or motor) is presented in Fig. 85.

One additional significant fact which should be stated is that interpoles act only in the interpolar zones. They can have absolutely no effect upon the armature mmf that distorts the main field, which is called the *cross-magnetizing* effect and can affect the operation of a loaded generator only slightly under average conditions. In some large machines, and in those in which load fluctuations are violent, the cross-magnetizing action can become severe enough to cause *flashover* between plus and minus brushes.

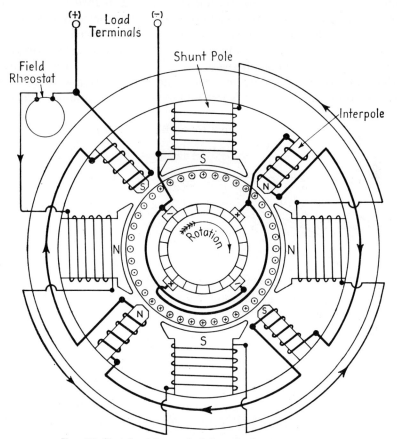

FIG. 85. Sketch of four-pole interpole shunt generator.

The method usually employed to overcome this trouble is to use *compensating windings*.

Compensating Windings for Direct-current Generators. *Compensating windings* are used for the purpose of neutralizing the effect of armature reaction in the zones outside the influence of the interpoles and particularly to maintain a uniform flux distribution under the faces of the main poles. They are special windings placed in slots or holes in the pole faces and carry, as do the interpole windings, the *total armature current*. Half of the conductors on the right side of a pole face, for example, are joined in series to half of the conductors on the left side of the adjoining pole face in such a manner that *the directions of the currents in these conductors are opposite to those in the wires in the armature winding directly below.* Figure 86 illustrates the arrangement of the pole-face conductors and their connections to form a compensating winding. The path of the current through the three turns (six conductors) of one of the coils may be considered to start at the rear of conductor a, then proceed consecutively through conductors b, c, d, and e, and end at the rear of conductor f. It is just as though a coil of wire were wound in a spiral with turn ab on the inside and turn ef on the outside. Since the compensating winding carries the *total* armature current, it is much more effective in producing flux per

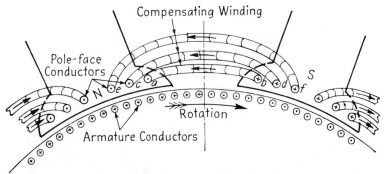

FIG. 86. Sketch representing compensating winding and relative directions of currents in pole-face conductors and armature conductors.

turn than is each armature winding turn. It will be remembered that the current per path (or any turn of wire) in the armature winding is equal to the total armature current divided by the number of armature paths. Thus the mmf produced by six pole-face conductors of a compensating winding will neutralize the cross-magnetizing mmf of 48 armature conductors if the machine has eight poles and a simplex-lap armature winding.

In the actual design of compensating windings, only those conductors in the armature winding directly opposite the pole faces are neutralized by an equal number of ampere-turns. The interpoles, which such machines always have, take care of the armature reaction effect in the interpolar zones. An example should make this clear.

EXAMPLE 5. A 2,500-kw 600-volt 16-pole generator has a lap-wound armature with 2,360 conductors. If the pole faces cover 65 per cent of the entire circumference, calculate the number of pole-face conductors in each pole of a compensating winding.

Solution

$$\text{Total armature current} = \frac{\text{watts}}{\text{volts}} = \frac{2{,}500{,}000}{600} = 4{,}170 \text{ amp}$$

$$\text{Armature current per path} = \frac{4{,}170}{16} = 261$$

$$\text{Armature conductors under each pole} = \frac{2{,}360}{16} \times 0.65 = 96$$

Armature ampere-conductors under each pole face = $96 \times 261 = 25{,}000$

Compensating winding ampere-conductors per pole face = $4{,}170 \times C$

Equating $4{,}170 C = 25{,}000$,

$$C = \frac{25{,}000}{4{,}170} = 6 \text{ conductors in each pole face}$$

The use of compensating windings together with properly designed interpole windings in d-c machines (generators or motors) will provide sparkless commutation and eliminate any possibility of flashover. Such good operating characteristics take place because the resultant field is absolutely uniform at all loads; the effect of armature reaction is neutralized, regardless of the armature current. Figure 87 is a photograph clearly showing the compensating winding and the interpole field. A sketch representing the wiring connections for a compound generator with four fields, i.e., compensating, interpole, shunt, and series, is given in Fig. 88.

Commutation and Reactance Voltage. It was previously pointed out that the commutator and its accompanying brushes are very important parts of the d-c generator. It is here that two essential actions take place, namely, *the commutation process* and *the passage of current from a moving armature to a stationary load*. The commutation process involves the change from a *generated alternating current to an externally applied direct current*. The transfer of current from the rotating armature to the stationary brushes (and thence to the load) involves a continuously moving contact. Both these actions must be carefully controlled by the use of suitable materials, good design, and proper adjustments, for otherwise there would be serious arcing and possible breakdown of the machine.

As one of the armature coils rotates between a positive brush and the succeeding negative brush, current flows in that coil in one direction. Then this coil is short-circuited by a brush for an extremely short fraction of a second, after which it passes into a region between a negative brush and the succeeding positive brush, where the current reverses. During this short period, when the current reverses from full value in one direction to full value in the opposite direction, two important factors tend to pre-

Fig. 87. Yoke with compensating winding, interpoles, and shunt and series fields for 3,000-kw 600-volt 360-rpm generator. (*Allis-Chalmers Mfg. Co.*)

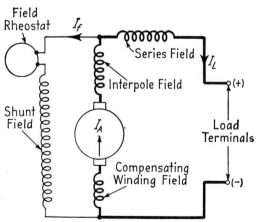

Fig. 88. Wiring diagram showing connections for four fields, namely, shunt series, interpole, and compensating winding.

vent smooth commutation: (1) a possible high current in the coil undergoing short circuit and (2) the inductance property of the coil (similar to inertia) that opposes a reversal of current.

Figure 89 should make this clear. In Fig. 89a, the commutator and one of the armature coils (lap winding for simplicity) are shown rotating clockwise. As the coil moves from position 1 to position 2, the current in the coil is to the left, to the positive brush. Then the coil is suddenly short-circuited by the negative brush, as indicated by Fig. 89b. If, in this position, the coil sides cut flux, a voltage will be generated, as a result of which a short-circuit current will flow. This is why it is necessary to use interpoles to offset the effect of armature reaction in order that the interpolar zones will be comparatively free of flux. Finally, the coil is relieved of the short circuit as it passes into the region between the negative and positive brushes, where the current reverses to flow to the right (see Fig. 89c). Note that there are three stages in this commutation process: (1)

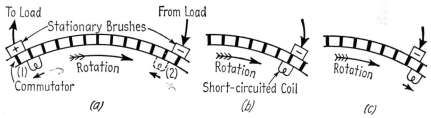

FIG. 89. Commutation process of one coil of an armature winding (a) before short circuit, (b) during short circuit, (c) after short circuit.

before short circuit, when the current is in one direction; (2) the short-circuit period, which is the most severe period; and (3) after short circuit, when the current begins to flow in the opposite direction.

Note also that the current must change from full value in one direction to full value in the reverse direction in an extremely short time interval. The inductance property of a coil containing several turns of wire and wrapped around good magnetic material tends to *oppose* such reversal, and this property gives rise to a voltage as a result of the inductance. It is called *reactance voltage* because it reacts to oppose a change of current from $+I$ to $-I$. If this reactance voltage is permitted to act, it will cause sparking, even though interpoles are used to neutralize the armature flux in the interpolar zone. To offset this reactance voltage, it is necessary to do one of two things: (1) shift the brushes, in noninterpolar machines, *beyond* the load magnetic neutral, so that a voltage will be generated in the coil in the direction in which the current is about to flow, or (2) strengthen the interpole more than would be necessary to neutralize the armature reaction flux, so that a voltage will be generated in the short-circuited coil in the direction in which the current is *about* to flow. In other words, in anticipation of the new current direction, *reactance voltage* is neutralized by making a coil generate enough emf to help the current reverse effectively and smoothly. This is why interpoles are always stronger than would be just necessary to overcome the armature reaction flux, or why brushes, when shifted, are moved a little beyond the exact magnetic neutral.

When generators are designed, it is generally difficult to make accurate calculations to determine the exact number of interpole-field turns for sparkless commutation because design equations involve terms that are, in part, based upon empirical data and certain necessary assumptions. It is for this reason, therefore, that such machines are generally constructed with a somewhat liberal use of commutating-pole windings and with pole cores whose radial lengths are shorter than required. Then, after the machine is constructed, it is customary to alter the reluctance of the interpole magnetic circuits by adjusting the air gaps under the poles. This is done, while the machine is undergoing tests, by inserting a combination of thin magnetic and nonmagnetic shims between the interpole cores and the yoke frame where the two are bolted together. The shims then make up for the short cores, so that correct air gaps are established for proper commutation.

Need for Operation of Generators in Parallel. Power plants will generally be found to have several small generators rather than large single units capable of taking care of the maximum peak loads. This is true of both d-c and a-c stations. The several units can then be operated singly or in various parallel combinations, on the basis of the actual load demand. Such practice is considered extremely desirable from the standpoint of efficiency, continuity of service, maintenance and repair problems, and additions to plant capacity as the service demands change.

As a rule, the load on an electric power plant fluctuates, usually having its peak sometime during the day and its minimum during the night hours. Since a generator operates most efficiently when delivering loads in the region of its rated capacity, it is logical to use a small unit delivering rated kilowatts when the demand is light. Then, as the load increases, a larger generator is substituted for the smaller one or another unit is connected in parallel with the one already in operation. Obviously, this must be done without disturbing the continuity of service, for the users of electric energy must not even be conscious of any switching operation. Further generator additions or changes can be made as the need arises. Exactly how a generator is "put on" or "taken off" a line without interruption to service will be discussed on p. 127.

One of the most important operating requirements of electrical systems is that there should be *continuity of service*. It would obviously be impossible if a power plant contained a single unit, because a breakdown of the prime mover or the generator would necessitate complete shutdown of the entire station. This matter of uninterrupted service has become so important in recent years, especially in factories and in the home, that it is now recognized as an *economic necessity*.

It is considered good practice to inspect a generator carefully and periodically to forestall the possibility of failure. This can, of course, be done only when the unit is at rest, which means that other machines must be in operation to take care of the load. Furthermore, when a generator does break down, it can be repaired with care and without rush if other machines are available to maintain service continuity.

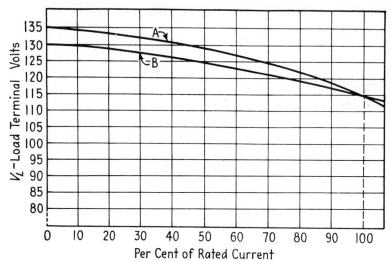

Fig. 90. Characteristic *voltage vs. per cent rated current* of two shunt generators operated in parallel.

Finally, additions are frequently made to power plants as they are called upon to deliver increasingly greater loads. In fact, engineers usually make plans for future extensions when they design power plants. Such extensions are made only as existing equipment begins to prove inadequate and when the installation of new machines has been proved economically advisable.

Operation of Shunt Generators in Parallel. It was previously shown that shunt generators have "drooping" voltage-vs.-load characteristics, i.e., that the voltage drops as the load increases. If two shunt generators have identical external characteristics, that is, if the voltage changes in both by exactly the same amount for the same per cent of change in load, then the two machines will divide the total load in proportion to their relative capacities. Thus, if a 100-kw shunt generator is in parallel with a 200-kw shunt generator, a total load of 240 kw will divide so that the first machine will supply 80 kw while the second will deliver 160 kw. However, if the external characteristics are *not* similar, as indicated by curves *A* and *B* of Fig. 90, then if both have been adjusted for the same voltage at rated load currents, generator *A* will always deliver a larger per cent of *its* rated capacity than will generator *B* of *its* rated capacity. For example, assume that generator *A* has a rating of 50 kw at 115 volts and that generator *B* has a rating of 75 kw at 115 volts. At 115 volts, generator *A* will deliver 50,000/115 = 435 amp, while generator *B* will deliver 75,000/115 = 652 amp. When the voltage is 125, generator *A* will deliver 68 per cent of 435 amp, or 296 amp, while generator *B* will supply 48 per cent of 652 amp, or 313 amp.

EXAMPLE 6. In Fig. 90, generator *A* has a rating of 150 kw and generator *B* has a rating of 200 kw, both at 230 volts. Calculate the kilowatt output of each machine and the total kilowatt load when the terminal voltage is 240 volts.

Solution

The voltage scale must first be multiplied by 2. At 240 volts, generator *A* delivers 86 per cent of its rated current. Therefore

$$0.86 \times \frac{150,000}{230} = 561 \text{ amp}$$

At 240 volts, generator *B* delivers 76 per cent of its rated current. Therefore

$$0.76 \times \frac{200,000}{230} = 661 \text{ amp}$$

Generator *A* delivers

$$\frac{240 \times 561}{1,000} = 134.6 \text{ kw}$$

Generator *B* delivers

$$\frac{240 \times 661}{1,000} = 158.5 \text{ kw}$$

Total load = 134.6 + 158.5 = 293.1 kw

A circuit diagram illustrating how the connections should be made for operating two shunt generators in parallel is given in Fig. 91. Assume that generator *A* is connected to the bus bars through switch S_A and that it carries a load. As the load increases, it will ultimately become necessary to (1) connect a larger generator than *A* in parallel with the latter, after which the smaller machine, when gradually unloaded, is disconnected from the line, or (2) connect another generator in parallel with *A*

DIRECT-CURRENT GENERATOR CHARACTERISTICS

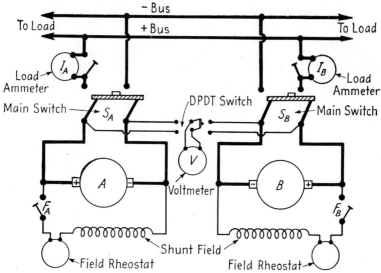

FIG. 91. Wiring connections for the operation of two shunt generators in parallel.

and have the two machines operate jointly to supply the total load. The procedure for accomplishing this is as follows:

1. Generator B is brought up to speed by its prime mover.
2. Field switch F_B is closed, whereupon the voltage will build up. With the double-pole double-throw (DPDT) switch closed to the left, the bus voltage (generator A in this case) is observed. This switch is then closed to the right, and the voltage of B is adjusted by means of its field rheostat until it equals the voltage of the line. It is important that the polarity of the incoming generator be exactly the same as that of the line polarity, i.e., plus to plus and minus to minus.

Caution! If this is not the case, a serious short circuit will occur when switch S_B is closed.

3. After proper adjustments are made, quickly close switch S_B. This places generator B in parallel with generator A. Generator A will still be supplying the entire load, while generator B will be running idle. (It is said to be "floating.")

4. To shift the load from A to B, it is necessary merely to manipulate both field rheostats simultaneously, cutting *in* resistance in the field of A and *at the same time* cutting *out* resistance in the field of B. Any degree of load shifting can be readily accomplished in this way; in fact, the entire load can be shifted to B and the main switch S_A opened to remove A from the line. While the load is shifted from one generator to the other, load ammeters I_A and I_B should be carefully watched to make sure that overloading does not occur. It is also important that the field rheostats are

not manipulated *beyond* the point where A is carrying no load because, if they are, generator A will attempt to operate as a motor and thus drive its prime mover.

When two shunt generators are operating in parallel, each one carrying its proper share of the total load, they will continue to do so even though one of them temporarily tries to deliver more current at the expense of the other. Such a situation might arise if the system load should suddenly change or if the speed of the prime mover of one machine should change temporarily. After the transient condition has passed, generator A might, for example, try to carry more than its proper share of the total load, while generator B would assume a smaller portion. Under this condition (see Fig. 90), the voltage of A would tend to drop, while that of B would tend to rise. This action would immediately result in a new series of tendencies in which A would try to drop load while B is taking it on. Since the total load is constant and the terminal voltage for both machines *must* be the same, a tendency on the part of one generator to take a greater share of the total load than it should is immediately accompanied by a series of actions that try to prevent it from doing so. It follows, therefore, that shunt generators operating in parallel are in *stable equilibrium*, that is, a load change is always shared proportionately by all.

Operation of Compound Generators in Parallel. When two compound generators are to be operated in parallel, it is necessary to use essentially the same wiring as that employed for shunt machines, except that an *equalizer connection* must be added. If the latter connection is not used, the two generators will not operate satisfactorily in parallel. An *equalizer* is a very low-resistance copper wire that joins together identical ends of the series fields, the other ends of which are connected together after the main switches have been closed. In Fig. 92, the equalizer is connected to points a and b, the negative terminals of the short-shunt compound generators. When A and B are operating in parallel, the two series fields are joined in parallel by equalizer ab and a portion of the negative bus $a'b'$. Should the equalizer be removed, instability would result if the machines are overcompounded (see Fig. 76) because any tendency on the part of one machine to assume a larger share of the total load than it should is immediately accompanied by an increase in its generated emf; simultaneously, the other machine, in dropping load, generates less voltage because the series-field flux is increased in the first and decreased in the second. The result of these actions is to make A take a still larger part of the total load, while B is losing more of its portion. If this condition is permitted to continue, A will not only assume the entire load but also attempt to drive B and its prime mover as a motor. Before this could happen in the practical installation, however, the circuit breaker would open and separate the machines electrically. It should be emphasized that what has been said applies to overcompound generators and not to those that have "drooping" characteristics, such as undercompound machines.

As was pointed out in the preceding paragraph, the equalizer connects the series fields in parallel, and this results in *a division of the total current in a ratio inversely proportional to the resistance values of the two fields.* Obviously, the voltage across the two series fields is always the same for a given load, which means that there can be no change in flux, the primary cause of the instability. This is to say that any tendency on the part

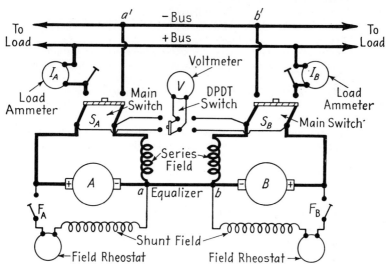

FIG. 92. Wiring connections for the operation of two compound generators in parallel.

of one series field to take more current is simultaneously accompanied by an identical tendency on the part of the other series field to do so. Instability is unlikely, therefore, when an equalizer connection is used.

The method for connecting one compound generator in parallel with another already supplying load current is exactly the same as for shunt machines. Load transfer, or disconnecting one machine from the line, may be accomplished similarly.

Questions

1. Name the three general types of d-c generator.
2. How does the construction of the fields differ in the three types of generator?
3. What current does the series field carry in the long-shunt compound generator? the short-shunt compound generator?
4. In a compound generator, why is the series-field coil wound *over* the shunt coil?
5. What voltage is developed by a series generator at no load? Why?

6. What voltage is developed by a shunt or compound generator with the shunt field open? Why?
7. Distinguish between a separately excited and a self-excited shunt generator.
8. Upon what two factors does the no-load voltage of a shunt generator depend?
9. How is the field of a shunt generator usually controlled?
10. Why is it not desirable to operate a generator at a speed lower than the one for which it is designed?
11. In normal operation of a shunt generator, why is the flux created by a shunt field not directly proportional to the field current?
12. What is meant by the *saturation curve?* What test must be performed to determine data for its construction?
13. What is the relation between no-load volts and field current below the *knee* of the magnetization curve? Why?
14. What does the saturation curve emphasize?
15. How does the no-load voltage of a shunt generator vary with changes in speed if the field current is kept constant?
16. What is meant by the *building up* of a self-excited shunt generator?
17. What conditions must be fulfilled before a self-excited shunt generator will build up?
18. What determines the voltage to which a self-excited generator will build up?
19. What simple test can be performed to show that a self-excited shunt generator will not build up because the field is reversed?
20. If a self-excited shunt generator does build up, will it do so if the field is reversed? if the armature leads are reversed? if the direction of rotation is reversed? if the residual field is reversed? Give reason for your answers.
21. What is meant by *flashing the field?* When must this be done?
22. What causes the voltage of a separately excited shunt generator to drop when it is loaded?
23. What causes the voltage of a self-excited shunt generator to drop when it is loaded?
24. Why is it important to keep the armature resistance of a generator as low as possible? In designing the winding, how could this be accomplished?
25. What is meant by *voltage regulation* of a shunt generator?
26. Assuming a certain regulation for a self-excited shunt generator, would this value be increased or decreased if the same machine were operated as a separately excited shunt generator? Give reasons for your answers.
27. How is the terminal voltage of a shunt generator usually controlled?
28. What is the principle of operation of the Tirrill regulator?
29. What types of voltage regulator are generally used at present for d-c generators?
30. Explain why the regulation of a self-excited shunt generator can be improved by operating the machine at a lower speed. Why is the lower speed practically inadvisable?
31. Make two simple sketches showing a compound generator connected long shunt and short shunt.
32. What are meant by the following terms when applied to compound generators: *cumulative? overcompound? undercompound? flat-compound? differential-compound?*
33. How can the degree of compounding be most readily adjusted?
34. What is meant by a *diverter?* What material is generally used for its construction? Why?
35. Why is no flux produced by the diverter?
36. What would be the objection to constructing the diverter of copper?

DIRECT-CURRENT GENERATOR CHARACTERISTICS

37. Explain why the terminal voltage of a series generator rises from its residual value at no load to some maximum value as load is applied.
38. What causes the terminal voltage of a series generator to level off to some maximum value as load is increased and then to decrease rapidly to zero with further applications of load?
39. Name several practical applications of the series generator.
40. What is meant by *armature reaction?*
41. What is the general direction of the armature-reaction flux with respect to the main-field flux?
42. What two important detrimental effects are produced by armature reaction?
43. Explain why armature reaction tends to reduce the total flux in a generator.
44. Why is it necessary to shift the brushes in noninterpole generators when the load changes?
45. In noninterpole generators, how can the main-pole tips be constructed to minimize the effect of armature reaction?
46. In what direction must the brushes be shifted as the load is increased on noninterpole generators?
47. What are *interpoles?*
48. By what electrical circuit must the interpole winding be excited? Why?
49. In what *zone* is the interpole effective?
50. Do interpoles prevent flux distortion? Explain.
51. What are *compensating windings?*
52. Explain why compensating windings correct flux distortion.
53. What current is carried by the compensating winding? Why?
54. Why is each turn of a compensating winding more effective in producing flux than each turn of the armature winding? What is the relation between them?
55. What is meant by *flashing over?*
56. What two important functions are performed by the commutator and the brushes in a d-c generator?
57. What are the three stages in the commutation process?
58. What is meant by *reactance voltage?* To what property of the armature coil does it owe its existence?
59. Explain why the brushes must be shifted *beyond* the magnetic neutral in noninterpole generators if sparkless commutation is to result.
60. When interpoles are used, why must they be somewhat stronger than would be necessary to neutralize the effect of armature reaction if sparkless commutation is to result?
61. Why do power plants usually have several generators to provide electrical service?
62. Why is service continuity so important in industrial plants? in the home? on the farm?
63. What important conditions must be fulfilled before a shunt generator is connected in parallel with another?
64. Under what conditions will two shunt generators operating in parallel divide the total load in exact proportion to the machine ratings?
65. Under what conditions will two shunt generators operating in parallel *not* divide the total load in exact proportion to the machine ratings?
66. Explain carefully the exact procedure for connecting a shunt generator in parallel with others already supplying a load.
67. How should a generator, already in parallel with others and supplying a load, be disconnected from the bus?
68. Why will two shunt generators operating in parallel always be in stable equilibrium?

69. What is an *equalizer?* When must it be used?

70. Why will two compound generators operating in parallel be in unstable equilibrium when no equalizer is used?

71. Explain carefully why an equalizer connection makes it possible for two compound generators to operate in parallel in stable equilibrium.

72. Make a wiring diagram showing how two *long-shunt* compound generators must be connected for operation in parallel.

Problems

1. Each shunt-field coil of a d-c generator produces 2,700 amp-turns. How many turns are there in the coil if the current is 1.5 amp?

2. A 50-kw 250-volt compound generator (short-shunt) has a series field in which each coil is wound with $8\frac{1}{2}$ turns. How many ampere-turns are produced by each coil?

3. If the curve in Fig. 67 was obtained for a speed of 1,500 rpm, calculate the points and draw a graph for a speed of 1,800 rpm.

4. Calculate the per cent voltage regulation of a shunt generator in which the no-load and full-load voltages are 135 volts and 120 volts, respectively.

5. The regulation of a 250-volt shunt generator is 6 per cent. Calculate the no-load voltage.

6. If the no-load voltage of a separately excited shunt generator is 110 volts at 1,350 rpm, what will be the voltage if the speed is increased to 1,600 rpm? is decreased to 1,100 rpm? (Assume constant field excitation.)

7. A self-excited shunt generator develops 230 volts when the field current is 3.6 amp. What will be the open-circuit voltage of this machine when the field resistance is reduced until the field current rises to 4.0 amp? (Assume that the flux changes half as much as the field current.)

8. A shunt generator has a no-load terminal voltage of 270 and a voltage of 240 when it delivers 180 amp. Assume a straight-line external characteristic and determine the voltage when the current is 120 amp. What is the equivalent load resistance under this condition?

9. Referring to Fig. 67, determine the total shunt-field resistance of a self-excited generator that will make it build up to 100 volts; 105 volts; 90 volts.

10. A 150-kw 250-volt compound generator is connected long-shunt. If the shunt-field resistance is 20 ohms, what is the series-field current at full load?

11. If the generator of Prob. 10 is connected short-shunt, what is the full-load series-field current?

12. A long-shunt compound generator has a shunt field with 1,200 turns per pole and a series field with $4\frac{1}{2}$ turns per pole. If the shunt-field and series-field ampere-turns are, respectively, 1,200 and 196, calculate the power delivered to a load when the terminal voltage is 230.

13. If the generator of Prob. 12 is connected short-shunt and delivers the same load at 230 volts, calculate the series-field and shunt-field ampere-turns per pole, assuming that each pole develops the same total mmf as before.

14. A short-shunt compound generator has a full-load current of 60 amp. If the series-field resistance is 0.04 ohm and a diverter carries 24 amp, what is the diverter resistance?

15. Determine how many ampere-turns are produced by each series-field coil of a compound generator, given the following particulars: rating = 100 kw; full-load volts = 600; series-field turns per coil = $8\frac{1}{2}$; resistance of series field = 0.025 ohm; diverter resistance = 0.068 ohm.

DIRECT-CURRENT GENERATOR CHARACTERISTICS

16. A 5-kw 120-volt compound generator has an armature resistance of 0.23 ohm, a series-field resistance of 0.04 ohm, and a shunt-field resistance of 57.5 ohms. Assuming a long-shunt connection and a voltage drop at the brushes of 2 volts, calculate the generated emf at full load.

17. A short-shunt compound generator has a shunt-field resistance of 77 ohms, a series-field resistance of 0.008 ohm, a commutating-pole winding resistance of 0.005 ohm, and an armature resistance, including brushes, of 0.02 ohm. When the armature current is 128 amp, the generated emf is 234.2 volts. Calculate the power delivered to the load.

18. The following information is given for a 300-kw 600-volt long-shunt flat-compound generator: shunt-field resistance = 75 ohms; armature resistance including brushes = 0.03 ohm; commutating-field winding resistance = 0.011 ohm; series-field resistance = 0.012 ohm; diverter resistance = 0.036 ohm. When the machine is delivering rated load, calculate the voltage and power generated by the armature.

19. In Prob. 18, calculate the power losses in: (a) the armature; (b) the commutating-pole winding; (c) the series-field winding; (d) the diverter; (e) the shunt-field circuit.

20. A 10-kw 250-volt compound (long-shunt) generator has a no-load voltage of 230. The shunt field has 800 turns per pole and the series field 8½ turns per pole. The shunt- and series-field resistances are 80 ohms and 0.07 ohm, respectively. In order to make the generator flat-compound, so that it will have a no-load voltage of 230 and a full-load voltage of 230, the series field must produce 225 amp-turns. Calculate: (a) the resistance of a diverter to accomplish this change; (b) the total number of ampere-turns produced by each pole at no load and at full load.

21. A 3,000-kw 500-volt 14-pole generator has a lap-wound armature with 2,340 conductors. The pole faces cover 67 per cent of the entire circumference. How many conductors are there in each pole face of a compensating winding?

22. Two shunt generators, A and B, are to be operated in parallel to supply a common load. Generator A has a no-load voltage of 240 and a voltage of 220 when it delivers a load of 60 amp. Generator B has a no-load voltage of 230 and a voltage of 220 when it delivers the same current as A. Assuming straight-line characteristics for both machines, calculate: (a) the line voltage and the total load in kilowatts when generator B is "floating"; (b) the total load delivered by both machines when the line voltage is 225.

CHAPTER 5

DIRECT-CURRENT MOTOR CHARACTERISTICS

Operating Differences between Motors and Generators. The study of electric generators in Chap. 4 involved discussions and problems relating to the conversion of mechanical energy to electrical energy. The purpose of this chapter will be to consider the machine that converts electrical energy into mechanical energy—*the electric motor*.

When a *generator* is in operation, it is driven by a mechanical machine such as an engine, a water turbine, or even an electric motor; the rotation through a magnetic field generates a voltage, which, in turn, is capable of producing a current in an electric circuit. When a *motor* is in operation, it is "fed" by an electric current from an electrical source of supply; the motor current then produces two stationary magnetic fields, one by the field poles and the other by the rotating armature, which react with each other to develop *torque*, which, in turn, produces mechanical rotation.

The *load on a generator* constitutes those electrical devices that convert electrical energy into other forms of energy; loads such as electric lighting, electric furnaces, electrical welding, electric motors, electric battery charging, etc., are well known. The *load on a motor* constitutes the force that tends to oppose rotation and is called a *countertorque*; such loads may be fan blades, pumps, grinders, boring mills, crushers, excavators, elevators, turntables, churns, drills, food mixers, and a host of other commonly used machines.

The *voltage of a generator* tends to change when the load changes; in shunt generators, a load increase is always accompanied by a drop in terminal voltage, while in compound generators, the voltage may fall, rise, or even remain constant as the load changes. In the case of the *motor*, the *speed of rotation* tends to change as the load varies; as will be pointed out later, an increase in load causes the speed of a shunt motor to drop slightly, that of a compound motor to drop considerably, and that of a series motor to drop very greatly.

The voltage of a generator can always be adjusted by doing either or both of two things: (1) changing the speed and (2) changing the strength of the magnetic field; in either case, an increase in speed or flux is accomplished by an increase in voltage. The speed of rotation of a d-c motor can be changed by varying either or both of two things: (1) the strength of the magnetic field and (2) the voltage impressed across the armature terminals; in general, an increase in flux decreases the speed, while a higher armature voltage raises the speed.

In practice, the speed of rotation of a generator is usually quite constant, since the speed of the mechanical prime mover is generally fixed by the governor controls. For all practical purposes, it is usually true that the impressed emf across the motor terminals is substantially constant, except in the case of special motors or applications in which the power supply constitutes a separate source; in such cases, the voltage of the generator is varied to change the speed of the motor.

Generators can be, and frequently are, operated in parallel with others to supply power to a common load; infrequently, they may be connected in series for the same purpose. Motors, on the other hand, usually operate as single independent units to drive their individual loads, although in special applications they may be connected in parallel or in series for the purpose of performing particular jobs at varying speeds; examples of series- and parallel-connected motors are electric excavators and traction equipment.

Generators are always started without electrical loads; the procedure is to bring them up to speed, adjust the voltage, and then close the main switch that permits the machine to deliver current. Motors, however, may or may not have a mechanical load when they are started; as a matter of practical significance, it is quite customary for a motor to start a load that is often equal to or greater than the rated name-plate value.

Structurally, d-c generators and motors are identical, except for minor differences that may permit them to function in accordance with known practical requirements. They may, in fact, often be operated either as generators or as motors with complete satisfaction if certain conditions, to be discussed later, are fulfilled. In some cases, as, for example, when dynamic or regenerative braking is employed, they operate as motors most of the time and as generators during the braking periods.

Classification of Direct-current Motors. In the study of d-c generators, it was learned that there are three general types, series, shunt, and compound, although only the last two are practically important. The series generator is used to a rather limited extent, usually in conjunction with shunt or compound types for "boosting" purposes; their principal fields of application are in connection with *boosters* and in the Thury high-voltage system for the transmission of power in Europe. There are also three general types of motor, namely, series, shunt, and compound, all of which are widely used in many applications. Each type of motor has very definite operating characteristics that differ markedly from those of the other two, so that it is important to know the load requirements before a proper selection is made.

When a d-c motor is loaded, it does more mechanical work per unit of time; under this condition, it always slows down. For example, an electrically powered bus runs more slowly going uphill than it does when operating on a level track; a drill press slows down when the drill begins to cut through heavy steel; when the pressure in the tank of an air compressor increases, the speed of the motor decreases. The amount by which a motor slows down for a given per cent increase in load, however, is different for the three types of motor. As will be pointed out later, the shunt motor slows down the least and the series motor the most, while the compound type is a sort of compromise between the first two. For the purposes of classification, it is convenient, therefore, to indicate how a motor behaves between no load and full load by using such terms as *constant* speed and *variable* speed. Understand that these terms refer to the *inherent* change in speed with change in load and that they do not apply to variations resulting from a manual or automatic adjustment of one kind or another. In general, if a change from no mechanical load to full load causes the speed to drop approximately 8 per cent or less, the motor is said to be of the constant-speed type; shunt motors fall into this classification. Motors in which the speed changes by greater values than indicated here are regarded as falling into the variable-speed classification; series and compound motors behave in this manner. It should be understood that the 8 per cent figure is merely an arbitrary value used as a guide to differentiate motors on the basis of speed classification. In some cases, a 5 per cent speed variation might be considered as excessive, while in other applications, a motor might be regarded as falling into the constant-speed class when the variation is as much as 10 or 12 per cent. The point is that the terms "constant speed" and "variable speed" are relative, like "tall" and "short," and must be applied advisedly in each particular instance. In some cases, however, there is absolutely no question about the speed classification; series motors, or compound motors with strong series fields, are definitely variable-speed machines.

Whenever the speed of a motor can be controlled by an operator who makes a manual adjustment, it is said to be of the *adjustable-speed* type. Note the difference between variable-speed and adjustable-speed motors; in the former, the speed changes inherently as a result of a modification of the loading conditions, while in the latter, the speed changes only because an operator or automatic control equipment has made an adjustment of some sort. It is possible, therefore, to have a *constant-speed–adjustable-speed* motor; a shunt motor with a field-rheostat control would fall into such a classification. Or a *variable-speed–adjustable-speed* motor might be a series motor with a line rheostat; such an arrangement is used on a hoist.

DIRECT-CURRENT MOTOR CHARACTERISTICS

Counter Electromotive Force (Counter EMF)—Voltage Generated by a Motor. It was previously shown (Chap. 2) that motor action results when an armature is placed in a magnetic field and the winding is supplied with current. Under this condition, the armature produces a magnetic field of its own, which is displaced 90 electrical degrees with respect to the main field. The interaction of these two fields then gives rise to a resultant field (Fig. 19), somewhat distorted, but of such a nature that the strength of the magnetic field on one side of each of the conductors is greater than it is on the other side. With the armature rotating as a result of motor action, the armature conductors continually cut through the resultant stationary magnetic field, and because of such flux cutting, *voltages are generated* in the very same conductors that experience force

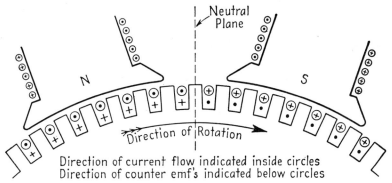

FIG. 93.—Sketch representing the relation between the direction of current flow and the direction of the counter emf in a d-c motor.

action. This can mean only that when a motor is operating, it is simultaneously acting as a generator. Obviously, motor action is stronger than generator action, for the direction of the flow of current in the armature winding is fixed by the polarity of the source of supply. The generated voltage does, however, oppose the impressed emf and, in this respect, serves to limit the current in the armature winding to a value just sufficient to take care of the power requirements of the motor.

Consider Fig. 93, which represents the armature conductors and two poles of a d-c motor. With the armature conductors carrying currents in the directions indicated by the crosses and dots *within the circles* and with the main pole polarities as marked, the direction of rotation of the armature resulting from motor action will be clockwise. As the armature rotates, the conductors cut the same flux that is responsible for the motor action; this flux cutting thereby causes voltages to be *generated* in the very same conductors that experience motor action. The *generated voltages* are indicated by crosses and dots *below the circles* and are in directions *opposite* to the *flow of current*. (This fact can be verified by the proper rule.) Since the generated voltage opposes the flow of current, it is called a *counter* electromotive force (counter emf). Clearly, *this counter emf can never be equal to, and must always be less than, the voltage impressed across the armature terminals*, because the direction in which the current flows determines first the direction of rotation and thus the direction of the counter emf. This can mean only that the armature current is *controlled* and *limited* by the counter emf. Therefore, by *Ohm's law*,

$$I_A = \frac{V_A - E_c}{R_A} \quad \text{amp} \tag{12}$$

where I_A = armature current
V_A = impressed voltage across armature winding
E_c = counter emf generated in armature
R_A = resistance of armature

EXAMPLE 1. A 115-volt shunt motor has an armature whose resistance is 0.22 ohm. Assuming a voltage drop across the brush contacts of 2 volts, what armature current will flow (*a*) when the counter emf is 108 volts? (*b*) if the motor load is increased so that the counter emf drops to 106 volts?

Solution

(a) $$I_A = \frac{(115 - 2) - 108}{0.22} = 22.7 \text{ amp}$$

(b) $$I_A = \frac{(115 - 2) - 106}{0.22} = 31.8 \text{ amp}$$

Since the counter emf is a generated voltage, it depends, for a given machine, upon two factors: (1) the flux per pole ϕ and (2) the speed of rotation S in revolutions per minute. That is, E_c is directly proportional to ϕ and S. If $E_c = k\phi S$, where k is a proportionality constant that depends upon the number of armature conductors, the type of armature winding, and the number of poles, Eq. (12) becomes

$$I_A = \frac{V_A - k\phi S}{R_A} \quad \text{amp} \tag{13}$$

EXAMPLE 2. A compound motor operates at a speed of 1,520 rpm when the voltage impressed across the armature terminals is 230. If the flux per pole is 620,000 maxwells and the armature resistance is 0.43 ohm, calculate: (*a*) the counter emf and (*b*) the armature current. (Assume a value of $k = 2.2 \times 10^{-7}$ and a brush drop of 2 volts.)

Solution

(a) $\quad E_c = 2.2 \times 10^{-7} \times 620{,}000 \times 1{,}520 = 207.5$ volts

(b) $\quad I_A = \dfrac{(230 - 2) - 207.5}{0.43} = 47.7$ amp

EXAMPLE 3. If the load on the motor of Example 2 is increased so that the armature current rises to 64 amp, what will be the speed of the motor, assuming that the flux increases by 6 per cent?

Solution

$$\phi = 620{,}000 \times 1.06 = 657{,}000 \text{ maxwells}$$

$$S = \frac{V_A - I_A R_A}{k\phi} = \frac{(230 - 2) - (64 \times 0.43)}{(2.2 \times 10^{-7}) \times 657{,}000} = 1{,}390 \text{ rpm}$$

It should be noted, in Example 3, that there are two reasons for the drop in speed: (1) the increased load, which requires an increased armature current, and (2) an increase in the flux because of the series field of the compound motor.

As a matter of practical importance, it should be stated that the counter emf developed in the armature of a motor is usually between 80 and 95 per cent of the voltage impressed across the armature terminals; the higher percentages generally apply to the larger motors, while the lower percentages apply to those near or in the fractional-horsepower ranges. It is also significant that motors in which E_c is a high percentage of the armature terminal voltage V_A will operate most efficiently, while those in which E_c is small compared with V_A will have a low efficiency. This may be seen from the following analysis:

From Eq. (12)

$$V_A - E_c = I_A R_A \tag{12a}$$

Multiplying Eq. (12a) by I_A gives

$$V_A I_A - E_c I_A = I_A{}^2 R_A$$

$$E_c I_A = V_A I_A - I_A{}^2 R_A \tag{14}$$

Equation (14) states that the power in watts *developed* by the armature is $E_c I_A$ because it is equal to the power in watts *supplied* to the armature $V_A I_A$ minus the copper loss in the armature $I_A{}^2 R_A$. Therefore, for a given load current I_A, it should be clear that *a motor will develop the greatest power when the counter emf E_c is a maximum.*

EXAMPLE 4. The armature of a 230-volt motor has a resistance of 0.312 ohm and takes 48 amp when operating at a certain load. (a) Calculate the counter emf and the power developed by the armature. (b)

If the armature resistance had been 0.417 ohm, the other conditions remaining the same, what would have been the values of E_c and the power developed in the armature? (Assume a brush drop of 2 volts in both cases.)

Solution

(a) $\quad E_c = (230 - 2) - (48 \times 0.312) = 213$ volts

Power developed $= E_c \times I_A = 213 \times 48 = 10{,}220$ watts

(b) $\quad E_c = (230 - 2) - (48 \times 0.417) = 208$ volts

Power developed $= 208 \times 48 = 10{,}000$ watts

Note, in Example 4, that a decrease of E_c from 213 to 208 volts reduces the developed power by 220 watts.

Starting a Direct-current Motor. At the instant a d-c motor is started, the counter emf E_c is zero because the armature is not revolving. As the armature accelerates to full speed, the value of E_c rises to a value that causes the proper value of armature current I_A to flow; the proper armature current is that required by the armature to permit it to drive its load at speed S, as fulfilled by Eq. (13). Since the counter emf E_c limits the current in the low-resistance armature winding, it should be understood that at the *instant of starting*, when E_c is zero, the armature current would be extremely high unless some resistance were added to offset the lack of E_c. In other words, if E_c is zero, or very small, as the motor is coming up to speed, a resistance must be inserted to take the place of E_c; as the speed increases, resistance may be cut out gradually because E_c rises; finally, when the motor has attained normal speed, all resistance can be cut out of the armature circuit. The following example should clarify this reasoning:

EXAMPLE 5. The armature of a 220-volt shunt motor has a resistance of 0.18 ohm. If the armature current is not to exceed 76 amp, calculate: (a) the resistance that must be inserted in series with the armature at the instant of starting; (b) the value to which this resistance can be reduced when the armature accelerates until E_c is 168 volts; (c) the armature current at the instant of starting if no resistance is inserted in the armature circuit. (Assume a 2-volt drop at the brushes.)

Solution

(a) $\quad I_A = \dfrac{V_A - E_c}{R_A + R}$

$R = \dfrac{V_A - E_c}{I_A} - R_A = \dfrac{(230 - 2) - 0}{76} - 0.18 = 2.82$ ohms

(b) $R = \dfrac{(230-2)-168}{76} - 0.18 = 0.61$ ohm

(c) $I_A = \dfrac{(230-2)}{0.18} = 1{,}265$ amp (a very dangerous value)

Exactly how a variable resistance should be inserted in the armature circuits of the three types of motor is shown in Fig. 94. In all cases, it

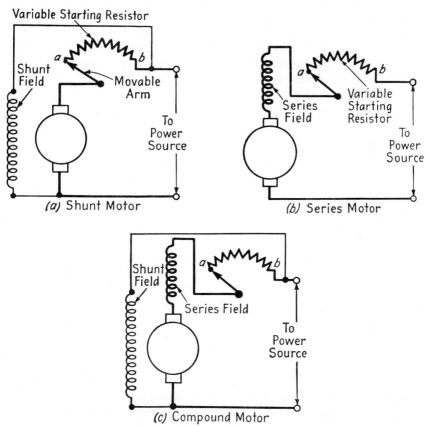

FIG. 94. Elementary connections showing how the variable starting resistor is inserted in the armature circuit of shunt, series, and compound motors.

should be recognized that the cross-sectional area of the material used in such resistors must be sufficient to carry the maximum current during the periods of time that they are in the circuit. Also, motors must be started with the movable arm at a, to be gradually moved to b as the armature accelerates to full speed.

In the case of very small motors, usually the fractional-horsepower sizes up to about ¾ hp, no starting resistor is necessary. Such motors may be started by simply closing the line switch. There are two reasons for this practice:

1. The resistance and the inductance of the armature winding are generally sufficiently high to limit the initial rush of current to values that are not particularly serious.

2. The inertia of a small armature is generally so low that it comes up to speed very quickly, thereby minimizing the serious effect that might otherwise result from a high *sustained* current.

Starters for Shunt and Compound Motors. Motor starters are generally manufactured in convenient sizes and styles for use as auxiliaries with d-c shunt and compound motors. *Their primary function is to limit the current in the armature circuit during the starting or accelerating period.* They are always rated on the basis of horsepower and voltage of the motors with which they are to be used. There are two standard types of motor starter for shunt and compound motors: the *three-point* type and the *four-point* type. Three-point starters are not completely satisfactory when used with motors whose speeds must be controlled by inserting resistance in the shunt-field circuit. (It will be learned later that the speed of a shunt or compound motor increases when resistance is inserted in the shunt-field circuit.) However, when applications require little or no speed control, either type may be employed.

In studying a wiring diagram of any starter, one should always remember that the *major circuit* (there may be several circuits in parallel) consists of a variable resistor of heavy wire in series with the armature circuit of the motor. Figure 95 represents a three-point starter, with its internal wiring, connected to a shunt motor.

Note that the starter has three terminals labeled L, F, and A. The line terminal L must be connected to either side, positive or negative, of the d-c source on the main switch (wire a); the field terminal F is connected to one field terminal on the motor (wire b); the armature terminal A must be connected to either one of the motor armature terminals (wire c). The final connection must then be made from the second line terminal on the main switch to a junction of the remaining two armature and field terminals of the motor. If it is desired that the speed of the motor be controlled, a field rheostat should be inserted in series between the field terminal F on the box and motor field terminal (wire b'). It is well to remember that *under no circumstances must any other wires be connected to the terminal ends of wires a, b, b', and c*. When the motor is at rest, the starter arm is held in the OFF position by a strong spiral spring.

To start the motor, one hand is held on the handle of the open main switch while the starter arm is moved to the first stud with the other hand; then the main switch is closed. If all the wiring is correct and the armature is free to turn, the motor will start. After the armature has accelerated sufficiently on the first stud, the starter arm is moved slowly to studs 2, 3, 4, 5, 6, etc., until the soft iron keeper rests firmly against the iron poles of the *holding-coil* electromagnet. The entire starting process should take from 5 to 10 sec. In the final position, the electromagnetic pull exerted by the holding coil will be greater than the force exerted by

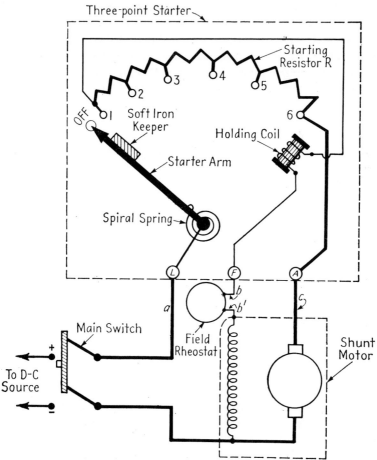

FIG. 95. Three-point starter connected to a shunt motor.

the spiral spring. Should there be a power failure or should the field circuit be opened accidentally, the starter arm will fall back to its OFF position. This function of the starter is particularly important because (1) if the power fails and the starter arm is not restored to the OFF position, the motor might be damaged should the power come on again, and, (2) if the shunt field circuit were opened accidentally and the starter arm did

144 ELECTRICAL MACHINES—DIRECT CURRENT

not return to the OFF position, the motor speed might become dangerously high. This will be discussed under Speed Control of Direct-current Motors, p. 165.

Figure 96 is a greatly simplified schematic wiring diagram clearly showing the two electric circuits of a shunt motor and its starter. Note that the main circuit, in heavy lines, consists of the variable resistor R and the armature. The second circuit includes the shunt field, the holding coil, and the field rheostat. In the last circuit, it should be noted that the current through the field is the *same* current that flows through the holding coil. Thus, if sufficient resistance is *cut in* by the field rheostat, so that the holding coil current is no longer able to create sufficient electromagnetic pull to overcome the spring tension, the starter arm will fall

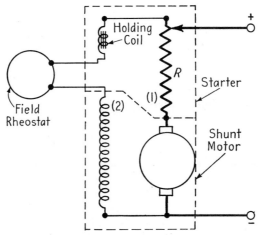

FIG. 96. Schematic wiring diagram of three-point starter connected to a shunt motor. Note two electrical circuits.

back to the OFF position. It is this undersirable feature of the three-point starter that makes it unsuitable for use with speed-controlled motors and that has resulted in the widespread application of four-point starters. Figure 97 shows a three-point enclosed type of starter with the cover removed. With the cover in place, no live parts are exposed. Note particularly the eight studs on the plate face plate, the electromagnet holding coil, and the three terminals at the bottom.

A four-point starter with its internal wiring is shown connected to a long-shunt compound motor in Fig. 98. If this starter is compared with the three-point type (Fig. 95), it will be observed that one important change has been made: the holding coil has been removed from the shunt-field circuit and, in series with a current-protecting resistor r, has been placed in a separate circuit in parallel with armature and field. Figure

Fig. 97. Enclosed standard-duty three-point starter for d-c motor, shown with cover removed. (*Cutler-Hammer, Inc.*)

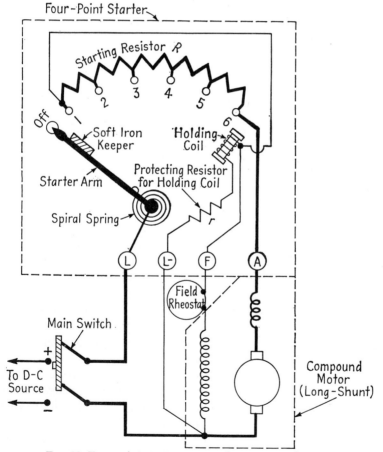

Fig. 98. Four-point starter connected to compound motor.

99 shows schematically how this change is accomplished. Thus, when the starter arm is on the first stud, the line current divides into *three* parts:

1. The main circuit is through the starting resistor R, the series field, and the armature.
2. The second circuit is through the shunt field and its field rheostat.
3. The third circuit is through the holding coil and a current-protecting resistor r.

Note particularly that this arrangement permits any change in current in the shunt-field circuit without affecting the current through the holding coil; in this regard it overcomes the objection to the three-point starter because the electromagnetic pull exerted by the holding coil will always be sufficient and will prevent the spiral spring from restoring the arm to the OFF position, no matter how the field rheostat is adjusted.

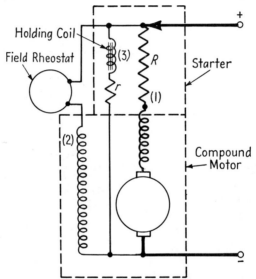

FIG. 99. Schematic wiring diagram of four-point starter connected to a compound motor. Note three electrical circuits.

Controllers for Series Motors. Whenever a starter is equipped with some means for varying the speed of the motor to which it is connected, it is called *a controller*. Controllers may also be designed to permit reversing the direction of rotation and may include protective features such as overload relays, undervoltage relays, and open-field devices. In this respect, a device used in connection with the starting of a series motor is called a *controller* because it usually serves also for reversing and speed-control purposes. As manufactured, they are generally designated as *drum* controllers, one such type being shown in Fig. 100. The variable series resistor in the armature circuit has a continuous duty rating, unlike those used in three- and four-point starters, which are designed for short-duty starting service only. A series-motor controller therefore has no spring that tends to restore the movable arm to the OFF position. The wiring connections are comparatively simple and are shown in Fig. 94b.

In construction, the drum controller consists of a central shaft to which are attached a group of copper cams which make contact with copper fingers or contactors as the drum is revolved. Since the currents are generally high, the breaking of the contacts is usually accompanied by much arcing; for this reason, fire-proof baffles are placed between successive sets of contacts to prevent dangerous short circuits that result from arcing. To improve further the operation of such controllers, it is customary to employ magnetic *blowout* devices, which blow out the arcs quickly when the contacts open.

Controllers for Shunt and Compound Motors. There are two objections to the use of a standard four-point starter when a field rheostat must be provided to control the speed of a motor. Refer again to Fig. 98 and note that a field rheostat is provided in the shunt-field circuit. As resistance is cut in, the speed increases; also, at a comparatively high speed, the field must be weakened considerably. Should the motor be stopped with a high value

Fig. 100. Drum-type controller used principally with speed-controlled motors employing resistance in the armature circuit. (*Cutler-Hammer, Inc.*)

of field resistance and then started again *before the rheostat is set at the all-out position*, the motor would attempt to start too rapidly; furthermore, the motor would draw an excessive armature current to compensate for the low field current because the required load torque depends upon the product of both the flux and the armature current. Under certain conditions, the high starting current might be sufficient to blow a fuse or open a circuit breaker. In addition, a minor disadvantage is the need for providing a separate mounting for the field rheostat apart from that required for the starter itself.

148 ELECTRICAL MACHINES—DIRECT CURRENT

To eliminate both these objections, controllers are available that incorporate both starting resistors and field rheostats in single, panel-mounted units and make it impossible to start shunt or compound motors with *any* resistance cut into the field rheostat. Since such units have two functions, starting and speed-control, they are properly called *controllers*. A

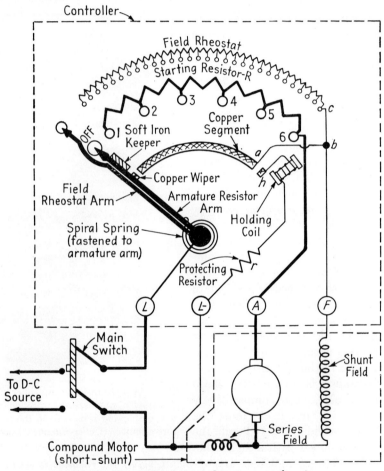

Fig. 101. Controller connected to compound motor.

wiring diagram of such a controller connected to a compound motor, short-shunt, is shown in Fig. 101.

To understand how this controller operates, it is necessary to recognize the following details of construction:

1. There are two arms, the longer one moving over a set of field-rheostat contact points (upper) and the shorter one moving over a set of **armature-resistor contacts**.

2. The handle for moving both arms clockwise simultaneously is on the upper arm.

3. The spiral spring is fastened to the armature-resistor arm only.

4. A copper wiper is mounted on the armature-resistor arm and wipes over a copper segment as it moves forward.

5. In the final position of the armature-resistor arm, the copper wiper makes contact with one end of the holding coil at point h, the copper wiper leaving the copper segment.

6. In the final position, the armature-resistor arm is held by the holding coil, while the field-rheostat arm is free to be moved counterclockwise to any point on the field rheostat.

When the motor is started, the two arms move forward simultaneously, the long one pushing the short one. Resistance is first inserted and then cut out of the armature circuit in the usual way as the armature accelerates. The field is excited without any rheostat resistance in the field circuit because the current passes directly from L to a to b to F. When the final position of the armature-resistor arm is reached, the arm is held by the electromagnet; at the same time, the connection between a and b is broken, so that the field current must now pass up to point c before reaching F. Obviously, when the field-rheostat arm is moved backwards, resistance is cut into the field circuit, which causes the motor to speed up.

It should be clear that the motor is always started automatically without field-rheostat resistance. Furthermore, a compact single unit is available for mounting.

The Automatic Starter for Shunt and Compound Motors. Automatic starting of motors is preferable to manual operation because, when properly designed and adjusted, the starting resistors are timed to be cut out so that the acceleration is uniform and the maximum allowable armature current is not exceeded. Manual starters, although cheaper, may be operated improperly at times, in which case damage may be done to both motor and starter.

Figures 102, 103, and 104 illustrate the wiring connections for three types of automatic starter that are designed to control automatically the acceleration of shunt or compound motors. Provided with a push-button station, such a motor is started by merely pressing a START button for an instant, whereupon acceleration proceeds smoothly to the proper speed; even the field rheostat may be preset (Fig. 103), although this resistance is not cut in until all of the starting resistors have been short-circuited by contactors. The motor is stopped by pressing the STOP button or when there is a power failure or an overload; this deenergizes the main contactor and opens the motor circuit. The push-button station is usually located conveniently, often at some distance from the motor, thus affording remote control.

When the automatic starter functions to bring a motor up to speed, two or more sets of contactors are generally made to short-circuit resistor units in a definite sequence, the latter being, of course, connected in series in the armature circuit. Several methods are used to accomplish this, but all depend upon the operation of relays, timing devices, contactors, and other auxiliary equipment. In one scheme, the counter-emf type of starter, a number of relays are connected across the armature where the counter emf increases as the motor accelerates, and the former

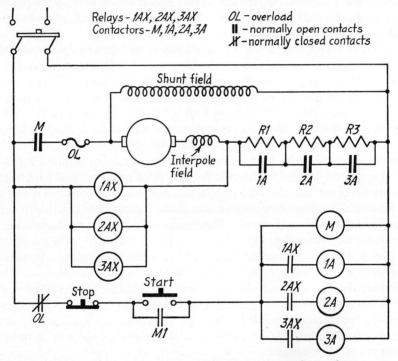

Fig. 102. Wiring diagram of a counter-emf starter connected to a shunt motor.

are adjusted to pick up at predetermined values of voltage. In a second type, the *time-limit starter*, a group of relays are *timed* to operate at preset intervals of time, by means of devices that function mechanically, pneumatically, or electrically. In still another method, the *current-limit starter*, the relays are designed so that they are sensitive to current changes in the armature circuit.

Referring first to Fig. 102, the *counter-emf* method, the shunt motor is started by pressing the START button. This energizes the main contactor M, which instantly closes the auxiliary contacts $M1$ (to seal the START button) and the main contacts M. The motor then starts with resistors

$R1$, $R2$, and $R3$ in series in the armature circuit. Note that relays $1AX$, $2AX$, and $3AX$ are connected *across* the armature terminals, where the voltage drop changes as the motor accelerates; since these relays are adjusted to pick up at preset and increasingly larger values of voltage, contacts $1AX$, $2AX$, and $3AX$ will close in a definite sequence. For example, at about 40 per cent armature voltage relay $1AX$ picks up, which, in turn, closes the $1AX$ contacts; this energizes contactor $1A$ and closes contacts $1A$ to short-circuit the resistor section $R1$. Next, at about 60 per cent armature voltage, relay $2AX$ picks up to close contacts $2AX$; this energizes contactor $2A$ and closes contacts $2A$ to short-circuit resistor $R2$. Finally, at about 80 per cent armature voltage the last resistor $R3$ is short-circuited when relay $3AX$ initiates a similar set of actions. The motor then runs normally until the STOP button is pressed, there is a power failure, or an overload opens the OL contacts; the main contactor M is then deenergized, the M contacts open, and the motor comes to a stop.

Figure 103 is a circuit diagram for a *time-limit acceleration starter* connected to a compound motor. In this design there are a group of three contactors $1A$, $2A$, and $3A$, each of which has one pair of instantaneously closing contacts across a block of armature resistance and another pair of *timed contacts that close with a time delay after the coil is energized*. When the START button is pressed, the control relay CR picks up; this closes the sealing contacts $CR1$ and the contacts $CR2$ that feed the four contactors $M, 1A, 2A$, and $3A$. The closing of $CR2$ energizes the M contactor, which opens contacts $M1$ to insert resistance in the M-coil circuit (to keep the current down) and closes the main contacts to start the motor. *After a definite time delay*, contacts $M2$-$T.C.$ close; this energizes the $1A$ contactor which, in turn, short-circuits resistor section $R1$. The operation of the $1A$ contactor also closes the $1A$-$T.C.$ contacts after a time delay and permits contactor $2A$ to pick up, which, in turn, closes the second pair of contacts $2A$ to short-circuit $R2$. Finally, the $2A$-$T.C.$ contacts close after a time delay to energize the $3A$ contactor, and this closes the $3A$ contacts to short-circuit the last resistor $R3$. The operation of contactor $3A$ also closes the $3A$-$T.C.$ contacts for sealing purposes and opens the normally closed $3A$ contacts across the shunt-field rheostat; the latter contacts are kept closed until the motor reaches full speed, after which the preset field rheostat resistance is inserted further to accelerate the motor to a desired higher speed. Stopping of the motor is accomplished in the same way as in the counter-emf starter.

The *current-limit acceleration starter* of Fig. 104 functions in still another way, depending for the motor's increase in speed upon the current taken by the armature circuit; this permits the motor to start more slowly when the load is heavy and more rapidly under light-load conditions. Note that there are three series relays, $SR1$, $SR2$, $SR3$, and three contactors, $1A$, $2A$, $3A$. When the START button is pressed, the M contactor is energized and this causes contacts $M1$ and M to close. The motor now starts as current passes through $R1$, $R2$, $R3$, and the $SR1$ series relay which open the normally closed contacts $SR1$ *instantly* and

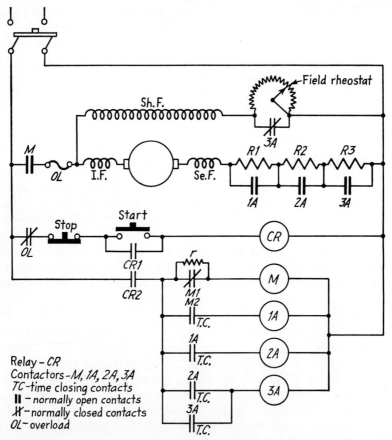

FIG. 103. Wiring diagram of a time-limit acceleration starter connected to a compound motor.

before the $M2$ contacts close; the latter is an important feature of the starter. When the current in relay $SR1$ drops to normal after the motor accelerates sufficiently, contacts $SR1$ close to energize contactor $1A$; this permits the $SR2$ relay to bypass $R1$ and the $SR1$ relay, and current, a greatly increased value, passes through relay $SR2$. The operation of relay $SR2$ opens the $SR2$ contacts, and an instant later the $1A$ contacts close; then, when the current through $SR2$ drops sufficiently, its contacts close again, contactor 2A is energized, and contacts 2A close. Again the motor increases speed, the armature current drops, and the same procedure is followed by relay SR3, normally closed contacts SR3 and contacts 2A, contactor 3A, and contacts 3A. The motor now runs with all of the armature-circuit resistance cut out. The motor is stopped in the same way as was previously described.

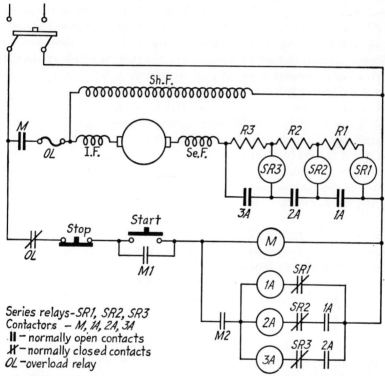

Fig. 104. Wiring diagram of a current-limit acceleration starter connected to a compound starter.

Loading a Motor—Effect upon Speed and Armature Current. When a *generator* delivers *electrical power* to a load, its terminal voltage *tends* to change. And, as was learned in Chap. 4, the operator (or designer of the machine) has practical control over this tendency on the part of the generator to change its terminal voltage. Briefly stated, *voltage control of a generator is generally exercised through the medium of flux adjustment or control.*

In practice, an electric motor generally receives its electrical power ($E \times I$) at substantially constant voltage. It then converts this electrical power into mechanical power, doing so by developing *torque* as it rotates its mechanical load. If the mechanical load on the motor changes, either the torque or the speed, or both, must change. And here again it will be learned that *the control of speed of a motor is generally exercised through the medium of flux adjustment or control.*

When a load is applied to a motor, the natural tendency of the latter is to slow down because the opposition to motion (i.e., the countertorque) is increased. Under this condition, the counter emf decreases, for the reason that E_c is proportional to the speed. This reduction in the speed immediately results in an increase in armature current (see Examples 2 and 3), since, by Eq. (12), I_A equals $(V_A - E_c)/R_A$. Obviously, this increase in armature current must be exactly that required by the motor to drive the increased load because *any increase in mechanical driving power must be met by a corresponding increase in electrical power input to the armature.* Moreover, since the electrical power input to the *armature* equals $V_A \times I_A$, it follows that only I_A must increase, for the reason that V_A is, as stated above, substantially constant. Thus it is seen that loading a motor always results in two changes: (1) a reduction in speed and (2) an increase in *armature* current. On the other hand, reducing a load on a motor causes its *armature* to take less current while it speeds up. Note that the word *armature* has been stressed in this discussion because it is here that the mechanical power is developed.

The speed at which a motor operates when it is driving its rated load, its so-called *rated horsepower*, is called the *normal speed*. If the mechanical load is completely removed from a *shunt* motor, so that it is merely overcoming its own bearing, brush, and wind friction, it will operate at a speed only slightly higher than the normal speed; this will generally be between 2 and 8 per cent higher than the normal speed. Doing the same thing to a *compound* motor will result in a rise in speed of about 10 to 25 per cent. In both shunt and compound motors, the no-load speed is very definite and stable; these types of motor do not attempt to operate at excessive unsafe speeds when running idle. The *series* motor, on the other hand, does attempt to race, or operate at a very high speed, when the load is removed. This fact is so well recognized in practice that a series motor is always geared or coupled to its load so that a countertorque will always exist; it is never belted to the load, because the accidental "throwing" of the belt will instantly result in a dangerous racing motor, a motor that is said to "run away."

Torque Characteristics of Direct-current Motors. The *torque* developed by a motor, i.e., the *tendency of a motor to produce rotation*, depends upon two factors: (1) the flux created by the main poles and (2) the current flowing in the armature winding. The torque is independent of the speed of rotation. This statement can be written in equation form thus:

$$T = k \times \phi \times I_A \quad \text{lb-ft} \tag{15}$$

DIRECT-CURRENT MOTOR CHARACTERISTICS

where T = torque (usually in pound-feet)
ϕ = flux per pole in maxwells
I_A = total armature current
k = proportionality constant (to be determined in the following equations)

To derive Eq. (15), the following analysis is given: the *power developed** by the motor is

$$P_d = E_c \times I_A \quad \text{watts} \quad \text{[by Eq. (14)]}$$

But
$$E_c = \frac{\phi \times P \times Z \times \text{rpm}}{a \times 60 \times 10^8} \quad \text{[by Eq. (2)]}$$

Therefore
$$P_d = \frac{\phi \times P \times Z \times \text{rpm}}{a \times 60 \times 10^8} \times I_A$$

Also
$$\text{hp} = \frac{2\pi \times \text{rpm} \times T}{33,000} \quad (16)$$

where T is in pound-feet and 1 hp = 746 watts.

Therefore
$$P_d = \frac{2\pi \times \text{rpm} \times T}{33,000} \times 746 \quad \text{watts} \quad (17)$$

Equating the two values of P_d

$$\frac{\phi \times P \times Z \times \text{rpm}}{a \times 60 \times 10^8} \times I_A = \frac{2\pi \times \text{rpm} \times T}{33,000} \times 746$$

$$T = \left(\frac{33,000}{2\pi \times 60 \times 746 \times 10^8}\right)\left(\frac{\phi \times P \times Z}{a}\right) \times I_A$$

$$= \left(\frac{0.1173}{10^8}\right)\left(\frac{P \times Z}{a}\right) \times \phi \times I_A \quad \text{lb-ft} \quad (15a)$$

In a given machine, the number of poles P, the total number of armature conductors Z, and the number of armature winding paths a are fixed. Hence, if

$$\left(\frac{0.1173}{10^8}\right)\left(\frac{P \times Z}{a}\right) = k$$

it follows that
$$T = k\phi I_A \quad \text{lb-ft}$$

A study of Eq. (15a) in connection with the three types of motor represented by Fig. 105 should make it clear how the torques of such machines behave under varying conditions of load. In the shunt and compound motors, the current through the shunt field I_{SH} is contant and is fixed only

* P_d, the power developed in watts, is the power to drive the mechanical load *and* the power necessary to overcome the motor's own rotational losses.

by the shunt-field resistance R_{SH} and the terminal voltage E_T; that is, $I_{SH} = E_T/R_{SH}$. This means that the shunt-field flux is independent of the load and is substantially constant because the flux depends only upon the field current. In the series and compound motors, however, the current through the series field changes with load because, as was pointed out previously, the armature current is determined by the load; at light loads, the armature current is small, whereas at heavy loads, the armature current is high. It follows, therefore, that the series-field flux will change with the load. This reasoning leads to the following conclusions concerning the manner in which the developed torque of a motor varies with the armature current I_A of a motor:

1. The torque of a *shunt motor* depends only upon the armature current; assuming that the shunt-field current is not changed by field-rheostat adjustment, the torque is independent of the flux. Therefore a graph

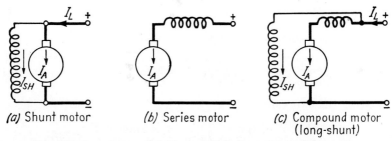

Fig. 105. Armature and field currents in three types of d-c motor.

indicating the relation between torque and load should be a straight line ($T = k_1 I_A$).

2. The torque developed by a *series motor* depends upon the armature current *and* the flux that this current produces in passing through the series field. At light loads, when the magnetic circuit iron is not saturated, the field flux is directly proportional to the load current. Under this condition, $T = k \times (k_2 I_A) \times I_A = k_3 I_A^2$, the equation of a parabola. At heavy loads, when the magnetic circuit iron is saturated, the flux will change very little or not at all with changes in load; under this condition, the graph will tend to become a straight line. Thus a complete graph of *torque vs. armature current for a series motor* will start out as a parabola and eventually become a straight line.

3. The torque of a *compound motor* (cumulative only, where the shunt- and series-field ampere-turns aid each other) combines the torque-load characteristic of the shunt and series motors. As the load on the motor increases, the armature, or load, current passing through the series field creates flux that adds to the constant shunt-field flux. The resultant flux thus tends to give the motor a rising torque curve (concave upward) at

DIRECT-CURRENT MOTOR CHARACTERISTICS

light loads when the iron is not greatly saturated; when the iron becomes saturated at heavy loads, the graph tends to become a straight line because no further increase in flux takes place.

Characteristic *torque vs. armature current* curves are shown in Fig. 106 for the three foregoing types of motor. For comparative purposes, these curves were drawn for motors having the same torque at some full-load

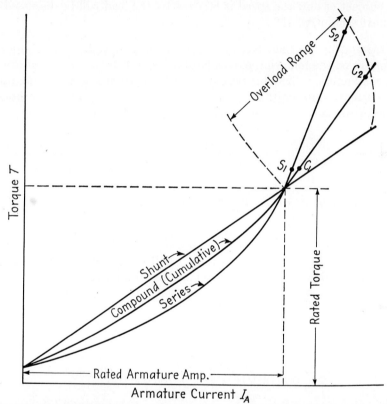

FIG. 106. Characteristic *torque vs. armature current* curves for three types of d-c motor.

value of I_A. Examination of the curves indicates that between no load and full load the shunt motor develops the greatest torque, while the series motor develops the least; the torque developed by the compound motor falls between these two. However, it is at overloads, i.e., above rated armature current, that the differences between these motors show up most prominently. Observe first that the series- and compound-motor curves become straight lines between S_1 and S_2 and between C_1 and C_2. Second, it is particularly significant that the overload torque of a series motor is considerably higher than that developed by a shunt motor; the

158 ELECTRICAL MACHINES—DIRECT CURRENT

compound motor again falls between the other two. In fact, it is generally recognized in practice that the series motor is capable of developing such high values of overload torque that it is difficult to stall. It is true, of course, that as the torque increases, the speed drops; but whereas the series motor will slow down considerably under heavy loading, the shunt motor will be unable to develop sufficient torque and will stall. The subject of how the speed is affected by the load will be discussed in the next section, p. 159.

EXAMPLE 6. (a) Calculate the torque in pound-feet developed by a d-c motor, given the following particulars: poles = 4; total number of armature conductors = 828; flux per pole = 1.93×10^5 maxwells; total armature current = 40 amp; winding = wave. (b) What will be the horsepower of the motor when operating at a speed of 1,750 rpm?

Solution

(a) $\quad T = \dfrac{0.1173}{10^8} \times \dfrac{4 \times 828}{2} \times 1.93 \times 10^5 \times 40 = 15$ lb-ft

(b) $\quad \text{hp} = \dfrac{2\pi \times \text{rpm} \times T}{33,000} = \dfrac{2\pi \times 1,750 \times 15}{33,000} = 5$

EXAMPLE 7. A four-pole shunt motor develops 20 lb-ft of torque when the flux per pole is 700,000 maxwells. If the armature winding has 264 conductors and is wound wave, calculate the total armature current.

Solution

$$20 = \dfrac{0.1173}{10^8} \times \dfrac{4 \times 264}{2} \times 700,000 \times I_A$$

$$I_A = \dfrac{20 \times 10^8 \times 2}{0.1173 \times 4 \times 264 \times 700,000} = 46 \text{ amp}$$

EXAMPLE 8. To change the shunt motor of Example 7 into a compound motor, the main poles are wound with a series field. What torque will this motor develop when the armature carries 65 amp if the series field increases the flux by 22 per cent?

Solution

$$T_2 = T_1 \times \dfrac{I_{A_2}}{I_{A_1}} \times \dfrac{\phi_2}{\phi_1} = 20 \times \dfrac{65}{46} \times \dfrac{1.22 \times 700,000}{700,000} = 34.5 \text{ lb-ft}$$

EXAMPLE 9. A series motor develops 62 lb-ft of torque when the current is 48 amp. Assuming that the flux varies directly with the current, calculate the torque if the load increases so that the motor takes 56 amp.

Solution

$$T_2 = T_1 \times \frac{I_2}{I_1} \times \frac{\phi_2}{\phi_1} = 62 \times \frac{56}{48} \times \frac{k \times 56}{k \times 48} = 84.4 \text{ lb-ft}$$

The torque developed by a motor at the instant of starting is called the *starting torque*. Some applications, such as lathes, grinders, centrifugal pumps, and fans, need have sufficient starting torque to overcome bearing and brush friction only. On the other hand, motors that must start heavy loads with high-inertia moving parts must possess high-starting-torque characteristics. Since torque is a function of both flux and armature current, it should be clear that the series motor would develop the highest starting torque because its series field is usually quite strong. The shunt motor would develop the least starting torque because the high armature current at the start has no effect upon the flux. The compound motor, on the other hand, usually wound with comparatively few series-field turns, will have more starting torque than the shunt motor but less than the series motor. In all this discussion, the terms "more" and "less" are relative; in fact, it is usually proper to designate the starting torque as a percentage of the full-load torque. Thus a series motor might have 500 per cent starting torque, a compound motor, 250 per cent starting torque, and a shunt motor, 125 per cent starting torque. In all cases, these percentages of torque would drop after the motors have attained normal speeds and are driving normal loads.

Speed Characteristics of Direct-current Motors. It was previously pointed out that (1) the speed of a shunt motor rises about 2 to 8 per cent when the rated load is completely removed; (2) the speed of a compound motor rises approximately 10 to 25 per cent when the rated load is completely removed; (3) the speed of a series motor rises very rapidly when the load is removed and must, therefore, always drive *some* load if it is to be prevented from racing dangerously, i.e., "running away." In order to understand why these conditions prevail, it will be desirable to study Eq. (13) by rewriting it in the form

$$S = \frac{V_A - I_A R_A}{k\phi} \quad \text{rpm} \tag{18}$$

where the meanings of the various terms are as already given on p. 138. Assuming that V_A (very nearly equal to the impressed voltage E_T) is constant, it is seen at once that the only factor affected when the load upon a *shunt motor* changes is I_A. And since the maximum change in the product of $I_A R_A$ between no load and full load is about 2 to 5 per cent of V_A, it follows that the maximum change in speed must be of the same order of magnitude. The following example makes this clear:

EXAMPLE 10. The armature of a 230-volt shunt motor has a resistance of 0.30 ohm and takes 50 amp when driving its rated load at 1,500 rpm. At what speed will the motor operate if the load is completely removed, i.e., when it is running idle, a condition under which the armature current drops to 5 amp? Assume that the flux remains constant and that the brush drops at full load and no load are 2 volts and 1 volt, respectively.

Solution

$$1{,}500 = \frac{(230 - 2) - (0.30 \times 50)}{k\phi}$$

$$S_{NL} = \frac{(230 - 1) - (0.30 \times 5)}{k\phi}$$

Dividing S_{NL} by S_{FL} gives

$$\frac{S_{NL}}{1{,}500} = \frac{(229 - 1.5)/k\phi}{(228 - 15)/k\phi} = \frac{227.5}{213}$$

Therefore $$S_{NL} = 1{,}500 \times \frac{227.5}{213} = 1{,}600 \text{ rpm}$$

In this typical shunt-motor example, it is seen that the rise in speed is 100 rpm, or about 6.7 per cent, when the rated load is completely removed from the motor so that it runs idle. It should also be pointed out that in motors like these, in which the flux does not change appreciably, the speed is directly proportional to the counter emf (i.e., $E_c = V_A - I_A R_A$).

When the rated load is removed from a *compound motor*, two factors are affected, namely, the armature current I_A and the flux ϕ. Unlike the shunt motor, in which the flux remains substantially constant for all conditions of loading, the effect of the series field is to cause the total flux in a compound motor to drop when the load is removed. Thus at full load the total flux results from the constant shunt field and an added strong series field, while at no load the total flux results from the same shunt field and a considerably weakened series field. It follows, therefore, that the speed of a compound motor varies much more between full load and no load than does the speed of a shunt motor, because (1) $I_A R_A$ changes and (2) the flux changes. The following example is given to illustrate this reasoning:

EXAMPLE 11. A 220-volt long-shunt compound motor (Fig. 105c) has an armature resistance of 0.27 ohm and a series-field resistance of 0.05 ohm. The full-load speed is 1,400 rpm when the armature current is 75 amp. At what speed will the motor operate at no load if the armature current drops to 5 amp and the flux is reduced to 90 per cent of its full-load value? Assume brush drops to be the same as in Example 10.

Solution

$$1{,}400 = \frac{(220 - 2) - 75(0.27 + 0.05)}{k\phi_{FL}}$$

$$S_{NL} = \frac{(220 - 1) - 5(0.27 + 0.05)}{k(0.90 \times \phi_{FL})}$$

$$\frac{S_{NL}}{1{,}400} = \frac{217.4/k(0.90 \times \phi_{FL})}{194/(k \times \phi_{FL})}$$

Therefore $\quad S_{NL} = 1{,}400 \times \dfrac{217.4}{194 \times 0.90} = 1{,}745$ rpm

Note that in this typical compound-motor example, the speed rises from 1,400 to 1,745 rpm; i.e., it changes 345 rpm between full load and no load. This change represents a rise of nearly 25 per cent, as compared with 6.7 per cent for the typical shunt-motor example previously given.

In operating a *series motor*, great care must be taken not to permit the load to be reduced to such an extent that the speed becomes excessive. To understand why this is so, the following points should be recognized:

1. The speed of any motor is *inversely* proportional to the flux [see Eq. (18)].
2. The flux produced in a given series motor depends entirely upon the load current.
3. When the load is heavy, the current is proportionately large and as a consequence, the flux is high; this results in a low speed.
4. When the load is lightened (*not completely removed*), the current drops, and this reduces the flux to increase the speed.

Moreover, the voltage drop in the armature circuit consisting of the armature resistance and the series-field resistance also affects the speed in the same way, although not to the same extent. Thus, when the load is heavy, the $I_A(R_A + R_{SE})$ drop is relatively high; this causes the speed to drop. When the load is lightened, the $I_A(R_A + R_{SE})$ drop is reduced; this causes the speed to rise. The extremely large change in speed of a series motor with variation in load is illustrated by the following example:

EXAMPLE 12. A 25-hp 240-volt series motor takes 93 amp when driving its rated load at 800 rpm. The armature resistance is 0.12 ohm, and the series-field resistance is 0.08 ohm. At what speed will the motor operate if the load is *partially* removed so that the motor takes 31 amp? Assume that the flux is reduced by 50 per cent for a current drop of 66⅔ per cent and that the brush drop is 2 volts at both loads.

Solution

$$800 = \frac{(240-2) - 93(0.12 + 0.08)}{k\phi_{FL}}$$

$$S_x = \frac{(240-2) - 31(0.12 + 0.08)}{k(0.5 \times \phi_{FL})}$$

$$\frac{S_x}{800} = \frac{232.8/k(0.5\phi_{FL})}{219.4/k\phi_{FL}} = \frac{231.8}{219.4 \times 0.5}$$

Therefore $\quad S_x = 800 \times \dfrac{231.8}{219.4 \times 0.5} = 1{,}690$ rpm

This represents a rise of 890 rpm, or a 111 per cent increase. A further reduction in load would cause the speed of the motor to increase by an even greater percentage because, with the iron unsaturated at light loads, the reduction in flux would be practically proportional to the drop in load.

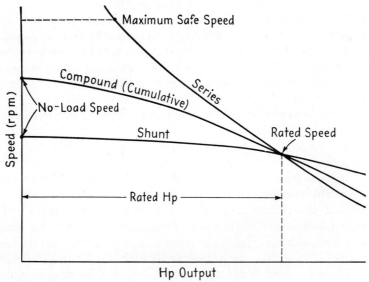

FIG. 107. Characteristic *speed vs. horsepower output* curves for three types of d-c motor.

In order to emphasize the importance of the *speed vs. load* characteristics of the three types of motor, typical curves are shown in Fig. 107. For purposes of comparison, the three motors are assumed to have the same speed when delivering the same horsepower output. (Per cent speed vs. per cent full-load horsepower would yield similarly shaped curves.) In studying these curves, note particularly that:

1. The speed of a shunt motor is substantially constant and has a very definite no-load value.

2. The speed of a compound motor varies considerably and also has a very definite no-load value.

3. The series motor operates over an extremely wide speed range and tends to "run away" at light loads—it should never be used with a belt drive or when the load is such that the torque might drop to approximately 15 per cent of the full-load torque.

Speed Regulation of Direct-current Motors. It is customary to speak of the natural, or inherent, change in speed of a shunt or compound motor between full load and no load as the *regulation*. And when this regulation is referred to the full load or rated speed of the motor expressed in per cent, it is called the *per cent speed regulation*. (It is a term similar to the one used in connection with shunt generators in Chap. 4, except that voltage values were implied there.) In equation form, this becomes

$$\text{Per cent speed regulation} = \frac{S_{NL} - S_{FL}}{S_{FL}} \times 100 \qquad (19)$$

In the shunt motor of Example 10, the speed change was 100 rpm in 1,500 rpm, or the per cent speed regulation was 6.67 per cent. For the compound motor of Example 11, the speed change was 345 rpm in 1,400 rpm, or the per cent speed regulation was 24.6 per cent. It should be thoroughly understood that these speed changes are the result only of the load changes imposed upon the motors and not the result of any manual or automatic adjustment on the part of the operator or a control mechanism. The greater the countertorque, the lower the speed. *Shunt* motors are generally regarded as *constant-speed motors* because their per cent speed regulation is very small. Of course, they are not constant-speed motors in the strictest sense, but their speed varies so little between full load and no load that they are called so for practical purposes. *Compound* motors are properly considered to be *variable-speed motors* because their per cent speed regulation is comparatively high. To be sure, there may be considerable variation between the regulations of motors of different design, but this is generally because of the number of series-field ampere-turns as compared with the shunt-field ampere-turns. If very few turns are used for the series field, the regulation will tend to approach that of a shunt motor; if many series-field turns are used, the regulation tends to depart greatly from the shunt-motor characteristic, although it never approaches the speed variations exhibited by series motors. The *series motor* certainly has a *variable-speed* characteristic, and it too is classed as such.

EXAMPLE 13. The full-load speed of a 10-hp shunt motor is 1,600 rpm. (*a*) If the per cent speed regulation is 4 per cent, calculate the no-load speed. Assuming a straight-line speed-load variation (approximately correct), determine the speed of the motor when it delivers (b) 6 hp and (c) 3.5 hp.

Solution

(a) $\quad 0.04 = \dfrac{S_{NL} - 1{,}600}{1{,}600}$

$\quad S_{NL} = 1{,}600 + (0.04 \times 1{,}600) = 1{,}664 \text{ rpm}$

(b) $\quad S_6 = 1{,}600 + \left(\dfrac{4}{10} \times 64\right) = 1{,}626 \text{ rpm} \qquad$ (by similar triangles)

(c) $\quad S_4 = 1{,}600 + \left(\dfrac{6.5}{10} \times 64\right) = 1{,}642 \text{ rpm} \qquad$ (by similar triangles)

EXAMPLE 14. The no-load speed of a compound motor is 2,200 rpm. At what speed will it operate at full load if the per cent speed regulation is 18 per cent?

Solution

$$0.18 = \dfrac{2{,}200 - S_{FL}}{S_{FL}}$$

$$1.18 S_{FL} = 2{,}200$$

$$S_{FL} = \dfrac{2{,}200}{1.18} = 1{,}865 \text{ rpm}$$

The arbitrary classification of motors on the basis of how the speed varies with load is especially important in connection with the problem of selection and application. In some applications, such as wood planers, circular saws, grinders, polishers, and line shafts, it has been found that constant-speed shunt motors perform most satisfactorily. Variable-speed compound motors should, however, be applied to loads requiring considerable torque upon starting or to loads that are subject to rapid change. Good examples of such applications are compressors, pumps, and pressure blowers. Series motors are most desirable in those applications requiring considerable starting torque or severe accelerating duty or when very high-speed variations are advantages. Some common applications of this type of motor are streetcars, turntables, cranes, bucket and mine hoists, and the operation of large valves.

Differential-compound Motors. Whenever the variable series-field ampere-turns of a compound motor "buck" the constant shunt-field ampere-turns, the total flux tends to diminish with increasing values of load. At light loads, the series-field current is low, so that it has little demagnetizing effect upon the shunt field; at heavy loads, the series-field current is comparatively high, which means that the demagnetizing action may be considerable. In any event, it should be recognized that this differential action of the two fields tends to make a motor run faster with increasing values of load [refer to Eq. (18)]. If the motor is designed with the proper number of series-field turns, it is possible to have a speed characteristic that is almost flat within a good part of the operating range. These motors tend, however, to become unstable when the load becomes appreciable; that is, increased load means an increase in speed, which, in turn, results in a further rise in speed, and so on until the motor "runs away" or the circuit breaker opens. *Differential-compound motors*, as they are called, have few practical applications; they may be used in special cases in which it is desirable to have a better constant-speed characteristic than has the shunt type of motor. Differential-compound motors must be started with caution, preferably with the series field short-circuited, because a large starting series-field current may be sufficient to reverse the *normal* magnetic polarities and cause the motor to start up in the wrong direction. Since the shunt field, with a comparatively large number of turns, has a high inductance, its current builds up very slowly in contrast with the rapid current rise in the low-inductance series-field circuit; this produces a resultant field that is largely influenced by the series field. The motor thus speeds up quickly and, in doing so, generates a considerable counter emf, takes a progressively smaller current, and develops diminishing values of torque. Meanwhile, the shunt-field current has been rising, and, when the motor eventually comes to a stop because of reduced torque, the counter emf is zero; a heavy surge of current now passes through the armature and series field. The machine finally starts up in the proper direction, but, if the new inrush of armature and series-field currents is high enough, the series-field mmf may again predominate and bring about a second reversal. These actions of instability may be repeated indefinitely or until the circuit breaker is opened or the fuses blow.

Figure 108 represents the typical comparative characteristics of shunt and differential-compound motors.

Speed Control of Direct-current Motors. It is frequently necessary to adjust the speed of a motor to some value other than the one at which it normally operates its load. This can readily be done with d-c motors in one or more of three different ways. These are (1) inserting a field rheostat in the shunt-field circuit of a shunt or compound motor; (2) inserting a resistance in the armature circuit of a shunt, compound, or series motor; (3) varying the voltage across the armature circuit of a shunt or compound motor while, at the same time, maintaining constant the voltage across the shunt field.

Method 1 is illustrated in Figs. 95, 98, 101, and 103. In this method, *the speed increases as resistance is cut in* by the field rheostat; this is true because, by Eq. (18), the speed rises as the flux is reduced.

Method 2 is illustrated by Fig. 94*b* for a series motor in which the starting resistor might also be used for speed-adjustment purposes. The method is also represented by Fig. 94*a* for a shunt motor and by Fig. 94*c* for a compound motor if the starting resistor is here, again, employed to control the speed of the motor. In the more practical case, a special continuous-duty rheostat having a high current capacity would be inserted in lead *c* of Fig. 95 or in series in the wire connecting the starter terminal *A* and the series field of Fig. 98. In this method, *the speed decreases as resistance is inserted in the armature circuit.* Referring to Eq.

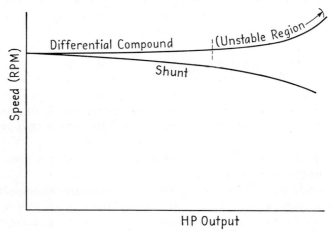

Fig. 108. Characteristic *speed vs. horsepower output* curves for shunt and differential compound motors.

(18), this is equivalent to increasing the second term in the numerator because it becomes $I_A \times (R_A + R)$; the greater the value of the inserted resistance R, the lower becomes the speed.

In method 3, it is necessary to have *two* sources of direct current for the *controlled motor*. The shunt field must be connected to a constant potential supply so that flux of unvarying intensity is created, while the armature is permanently connected to the armature terminals of a special generator whose voltage can be varied; the latter is usually separately excited from the same source that "feeds" the controlled motor. To drive both the *controlling generator* and the *exciter*, a prime mover, frequently a constant-speed a-c motor, is necessary. Figure 109 is a wiring diagram for this system of control, known as the *Ward Leonard method*. Obviously, the arrangement involves the expense of a three-machine set for the control of a motor.

In Fig. 110, the *motor-generator control set* is represented by the three machines mounted on a common base. Note that the exciter supplies current to the controlled motor and the controlling generator, while the armature terminals of the generator are directly connected to the controlled motor. Variation of the generator voltage is accomplished readily by

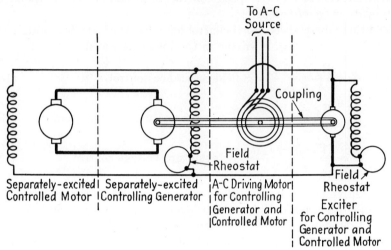

Fig. 109. Wiring connections of a Ward Leonard variable-voltage system of control for a shunt motor.

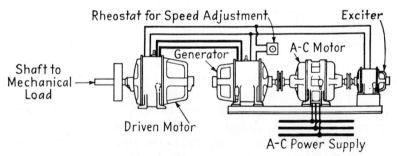

Fig. 110. Arrangement of machines and wiring connections of Ward Leonard method of control.

means of a field rheostat in the shunt field of the controlling generator; the variable armature voltage provides the means by which the motor speed is controlled. When the generator voltage is low (high field resistance), the motor speed is low; when the generator voltage is high (low field resistance), the motor speed is high. This variable-voltage control system has many important applications when extremely wide speed ranges, often as much as 10:1, are desired. It is frequently found on electric excavators, on freight-handling ships, and in blooming and paper mills and for the operation of passenger elevators in tall buildings. The chief disadvantage of the Ward Leonard system is its high first cost and its low over-all efficiency; to offset these disadvantages, it must be said that it provides excellent stepless speed control for a motor which must have a very wide range of speed.

The three methods of speed control here discussed are best understood by analyzing Eq. (18), rewritten in the following ways:

$$S = \frac{V_A - I_A R_A}{k \times \phi_{\text{var}}} \qquad \text{field-resistance control} \qquad (18a)$$

$$S = \frac{V_A - I_A(R_A + R_{\text{var}})}{k \times \phi} \qquad \text{armature-resistance control} \qquad (18b)$$

$$S = \frac{(V_{A\text{var}}) - I_A R_A}{k \times \phi} \qquad \text{armature voltage control} \qquad (18c)$$

If rated voltage is impressed across a motor and no rheostats (field or armature) are inserted, the machine is said to operate at *normal speed*. In the field-resistance method of control [Eq. (18a)], the speed is inversely proportional to the flux; it follows that the speed can only be increased *above normal* by this method. In the *armature-resistance* method of control [Eq. (18b)], the speed can only be lowered *below normal* by the insertion of resistance R. The armature voltage method, however, permits speed variations both above and below normal, assuming, of course, that the value of V_A can be widely adjusted.

The insertion of a resistance in the field or armature circuit of an *adjustable-speed* motor always involves a power loss. This power loss in watts is generally a small per cent of the total power input to the motor if the field-resistance method of control is used because the shunt-field current is about 2 to 8 per cent of the total motor current. On the other hand, if the armature-resistance method of control is employed, the power loss may be quite large because the current in the armature circuit is nearly equal to the line current. Two examples should make this point clear.

EXAMPLE 15. A 5-hp 115-volt shunt motor takes 36 amp when operating at full load at a speed of 1,500 rpm. The resistance of the shunt field is 52 ohms. To increase the speed of the motor to 1,700 rpm, it is necessary to "cut in" 28 ohms with the field rheostat. Under this condition, the motor takes 42 amp. Calculate: (a) the power loss in the field, and its percentage of the total power input for the 1,500 rpm speed; (b) the power loss in the field rheostat and in the entire field at 1,700 rpm; (c) the percentage power loss in the field rheostat.

DIRECT-CURRENT MOTOR CHARACTERISTICS

Solution

(a) $$I_{SH} = \frac{115}{52} = 2.21$$

$$P_f = 115 \times 2.21 = 254 \text{ watts}$$

and $$P_m = 115 \times 36 = 4{,}140 \text{ watts}$$

Per cent loss in field $= \dfrac{254}{4{,}140} \times 100 = 6.14$

(b) $$I_{SH} = \frac{115}{52 + 28} = 1.44$$

$$P_m = 115 \times 42 = 4{,}830 \text{ watts}$$

$$P_f = 115 \times 1.44 = 166 \text{ watts}$$

$$P_{\text{rheo}} = (1.44)^2 \times 28 = 58 \text{ watts}$$

(c) Per cent loss in field rheostat $= \dfrac{58}{4{,}830} \times 100 = 1.2$

EXAMPLE 16. To reduce the speed of the motor in Example 15 to 1,300 rpm, it is necessary to add 0.8 ohm in the armature circuit. If the total current is 32 amp and the shunt-field current is 2.2 amp, calculate the power loss in the armature rheostat and its percentage of the total power input.

Solution

$$I_A = 32 - 2.2 = 29.8 \text{ amp}$$

$$P_{\text{rheo}} = (29.8)^2 \times 0.8 = 710 \text{ watts}$$

and $$P_m = 115 \times 32 = 3{,}680 \text{ watts}$$

Per cent loss in armature rheostat $= \dfrac{710}{3{,}680} \times 100 = 19.3$

An interesting modification of the Ward Leonard system of control involves the use of *two* exciters that are electrically connected *in series bucking*, so that one of them excites the field of the main generator while the series-connected combination excites the field of the controlled motor. Both exciters are mechanically coupled to the controlling generator, and the three units are driven by a prime mover, usually a large a-c motor. In practice, the field of the main exciter is permanently adjusted, while the so-called *intermediate* exciter is used to control the field currents in both the main generator and the controlled motor, in such a manner as simultaneously to increase the strength of one field and to decrease that of the other. The intermediate exciter therefore exercises a dual-control function that makes it possible, for example, (1) to increase the speed of the controlled motor by strengthening the main-generator field and (2) to decrease the speed of the controlled motor by weakening the controlled-motor field.

A wiring diagram of this unique scheme of connections is given in Fig. 111. Note first that the fundamental interconnection of the two armatures of the main machines—the so-called *loop-circuit*—is identical with that of Fig. 109. However, the important departure from the original Ward Leonard system is that a controlled intermediate exciter, connected directly across the field of the main generator, is used not only to adjust the voltage in the loop [the V_A term in Eq. (18c)] but also to vary the flux

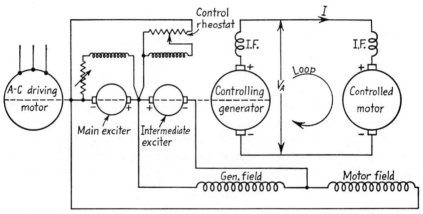

Fig. 111. Modified Ward Leonard speed-control system using two exciters.

produced by the field of the controlled motor [ϕ_{var} in Eq. (18a)]; the latter is accomplished by having the two exciters in series bucking and directly across the field of the controlled motor. Thus, when the voltage of the intermediate exciter is increased to raise V_A, the voltage across the series-opposed exciters is simultaneously lowered; the result is that the main motor is affected by two actions, both of which tend to increase the speed; conversely, when the voltage of the intermediate exciter is reduced to lower V_A, the voltage across the series-opposed exciters is increased; the controlled motor is again affected by two actions, both of which tend to reduce the speed.

Another modification of the basic Ward Leonard method of speed control makes use of two series machines in which the loop comprises a *series generator* and a *series motor*. This greatly simplified scheme of connections dispenses with the need for separate excitation as in Figs. 109 and 111 and thereby reduces the cost of the installation somewhat. As illustrated by Fig. 112, the controlling generator is driven by a prime mover, usually a constant-speed a-c motor, and speed control of the controlled motor is effected by shunting the series field of the generator with a variable resistance. To understand how such a system of control operates, it should be remembered that (1) the terminal voltage of a series generator (operating at constant speed) depends upon the series-field current or excitation, Fig. 78, and this, in turn, is a function of the loop (or load) current, or, as here, of that part of the current that is not shunted by the series-field rheostat, and (2) the speed of a series motor varies inversely as the load (Fig. 107), which, in turn, also depends upon the loop current. Since the characteristic curves of the two series machines are complementary, in the sense that a generator-current rise

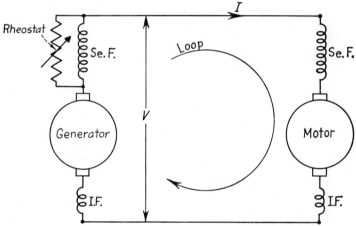

FIG. 112. Modified, and simplified, Ward Leonard speed-control system using two series machines.

attempts to increase the motor speed (the voltage is higher), while the same motor-current rise has an inverse effect upon the speed, the resulting action is to keep the motor speed constant, *for a given rheostat setting*.

The foregoing discussion may be further verified by the following mathematical analysis:

The motor speed is given by the equation

$$S_M = \frac{V - IR_M}{k\phi_M}$$

where R_M = equivalent resistance of motor
ϕ_M = flux produced by motor

But
$$V = E_G - IR_G$$

where E_G = generated voltage of generator
R_G = equivalent resistance of generator

Assuming that the two series machines are designed so that the fluxes are directly proportional to the loop current, i.e., that the magnetic circuits are never saturated,

$$\phi_M = k_1 I \quad \text{and} \quad E_G = k_2 \phi_G = k_3 I$$

where ϕ_G = flux produced by generator and k_1, k_2, k_3 = constants of proportionality, it follows that

$$S_M = \frac{k_3 I - k_4 I - k_5 I}{k_1 I} = \frac{k_6 I}{k_1 I} = k \quad \text{(a constant)}$$

where
$$k_4 = R_G$$
$$k_5 = R_M$$
$$k_3 + k_4 + k_5 = k_6$$

Armature Reaction in Direct-current Motors. As in the case of d-c generators, the magnetic neutral tends to shift when a motor is loaded. The reason for this tendency is the fact that the armature current creates a magnetic field of its own, apart from that created by the stationary poles, the magnetic axis of which is exactly halfway between the centers of the main poles. The two fields then react with each other to produce a resultant magnetic field whose axis lies somewhere between the center of the main pole and the axis of the armature field. And since the armature current in a motor is opposite to that of a generator *for the same direction of rotation*, it follows that the shift of the magnetic axis is opposite to the shift produced in the generator. The result is that *the magnetic neutral in a motor always tends to shift in a direction opposite to that of the armature rotation*. Figure 113 represents a simple two-pole motor with rotation *clockwise*. Indicated in Fig. 113a are the directions of the currents in the conductors, the main field flux in continuous lines and the armature field flux in broken lines. Figure 113b shows the resultant field and clearly emphasizes the fact that the magnetic neutral is shifted *counterclockwise*.

For good commutation, therefore, the brushes should be shifted to a new neutral plane with every change in load. Obviously, such brush-shifting practice is inconvenient and unsatisfactory because load changes generally occur quite frequently in the operation of motors. As in the case of generators, a more satisfactory method is to equip motors with commutating poles (often called *interpoles*). It will be recalled that their windings are always in series with the armature because it is necessary for interpoles to create fluxes that are directly proportional to the varying load. And note, too, that the interpole polarities are opposite to those in generators (see Fig. 85); that is, *the polarities of interpoles of d-c motors are*

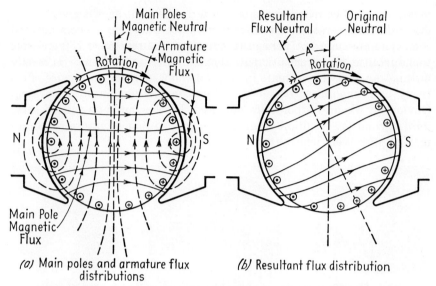

FIG. 113. Sketches illustrating flux distributions of loaded two-pole motor.

always the same as the preceding main poles with respect to the direction of rotation. Figure 114a shows how the interpoles counteract the effect of armature reaction, and Fig. 114b illustrates how their polarities are related to those of the main poles.

Interpoles for d-c motors are always made somewhat stronger than would be required to neutralize the armature reaction flux in the inter-

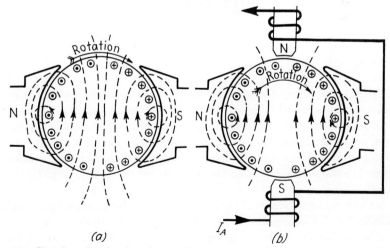

FIG. 114. Sketches showing how interpoles counteract effect of armature reaction in interpolar zones of d-c motor.

polar zones. The reason for this is exactly the same as was given in the discussion of generators; that is, the interpoles must help the commutated coils generate sufficient voltage to overcome the reactance voltage (due to inductance) and thus anticipate the new current directions. It is only in this way that the currents in the coils may be made to reverse effectively and smoothly without the objectionable commutator sparking. Obviously, when interpoles are not used, brushes must be shifted *backward* beyond the resultant magnetic neutral, so that armature reactance may be effectively neutralized.

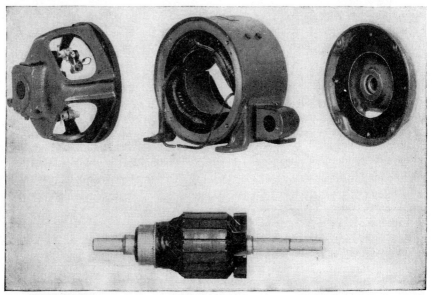

Fig. 115. Dismantled small two-pole motor in which only one interpole is used. (*General Electric Co.*)

In small machines, it is often found possible to use half as many interpoles as main poles, thus reducing the cost of manufacture. The reason for this practice is that the span of every coil is 180 electrical degrees. Therefore, if an interpole is made doubly strong, its effect on one coil side is equivalent to the action of two interpoles, each acting on one coil side. Figure 115 shows a dismantled two-pole motor in which a single interpole is used.

Compensating windings are also used in large d-c motors, the principles discussed under generators being quite similar to those of motors and equally applicable. The need for compensating windings on motors is often more important than on generators because of sudden violent load changes. This is particularly true in steel-mill applications.

Reversing the Direction of Rotation of Direct-current Motors. There are two general methods for reversing the direction of rotation of a d-c motor. These are (1) changing the direction of current flow through the armature and (2) changing the direction of current flow through the field circuit on circuits. The direction of rotation of a d-c motor *cannot* be reversed by interchanging the connections to the starting switch, because this reverses the current flow through *both* the armature and the field. The principle underlying the *direction* of force action exerted upon a current-carrying conductor when placed perpendicularly to a magnetic field was discussed in Chap. 2 (see Figs. 18 to 20). Figure 116 shows how the reversal of the direction of rotation of a shunt motor may be accomplished by both methods, using a DPDT reversing switch. In Fig. 116a, the reversing switch is connected to the shunt field, while in Fig. 116b, it is

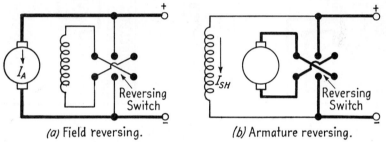

Fig. 116. Wiring diagrams showing methods for reversing the direction of rotation of a shunt motor.

connected to the armature. When the switch is closed to the left, the current will be *down* through both field and armature. When the switch is closed to the right, the current will be *up* through one of the elements and *down* through the other.

To reverse the direction of rotation of a compound motor, it is necessary to reverse the current flow through the armature winding only (Fig. 117a) or through *both* the series and shunt fields (Figs. 117b and 117c). Note that the wiring is much simpler when the first of the two methods is used because the reversal of the current through a single element is involved; a reversing switch can be used if frequent reversals are desired. If the second of the two methods is used, it is necessary to reverse the current through two field windings, the series field and the shunt field, as shown in Figs. 117b and 117c.

When motors have interpoles, the brushes are on the magnetic neutral, regardless of the direction of rotation; it will be recalled that one of the important advantages of interpoles is that the magnetic neutral is always the mechanical neutral, which, of course, is unaffected by load changes. However, when there are no interpoles in the motor, it is very important that the brushes be located properly with respect to the direction of rotation. Thus, if the brushes are correctly set at an angle of $A°$ clockwise from the mechanical neutral for a counterclockwise rotation, then they should be shifted to a point $A°$ counterclockwise from the mechanical neutral for a clockwise rotation.

A wiring diagram of a *reversing* time-limit automatic starter connected to a compound motor is given in Fig. 118. Note particularly that:

1. It is provided with two acceleration contactors and resistors, designated by $1A$, $2A$ and $R1$, $R2$.

2. Arrangement is made for *armature reversing* through forward contacts F and reversing contacts R.

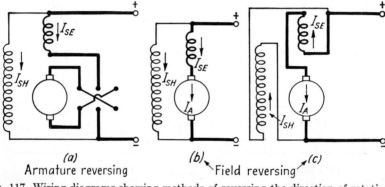

(a) Armature reversing (b) (c) Field reversing

FIG. 117. Wiring diagrams showing methods of reversing the direction of rotation of a compound motor.

3. The push-button station is equipped with FOR and REV buttons, each of which, when pressed, closes one set of contacts and simultaneously opens another set.

When the FOR button is pressed, the F contactor is energized and the R circuit is opened at f as a safety measure; this seals the FOR button at $F1$, closes the F contacts, and the current passes through the armature circuit from a to b. At the same instant the $F2$ contacts close to energize the timing relay T; after the latter times out, the T-$T.C.$ contacts close to energize the $1A$ contactor, which, in turn, closes the $1A$ contacts to short-circuit resistor $R1$. Contactor $1A$ also has a set of timed contacts $1A$-$T.C.$, which, after a certain delay, close to make contactor $2A$ pick up; the operation of the latter closes contacts $2A$ to short-circuit the second resistor so that the motor can run normally.

To reverse the motor, the STOP button is first pressed and the motor is permitted to come to rest; then, the REV button is pressed. This energizes the control relay CR, which opens the normally closed contacts $CR1$ to prevent the F contactor from being operated accidentally and momentarily opens the r contacts as a further safety measure. The normally

DIRECT-CURRENT MOTOR CHARACTERISTICS

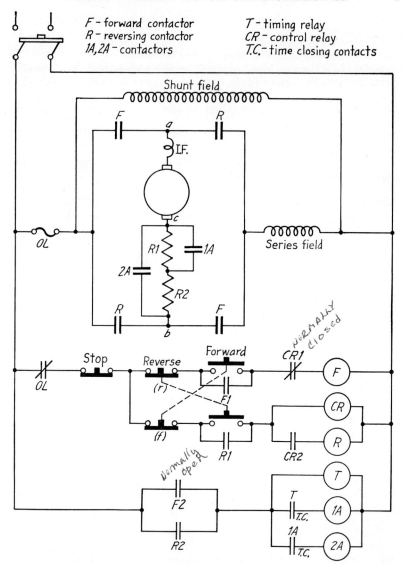

Fig. 118. Wiring diagram of a reversing time-limit acceleration starter connected to a compound motor.

open contacts $CR2$ then close to energize the R contactor, and the operation of the latter seals the reversing circuit at $R1$, closes contacts $R2$, and causes the main R contacts to close. The motor now picks up speed in the opposite direction since current passes through the armature circuit from b to a. Further actions of the starter proceed as explained for forward rotation.

Questions

1. What is meant by a *load on a generator?* a *load on a motor?*
2. List several practical types of loads applied to motors.
3. When the load changes, what tends to change in a generator? in a motor?
4. What methods are usually employed to adjust the voltage of a generator? the speed of a motor?
5. Generally speaking, what factor is kept constant when a generator is in operation? when a motor is in operation?
6. Are generators started under load? Explain.
7. Are motors started under load? Explain.
8. Is it possible to operate a d-c generator as a motor and vice versa? Explain.
9. Name the three general types of d-c motor.
10. Indicate, in a general way, how the speeds of the three types of d-c motor are affected by an increase in load.
11. What is meant by a *constant-speed motor?* What type of motor exhibits constant-speed characteristics?
12. What is meant by a *variable-speed motor?* What types of motor exhibit this characteristic?
13. What is meant by an *adjustable-speed motor?*
14. Under what condition would a motor be called a *constant-speed–adjustable-speed* motor? a *variable-speed–adjustable-speed* motor?
15. Why is the generated emf in a d-c motor called a *counter emf?*
16. Can the counter emf ever be equal to the impressed voltage in a motor? Give reasons for your answer.
17. How does the counter emf control the value of the armature current?
18. Upon what two factors does the counter emf depend in a given motor?
19. When the load upon a shunt motor is increased, what electrical factor affects the speed?
20. When the load upon a compound or series motor is increased, what two factors affect the speed?
21. Approximately, what percentage of the impressed voltage is the counter emf in a d-c motor?
22. Explain why the power developed by a d-c motor is determined by the value of the counter emf.
23. What limits the armature current in a d-c motor at the instant of starting?
24. How is it possible to keep the armature current down to a reasonable value when a d-c motor is started?
25. Why is it not particularly serious to start a small motor directly from the line without the use of external resistors?
26. What is the primary function of a *starter* for a d-c motor?
27. What are the two general types of manual starter for d-c motors?
28. How are starters rated?
29. When starting a d-c motor with a manually operated starter, why is it *not* permissible to hold the handle on an intermediate stud for a considerable length of time? Give reasons for your answer.
30. Explain exactly how a d-c motor should be properly started with a manual starter.
31. How many electrical circuits are there in a three-point starter? a four-point starter?
32. What is the disadvantage of the three-point starter? How is this disadvantage overcome in the four-point starter?

DIRECT-CURRENT MOTOR CHARACTERISTICS

33. What is the function of the *holding coil?*
34. What is a controller? What functions, other than starting, can it perform?
35. What two important advantages are possessed by a manual controller for a shunt or compound motor?
36. Explain the operation of a four-point controller for a shunt or compound motor. Refer to the wiring diagram of Fig. 101 in doing this.
37. In an automatic starter distinguish between: a *relay* and a *contactor;* normally *open* and normally *closed* contacts.
38. What is a *timing relay?* Explain its operation.
39. In what respects do the counter-emf, time-limit, and current-limit automatic types of starter differ from one another in operation? What advantages are possessed by each?
40. Describe the operation of the counter-emf automatic starter of Fig. 102.
41. Describe the operation of the time-limit automatic starter of Fig. 103.
42. Describe the operation of the current-limit automatic starter of Fig. 104.
43. In the push-botton automatic starter, is the START button normally open or closed? is the STOP button normally open or closed? are the overload relay contacts normally open or closed?
44. Explain why the armature of a d-c motor automatically draws more current from the source when the load is increased.
45. What is meant by the *normal speed* of a motor?
46. What general statements can be made with regard to the change in speed with load for shunt motors? compound motors? series motors?
47. What is meant by *torque?* In what units is it usually expressed?
48. In general, upon what two factors does the torque of a motor depend?
49. How much power must a motor develop? Be explicit.
50. Why is the torque of a shunt motor directly proportional to the armature current?
51. Explain how the torque varies with increased load upon a compound motor; a series motor.
52. Under what operating conditions is it desirable to use a shunt motor? a series motor? a compound motor?
53. Are shunt and compound types of motor stable at no load?
54. What precautions must be taken in operating a series motor, the load on which varies over wide limits?
55. Two similar shunt motors are changed to compound machines. If one of them is wound with twice as many series-field turns per pole as the other, which will have: (*a*) the greater speed change with load? (*b*) the greater starting torque? (*c*) the greater overload torque?
56. Define *speed regulation.*
57. What approximate values of speed regulation can be assigned to shunt and compound motors?
58. Why is it improper to speak of the speed regulation of a series motor?
59. Distinguish between the terms *speed regulation* and *voltage regulation.*
60. Why are shunt motors generally referred to as *constant-speed* motors?
61. Why are compound and series motors generally referred to as *variable-speed* motors?
62. Give several practical applications for shunt (*constant-speed*) motors.
63. Give several practical applications for compound motors.
64. Give several practical applications for series motors.
65. Why are differential-compound motors unstable at heavy loads? Explain carefully.

66. Under what conditions is it permissible and desirable to use differential-compound motors?
67. What precaution must be taken when starting a differential-compound motor?
68. Distinguish between *speed control* and *speed regulation*.
69. How does the speed vary when the shunt-field rheostat is adjusted?
70. How does the speed vary when the armature rheostat is adjusted?
71. How does the speed vary when the armature voltage is adjusted?
72. Using the fundamental equation for the speed of a d-c motor [Eq. (18)], justify the answers to Questions 69 to 71.
73. Describe the Ward Leonard system of control for a shunt motor. Refer to Fig. 109 in doing this.
74. Compare the power losses in the rheostats of field- and armature-resistance methods of control.
75. What advantages are possessed by the Ward Leonard system of control? What are its disadvantages?
76. List several practical applications of the Ward Leonard control system.
77. Describe the operation of the modified Ward Leonard system of control of Fig. 111, in which two exciters are used.
78. What advantages are possessed by the modified Ward Leonard control system of Fig. 111?
79. Explain why the voltages of the main and intermediate exciters of Fig. 111, in the modified Ward Leonard system of control, must never be equal. What would happen if they were?
80. Describe the operation of the simplified Ward Leonard system of control of Fig. 112, in which two series machines are used. What important magnetic design features must the machines possess for good operation?
81. How does the effect of armature reaction in motors differ from its effect in generators?
82. If no interpoles are used, how must the brushes be shifted in generators? In motors?
83. What are the polarities of the interpoles with respect to the main poles in generators? in motors?
84. Why is it possible to use half as many interpoles as main poles in some small motors? What advantage would this have?
85. What two fundamental methods may be used to reverse a d-c motor? Which is preferable in compound machines?
86. When a DPDT switch is used in the field circuit of a shunt motor for reversing purposes, what precaution must be taken when the motor is started? Is this precaution necessary if the DPDT switch is placed in the armature circuit?
87. Describe the operation of the automatic reversing starter of Fig. 118.
88. Carefully explain why the control relay *CR* in Fig. 118 has a normally closed contact in the *F* contactor circuit.
89. In Fig. 118, explain why each of the push buttons, FOR and REV, has one normally open and one normally closed set of contacts.
90. What would happen in Fig. 118 if the FOR and REV buttons were pressed simultaneously?

Problems

1. A 230-volt shunt motor has an armature resistance of 0.26 ohm. Assuming a 2-volt brush drop, calculate the counter emf when the armature current is 35.4 amp.

DIRECT-CURRENT MOTOR CHARACTERISTICS 181

2. A 240-volt shunt motor has an armature resistance of 0.38 ohm. What current will flow in the armature when the counter emf is 227.5 volts? (Assume a 2-volt brush drop.)

3. In Prob. 2, what is the *total* voltage drop in the armature circuit?

4. A 500-hp 600-volt compound motor operates at a speed of 495 rpm at full load. If the flux per pole is 9.1×10^6 maxwells and the armature resistance is 0.015 ohm calculate: (a) the counter emf; (b) the armature current. (Assume a value of $k = 1.3 \times 10^{-7}$ and a brush drop of 5 volts.)

5. If the load on the motor of Prob. 4 is increased so that the armature current rises to 800 amp, at what speed will the motor operate, assuming that the flux increases 9 per cent?

6. When the load in Prob. 4 is removed, the speed rises to 560 rpm. What is the regulation of the motor?

7. A 10-hp 1,750-rpm 550-volt shunt motor has an armature resistance of 1.55 ohms. If the armature takes 14.8 amp at full load, calculate: (a) the counter emf developed by the motor; (b) the power developed by the motor, in watts and in horsepower. (Assume a 5-volt brush drop.)

8. When a motor is operating under load, the armature takes 8,280 watts and its current is 36 amp. If the armature-circuit resistance, including brushes, is 0.4 ohm, what horsepower is developed by the motor?

9. A 5-hp 230-volt shunt motor takes 18 amp when operating at full load. The shunt-field resistance is 115 ohms and the armature resistance is 0.25 ohm. Calculate the value of the starter resistor if the armature current is limited to 1.5 times the rated value at the starting instant. (Assume a 3-volt brush drop.)

10. In Prob. 9 calculate the resistance which is *cut out* of the starter in moving to the second stud, assuming that this is done when the counter emf is 74 volts and that the armature current immediately jumps to a value 1.5 times its rated value.

11. What would be the armature current upon starting if the motor in Prob. 9 were connected directly to the line without a starting resistor?

12. A 60-hp 230-volt shunt motor has a shunt-field resistance of 38.3 ohms and an armature resistance of 0.04 ohm. If the resistance in the starter is 0.66 ohm, calculate the current input to the motor at the instant it is started. (Assume a 3-volt brush drop.)

13. A four-pole shunt motor has an armature with a total of 282 conductors. If each pole face has an area of 25 sq in. and the flux density is 50,000 lines per square inch, calculate the torque in pound-feet for an armature current of 85 amp. The armature winding is wave.

14. What horsepower will be developed by the motor of Prob. 13 for a speed of 1,420 rpm?

15. A 220-volt shunt motor has an armature resistance of 0.32 ohm and a field resistance of 110 ohms. At no load the armature current is 6 amp and the speed 1,800 rpm. Assume that the flux does not vary with load and calculate: (a) the speed of the motor when the rated line current is 62 amp (assume a 2-volt brush drop); (b) the speed regulation of the motor.

16. A 50-hp 550-volt shunt motor has an armature resistance, including brushes, of 0.36 ohm. When operating at rated load and speed, the armature takes 75 amp. What resistance should be inserted in the armature circuit to obtain a 20 per cent speed reduction when the motor is developing 70 per cent of rated torque? Assume that there is no flux change.

17. A 240-volt series motor has an armature resistance of 0.42 ohm and a series-field resistance of 0.18 ohm. If the speed is 500 rpm when the current is 36 **amp,**

what will be the motor speed when the load reduces the line current to 21 amp? (Assume a 3-volt brush drop and that the flux is proportional to the current.)

18. A 20-hp 220-volt 540-rpm shunt motor has an armature and a field resistance of 0.24 ohm and 157 ohms, respectively. If the starter resistance is 1.7 ohms, what current does the motor take at the instant of starting? (Assume a 2-volt brush drop.)

19. A 25-hp 230-volt 1,200-rpm shunt motor has an armature resistance, including brushes, of 0.2 ohm, and the armature current is 92 amp when the machine is operating at rated load and speed. To reduce the speed, a 6.8-ohm resistance is shunted across the armature and a resistance of 1.6 ohms is connected in series with the parallel combination of armature and shunted resistor. If, under the latter conditions, the armature current is 20 amp, calculate the speed of the motor. Assume that there is no flux change.

20. What should be the ohmic value of an armature shunt which would produce the speed in Prob. 19 with an armature current of 10 amp?

21. A series motor develops 164 lb-ft of torque when the current is 94 amp. If the load increases the current 50 per cent with a flux increase of 20 per cent, calculate the torque at the second value of current.

22. A 550-volt long-shunt compound motor has an armature resistance of 0.815 ohm and a series-field resistance of 0.15 ohm. The full-load speed is 1,900 rpm when the armature current is 22 amp. (a) At what speed will the motor operate at no load if the armature current drops to 3 amp with a corresponding drop in flux to 88 per cent of the full-load value? (Assume a brush drop of 5 volts at full load and 2 volts at no load.) (b) Calculate the per cent speed regulation of the motor.

23. A 15-hp 440-volt compound motor has a shunt-field resistance of 293 ohms and takes 28.5 amp at rated load when it operates at 1,150 rpm. What will be the horsepower output of the motor when its speed is 1,070 rpm, under which condition it takes 34.5 amp and the flux is increased by 10 per cent?

24. A 550-volt series motor takes 112 amp and operates at 820 rpm when the load is 75 hp. If the effective armature-circuit resistance is 0.15 ohm, calculate the horsepower output of the motor when the current drops to 84 amp, assuming that the flux is reduced by 15 per cent.

25. A 50-hp 230-volt shunt motor has a field resistance of 17.7 ohms and operates at full load when the line current is 181 amp at 1,350 rpm. To increase the speed of the motor to 1,600 rpm, a resistance of 5.3 ohms is "cut in" the field rheostat; the line current then increases to 190 amp. Calculate: (a) the power loss in the field and its percentage of the total power input for the 1,350-rpm speed; (b) the power losses in the field and the field rheostat for the 1,600-rpm speed; (c) the per cent losses in the field and in the field rheostat at 1,600-rpm speed.

26. To reduce the speed of the motor of Prob. 22 to 1,200 rpm, it is necessary to insert a resistance of 1.2 ohms in the armature circuit. Calculate the power loss in this resistor if the current is 20 amp.

CHAPTER 6

EFFICIENCY, RATING, AND APPLICATIONS OF DYNAMOS

Power Losses in Dynamos. A dynamo is a machine that converts energy from one form to another. In the generator, it is from mechanical energy to electrical energy; in the motor, it is from electrical energy to mechanical energy. When this conversion takes place at a uniform rate, that is, when the energy *received* by the machine *per unit of time* and the energy *delivered* by the machine in the same unit of time are *constant*, then it is proper to say that a dynamo converts *power* from one form to another. In this connection, it should be recognized that a dynamo must be in operation, i.e., rotating, if the energy it receives is to be converted into some other form; a dynamo cannot store energy. (A storage battery, on the other hand, can store chemical energy and later release it as electrical energy.)

The power *received* by a dynamo is called its *input;* in a generator it is mechanical power, while in a motor it is electrical power. The power *delivered* by a dynamo is called its *output;* in a generator, it is electrical power, while in a motor it is mechanical power. Although mechanical power is usually expressed in horsepower (1 hp = 33,000 ft-lb of energy per min), it is customary to change the mechanical unit to its equivalent electrical unit of watts when making calculations involving power losses and efficiency (1 hp = 746 watts).

The power input to a dynamo is always more than its power output; that is, a generator or motor cannot convert all the power it receives into useful output. This implies, of course, that some of the power input is used to perform functions that do not show up as power output. Understand that they may be *necessary* functions, but not *useful* in the sense that they represent electrical (generator) or mechanical (motor) loads.

The difference between the power input to a machine (in watts) and its power output (in watts) is called the *power loss* because it is unavailable to drive a mechanical load in a motor or to supply electrical power in a generator. This power loss always produces heating in the dynamo; therefore, the greater the power loss, as a percentage of the power input, the hotter will the machine tend to become. If this loss should reach an excessive value, the temperature rise might be high enough to cause failure.

There are two general classifications of power losses in electric machines: (1) those that are caused by the *rotation of the armature* and (2) those that result from a *current flow* in the various parts of the machine. The former are generally called *rotational losses* (sometimes *stray-power* losses), while the latter are termed *electrical losses*. Obviously, the rotational losses will vary only if the speed changes, while the electrical losses are affected by the current values through the various electrical elements such as the armature winding, the fields, and the brush contacts.

For the purposes of calculation, as well as for understanding what takes place in a dynamo, it is desirable to separate each of the two general losses into its components and then determine what values they have as well as the way in which they vary with speed and load. The rotational losses may be divided into five parts: (1) bearing friction, (2) brush friction, (3) wind friction, usually called *windage*, (4) hysteresis, and (5) eddy currents. The first three are obviously mechanical losses resulting from rotation.

The *hysteresis loss* takes place in the revolving armature core because the magnetic polarity in the iron changes in step with the changing positions of the magnetic material under the various poles. Thus, when an armature-core tooth is passing under a *north* pole, its polarity will be *south;* the iron particles are then oriented with their north ends pointing toward the shaft center. When this same tooth moves under a *south* pole, its polarity will be *north* and the iron particles will then be directed so that their north ends point away from the shaft center. The rapid "jerking" around of the tiny magnetic molecules in the armature-core iron as it revolves rapidly causes a sort of magnetic particle friction and produces heating. This hysteresis loss is magnetic in character but results only because the armature core is turning; it may therefore be properly classified as a rotational loss. In the modern dynamo, it depends upon the flux density in the armature-core iron, the speed of rotation, and the quality of the magnetic iron. Hysteresis loss is unaffected by whether or not the core is laminated. In generators, whose speeds are usually constant, and in shunt motors that operate at substantially constant speed, the hysteresis loss is fairly constant because the flux density in the core iron changes very little with load. In series and compound motors, the hysteresis loss does vary somewhat with changes in load because both the speed and flux density are affected by the load.

As the armature core revolves, voltages are generated in the iron exactly as they are in the copper wires. These voltages are objectionable, however, because they create a flow of current in the iron core in "eddies." These *eddy currents*, as they are called, result because the generated voltages in the iron near the outside surface are greater than those closer to the center of the shaft because of the higher speed; the difference in potential then causes currents to flow in the iron. Since these eddy currents have paths mostly parallel to the shaft, the logical way to minimize their magnitude is to introduce high resistances in the form of air spaces in direct line to such paths. This is readily accomplished by slicing or laminating the armature core and then coating each lamination with a high-resistance varnish. Eddy currents are electromagnetic in character but result only because of the rotation of the armature core; these, too, may therefore be regarded primarily as a rotational loss. Its value in watts depends upon the core flux density, the speed of rotation, and the thickness of the laminations; it is independent of the quality of the magnetic iron. The eddy-current loss is affected in much the same way by the load as is hysteresis; if the speed and flux densities remain constant, the eddy-current loss does not change; if they vary, as they must in series and compound motors, the eddy-current loss will change.

Copper losses always occur when there is a current flow through the various copper circuits. One of them takes place in the armature winding and is equal to $I_A^2 R_A$. There will be another loss at the brush contacts between the copper commutator and the carbon brushes; in practice, this loss depends upon the *brush-contact voltage drop* and the armature current I_A. In low-voltage machines (115 to 230 volts) the brush drop E_B varies between 1 and 3 volts, while in the higher-voltage machines (550 to 1,100 volts) E_B may be as much as 6 volts. From this it may be seen that the armature-resistance copper loss is approximately proportional to the *square* of the load, while the brush contact loss is nearly *directly* proportional to the load.

The various field copper losses are (1) the shunt field, (2) the series field, (3) the interpole field, and (4) the compensating-winding field, if there is one. The shunt-field loss is the only one that remains nearly constant (except for minor line-voltage changes) and is equal to $E_T \times I_{SH}$, where E_T is the terminal voltage and I_{SH} is the shunt-field current. Since the other fields are always connected in the armature circuit, their losses are almost proportional to the square of the load.

Another loss, very difficult to determine, is known as the *stray-load loss*. It results from such factors as (1) the distortion of the flux because of armature reaction, (2) lack of uniform division of the current in the armature winding through the various paths and through individual conductors of large cross-sectional area, and (3) short-circuit currents in the coils undergoing commutation. The indeterminate nature of the stray-load loss makes it necessary to assign it a reasonable value arbitrarily; it is usually assumed to be 1 per cent of the output of the machine when the rating is about 150 kw (200 hp) or more; for the smaller ratings, the stray-load loss is generally neglected when efficiency calculations are made, without much loss of accuracy.

Table 2 summarizes the points made in connection with losses in dynamos.

TABLE 2. DYNAMO LOSSES

Losses	Affected by	How determined

Rotational or stray power:
   Friction:
     Bearing ⎫
     Brush   ⎬ ....
     Windage ⎭
   Armature core:
     Hysteresis ⎫ ....
     Eddy current ⎭
	Speed changes
	
	 Speed and flux changes
		Usually by test

Copper:
   Armature winding ⎫
   Interpoles
   Series field   ⎬ ..
   Compensating winding
   Brush contact   ⎭
   Shunt field.............
	Load
	
	 Terminal voltage changes
		$I_A^2 \times R_A$
		 $I_A^2 \times R_I$
		 $I_A^2 \times R_{SE}$ or $I_L^2 \times R_{SE}$
		 $I_A^2 \times R_C$
		 (1 to 6) $\times I_A$
		 $E_T \times I_{SH}$

Stray load................	Flux distortion and commutation	1 per cent of output for machines 150 kw (200 hp) and over

Efficiency of Direct-current Generators. The efficiency of a d-c generator is the ratio of the electrical power output $E_T \times I_L$ to the mechanical power input, converted to watts. As a percentage, this statement may be written in equation form

$$\text{Per cent efficiency} = \frac{\text{watts output}}{\text{watts input}} \times 100 \qquad (20)$$

Since watts input = watts output + watts losses,

$$\text{Per cent efficiency} = \frac{\text{watts output}}{\text{watts output} + \text{watts losses}} \times 100 \qquad (21)$$

Another way to write Eq. (21) for more accurate calculations when a slide rule is used is

$$\text{Per cent efficiency} = \left(1 - \frac{\text{watts losses}}{\text{watts output} + \text{watts losses}}\right) \times 100 \qquad (22)$$

EFFICIENCY, RATING, AND APPLICATIONS OF DYNAMOS 187

EXAMPLE 1. A 5-kw generator has a total loss of 700 watts when operating at full load. Calculate the per cent efficiency.

Solution

$$\text{Per cent efficiency} = \left(1 - \frac{700}{5{,}000 + 700}\right) \times 100 = 87.7$$

EXAMPLE 2. A 50-kw generator has full-load efficiency of 90.5 per cent. Determine the total losses.

Solution

$$90.5 = \frac{50{,}000}{50{,}000 + \text{losses}} \times 100$$

$$\text{Losses} = \frac{50{,}000 \times 100}{90.5} - 50{,}000 = 5{,}250 \text{ watts}$$

The two methods for determining the efficiency of a generator are (1) by directly measuring the total power output and the total power input and (2) by making certain necessary tests from which the various power losses are determined, and then applying Eq. (22). The first of these involves an actual test upon the generator in which electrical instruments (voltmeter and ammeter) measure the output, while a *calibrated motor* drives the machine under test. A calibrated motor is one the outputs of which are known for all values of input (the calibrated motor output equals the generator input). Ordinarily, direct efficiency tests are both difficult to perform, especially upon large machines, and somewhat inaccurate. Furthermore, there is a considerable waste of power when actual loads must be applied. For these reasons, it is usually customary to determine the efficiency by the so-called *conventional* method, i.e., method 2, by which the various power losses are determined by calculation. To illustrate why the *conventional* efficiency determination is more accurate than one measured directly, assume an efficiency of 90 per cent. Under this condition, the losses will be about 11 per cent of the output [losses = (output/efficiency) − output]. Thus an error of even as much as 5 per cent in calculating the losses will result in an error in the over-all efficiency of less than 0.6 per cent. An error of 5 per cent in making measurements by the direct method would yield worthless results.

In order to determine the conventional efficiency of a generator, the following procedure may be followed.

1. Measure the resistance of the *armature*. Do this by connecting the armature in series with a current-limiting rheostat of sufficient current-carrying capacity to a d-c source. Adjust the current I_A to approximately rated value (name-plate value); then measure the voltage drop across the

armature winding by touching the voltmeter leads to commutator segments directly below two adjacent sets of brushes (see Fig. 119a); $R_A = E_A/I_A$.

2. Measure the resistance of the *interpole field*. With the same connections as in test 1, merely shift the voltmeter leads to the interpole winding (Fig. 119a); $R_I = E_I/I_A$. (If there is a compensating winding, the resistance measurement will include both the interpole and the compensating windings.)

3. Measure the resistance of the *series field*. Do this by simply shifting the leads in tests 1 and 2 from the armature terminals to the series field (see Fig. 119b); $R_{SE} = E_{SE}/I_{SE}$.

4. Measure the *shunt-field* resistance. Connect this field directly across a source of direct current whose voltage is the same as the name-plate value. Record E_{SH} and I_{SH}; $R_{SH} = E_{SH}/I_{SH}$.

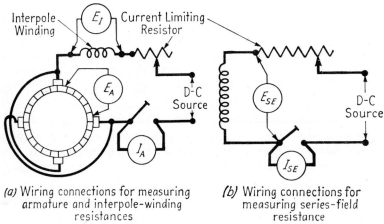

(a) Wiring connections for measuring armature and interpole-winding resistances

(b) Wiring connections for measuring series-field resistance

Fig. 119. Method of measuring armature-circuit and series-field resistances.

5. Measure the rotational (stray-power) loss. To do this, it is necessary to operate the generator as a shunt motor without load at *rated speed* and with a voltage across the armature winding equal to the normal full-load *generated* voltage. The power input to the armature then equals the friction and windage loss plus the core loss. The fact that the machine operates at rated speed means that the frictional losses (bearing, brush, and windage) are duplicated. Also, since the core loss is proportional to the flux and $E_g = k\phi S$, it follows that the core loss will be proportional to the generated emf if the speed is correct. Proper speed adjustment is made with the field rheostat. The correct armature voltage to be applied is

$$E_A = (E_T + E_B) + I_A R_A$$

EFFICIENCY, RATING, AND APPLICATIONS OF DYNAMOS

where E_T = rated name-plate voltage
 E_B = brush drop
 I_A = rated armature current
 R_A = armature resistance determined by test 1

Figure 120 is a diagram of connections for this test. The product of the impressed emf across the armature E_A and the current I_A in the armature,

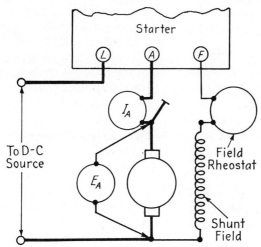

Fig. 120. Wiring connections for measuring the rotational (stray power) loss. NOTE: I_A = no-load current at rated speed; E_A = adjusted full-load generated voltage.

with the generator operating as a motor at no load at rated speed, is equal to the rotational (stray-power) loss; i.e., $E_A \times I_A$ = stray-power loss.

EXAMPLE 3. The following information is given in connection with a 10-kw 250-volt long-shunt flat-compounded generator: shunt-field resistance = 125 ohms; armature and interpole resistance = 0.4 ohm; series-field resistance = 0.05 ohm; stray-power loss = 540 watts; brush drop at full load (assumed) = 2 volts. Calculate the efficiency of the generator at full load. All resistances are given for a temperature of 75°C.

Solution

$$I_{L_{FL}} = \frac{10{,}000}{250} = 40 \text{ amp}$$

$$I_{SH} = \frac{250}{125} = 2 \text{ amp}$$

and $I_{A_{FL}} = 40 + 2 = 42$ amp

Losses	Watts
Stray-power............................	540
Armature............... $(42)^2 \times 0.4$ =	705
Series-field............. $(42)^2 \times 0.05$ =	88
Shunt-field.............. 250×2 =	500
Brush contact............ 2×42 =	84
Total...............................	1,917

$$\text{Per cent efficiency} = \left(1 - \frac{1,917}{10,000 + 1,917}\right) \times 100 = 83.9$$

To determine the efficiencies of a generator at loads other than that at full load, it is necessary merely to use the proper values of current for the several resistances. The stray-power loss in each case must be calculated for each load on the basis of the proper generated emf, where the stray-power loss is assumed to be proportional to this generated voltage. A table illustrating the method of solution for the data of Example 3 is given in Example 4.

EXAMPLE 4. Determine the efficiencies of the generator of Example 3 for load currents of 10, 20, 30, and 50 amp. Plot a curve of *efficiency vs. kilowatt output* from the calculated data.

Solution

Line volts..................	250	250	250	250	250
Current, amperes:					
Line.....................	10	20	30	40	50
Shunt field...............	2	2	2	2	2
Armature.................	12	22	32	42	52
Generated volts $= V_a + I_a R_a + V_b$	257.4	261.9	266.4	270.9	275.4
Losses, watts:					
Stray power.............	513	522	531	540	548
Armature...............	58	193	410	705	1,082
Series field.............	7	24	51	88	135
Shunt field.............	500	500	500	500	500
Brush contact..........	24	44	64	84	104
Total..................	1,102	1,283	1,556	1,917	2,369
Output, watts $V \cdot I_e$	2,500	5,000	7,500	10,000	12,250
Input, watts... $P_o + L_{osses}$	3,602	6,282	9,056	11,917	14,619
Efficiency, per cent. $\frac{P_o}{P_{in}}$	69.4	79.6	82.8	83.9	83.8

The graph of *efficiency vs. kilowatt output* is given in Fig. 121.

Efficiency of Direct-current Motors. The efficiency of a d-c motor is the ratio of the mechanical power output, converted to watts, to the electrical power input. As a percentage, this statement may be written

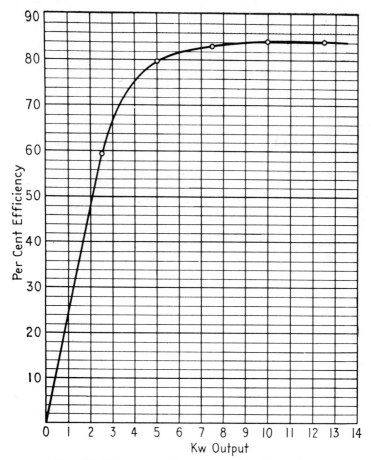

Fig. 121. *Efficiency vs. kilowatt output* for Example 4.

in equation form

$$\text{Per cent efficiency} = \frac{\text{hp output} \times 746}{\text{watts input}} \times 100 \qquad (23)$$

Since watts input = (hp output × 746) + watts losses,

$$\text{Per cent efficiency} = \left[\frac{\text{hp output} \times 746}{(\text{hp output} \times 746) + \text{watts losses}}\right] \times 100 \qquad (24)$$

or written in another way,

$$\text{Per cent efficiency} = \left[1 - \frac{\text{watts losses}}{(\text{hp output} \times 746) + \text{watts losses}}\right] \times 100 \qquad (25)$$

EXAMPLE 5. A 15-hp motor has a total loss of 1,310 watts when operating at full load. Calculate the per cent efficiency.

Solution

$$\text{Per cent efficiency} = \left[1 - \frac{1{,}310}{(15 \times 746) + 1{,}310}\right] \times 100 = 89.5$$

EXAMPLE 6. A 25-hp motor has an efficiency of 84.9 per cent when delivering three-quarters of its rated output. Calculate the total losses at this load.

Solution

$$84.9 = \frac{(0.75 \times 25 \times 746)}{(0.75 \times 25 \times 746) + \text{losses}} \times 100$$

$$\text{Losses} = \frac{14{,}000 \times 100}{84.9} - 14{,}000 = 2{,}500 \text{ watts}$$

The conventional efficiency of a motor may be determined in exactly the same way as for a generator, with one exception. In making the test for the rotational loss, the impressed voltage across the armature must equal the terminal emf *minus* the brush contact and armature-resistance drops at full load. Thus

$$E_A = (E_T - E_B) - I_A R_A$$

where the terms are as previously defined on page 189.

An example will now be given to illustrate the method for finding the efficiency of a motor.

EXAMPLE 7. Determine values for, and plot, the *efficiency-vs.-horsepower-output* curve for a 25-hp 230-volt long-shunt compound motor, given the following information: shunt-field resistance = 57.5 ohms; armature and interpole resistance = 0.18 ohm; series-field resistance = 0.03 ohm; brush drop (assumed) = 2 volts; stray-power loss = 1,088 watts. All resistances were measured at room temperature but were corrected to 75°C.

Determine the efficiency when the motor is delivering 25 hp.

Solution

Since the rated line current is unknown, a preliminary calculation for an approximate value is made based upon an *assumed* efficiency at full load of 88 per cent:

$$I_L = \frac{25 \times 746}{230 \times 0.88} = 92 \text{ amp}$$

EFFICIENCY, RATING, AND APPLICATIONS OF DYNAMOS 193

A table can now be constructed and calculations made for assumed values of armature current, based upon the line current given:

Line volts...............	230	230	230	230	230
Current, amperes:					
Line...................	26	48	70	92	114
Shunt field............	4	4	4	4	4
Armature..............	22	44	66	88	110
Generated volts..........	223.4	218.8	214.1	209.5	204.9
Losses, watts:					
Stray power...........	1,168	1,135	1,110	1,088	1,064
Armature..............	87	348	783	1,400	2,180
Series field............	15	58	131	233	363
Shunt field............	920	920	920	920	920
Brush contact.........	44	88	132	176	220
Total.............	2,234	2,549	3,076	3,817	4,747
Input, watts.............	5,980	10,040	16,100	21,160	26,220
Output, watts............	3,746	7,491	13,024	17,343	21,473
Output, horsepower......	5.0	10.0	17.5	23.3	28.8
Efficiency, per cent......	62.7	74.6	80.9	82.0	81.9

The graph of *efficiency vs. horsepower output* is given in Fig. 122. The full-load efficiency from the curve is 83.4 per cent.

Importance of Efficiency. A dynamo that operates at a comparatively high efficiency loses little power; on the other hand, one that operates at a low efficiency loses much power. Since all power losses are converted into heat, it follows that the temperature rise of a machine is affected very definitely by the efficiency. If the temperature to which a generator or motor rises exceeds well-established practices, it tends to cause insulation failure and eventual breakdown. Obviously then, from the standpoint of heating, it is very desirable that machines shall have efficiencies that are as high as possible. Furthermore, if a machine has a low efficiency and the temperature rise tends to become excessive, it usually becomes necessary to use cooling fans mounted on the shaft; such fans require additional power, so that the efficiency becomes still lower.

Another factor of importance is the cost of operation. The energy cost of the losses in a dynamo, as a percentage of the total operating cost, goes up as the efficiency becomes less. Therefore, in a machine of a given output, the energy charge increases as the efficiency drops because the operator pays for input, which is the sum of the output and the losses. A simple example should make this clear. Suppose that a 10-hp motor has an efficiency of 90 per cent. The input at full load would then be $(10 \times 746)/0.9 = 8,290$ watts, and the losses would be $8,290 - 7,460 = 830$ watts. The cost of this wasted heat energy for 8 hours a day and 25 days a month at 2 cents per kilowatt-hour would be $(830 \times 8 \times 25)/1,000 \times 0.02 =$

$3.32 per month, or $39.84 per year. However, if the efficiency had been 82 per cent, the cost of the energy loss would have been almost double the figures given above, i.e., $6.64 per month, or $79.68 per year. It is possible, of course, that a highly efficient motor may be more expensive than one whose efficiency is low and that it therefore adds to the fixed

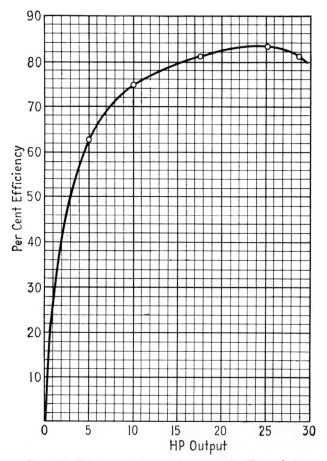

FIG. 122. *Efficiency vs. horsepower output* for Example 7.

charges such as interest, taxes, and insurance. But it is also significant that the better machine is likely to be more reliable and less subject to breakdown, and thus involve fewer maintenance costs, than one whose efficiency is comparatively low.

A well-designed machine in which good materials are used and high standards of manufacturing are employed is likely to perform well. Such practice requires careful planning, from design calculations and the selection of the raw materials to the various steps in production. In connection with the latter point, much inspection is necessary as the machine goes from one operation to the next.

In general, it is usually true that large machines are more efficient than small ones. Also, high-speed dynamos are likely to perform more satisfactorily, from the standpoint of efficiency, than are low-speed machines because the losses, as a *percentage* of the total input, tend to become smaller for the higher capacity high-speed generators and motors. It should be understood that these are general statements and that they may not always apply to differences between individual machines. Another point that should be mentioned is that mass-production methods have made it possible to produce uniformly high-efficiency dynamos at costs that are much lower than they would be for production requiring a considerable amount of hand work. This is especially true in connection with the manufacture of fractional-horsepower motors.

Rating of Generators and Motors. Generators are rated in terms of kilowatt output at a given speed and voltage. Motors are rated in terms of horsepower output at a certain speed and voltage. Such ratings imply that the temperature rise above the ambient room temperature will not exceed definite well-established practices when operation proceeds in accordance with the indicated name-plate values. Overloads are generally permissible for short periods of time, but when machines are required to carry greater loads than those specified, they must be kept under inspection to see that the temperature does not rise too much and that severe sparking at the commutator does not occur.

The type of service to which a machine is subjected is of great importance. Those that operate continuously at rated (or near rated) load are physically larger than machines whose loads are intermittent. Also, generators and motors that are not enclosed and are, in addition, well cooled by fans are likely to have higher ratings than similar machines that must be covered up or that are located where air does not circulate freely through and over them. Since temperature rise is so important, it should be clear that high-speed machines and those manufactured with insulations such as mica, glass tapes, and the new silicones can generally be physically smaller, in given ratings, than the low-speed dynamos constructed with the standard insulations. Obviously, in d-c generators and motors, commutation is extremely important; this factor often limits the output, even though heating may not have proceeded to permissible values.

In the operation of any d-c generator or motor, therefore, it is well to remember that there are two danger signals: temperature rise and commutator sparking. Usual temperature rise values are 40 and 50°C; these limits are often specified on the name plate in accordance with the type of service, i.e., *continuous duty* or *intermittent duty*. If a machine is equipped with cover plates on the end bells, removing the former makes better air circulation possible. Also, interpoles, compensating windings, and equalizer connections are construction features that improve commutation materially.

Selection of Generators and Motors. When a d-c generator or motor is selected for a given application, there are a great many factors that must be borne in mind if the proper choice is to be made. First of all, what should be the rating in kilowatts or horsepower, voltage, and speed? The type of machine, whether shunt, series, or compound, is the next consideration. Then, for a generator, it is also necessary to know such things as the degree of compounding, the way it is to be mounted, the kind of prime mover it will have, where it will be located, the type of control it will require, whether or not it will be paralleled with other machines, the general service conditions, and whether it is belted, geared, or coupled. In the case of a motor, selection should also be made on the basis of such factors as the following: starting-torque requirements, overload possibilities, type of service (such as continuous duty or intermittent duty), the regulation, whether or not the motor will be reversed, type of speed control, how it is to be mounted (i.e., on the floor, side wall, or ceiling, horizontal or vertical), the surrounding conditions (whether damp or hot), the altitude, and the kind of driver (i.e., belt, coupling, gear, etc.). All these factors must be carefully considered if the installation is to be satisfactory and efficient. In general, it is well to consult experienced application engineers before a proper choice is made because this field of service requires a great deal of intimate practical information learned only by a long period of actual work with such problems. An improper installation may mean poor service, inefficiency, high cost for energy and fixed charges, early breakdown, and much loss of time, as well as other more or less serious matters.

Special Dynamos and Applications. *Three-wire Generators.* One of the most serious disadvantages of d-c systems is that the generated voltage cannot be readily raised or lowered efficiently for long-distance transmission purposes. (A-c systems using transformers make voltage changes efficiently possible.) The three-wire generator is an attempt to overcome this difficulty to a limited extent, in that it makes available two voltages, one twice as much as the other. Usually, the generator is designed for 230-volt service, but it has provision for a *neutral*, which serves to provide 115 volts between it and either side of the line. The method was first proposed by Dobrowolsky.

To understand how such a machine operates, it is first necessary to remember that a d-c generator generates *alterating* current in its armature winding; in fact, if slip rings are provided (see Fig. 13) instead of a com-

EFFICIENCY, RATING, AND APPLICATIONS OF DYNAMOS

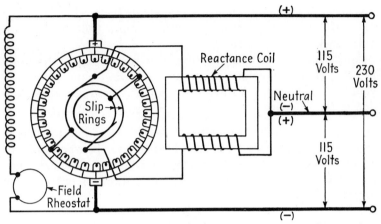

Fig. 123. Simple two-pole three-wire generator for 230/115-volt operation.

mutator, the current will be alternating. Therefore, if a high-inductance coil is connected to points on the armature winding that are exactly 180 electrical degrees apart, the *midpoint* on this reactance coil, usually called a *balance coil*, will always have a potential exactly halfway between the d-c voltage existing between positive and negative brushes. Figure 123 represents a simple two-pole three-wire generator for 230/115-volt operation. Note that the exact center of the reactance coil serves as the neutral, the wire connected to this point usually being grounded for safety purposes. In practice, the 230-volt source operates the high-current devices, such as motors, while the 115-volt supplies "feed" incandescent lighting and small appliances. Should there be any unbalance on the two 115-volt sides of the line, the neutral carries the difference in the two currents. Thus, if the current in a load between the positive line and neutral is 50 amp and the current in a load between the negative line and neutral is 30 amp, the neutral wire will carry 20 amp *toward* the reactance

Fig. 124. Balance coil for a 100-kw two-ring three-wire 120/240-volt d-c generator. (*Allis-Chalmers Mfg. Co.*)

coil. When the two 115-volt loads are equal, there will be no neutral current. Moreover, any current resulting from a 230-volt load *cannot* flow through the neutral wire.

A balance coil for a 100-kw two-ring three-wire d-c generator is shown in Fig. 124. Note particularly that it has two series-connected coils, each one wound around one leg of what looks like a simple transformer, and the junction of the coils brought out to function as the neutral terminal.

EXAMPLE 8. A 250-kw 230/115-volt three-wire generator delivers a 150-kw 230-volt load, a 30-kw 115-volt load between plus and neutral and a 50-kw 115-volt load between negative and neutral. Calculate the current in each line wire and in the neutral and the direction of the current in the neutral.

Solution

$$I_{230} = \frac{150,000}{230} = 652 \text{ amp}$$

$$I_{115+} = \frac{30,000}{115} = 261 \text{ amp}$$

and

$$I_{115-} = \frac{50,000}{115} = 435 \text{ amp}$$

$$I_{(+)} = 652 + 261 = 913 \text{ amp}$$

$$I_{(-)} = 652 + 435 = 1{,}087 \text{ amp}$$

and

$$I_N = 435 - 261 = 174 \text{ amp } \textit{away} \text{ from reactance coil}$$

For better performance, three-wire generators are usually equipped with two or three reactance coils connected to four or three slip rings, respectively. Figure 125 is a sketch showing how this is done when three reactance coils are used in a four-pole generator. The reactance or balance coils are said to be connected in *star* (*Y*) in this arrangement.

Third-brush Generators. Small generators that operate over a considerable range in speed, such as automobile generators used to charge storage batteries, utilize the effect of armature reaction to maintain substantially constant voltage. It will be recalled that wide variations in speed of *ordinary shunt and compound generators* have a marked effect upon the terminal voltage; if used for such service as indicated above, the voltage and charging current would change greatly between low and high automobile speeds. In addition, at high speed, the generator emf might reach values that would cause the low-voltage appliances (lights, small motors, radios, heaters, etc.) to burn out.

A method that was widely used in automobiles (now generally superseded by specially designed electromagnetic relays) is the *third-brush* type of generator. In construction it is a simple shunt generator, with the exception that the shunt field is connected to the positive brush and an auxiliary brush about 120 electrical degrees away in the direction of rotation. (In the standard shunt generator, the field is connected to the positive and negative brushes, 180 electrical degrees apart.) Figure 126 shows the arrangement for a two-pole generator in which the flux distributions are indicated for light load and high load. At light load (Fig. 126a), the armature current is small, so that the armature-reaction flux has little effect upon the main field. Under this condition, the voltage between the positive and negative brush will be determined by the number of series conductors in the angle between these two brushes. This will

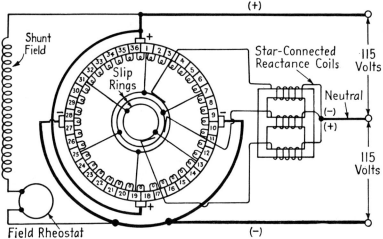

Fig. 125. Simple four-pole, three-wire generator for 230/115-volt operation. Note three slip rings and Y-connected reactance coil.

make the field voltage a maximum, which, in turn, means maximum flux and a given voltage between plus and ground at low speed. At high speed and high load (Fig. 126b), the armature-reaction flux greatly distorts the main flux, shifting the field in the direction of rotation (clockwise in this case). The result is that the weak field at pole tips a and b causes a drop in the generated voltage in those conductors that pass the weakened pole tips, so that the emf between the plus and third brushes is reduced. This, in turn, reduces the field flux and with it the voltage between the plus and ground brushes. In other words, since $E = k\phi S$, at low speeds ϕ is comparatively high, while at high speeds ϕ is correspondingly lower; for a wide range in speed, therefore, the terminal voltage remains nearly constant. The battery-charging current may be readily adjusted by simply shifting the third brush; to increase the charging rate, the third brush is shifted toward the grounded brush in the direction of rotation.

Another factor that tends to stabilize the current output of a third-brush generator at varying speeds is the short-circuit current in an armature coil being commutated. Since, in this type of machine, a commutated coil cuts flux in a strong field at heavy load, the generated emf will be accompanied by an opposing action that tends further to distort the main field, thus limiting the current delivered to the load.

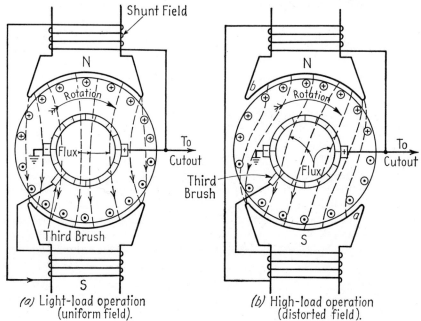

Fig. 126. Third-brush generator, showing field distribution under two conditions of loading.

Third-brush generators are generally designed to produce a weak shunt field (few ampere-turns) and a strong armature-reaction field. In this way, the latter will have a substantial effect upon the main-field flux. (A weak field is more easily distorted by armature reaction than is a strong field.)

Diverter-pole Generators. For battery-charging service, standard shunt and compound generators have certain limitations. In the case of the shunt machine, the terminal voltage drops with increased load. This characteristic acts to charge a battery at a constant rate, rather than at a high rate when the battery is discharged and at a low rate as the battery approaches full charge. When a battery is discharged, its terminal voltage is low; as a shunt generator attempts to charge the battery at a high rate, its voltage drops, which, in turn, keeps the charging current down. When the battery reaches full charge, its voltage rises. As the shunt generator attempts to charge the battery at a low rate, its voltage rises, which, in turn, keeps the charging rate up. For practical reasons, however, it is better to start out at a high charging rate and taper down as the battery approaches full charge, to prevent "gassing."

A compound generator with a *flat* characteristic would be better than the shunt type of machine but is open to the objection that it can accidentally become a motor (should the battery voltage exceed the generator emf) and thus reverse the current through the series field. As a differential-compound motor, it would be unstable and attempt to "run away."

A special type of generator, which has the flat characteristic of a compound machine but which cannot "motor," is the so-called *diverter-pole generator*. In construction, it looks like the standard generator with interpoles, except that a magnetic bridge joins each of the interpoles with the *succeeding* main pole in the direction of rotation. The winding on the *diverter poles* (which are actually interpoles) is connected in the armature circuit and tends to make the interpole polarities the same as the succeeding main poles in the direction of rotation.

Figure 127 is a sketch representing a two-pole diverter-pole generator. Note that a slot is cut in each of the magnetic bridges so that the latter tend to become easily saturated. At no load, the diverter-pole current is zero, under which condition it produces no mmf; a good part of the main-pole flux is thus diverted through the magnetic shunts to the diverter poles. The amount of this flux diversion is limited by the saturation of the bridges the areas of which are reduced by the slots. This flux distribution is shown approximately in Fig. 127a. Under load, however, the mmf of the diverter poles (acting to oppose flux diversion) prevents the main-pole flux from passing into the diverter poles. Thus the diverted flux decreases with increase in load, the result being that the emf of the machine tends to remain constant for all values of load. By proper design of the bridge areas and their slots, it is possible to have a characteristic that is practically flat. Figure 127b represents the diverter-pole generator under load.

Another advantage of this type of generator is that the flux that enters the armature from the bridge is in exactly the right direction to neutralize the armature-reactance effect; the result is that excellent commutation takes place. Moreover, this type of generator is self-protecting at heavy overloads and will not "motor," because the diverter-pole mmf cannot reverse the main-field flux; the small cross-sectional areas of the bridges become saturated at heavy loads and thus prevent the passage of flux from the diverter poles to the main poles.

Figure 128 shows a typical *voltage vs. load* characteristic of such a

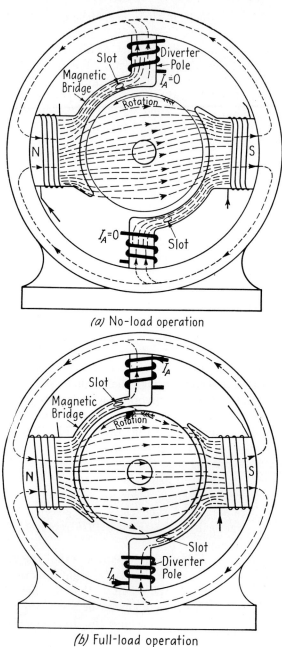

(a) No-load operation

(b) Full-load operation

FIG. 127. Diverter-pole flux distributions.

machine. Note particularly how the voltage remains practically constant up to a load a little above full-load value and then drops off suddenly. As mentioned above, this type of generator is used for charging batteries, doing so at a high rate when the battery voltage is low and tapering off to a low value as the battery voltage increases with charge.

Dynamotors. In some applications, it is necessary to have a d-c voltage much different from that available from the supply source. For radios that are served by storage batteries or 32-volt farm-lighting plants, for example, a B-battery voltage of 135 to 180 volts must be available. Also, in some 600-volt systems, a 115-volt lighting service has been used

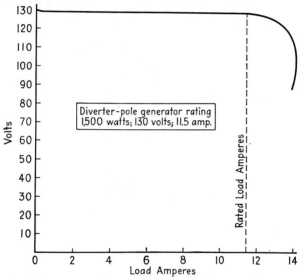

FIG. 128. Typical *voltage vs. load ampere* curve for diverter-pole generator used for battery-charging purposes.

for lighting. In such cases, it is possible to use a motor-generator set in which the motor is supplied with power from the available source and the driven generator develops the required voltage. Such an arrangement is, of course, expensive and inefficient, since two machines must be used.

A more desirable arrangement for comparatively small amounts of power is a single machine that functions both as a motor *and* a generator. Called a *dynamotor* (from the words *dynamo* and *motor*), it has a single field structure and an armature with two armature windings and two commutators. The two independent windings are usually placed in the same slots, one on top of the other (in some cases alternate sets of slots are used for each winding), and are connected to separate commutators, one at each end of the shaft. Special dynamotors have been constructed with as many as four armature windings and four commutators, two at each end of the shaft; in such arrangements, one winding is "fed" by the available source and provides the motor action, while the other three windings develop three different voltages needed. Dynamotors are less expensive and more efficient than motor-generator sets; the higher efficiency results because the rotational (stray-power) losses, consisting of the friction, windage, and iron losses, and the field losses are from a single machine. They have the disadvantage, however, that the generated voltage cannot be controlled and is fixed by the emf supplied to the "motor end."

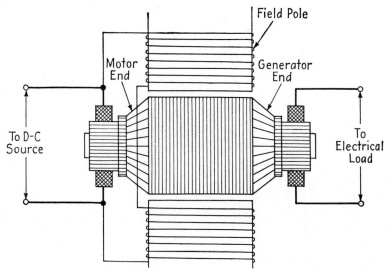

Fig. 129. Sketch illustrating dynamotor construction.

Briefly analyzing the dynamotor it is found that the voltage generated *in each conductor* is the same for either winding because the same flux is cut at the same speed. If the motor and generator windings are assumed to be similar (both lap or wave), the counter emf in the motor winding and the generated emf in the generator winding will be directly proportional to their respective numbers of turns. This implies that any change in flux by rheostat adjustment or any change in speed affects the two windings similarly. Therefore, if the $I_A R_A$ drops in the windings are neglected, there will always be a definite ratio of the input voltage (to the motor) to the output voltage (from the generator). To put it another way, if the field is strengthened, the motor slows down, but, since the generator armature winding cuts *more* flux, the product of $\phi \times S$ remains constant, so that the generator voltage is, in turn, kept constant. On the other hand, if the field is weakened, the motor speeds up, but since the generator armature winding cuts *less* flux, the product of $\phi \times S$ remains unchanged, so that the generator voltage is, in turn, kept constant.

Commutation in dynamotors is usually excellent because the armature-reaction effect results from two sets of ampere-turns in opposite directions. This is true because the current in the motor winding is always opposite to that in the generator winding; the net armature ampere-turns are exactly those needed to provide sufficient torque for the rotational losses. Strictly speaking, a dynamotor is a *rotary converter*, since it is a rotating machine that converts d-c energy at one voltage to d-c energy at another voltage. The term *rotary converter* is generally used, however, to imply a change from direct to alternating current or from alternating to direct current.

Figure 129 is a sketch illustrating dynamotor construction.

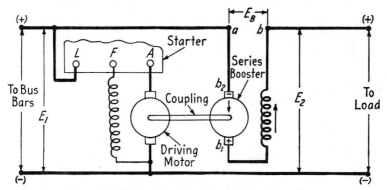

Fig. 130. Wiring diagram of series booster system.

Booster systems. *Series Booster.* The voltage of a *series generator* depends upon the current through it, that is, its load, because this same load current must pass through the series field to produce the flux. If such a generator is therefore connected properly in series in one of the line wires of a transmission system in which the line potential varies considerably with wide load changes, it is possible to make it compensate for the IR drop in the wires connecting the generator and the load. Furthermore, if the series generator is operated at low flux densities, so that its voltage is directly proportional to the load current ($E = k\phi S$, where S is constant), it may be made to function to keep the voltage at the far end of a transmission system practically constant regardless of the load; under this condition, the voltage of the series generator is always exactly equal to the line drop. Used in this way, a series generator is called a *series booster*. Figure 130 is a wiring diagram for a system such as the one described. Note that the series booster must be driven by a motor, but this is a simple matter, since the motor is merely a shunt machine connected to the bus bars. In operation, such a motor-generator set must be protected against the possibility that the driving motor might accidentally be disconnected from its source of power. Should this happen, the series generator would immediately function as a series motor, reversing its direction of rotation, and, operating without load, attempt to *race*. A study of the wiring diagram will indicate why this will happen. When the machine is a *generator*, booster voltage E_B adds to E_1 to yield E_2; brush b_1 is then positive and brush b_2 is negative. However, when the booster acts as a *motor* because it has lost its motive power, brush b_2 becomes positive and brush b_1 negative. Finally, since the field current has not changed direction, the booster reverses direction and attempts to "run away" as an unloaded series motor.

A more frequent application of the series booster than its use to compensate for line drop is to prevent electrolysis of water and gas mains and

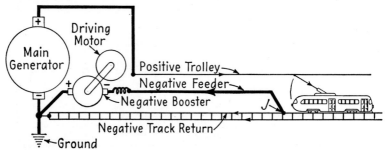

Fig. 131. Wiring diagram of negative-feeder booster system.

lead-covered telephone and telegraph cables. In street-railway systems, the negative or *return* circuit consists of the tracks, the surrounding earth, and any metallic structure capable of carrying the current back to the grounded negative bus of the main generator. Even if the tracks are carefully bonded together, some of the current will always leave the tracks to enter gas and water mains and lead-covered cables. Eventually, when the current leaves these to return through the earth to the tracks, electrolysis, i.e., "eating away" of metal, takes place. The damage to underground metallic structures has been so serious that several methods have been used to prevent electrolysis. Series boosters connected in series between the tracks and a negative feeder have proved very successful; when used in this way, they are called *negative* or *track-return boosters*.

Figure 131 is a wiring diagram of a negative-feeder booster circuit. Note that the polarity of the booster is to force current in a direction from a distant point J on the tracks through the feeder and back to the grounded negative terminal of the generator. If the booster voltage

EFFICIENCY, RATING, AND APPLICATIONS OF DYNAMOS

exactly equaled the negative-feeder voltage drop, no current would flow in the track system out to the point of juncture J, because point J would then be at exactly the same potential as the generator ground. In the actual installation, however, the negative-booster voltage is made only large enough to keep the leakage current from doing minor damage to underground structures.

Constant-current Booster. In large office buildings and hotels that have individual d-c power plants, two general types of loads must be serviced: (1) fairly constant lighting and appliance loads and (2) violently changing elevator motor loads. To prevent the lights from flickering when elevators are started, it is necessary to use a storage battery in conjunction with a booster to smooth out the wide current fluctuations on the main generator. A satisfactory scheme is illustrated in Fig. 132; it consists of a

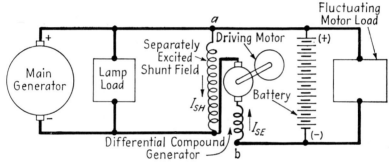

FIG. 132. Constant-current booster system.

differential-compound generator booster and a storage battery. In operation, the main generator "feeds" the lighting load directly. When the elevator load is zero or light, the current I_{SE} in the series field, the ampere-turns of which "bucks" the shunt-field ampere-turns, is zero or small; under this condition, the line voltage between a and b, being higher than the battery emf, causes the battery to charge. When the elevator motor load is heavy, the series-field current I_{SE} causes the booster to lose voltage to such an extent that the battery is forced to discharge into the motor load. Thus the main generator load current is kept fairly constant, since it "feeds" the lighting load and part of the motor load *or* the lighting load and battery-charging load. Such an arrangement is known as a *constant-current booster system.*

Electrical Braking of Direct-current Motors. Several methods are used to bring a motor quickly to a stop. The mechanical brake is, of course, well known; it is open to the objection, however, that a smooth stop is difficult because it depends upon the condition of the braking surfaces as well as the skill of the operator. Three excellent electrical methods find wide application in practice, eliminating, as they do, the need for brake linings, levers, and the other mechanical devices.

Plugging. This method uses the principle of applying power to the motor in such a direction that the latter attempts to reverse. Obviously, it must come to rest before it can reverse; the power to reverse the motor must first bring it to a stop, at which instant the circuit is opened. In some applications, such as in rolling-mill service, the plugging action is used to bring the motor to rest quickly and then permit reversal.

At the instant that a motor is plugged, the armature impressed emf and its counter emf are nearly equal and *in the same direction*. Therefore, to limit the armature current to a reasonable value, it is necessary to

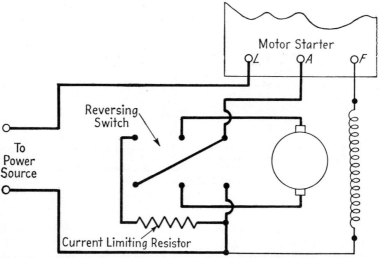

Fig. 133. Simple sketch illustrating the *plugging* method of stopping a shunt motor.

insert a resistor in this circuit when the power to the armature is reversed. Figure 133 represents a simple circuit illustrating the plugging method. When the reversing switch is closed to the right, the motor rotates in the proper direction. To stop the motor, the switch is quickly thrown to the left; at the instant the motor stops, the switch is opened. The current-limiting resistor is shown connected to the diagonal terminals on the reversing switch.

EXAMPLE 9. A 15-hp 230-volt motor has an armature resistance, including brushes, of 0.38 ohm and, when operating at rated load, the armature current is 54 amp. Calculate the value of a plugging resistor R_P if the armature current is to be limited to $1\frac{3}{4}$ times the full-load value at the instant the motor is plugged, assuming that the counter emf is $0.9V$ at that instant.

EFFICIENCY, RATING, AND APPLICATIONS OF DYNAMOS   209

Solution

$$R_P = \left[\frac{230 + (0.9 \times 230)}{1.75 \times 54}\right] - 0.38 = 4.25 \text{ ohms}$$

A complete wiring diagram showing a compound motor connected to an automatic starter that performs a plugging function when the STOP button is pressed is shown in Fig. 134. In addition to the usual accelera-

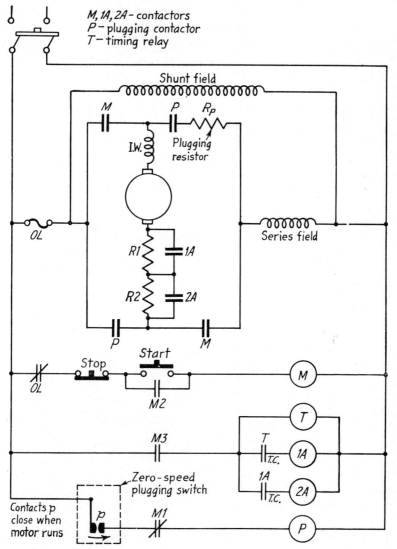

FIG. 134. Wiring diagram of time-limit acceleration starter, with provision for *plugging*, connected to a compound motor.

tion contactors, a time-delay relay, numerous contacts, and other parts shown in Fig. 118, it is equipped with a special type of switch which is designed to prevent the motor from reversing *after* it is plugged to a stop; called a *zero-speed plugging switch*, it is mounted on the motor shaft, operating to close a pair of contacts when the rotor is turning and opening them at the very instant the motor stops.

Referring to Fig. 134, the motor is started in the usual way by pressing the START button. This energizes the M contactor, which instantly opens the normally closed $M1$ interlock in the plugging-switch circuit *before* the $M2$ contacts close (to seal the M contactor) and close the main contacts M to start the motor. When the M contacts do close, current flows downward in the armature circuit and the machine accelerates in its normal direction as explained for Fig. 118. Note particularly that the plugging contactor P is prevented from picking up even though the plugging-switch contacts p are closed; this is because the normally closed interlock $M1$ was opened by the M contactor. However, when the STOP button is pressed, the M contactor and the accelerating contactors $1A$ and $2A$ drop out, and the $M1$ contacts close. Since the plugging-switch contacts p are already closed by the rotation of the rotor, the P contactor is energized; this causes the P contacts to close which, in turn, reverses the current through the armature circuit, now upward, and attempts to reverse the motor. Observe also that this plugging action is accompanied by the insertion of plugging resistor R_P and accelerating resistors $R1$ and $R2$. The motor now slows down quickly to a stop, and at the very instant it tries to reverse *the plugging contacts open;* this causes the P contactor to drop out to open the P contacts and prevents reversal of the motor.

Dynamic Braking. This method makes use of the generator action in a motor to bring it to rest. If the armature terminals are disconnected from the source *while the field is kept energized*, the motor will coast to a stop; the time it will take to come to rest will depend upon the inertia of the moving parts and the bearing, brush, and wind friction. However, if the armature terminals are discontinued from the source and immediately connected across a resistor, *the field being kept energized*, the motor will come to rest very quickly. The reason for this braking action is that the counter emf immediately sends a current into the resistor, the electrical energy for the resistor, $I_2^2 R$, coming from the mechanical energy stored in the moving parts. When the mechanical energy is expended, the motor comes to rest. If the value of the resistor is made low enough, the energy is dissipated very rapidly, which means that the motor will stop quickly. Figure 135 is a simple wiring diagram of this *dynamic braking* method. When the DPDT switch is closed to the right, the motor operates normally. Closing the switch to the left brings the motor to a quick stop.

Since the dynamic-braking resistance determines the *maximum* value of the current in the armature circuit as well as the stopping time of the motor, it is generally necessary to calculate its ohmic value on the basis of the permissible current at the instant the braking action is initiated; this is because the counter emf, which is responsible for the current, is a maximum at that instant and diminishes rapidly to zero as the motor comes to

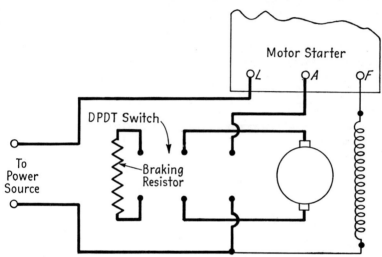

FIG. 135. Simple sketch illustrating the *dynamic-braking* method of stopping a shunt motor.

a stop. The following example illustrates how the dynamic-braking resistance is evaluated:

EXAMPLE 10. For the motor of Example 9 (hp = 15, V = 230, E_c = 0.9V, R_A = 0.38, I_A = 54), calculate the resistance of a dynamic-braking resistor R_{DB} if the armature current is to be limited to 1¾ times its rated value.

Solution

$$R_{DB} = \left(\frac{0.9 \times 230}{1.75 \times 54}\right) - 0.38 = 1.81 \text{ ohms}$$

Figure 136 represents a wiring diagram of a compound motor connected to an automatic starter that is designed to perform a braking function when the STOP button is pressed. Note particularly that the fundamental circuit, Fig. 103, is modified mainly by the addition of *a shunt path around the armature* consisting of a dynamic-braking resistor R_{DB} in series with a normally closed contact $M1$; the latter interlock is opened instantly when the START button is pressed for normal acceleration.

However, when the STOP button is pressed, the M contactor drops out and the M contacts open, after which the $M1$ contacts close. With the *shunt field* still energized, the armature proceeds to discharge into the resistor R_{DB}, and the motor is quickly brought to rest.

The reversing circuit of Fig. 118 may be readily arranged for dynamic braking from both directions of rotation. This is done by simply joining a dynamic-braking resistor in series with *two* normally closed contacts and connecting the series combination *across* the terminals marked a and c;

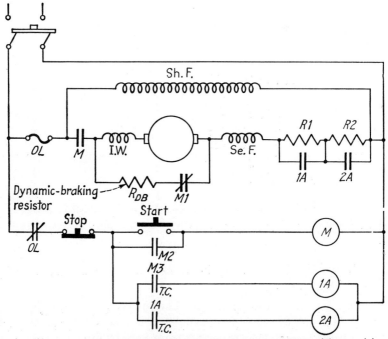

FIG. 136. Wiring diagram of a time-limit acceleration starter, with provision for *dynamic braking*, connected to a compound motor.

one of the normally closed contacts is then actuated by the forward contactor F and the other by the reversing contactor R. Thus, when the motor is running in either direction, the shunt path around the armature is open at one or the other of the two contacts. However, when the STOP button is pressed, both the F and R contactors drop out, the two interlocks in the dynamic-braking circuit close, and the armature discharges into R_{DB} to bring the motor to a quick stop.

Regenerative Braking. Although the general principle of operation of this method is similar to that of dynamic braking (in both schemes the motor develops a braking torque with respect to the negative torque exerted by the load), there are two important operating differences between these systems; in dynamic braking the load is always brought to a stop and the energy of rotation is dissipated in a resistor (see Fig. 136), whereas rotation continues and energy is restored to the electric circuit during the regenerative braking period. In practice, regeneration can be applied only when the load has an overhauling characteristic, as in the lowering of the cage of a hoist or the downgrade motion of a railroad train.

Regeneration takes place when the induced voltage in the armature, the counter emf, is greater than the source voltage, and this condition

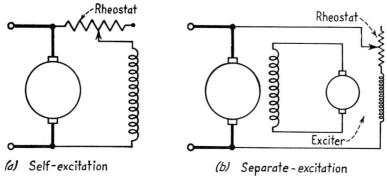

FIG. 137. Simple sketches illustrating regenerative braking arrangements for shunt motors.

results when the overhauling load, acting as a prime mover, drives the machine as a generator. In the case of the lowering of a load, for example, the field of the machine is properly overexcited to give a generated emf that is sufficiently greater than the source voltage to permit a return of power to the line as well as to maintain a constant speed slightly in excess of normal running speed of the motor. Speed variations can, however, be obtained by field-current adjustment; increasing the field strength lowers the speed, while decreasing the flux raises the speed. For protective purposes it is obviously necessary to have some type of mechanical brake so that the load may be held in the event of a power failure; in the absence of such a brake, the overhauling load would tend to "run away" because there would be no electrical braking.

A number of wiring arrangements have been developed to permit shunt, series, and compound motors to regenerate during the braking periods. These include schemes involving self- and separate-excitation, differentially connected machines with dual excitation, special voltage systems requiring separate motor-generator sets and booster generators, and others. Figure 137a illustrates a simple shunt machine with self-excitation, while Fig. 137b shows a shunt machine in which the field is energized by a separate exciter; the latter may be driven by its own motor directly connected to the common source or by the main motor through suitable gearing. The scheme of Fig. 137b can also be applied to the operation of series motors.

Series-Parallel Control of Railway Motors. In the operation of railway trains and cars, two or four series motors are mounted on trucks, and these drive the wheels through gears. The method generally used for such applications, which require frequent starting and stopping and efficient speed control, is *series-parallel control*. When two motors are used on small cars, one on each of the two trucks, the starting position of the drum

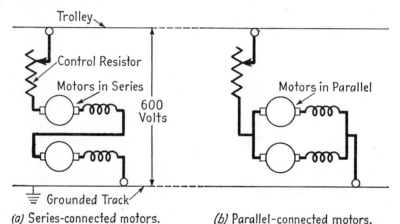

(a) Series-connected motors. (b) Parallel-connected motors.

Fig. 138. Two-motor, series-parallel control system for traction service.

controller connects the two machines in *series* and in series with a control resistor. As the car gains speed, the control resistor is cut out gradually until the two motors are in series across the 600-volt trolley-to-track circuit; under this condition, each motor receives 300 volts (see Fig. 138a). With further increase in car speed, the two motors are connected in *parallel* (Fig. 138b) and in series with the same control resistor. As acceleration progresses, the control resistor is gradually cut out until each motor is connected directly to the 600-volt source. Since the torque that is developed by a series motor is approximately proportional to the square of the line current, this scheme of control makes it possible for the two motors to develop values of torque up to the maximum attainable when full voltage is impressed across each machine.

When four motors are used in the larger cars, the control method is similar to that described above for the two-motor car, with the exception that there are always *two pairs* of parallel-connected motors that may be connected in series or parallel. During the initial accelerating period, the two pairs of motors are connected in series and in series with the control resistor (Fig. 139). As the car gains speed, the resistor is cut out until each motor receives 300 volts. With further acceleration, the drum-controller handle is moved to the position that connects the four motors in parallel and in series with the same control resistor. The latter is gradually cut out until, at maximum speed, all four motors receive the full line voltage.

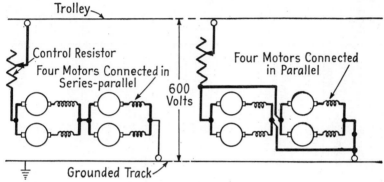

FIG. 139. Four-motor series-parallel control system for traction service.

Direct-current Dynamo Applications. *Generator Applications.* D-c generators serve a great many kinds of electrical loads. The type of generator used depends, of course, upon the requirements with regard to voltage, voltage regulation, kind of service, and other factors. The usual applications may be listed as follows: lighting service, stationary motor service, traction motor service, lifting electromagnets, heating (such as stoves, space heaters, furnaces, etc.), electric welding, battery charging, electrochemical processes, booster service in transmission systems, standby service for hospitals, hotels, and other public buildings, and communication systems.

Motor Applications. D-c motors are not used so widely as are those designed for a-c service. There are several reasons for this, the most important of which are the following:

1. D-c service is not generally available, because modern power and transmission systems are more economical and efficient when alternating current is generated and transmitted over long distances.

2. D-c motors have commutators that are subject to trouble resulting from sparking, brush wear, arc-over, and the presence of moisture and destructive fumes in the surrounding air.

3. D-c motors are generally more expensive than a-c machines for similar operating conditions.

D-c motors are nevertheless used in a large number of applications, especially when their excellent torque and speed operating characteristics

TABLE 3. APPLICATIONS OF SHUNT, COMPOUND, AND SERIES DIRECT-CURRENT MOTORS

Shunt				Compound		Series	
Constant speed	Adjustable speed			Variable-adjustable speed		Variable-adjustable speed	Line-resistance control
Plain	Field control	Armature control	Multivoltage control	Field control	Armature control		
Line shaft	....	....	....	Compressor	....	Crane	
Vacuum cleaner	....	....	....	....	Door lift	Coal and ore bridge	
....	Large blower	....	....	Stamping machine	....	Mine hoist	
Pressure blower	....	....	....	Pressure blower	....	Continuous conveyer	
Centrifugal pump Constant head	Centrifugal pump Variable head	....	....	Centrifugal pump Variable head	....	Valve operation	
Displacement pump	Displacement pump	....	....	Displacement pump	....	Turntable	
....	Compressor	....	....	Rotary press	....	Motor vehicle	
....	Small print press	....	....	Flat-bed press	....	Bucket hoist	
Circular saw	....	....	....	....	Circular saw		
Wood planer	....	....	....	Metal planer	....		
Grinder	....	....	....	Shearing machine	....		
Polisher	....	....	....	Punch press	....		
....	Milling machine	....	Milling machine	Rolling mill	....		
....	Lathe	....	Lathe	....	....		
....	Passenger elevator	....	....	....	Freight elevator		
....	Continuous conveyer	....	Continuous conveyer	Passenger elevator	....		
Laundry washing machine	....	....	....	Continuous conveyer	....		
....	....	....	....	Hydroextractor	....		

EFFICIENCY, RATING, AND APPLICATIONS OF DYNAMOS

cannot be duplicated by a-c motors. Moreover, the speed control that can be provided for d-c motors is far more flexible and satisfactory than that which can be provided for a-c motors. These requirements are particularly significant in connection with such services as the operation of traction equipment, hoists, and elevators. In other cases, the matter of speed control is so important that the existing a-c service is frequently rectified into d-c service so that d-c motors may be used. Good examples of such applications are motors used in electric excavators, steel mills, and cranes.

Before selecting the proper motor for a given application, one must know several important things. This information, fully discussed previously under Selection of Generators and Motors, p. 196, is briefly summarized as follows: duty cycle (that is, the number of times a motor must be started and stopped), whether or not the motor must be reversed, the required acceleration of the application, the required starting torque, the speed range, whether or not the motor must be stopped quickly, and whether the motor should be open or enclosed. In addition to these factors, it is also necessary to know the voltage of the power source, how the motor is to be mounted, whether it is to have a belt, gear, or coupled drive, and whether it should have manually operated or automatic starting equipment.

Obviously, such information can be obtained by those who not only are familiar with the various types of motor and their operating characteristics, but also have had actual experience with the mechanical machines to which they are applied. Table 3 illustrates a few of the more frequent applications and the types of motor employed. These should be carefully studied in connection with the points discussed here.

Questions

1. Define the term *efficiency*.
2. Why is the efficiency of a dynamo less than 100 per cent?
3. List the various losses in a dynamo.
4. Which of the losses listed in Question 3 are due to the rotation of the armature? to the flow of current in the electric circuits?
5. What happens to the power losses in a motor or generator?
6. Which of the losses are affected by the speed of rotation? by the value of the flux? directly by the load? by the square of the load?
7. What combination of losses is called the *stray-power* loss?
8. What losses are designated as the *stray-load* loss?
9. What two methods may be used to determine the efficiency of a dynamo?
10. Give several reasons for the preference of the *conventional* efficiency method over the method of *directly measured* efficiency.
11. Explain in detail the various tests that must be made *before* the conventional efficiency calculations can be made.
12. Why is the stray-power loss a function of the generated emf in the armature of a dynamo?

13. How is the generated emf calculated for a generator? for a motor?
14. Why is the efficiency of a generator or motor zero at no load?
15. Write the efficiency equations for the generator and the motor and discuss their differences.
16. Explain carefully why low efficiency affects the temperature rise of a dynamo; the cost of operation of a dynamo.
17. Why does the efficiency of a generator or motor influence the problem of reliability of electrical service?
18. In general, how is the efficiency of a dynamo affected by the speed? the size? Give reasons for your answers.
19. How are generators rated? motors?
20. What factors will influence the rating of a dynamo?
21. Discuss the question of cooling as it concerns the rating of a dynamo.
22. Distinguish between the terms *continuous duty* and *intermittent duty*.
23. Make a list of the important factors that must be considered when a generator is selected; when a motor is selected.
24. What are the objections to the selection of a motor that is larger than necessary for a given application?
25. What objections might be raised to the selection of a compound motor instead of a shunt motor? a shunt motor instead of a compound motor? a compound motor instead of a series motor?
26. Under what conditions is a flat-compound generator more desirable than an overcompound generator? an overcompound generator more desirable than a flat-compound generator?
27. What is a three-wire generator?
28. Give several advantages possessed by a three-wire generator.
29. What function does the balance coil have in the operation of a three-wire generator system?
30. Would it be possible to use a simple, center-tapped resistor instead of a reactor for three-wire service? Give reasons for your answer.
31. What current does the neutral wire carry in a three-wire system? Under what condition is the current in the neutral wire *toward* the reactor? *away* from the reactor? *zero?*
32. What is a *third-brush generator?* When is it used?
33. Why would it be unsatisfactory to use a standard shunt or compound generator in an automobile for charging and utilities services?
34. Explain the principle of operation of the third-brush generator.
35. Discuss the general constructional features of the *diverter-pole generator* and how it differs from the standard generator construction.
36. What circuit energizes the diverter-pole winding? Why?
37. What advantage, in operation, is possessed by the diverter-pole generator as compared with the flat-compound generator for battery-charging purposes?
38. What is the purpose of the slot in the bridge of the diverter-pole generator?
39. Explain how the diverter-pole generator, when used for battery-charging purposes, charges the battery at a high rate at the beginning and tapers off with increasing charge. Why is this not possible when a shunt generator is used?
40. What is a *dynamotor?*
41. Why is a dynamotor more efficient than a motor-generator set?
42. Explain why the ratio of the output voltage to the input voltage is fixed in a dynamotor and cannot be changed by flux adjustment.
43. Describe the constructional features of a dynamotor.
44. List several applications of dynamotors.

EFFICIENCY, RATING, AND APPLICATIONS OF DYNAMOS

45. Why is commutation usually good in dynamotors?
46. Distinguish between the *dynamotor* and the *rotary converter*.
47. Discuss the operation of the *series-booster* system, referring to Fig. 130.
48. Explain how the *negative-feeder booster* system functions to prevent electrolysis. Refer to Fig. 131 in your discussion.
49. Why is it desirable to have a *constant-current booster* system in large buildings where direct current is used?
50. Explain the operation of the constant-current booster system illustrated by Fig. 132.
51. What are the disadvantages of mechanical brakes for stopping electric motors quickly?
52. Explain the principle of the *plugging* braking method.
53. Carefully explain the operation of the circuit of Fig. 134, in which a compound motor may be *plugged* to a stop.
54. What is a *zero-speed plugging switch* and what is its function?
55. Is it possible to permit the motor of Fig. 134 to coast to a stop?
56. What is a *plugging resistor* and what determines its value?
57. Explain the principle of the *dynamic-braking* method.
58. Carefully explain the operation of the circuit of Fig. 136, in which a compound motor may be brought to a stop by the dynamic-braking method.
59. What is a *dynamic-braking resistor* and what determines its value?
60. In Fig. 134, are the acceleration resistors used when the motor is plugged to a stop?
61. In Fig. 136, are the acceleration resistors used when the motor is stopped by dynamic braking?
62. Explain the principle of *regenerative braking*.
63. Under what conditions would the regenerative braking method be advantageous?
64. In what respects are dynamic braking and regenerative braking similar? In what respects do they differ?
65. What types of load may employ the regenerative braking method?
66. In lowering a hoist by the regenerative braking method, how is the speed controlled? Explain carefully.
67. Why is it necessary to use a mechanical brake as a safety precaution when regenerative braking is employed?
68. Explain the operation of the two regenerative braking schemes illustrated in Fig. 137.
69. Referring to Fig. 139, describe the *series-parallel* control system for traction service.
70. Explain how it would be possible to employ dynamic braking on the series motors of the series-parallel system of control.
71. Referring to Table 3, lists several additional motor applications for each type of motor.
72. Exactly what factors influence a choice between shunt, compound, and series motors?
73. Distinguish between *constant-adjustable-speed* motors and *variable-adjustable-speed* motors.

Problems

1. What is the watts output of a 25-hp motor at full load?
2. A 150-kw generator has an efficiency of 91 per cent at full load. Calculate the input in kilowatts and the power loss.

3. A 15-hp motor operates at an efficiency of 87.5 per cent at full load. If the stray-power loss is approximately one-fourth of the total loss, calculate the copper loss.

4. The rotational loss in a generator was found to be 780 watts when the generated emf was 132 volts. Determine the rotational loss for generated voltages of 138 and 126 volts.

5. Referring to Examples 3 and 4, pages 189–190, make a table similar to the one there and calculate the efficiencies for line currents of 5, 10, 15, 20, and 25 amp, assuming the following changes: line volts = 500; shunt-field resistance = 500 ohms; armature and series-field resistance = 1.6 ohms; brush drop (assumed) = 3 volts.

6. The output torque of a motor is 69.2 lb-ft when it operates at 950 rpm. Calculate the losses in the machine and its efficiency if, under this condition, the power input is 10,900 watts.

7. What should be the full-load horsepower rating of a motor that drives a 50-kw generator whose efficiency is 89.5 per cent?

8. If the over-all efficiency of the motor-generator set of Prob. 7 is 78 per cent, calculate: (a) the efficiency of the motor; (b) the total losses in the motor-generator set; (c) the losses in each machine.

9. A 10-kw 220-volt compound generator is operated at no load at the proper armature voltage and speed, from which the stray-power loss calculations are determined to be 705 watts. The shunt-field resistance is 110 ohms, the armature resistance is 0.265 ohm, and the series-field resistance is 0.035 ohm. Assume a 2-volt brush drop and calculate the full-load efficiency.

10. If the maximum efficiency of the generator of Prob. 9 occurs when the sum of the copper loss in the armature and series field is equal to the sum of the stray-power loss and the shunt-field loss, calculate: (a) the armature current; (b) the line current; (c) the kilowatt output; (d) the maximum efficiency.

11. A 250-volt shunt generator has a rated armature current of 40 amp and the following losses at full load: friction and windage = 200 watts; core loss = 260 watts; shunt-field loss = 100 watts; brush contact loss = 80 watts; armature copper loss = 400 watts. Assuming that the maximum efficiency occurs when the approximated *constant* losses are equal to those losses that vary as the *square* of the load, calculate the armature current for the condition of maximum efficiency.

12. A 25-kw series generator has an efficiency of 85 per cent when operating at rated load. If the stray-power loss is 20 per cent of the full-load loss, calculate the efficiency of the generator when it is delivering a load of 15 kw, assuming that the stray-power loss is substantially constant and the other losses vary as the square of the load.

13. The following information is given in connection with a long-shunt compound generator: E = 220 volts; output = 20 kw; stray-power loss = 705 watts; R_{SH} = 110 ohms; R_A = 0.265 ohm; R_{SE} = 0.035 ohm; brush drop = 3 volts. Calculate the efficiency.

14. A 250-volt shunt generator has a full-load armature current of 40 amp, under which condition the losses are: friction + windage = 200 watts; shunt field = 100 watts; core = 260 watts; brush contact = 120 watts; armature copper = 400 watts. For operation at maximum efficiency, when the constant losses are equal to those losses that vary as the square of the load, calculate: (a) the armature and line currents; (b) the load power; (c) the maximum efficiency.

15. Referring to Example 7, p. 192, make a table similar to the one given and calculate the efficiencies for armature currents of 11, 22, 33, 44, and 55 amp, assuming the following changes: line volts = 460; shunt-field resistance = 230 ohms; armature and interpole resistance = 0.72 ohm; series-field resistance = 0.12 ohm; brush drop = 3 volts.

EFFICIENCY, RATING, AND APPLICATIONS OF DYNAMOS 221

16. A 20-hp motor has an efficiency of 88.5 per cent and operates continuously to drive a ventilator at full load. Calculate: (a) the power loss; (b) the energy loss per month, assuming operation for 200 hours during that period; (c) the cost of the energy loss at 1½ cents per kilowatthour.

17. What saving would be made per year in energy cost if the motor of Prob. 16 had an efficiency of 91.5 per cent?

18. An enclosed motor has a rating of 50 hp. The cover plates are removed, and the machine is located where it is capable of cooling itself extremely well. If tests show that it can carry 28 per cent more load without excessive heating, what rating should be given the motor?

19. A 230/115-volt three-wire generator delivers the following loads: 65 kw at 230 volts; 35 kw at 115 volts between the positive and neutral lines; 25 kw at 115 volts between the negative and neutral lines. Calculate: (a) the total kilowatt load delivered by the generator; (b) the current in the positive line; (c) the current in the negative line; (d) the current and its direction in the neutral line.

20. If the generator of Prob. 19 delivers the same total load, except that the two 115-volt loads are balanced with respect to each other, calculate the line and neutral currents.

21. The generator of Prob. 19 delivers a total load of 120 kw. If the current in the positive wire is equal to the sum of the currents in the negative and neutral wires, calculate the three kilowatt loads.

22. A dynamotor has an output rating of 0.16 amp at 500 volts, under which condition it takes 11 amp from a 12-volt storage battery. Calculate the efficiency

23. A dynamotor has low-voltage input and high-voltage output windings whose resistances are, respectively, 0.1 ohm and 4 ohms, and the high-voltage winding has 7½ times as many conductors as the low-voltage winding. If the machine delivers a current of 0.5 amp at 180 volts, under which condition the core + friction + windage loss is 35 watts, calculate the input current and voltage.

24. A 35-hp 230-volt shunt motor has a full-load armature-current rating of 135 amp and an armature resistance, including brushes, of 0.125 ohm. The accelerating resistors in the automatic starter to which the machine is connected have a total resistance of 0.73 ohm. Calculate the ohmic value of a plugging resistor that should be placed in series with the acceleration resistors to limit the inrush armature current to 1.5 times its full-load value at the instant the motor is plugged. Assume that the counter emf is 80 per cent of the impressed voltage.

25. For the motor of Prob. 24, calculate the ohmic value of a dynamic-braking resistor that will limit the inrush armature current to 2.5 times its rated value.

Part II

ELECTRICAL MACHINES
ALTERNATING CURRENT

CHAPTER 7

ALTERNATING-CURRENT GENERATORS

Alternator Construction. D-c and a-c generators are similar in one important respect—they both *generate* alternating emfs. In the d-c generator, the alternating voltage is rectified through the medium of a commutator and brushes, whereas the a-c generator has no rectifier and delivers a-c electric energy to its loads. It is, in fact, quite possible to use a d-c generator as an a-c generator by placing a set of collector rings on the shaft and connecting these rings to the proper points on the armature winding; brushes riding on the rings can then be connected to the electrical load.

A-c generators are usually called *alternators.* But unlike d-c generators, they must be driven at a very definite *constant* speed because the frequency of the generated emf is determined by that speed. The latter is usually referred to as the *synchronous speed,* for which reason these machines are frequently called *synchronous alternators* or *synchronous generators.* Remembering that generator action depends upon the *relative* motion of conductors with respect to lines of force, it should be clear that it is possible to construct an alternator with a stationary field and a moving armature or with a stationary armature and a moving field. In practice, the latter arrangement is the preferable one for several reasons:

1. The armature winding is more complex than the field and can be constructed more easily on a stationary structure.
2. The armature winding can be braced more securely in a rigid frame.
3. It is easier to insulate and protect the high-voltage armature windings common to alternators.
4. The armature winding is cooled more readily because the stator core can be made large enough and with many air passages or cooling ducts for forced air circulation.
5. The low-voltage field can be constructed for efficient high-speed operation.

In general, alternators are built in much larger sizes than are d-c generators—commutation is a serious limiting factor in the design of very large d-c machines, the reason being that voltages are comparatively low and currents high for large kilowatt ratings. Three common types of prime

226 ELECTRICAL MACHINES—ALTERNATING CURRENT

movers for alternators are steam turbines, steam engines, and water wheels, and, as stated above, the mechanical power is usually applied to the rotor field. Figure 140 shows a cutaway view of a vertical-type *waterwheel generator* of about 3,000-kw capacity for operation at *moderate speed*. Careful inspection of this picture will make clear how the armature and

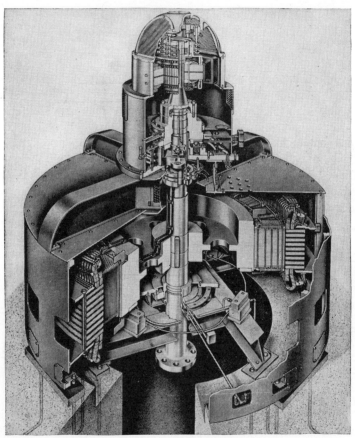

FIG. 140. Cutaway view of typical water-wheel synchronous alternator of about 3,000-kw capacity. (*Westinghouse Electric Corp.*)

field are arranged, how the small d-c *exciter* generator on top "feeds" direct current to the rotating field through collector rings, and how the great weight of the stator and rotor is supported by a strong concrete foundation and a carefully designed guide and thrust bearing assembly. Figure 141 shows a cutaway view of a steam-turbine-driven *alternator* for *high-speed* operation. Note particularly the long rotor field, several armature-winding coils, the high-voltage insulators at the bottom for

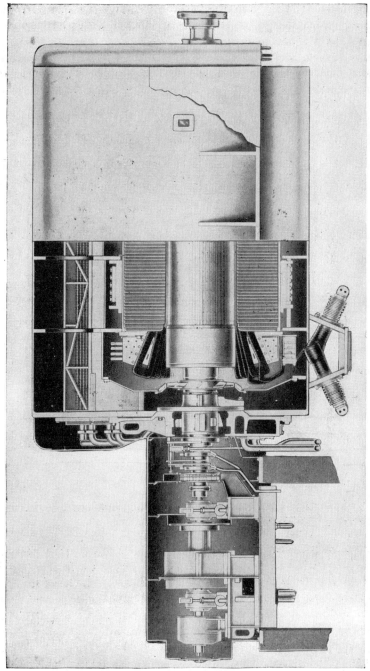

FIG. 141. Cutaway view of typical steam-turbine synchronous alternator for operation at high speed. (*Westinghouse Electric Corp.*)

bringing out line leads, and the two d-c exciter generators at the left. (The *pilot* self-excited d-c generator at the far left separately excites the alternator-field exciter to its right.)

The stator of an alternator consists of a laminated, slotted, good magnetic steel core and an armature winding placed in the slots in much the same way as is done in d-c generators. Figure 142 shows a completely wound

Fig. 142. Wound stationary armature for a 400-kw 2,400-volt 300-rpm 60-cycle synchronous alternator. (*General Electric Co.*)

stator. It has a rating of 400 kw at a speed of 300 rpm, delivering 60-cycle alternating current at 2,400 volts. The rotor for this machine has 24 poles. Note the pipe bracket to which are fastened the brush rigging and brushes that ride on the rotor collector rings. A typical revolving field for an alternator like that shown in Fig. 142 is depicted in Fig. 143. This is a 22-pole rotor that is designed for a 200-kw 480-volt 327.3-rpm 60-cycle machine. Clearly visible are the collector rings, the cooling-fan blades, the field coils and their connections, and the split spider. To illustrate alternator-rotor construction further, the rotor field and its water wheel for a 5,000-kw 6,600-volt 514-rpm 60-cycle synchronous generator are shown in Fig. 144.

Frequency of Alternating-current Generators. As the poles of a *two-pole* alternator revolve, the generated emf in the stationary armature winding changes direction every half revolution; therefore one complete positive and negative pulse, one cycle, will occur in one revolution. It follows, then, that the frequency in cycles per second will depend directly

Fig. 143. Wound revolving field for a 200-kw 480-volt 327.3-rpm 60-cycle synchronous alternator. (*General Electric Co.*)

upon the number of revolutions per second (rpm/60) made by the field. Moreover, if the generator is multipolar, i.e., if it has four, six, eight, or more poles, then the frequency per revolution will be, respectively, two, three, four, or more. Or, to put it more generally, the frequency *per revolution* is equal to the *number of pairs of poles*. Combining both facts into a single statement, it should be clear that the frequency of the emf in an alternator is proportional to (1) the speed in revolutions per second (rpm/60) and (2) the number of pairs of poles, $P/2$. The relationship

ELECTRICAL MACHINES—ALTERNATING CURRENT

may be written in the form of the equation

$$f = \frac{P}{2} \times \frac{\text{rpm}}{60} = \frac{P \times \text{rpm}}{120} \tag{26}$$

EXAMPLE 1. An alternator has six poles and operates at 1,200 rpm. (a) What frequency does it generate? (b) At what speed must the machine be operated if it is to have a frequency of 25 cycles? 50 cycles?

Solution

(a) $$f = \frac{6 \times 1{,}200}{120} = 60 \text{ cps}$$

(b) $$\text{rpm}_{25} = \frac{120 \times 25}{6} = 500$$

$$\text{rpm}_{50} = \frac{120 \times 50}{6} = 1{,}000$$

EXAMPLE 2. What is the maximum speed at which the field of an alternator can be operated to develop 60 cycles? 25 cycles? 50 cycles?

Solution

Since the maximum speed at which a generator must be operated for a given frequency occurs when the machine has the *fewest* number of poles, it follows that there must be two poles in the machine. Therefore

$$\text{rpm}_{60} = \frac{120 \times 60}{2} = 3{,}600$$

$$\text{rpm}_{25} = \frac{120 \times 25}{2} = 1{,}500$$

$$\text{rpm}_{50} = \frac{120 \times 50}{2} = 3{,}000$$

The three most widely used frequencies in this country are 60, 25, and 50 cps. The 60-cycle frequency is most general, wnile 50 cycles is found on one system in Southern California (and also in Mexico). At one time, 25 cycles was used most successfully in steel mills and in the hydroelectric districts, like Niagara Falls; it is for this reason that this lower frequency may still be found in such localities, although the general tendency is to standardize the entire country at 60 cycles. For the operation of specially designed a-c series motors for traction service, 25 cycles appears to be the most desirable. Table 4 gives the speed-pole combinations for 60-, 25-, and 50-cycle alternators. Note particularly that, for each frequency, the speed and the number of poles are *inversely* proportional to each other.

The speeds given in Figs. 142, 143, and 144 should be compared with those in Table 4 at 60 cycles.

ALTERNATING-CURRENT GENERATORS

FIG. 144. Rotor field and water wheel for a 5,000-kw 6,600-volt 514-rpm 60-cycle synchronous generator. (*Allis-Chalmers Mfg. Co.*)

TABLE 4. ALTERNATOR SPEED-POLE COMBINATIONS

Poles	Revolutions per minute		
	$f = 60$ cycles	$f = 25$ cycles	$f = 50$ cycles
2	3,600	1,500	3,000
4	1,800	750	1,500
6	1,200	500	1,000
8	900	375	750
10	720	300	600
12	600	250	500
14	514	214	429
16	450	187	375
18	400	167	333
20	360	150	300
22	327	136	273
24	300	125	250
26	277	115	231
28	257	107	214
30	240	100	200
32	225	94	187

The Revolving Field. Figures 143 and 144 show the general constructional details of one type of field structure. It consists essentially of an even set of laminated pole cores around each of which is placed an exciting winding. Each lamination has a shape approximately like that shown in Fig. 145. The outer surface of the pole almost follows the inner cylindrical surface of the stator core, while the inner part has a dovetail that fits into a wedge-shaped recess in the projecting spider. For high-speed turboalternators, the field structure does not have projecting poles. Instead, the slotted core is made up of several forged-steel sections or of heavy, thick disks cut from forged plates. The shaft does not pass through a hole in the core, but is made in two pieces, each of which is rigidly bolted to one of the end plates. A spiral type of field winding is placed in radial or parallel slots. Figure 146 shows an end view of a two-pole field structure, partially completed. The field winding is usually designed to be connected to a 115- or 230-volt d-c source, the latter being supplied by either a small d-c generator mounted on the end of the alternator shaft or one driven by a separate electric motor.

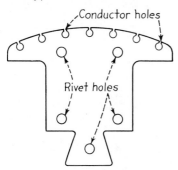

Fig. 145. Typical lamination for the poles of an alternator.

The Stator. The stator of an alternator consists of a laminated steel core slotted to receive the armature winding. The number of slots is generally such that a symmetrical polyphase (three-phase, as a rule) winding can be used. Such a winding is possible if the number of slots divided by poles times phases is an integer (i.e., slots/poles × phases = integer). In slow-speed large-diameter machines that have many poles (see Figs. 142 and 143), the axial length of the core is comparatively short. Since such machines are completely open, cooling is simple and effective. In high-speed machines of the turboalternator type (see Fig. 141), in which two or four poles are used, the axial length is many times the diameter. Since self-cooling would not be satisfactory, such alternators are completely enclosed, in order that large volumes of clean dry cool air may be forced through them. It is interesting to note that air velocities as high as 3,000 ft per min (fpm), which is about 35 miles per hour, are used to keep the temperature rise of such generators down to reasonable values. Also, for the purpose of providing enough air-duct area so that large volumes of air may be readily forced through them, the cores are built out radially with a great many holes lined up axially; the air is forced through these holes as well as through the rather wide air gap.

The voltage generated in the armature windings of alternators is usually much higher than that of d-c generators; voltages of 2,300, 4,600, 6,600, and 13,200 are quite general. Obviously, the same insulating materials that are used in d-c armature windings would be unsatisfactory in the high-voltage a-c armature windings. Also, higher temperatures are permitted in the a-c machine. Since mica is the only reliable material that will withstand high voltage and high temperature, it is generally employed. This material is given a special treatment to overcome its

FIG. 146. Partially completed field structure for a two-pole alternator. Note the slotted core and the spiral winding. (*Allis-Chalmers Mfg. Co.*)

poor mechanical properties. The mica is split into thin flakes and then lap-bonded together with adhesive varnishes; the large sheets thus formed to the proper thickness are then baked in ovens under extreme pressure. Mica in this form can be readily molded to any shape for slot and coil insulation.

Generated Voltage in an Alternator. The fundamental law of generator action may be stated as follows: *an average of 1 volt is generated in one turn of wire if the flux passing through that turn changes at the rate of 100,000,000* (10^8) *maxwells per second.* Referring to Fig. 147a one full-pitch turn of wire is shown surrounding a north pole. Assume that this pole produces 10^8 maxwells. Assume further that the turn of wire moves over to a position indicated by Fig. 147b in 1 sec; here the net flux passing *through* the coil is zero, because 50 per cent flux from the *north* pole neutralizes the 50 per cent flux from the *south* pole. Therefore the *average* generated voltage in the coil of one turn will be 1 volt. If the change of flux through the coil had taken place at a greater *rate*, or if more than one turn had been used, the generated voltage would have been in direct proportion to such increases. Thus, in equation form, these statements

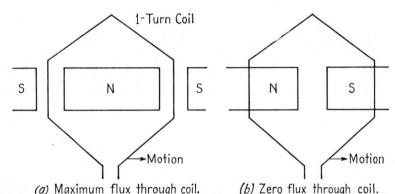

(a) Maximum flux through coil. (b) Zero flux through coil.

FIG. 147. Sketch illustrating change in flux through a one-turn coil as it moves from left to right.

become

$$E_{av} = N \frac{\phi}{t} \times 10^{-8} \quad \text{volts} \tag{27}$$

where E_{av} = average generated voltage
N = number of turns in coil
ϕ = flux per pole
t = time, in seconds, for flux to change by ϕ

Referring again to Fig. 147, note that there is a flux change of ϕ maxwells through a coil as the latter moves a distance equal to one-half of a pole pitch, or, more specifically, the distance a coil must travel to generate ¼ of a cycle of voltage. Since one cycle occurs in $1/f$ sec, the elapsed time for ¼ of a cycle will be $1/4f$ sec. Therefore, when $1/4f$ is substituted for t in Eq. (27), it becomes

$$E_{av} = N \frac{\phi}{1/4f} \times 10^{-8} = 4fN\phi \times 10^{-8} \quad \text{volts} \tag{28}$$

Equation (28) is quite general and is applicable to all alternators regardless of the flux-density distribution under the poles. However, if the alternator is so designed that the flux-density distribution is *sinusoidal*, the so-called *effective* value of the voltage, usually designated by the symbol E, is 1.11 times the average value. Thus

$$E = 4.44fN\phi \times 10^{-8} \quad \text{volts} \tag{29}$$

EXAMPLE 3. Calculate [RMS] the effective voltage in one phase of an alternator, given the following particulars: $f = 60$ cps; turns per phase $N = 240$; flux per pole $\phi = 2.08 \times 10^6$.

Solution

$$E = 4.44 \times 60 \times 240 \times 2.08 \times 10^6 \times 10^{-8} = 1{,}330 \text{ volts}$$

If the flux density distribution is not a sine function, the effective voltage will be a little different from that given by Eq. (29); the factor 4.44 will be modified to account for the so-called *harmonics*, a subject that is considered beyond the scope of this book. Practically speaking, however, such modification would be slight, so that quite accurate values are obtained even if harmonics are neglected.

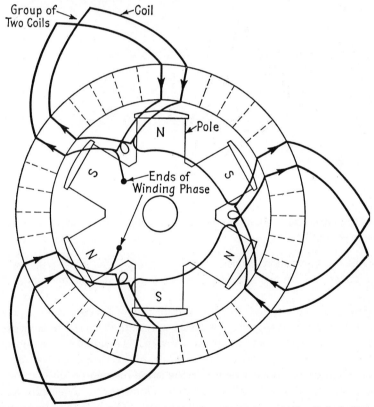

FIG. 148. Simple sketch showing one phase of a three-phase *half-coiled* winding for a six-pole alternator.

236   ELECTRICAL MACHINES—ALTERNATING CURRENT

Armature Windings for Alternators. The type of winding most generally used in alternators is very similar to the d-c *lap winding*. In d-c armatures, it will be recalled, the coils are connected together at the commutator; in a-c armatures they are joined together by merely connecting the proper coil ends in the correct sequence. Two general arrangements of coils are employed: (1) the *half-coiled* and (2) the *whole-coiled*. In the

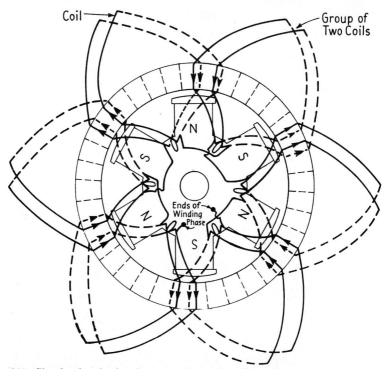

FIG. 149. Simple sketch showing one phase of a three-phase *whole-coiled* winding for a six-pole alternator.

half-coiled arrangement, there are half as many coils, each with twice as many turns, as in the whole-coiled arrangement, *for the same phase voltage*.

Figure 148 shows a simplified sketch of one of three phases of a half-coiled winding for a six-pole alternator. Note that there are *three* groups of coils ($P/2$ groups), two coils in each group. It should also be observed that the coils are connected so that they *all carry current in the same direction* (clockwise, for example) at the same instant.

Figure 149 shows a simplified sketch of one of the three phases of a whole-coiled winding for a six-pole alternator. The points to be noted are that, in contrast to the half-coiled winding, the whole-coiled type has six

ALTERNATING-CURRENT GENERATORS 237

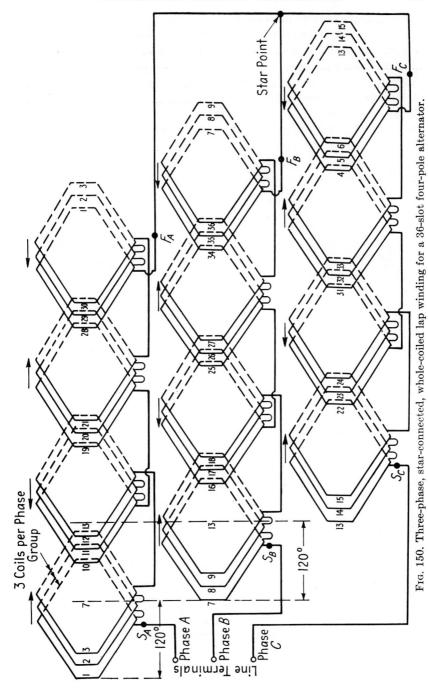

Fig. 150. Three-phase, star-connected, whole-coiled lap winding for a 36-slot four-pole alternator.

groups of coils (P groups) and that *successive groups carry currents in opposite directions* (clockwise and counterclockwise, for example). Also, there are two coil sides in each slot as against one coil side per slot in the half-coiled winding.

The whole-coiled winding is the more usual of the two arrangements; it is very similar to the d-c lap winding in that it is a double-layer winding.

Most alternators are wound for three-phase operation, i.e., there are three distinct and independent windings in the stator-core slots that are displaced from each other by exactly 120 electrical degrees. In practice, the three phases are ordinarily connected to form a *star* or *delta*. (Star Y, and delta Δ, connections are discussed in elementary texts dealing with

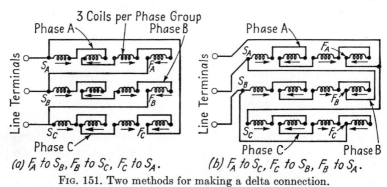

FIG. 151. Two methods for making a delta connection.

polyphase circuits.) As a rule, only three terminal leads are brought out from the interconnected three-phase winding, these being the ends of the Y or the junction points of the Δ. Frequently, the *neutral* point of the Y is brought out to be grounded for purposes of protection.

Figure 150 shows in more detail how a complete three-phase winding appears. It is represented spread out horizontally in three stages, one for each phase, so that it may be studied more easily and carefully than if drawn in the correct circular form. Each of the three *four-pole* Y-connected phases is lettered with an S and F (start and finish), i.e., $S_A F_A$, $S_B F_B$, $S_C F_C$. Note that each phase group has three coils in series $[36/(4 \times 3)]$; that all phase groups are connected in series, with successive groups carrying current in opposite directions; that, having arbitrarily assumed a direction *into* the winding at S_A, the same direction must be assumed for S_B and S_C; that the star point is the junction of F_A, F_B, and F_C; that phase A starts in slot 1, phase B in slot 7, 120 electrical degrees away from phase A, and phase C in slot 13, 120 electrical degrees away from phase B.

Figure 151 is a simplified sketch of this winding connected in delta. Note that the junctions of S and F are line terminals, with two possibilities shown.

ALTERNATING-CURRENT GENERATORS

Coil Pitch and Pitch Factor. The distance between the two sides of a coil is its *coil pitch*. When this is exactly equal to the distance between the centers of two adjacent poles, 180 electrical degrees, the coil is said to be *full pitch* and the winding a *full-pitch* winding. Under this condition, the generated voltages in both coil sides are exactly *in phase*. However, if the distance between the two sides of a coil is *less* than 180 electrical degrees, the coil is said to be *fractional pitch* and the winding a *fractional-pitch* winding. Fractional-pitch windings are more generally used than those that are full pitch, because their generated voltage can be made to approximate a sine wave more easily and the distorting *harmonics* may be reduced, or even eliminated, in this way. (Since the subject of *harmonics* is beyond the scope of this book, it is not treated here.) When fractional-pitch windings are employed, however, the generated voltages in the two sides of a coil are *not* in phase with each other, so that the resultant coil voltage is *less* than it would be for a full-pitch winding. The ratio of the voltage generated in the fractional-pitch coil to the voltage generated in the full-pitch coil is called the *pitch factor*.

It is customary to express the coil pitch as a per cent of full pitch. Thus, a 36-slot four-pole machine would have nine slots per pole; 100-per cent pitch would require a coil span from slot 1 to slot 10; a coil span from slot 1 to slot 9 would be an 88.9 per cent pitch; a coil span from slot 1 to slot 8 would be a 77.8 per cent pitch, etc.

As stated above, the voltage generated in a fractional-pitch winding is less than the voltage developed in a full-pitch winding. The factor, always a decimal, by which the full-pitch voltage is multiplied to give the fractional-pitch voltage is called the *pitch factor*. Its value may be calculated by the equation

$$k_p = \sin \frac{p°}{2} \qquad (30)$$

where k_p = pitch factor (always a decimal for a fractional-pitch winding)
$p°$ = span of the coil in electrical degrees

EXAMPLE 4. Calculate the pitch factors for the following windings: (a) 36 slots, four poles, span 1 to 8; (b) 72 slots, six poles, span 1 to 10; (c) 96 slots, six poles, span 1 to 12. Illustrate the three coil spans by sketches.

Solution

(a) $p° = \frac{7}{9} \times 180° = 140°$ $k_p = \sin \frac{140°}{2} = \sin 70° = 0.940$

(b) $p° = \frac{9}{12} \times 180° = 135°$ $k_p = \sin \frac{135°}{2} = \sin 67.5° = 0.924$

(c) $p° = \frac{11}{16} \times 180° = 124°$ $k_p = \sin \frac{124°}{2} = \sin 62° = 0.883$

240 ELECTRICAL MACHINES—ALTERNATING CURRENT

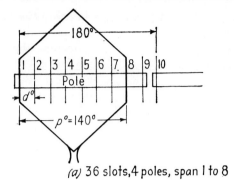

(a) 36 slots, 4 poles, span 1 to 8

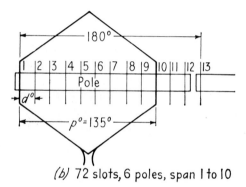

(b) 72 slots, 6 poles, span 1 to 10

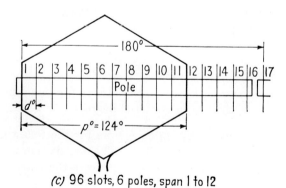

(c) 96 slots, 6 poles, span 1 to 12

FIG. 152. Sketches illustrating the solutions to Example 4.

Figure 152 illustrates the coil spans for the three parts of Example 4.

It should be understood that a pitch-factor value of less than 1 reduces the calculated voltage determined by Eq. (29). Thus, if the generated emf for a full-pitch winding is 1,330, it would be 1,250, 1,228, and 1,175 volts for pitch factors of 0.940, 0.924, and 0.883, respectively.

ALTERNATING-CURRENT GENERATORS   241

Table 5 lists all the possible pitch factors k_p for three-phase alternators having 3 to 15 slots per pole.

TABLE 5. PITCH FACTORS FOR THREE-PHASE ALTERNATORS

Slots per pole	$p°$										
	180	168	165	160	156	150	144	140	135	132	120
3	1.0	.....	.....	.....	.....	.....	.....	.....	.....	.....	0.866
6	1.0	.....	.....	.....	.....	0.966	.....	.....	.....	.....	0.866
9	1.0	.....	.....	0.985	.....	.....	.....	0.940	.....	.....	0.866
12	1.0	0.991	.....	.....	.....	0.966	.....	.....	0.924	.....	0.866
15	1.0	0.995	.....	.....	0.978	.....	0.951	.....	.....	0.914	0.866

Distribution Factor. When several coils in a group (see Fig. 150) are connected in series, the total voltage generated by that group is *not* equal to the voltage per coil multiplied by the number of coils in the group. It is always *less* than this because the coils are displaced from each other, which means that *the voltages* generated in the several coils *are not in phase with each other.* The factor by which the generated voltage E must be multiplied (because the coils are *distributed* in several slots under the poles instead of being concentrated in single slots under the poles) to obtain the correct value is called the *distribution factor;* it is designated by the symbol k_d.

Distributing the winding in many slots has the effect of improving the shape of the voltage wave, i.e., making it approach a sinusoidal function, as well as of adding rigidity and mechanical strength to the winding. The value of k_d is given by the equation

$$k_d = \frac{\sin(nd°/2)}{n \times \sin(d°/2)} \tag{31}$$

where k_d = distribution factor
 n = number of slots per phase per pole (slots/poles × phases)
 $d°$ = number of electrical degrees between adjacent slots

EXAMPLE 5. Calculate the distribution factor for a 36-slot four-pole three-phase winding.

Solution

Slots per pole = 36/4 = 9; $d°$ = 180°/9 = 20° between slots (see Fig. 152a); and n = 36/(4 × 3) = 3 slots per phase per pole.

$$k_d = \frac{\sin(3 \times 20°/2)}{3 \times \sin(20°/2)} = \frac{\sin 30°}{3 \times \sin 10°} = \frac{0.50}{3 \times 0.1736} = 0.960$$

Again let it be understood that a distribution factor value of less than 1 reduces the calculated voltage determined by Eq. (29). Thus, if the generated emf for a concentrated winding is 1,330, it would be 1,276 volts for a distribution factor of 0.960.

Table 6 lists the distribution factors k_d for three-phase alternators having 3 to 15 slots per pole.

TABLE 6. DISTRIBUTION FACTORS FOR THREE-PHASE ALTERNATORS

Slots per pole	n	$d°$	k_d
3	1	60	1.000
6	2	30	0.966
9	3	20	0.960
12	4	15	0.958
15	5	12	0.955

Corrected Voltage of an Alternator. Equation (29) will now be corrected by the pitch factor k_p and distribution factor k_d terms to take account of the fact that the generated emf is *less* than was given. In its final form it becomes

$$E = 4.44 f N \phi k_p k_d \times 10^{-8} \quad \text{volts} \tag{32}$$

where the meanings of the symbols are as already defined.

EXAMPLE 6. The following information is given in connection with an alternator: slots = 96; poles = 4; rpm = 1,500; turns per coil = 16; $\phi = 2.58 \times 10^6$; coil span = slots 1 to 20; winding = whole-coiled lap, three phase. Calculate the generated voltage per phase.

Solution

$$f = \frac{4 \times 1,500}{120} = 50 \text{ cps}$$

Since there are 96 slots, there will be a total of 96 coils for this winding.

$$\text{Coils per phase} = \frac{96}{3} = 32$$

Therefore

$$n = \frac{32}{4} = 8$$

$$\text{Slots per pole} = \frac{96}{4} = 24$$

Therefore

$$p° = \frac{19}{24} \times 180° = 142.5°$$

and
$$d° = \frac{180°}{24} = 7.5°$$
Turns per phase $= 32 \times 16 = 512$

$$k_p = \sin \frac{142.5}{2} = 0.95$$

$$k_d = \frac{\sin (8 \times 7.5°/2)}{8 \times \sin (7.5°/2)} = \frac{\sin 30°}{8 \times \sin 3.75°} = \frac{0.5}{0.513} = 0.955$$

$$E = 4.44 \times 50 \times 512 \times 2.58 \times 10^6 \times 0.95 \times 0.955 \times 10^{-8} = 2{,}660 \text{ volts}$$

Alternator Regulation. Before an a-c generator is ready to function to deliver electrical load, (1) it must be brought up to synchronous speed by its prime mover (see Table 4); (2) it must be separately excited from a d-c source; (3) it must have its terminal voltage adjusted to the correct value by proper manipulation of the field rheostat. In contrast, a d-c generator need not have a speed of exactly the name-plate value, nor is it usually separately excited.

Loading an alternator will affect its terminal voltage, just as it does a d-c generator, but the manner in which it does so will depend upon the *character* of the load. If the load consists of pure resistance units, such as incandescent lamps or heating devices (electric stoves, flatirons, toasters, and the like), the power factor will be unity; this type of load will cause the terminal voltage to drop about 8 to 20 per cent below its no-load value. A lagging power-factor load, such as induction motors (to be discussed subsequently), electrical welders, fluorescent lighting, and electromagnetic devices, will cause the terminal voltage of the alternator to drop as much as 25 to 50 per cent below the no-load value. Leading power-factor loads, however, such as capacitor devices or special types of synchronous motor, will tend to raise the terminal voltage of the alternator above the no-load value. Exactly how much the terminal voltage will drop or rise will depend upon (1) the magnitude of the load and (2) the actual over-all power factor of the combined loads. In general, it can be said that (1) the greater the load, the greater will be the drop or rise; (2) the lower the lagging power factor, the greater will be the voltage drop; and (3) the lower the leading power factor, the greater will be the voltage rise.

In the study of d-c generator operation, it was learned that two factors are responsible for the *change* in voltage as load is applied: (1) resistance drop in the armature circuit and (2) change in flux. The same two factors are responsible for the change in voltage of an alternator but, in addition, there is a third factor: the *reactance voltage drop*. This drop results from the fact that the armature winding possesses considerable inductance L, which, in a-c circuits, asserts itself as a reactance X_L, where $X_L = 2\pi f L$; the reactance drop, like a resistance drop, is then equal to IX_L. However, in applying the reactance drop as a calculation, it must be done not arithmetically but *vectorially*, as will be shown later. Moreover, the effect of the flux change, the *armature-reaction effect*, is generally interpreted as a voltage drop and it, too, must be applied *vectorially*.

In summary, then, the terminal voltage of an alternator is affected by three factors, namely, (1) the armature resistance, (2) the armature *reactance*, and (3) the effect of armature *reaction*. The manner in which these factors, treated as voltages, are to be determined and applied is discussed in the next section.

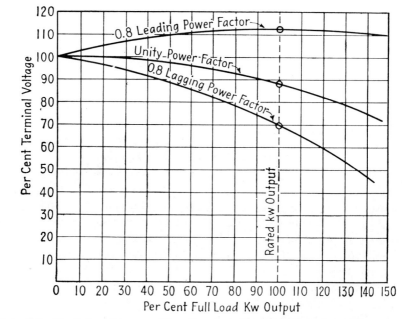

FIG. 153. Typical *voltage vs. kilowatt output* characteristics of an alternator at different power factors.

Figure 153 shows the typical relation between the percentage of terminal voltage and the percentage of kilowatt output for the three types of loading. The curves were plotted in terms of percentage because general operating conditions can be illustrated better in this way. Note that when 100 per cent rated kilowatt load is delivered by the alternator, the voltages are 88, 70, and 112 per cent of the starting no-load values for unity, 0.8 lagging, and 0.8 leading power factors, respectively. It should be understood that the data for such curves as are represented in Fig. 153 are obtained by starting at exactly the same no-load voltage for all three types of loading and maintaining the field-excitation direct current absolutely constant at all times as the alternator progressively delivers more power to the load.

Since the per cent regulation formula [Eq. (9), p. 103] applies equally well for alternators, it may be calculated in the same manner as before. It is repeated here:

$$\text{Per cent regulation} = \frac{V_{NL} - V_{FL}}{V_{FL}} \times 100$$

EXAMPLE 7. Referring to Fig. 153 calculate the per cent regulation for the three types of loading indicated by the unity power-factor, the 0.8 lagging power-factor, and the 0.8 leading power-factor curves.

Solution

$$\text{Per cent regulation (power factor} = 1.0) = \frac{100 - 88}{88} \times 100 = 13.65$$

Per cent regulation (power factor = 0.8 lagging)

$$= \frac{100 - 70}{70} \times 100 = 42.9$$

Per cent regulation (power factor = 0.8 leading)

$$= \frac{100 - 112}{112} \times 100 = -10.7$$

The minus (negative) regulation calculated for the leading power-factor load merely indicates that the full-load voltage is *more* than the no-load voltage. In other words, the terminal voltage rises with increasing kilowatt output as in the overcompound d-c generator.

It is much more important that automatic voltage regulators be used in connection with the operation of alternators than with d-c generators for several reasons. In the first place, an alternator cannot be compounded to make its *voltage vs. load* characteristics correspond to those of the d-c compound generator. Thus the inherent voltage change in the alternator, expecially with low lagging power-factor loads, cannot be compensated. Secondly, the voltage variations resulting from load changes and power factor are considerably greater than those displayed by d-c machines because of the greater effect of armature reaction and the additional influence of armature reactance. And thirdly, the alternator must generally "feed" a comparatively long transmission system consisting of wires and transformers whose resistances and reactances introduce additional voltage drops. The combined result of these factors acting simultaneously is to cause the load voltage to change very greatly with load changes. And, since large voltage fluctuations cannot be tolerated in otherwise satisfactory transmission systems, specially designed regulators must be employed to operate with alternators; these function to act upon the d-c field of the a-c generator so that a drop in the a-c terminal voltage is accompanied by an inverse adjustment of the flux.

Voltage Drops in Alternator Armatures. When the load upon a d-c shunt generator increases, the terminal voltage drops. There are two reasons for this: (1) the flux diminishes because of the effect of armature reaction and the fact that the field voltage is reduced and (2) a part of the generated voltage is used in overcoming the resistance of the armature *circuit*. To be sure, the armature-circuit resistance consists of the armature-winding resistance, the brush-contact resistance, the interpole-field resistance, and, in some cases, the compensating-winding resistance.

When the load upon an alternator increases, however, the terminal voltage may drop or rise, depending upon the *character* of the load. Three

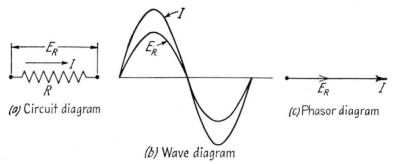

FIG. 154. Diagrams representing conditions for a pure resistance circuit.

factors control this tendency on the part of the alternator to lose or gain voltage: (1) the armature-winding resistance, (2) the armature reactance, and (3) the armature reaction. The first of these always causes a voltage *drop*, exactly as in the d-c generator. The second and third factors, which are the predominating ones, may produce either a voltage drop or a voltage rise. As will be explained, a zero *lagging* power-factor load causes a voltage *drop*, while a zero *leading* power-factor load produces a voltage *rise*; a power factor between the two extremes can have either effect upon the terminal emf.

Armature-resistance Voltage Drop. In elementary a-c circuit theory, it is proved that the current in a pure *resistance* circuit is *in phase* with the voltage required to cause *that* current to flow through *that* resistance. Figure 154 represents such a simple circuit; in (*a*) the voltage E_R is shown impressed across a resistor of R ohms, through which a current of I amperes flows; in (*b*) the voltage and current sine waves are shown *in phase* with each other; in (*c*) the same *in-phase* relationship of E_R and I is shown in the simplified phasor notation.

Since the alternator generated voltage E_G must first be sufficient to overcome the armature-winding resistance drop $I \times R_A (= E_R)$ and provide a terminal voltage V, it follows that, *vectorially*, $E_G = V + E_R$.

This is shown in Fig. 155 for unity, lagging, and leading power-factor loads. Note that in every case the generated voltage E_G is *greater* than the terminal voltage V.

Armature-reactance Voltage Drop. It is also proved in a-c circuit theory that the current in a pure *reactance* circuit (one possessing inductance but no resistance) *lags by 90 electrical degrees behind* the voltage required to cause *that* current to flow. Figure 156 represents such a simple circuit: in

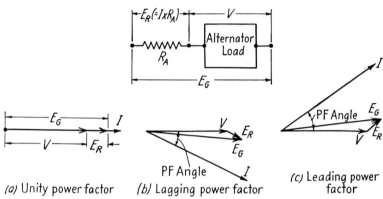

FIG. 155. Phasor diagrams representing alternator voltages for three types of load, assuming that the winding has resistance only.

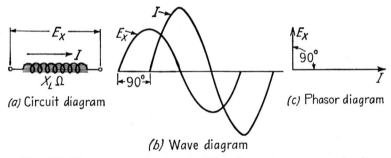

(b) Wave diagram

FIG. 156. Diagrams representing conditions for a pure reactance circuit.

(a) the voltage E_X is shown impressed across the reactor X_L ohms, through which a current of I amperes flows; in (b) the voltage and current sine waves are shown *out of phase* by 90 electrical degrees; in (c) the same *out-of-phase* relationship of E_X and I is shown in the simplified phasor notation. Note particularly that E_X leads I by 90° or that I lags behind E_X by 90°.

Again, since the alternator voltage E_G must be sufficient to overcome the armature-winding reactance drop $I \times X_L$ ($= E_X$) and provide a terminal voltage V, it follows that, *vectorially*, $E_G = V + E_X$. This is shown in Fig. 157 for unity, lagging, and leading power-factor loads. Note that for *unity* and *lagging* power-factor loads, the generated voltage E_G is *greater* than V; for the particular *leading* power-factor load, the generated voltage E_G is *less* than V.

Armature-reaction Voltage Drop. When load current flows through the armature winding of an alternator, the resulting mmf produces flux. This armature flux reacts with the main-pole flux, causing the total (resultant) flux to become either less than or more than the original main

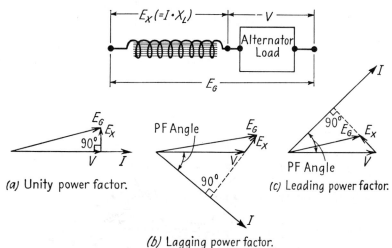

Fig. 157. Phasor diagrams representing alternator voltages for three types of load, assuming that the winding has reactance only.

flux. If it is assumed that the magnetic circuit iron is unsaturated, the amount of flux reduction or addition involves a voltage drop or rise, depending upon the magnitude of the armature current. In other words, the flux created by the armature current may be said to develop a voltage, called the *armature-reaction voltage* E_{AR}, that will vectorially add to or subtract from the generated voltage E_G.

It will now be shown that the flux created by the armature mmf subtracts directly from the main flux if the load power factor is zero lagging and that the flux created by the armature mmf adds directly to the main flux if the load power factor is zero leading. And since it is the *resultant flux* that is responsible for the actual generated voltage (neglecting, for the moment, an $I_A R_A$ drop), it should be clear that the terminal emf will depend upon whether the armature-reaction flux subtracts from or adds to the main-field flux.

Consider Fig. 158a. If the poles are assumed to be moving from left to right, the directions of the *voltages* in the armature conductors will be as

ALTERNATING-CURRENT GENERATORS 249

indicated by the crosses (in) and dots (out). For zero power factor lagging, the *currents* in the armature conductors would have the directions indicated in Fig. 158*b* *after* the poles have moved to a new position 90 electrical degrees beyond because the current lags behind the voltage by 90 electrical degrees. Careful study of Fig. 158*b* indicates that the flux

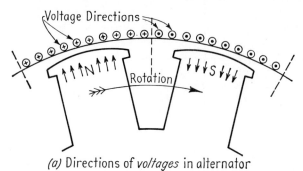

(a) Directions of *voltages* in alternator

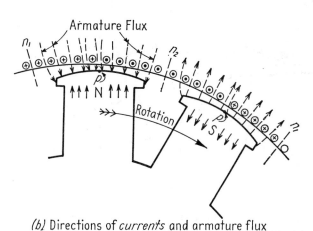

(b) Directions of *currents* and armature flux for zero power factor lagging

FIG. 158. Sketches illustrating how the armature flux directly *opposes* the main field flux in an alternator supplying a zero-lagging power-factor load.

created by the armature mmf directly *opposes* the main field flux. The latter may be understood if the right-hand rule is used for the conductors between n_1 and n_2, assuming that these conductors form a spiral about point p. Thus the armature flux reacts with the main-pole flux in such a manner as to *reduce* the latter in proportion to the value of the armature current *if the iron is unsaturated*. It follows, therefore, that for zero power factor lagging, the voltage created by this armature flux can be assumed to subtract directly from the generated voltage E_G to produce the lower terminal voltage V.

250 ELECTRICAL MACHINES—ALTERNATING CURRENT

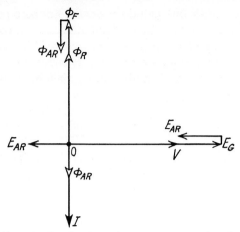

Fig. 159. Phasor diagram showing how the armature-reaction flux and voltage *subtract*, respectively, from the field flux and no-load voltage for a zero-lagging power-factor load.

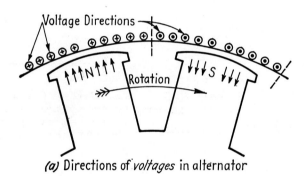

(a) Directions of *voltages* in alternator

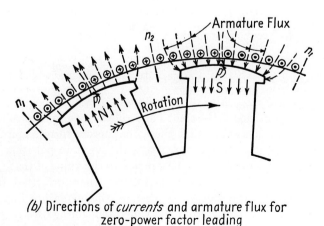

(b) Directions of *currents* and armature flux for zero-power factor leading

Fig. 160. Sketches illustrating how the armature flux directly *aids* the main field flux in an alternator supplying a zero-leading power-factor load.

ALTERNATING-CURRENT GENERATORS

A phasor diagram illustrating the relations for the zero-lagging power-factor load is given in Fig. 159. Note that the main-field flux ϕ_F is responsible for the no-load generated voltage E_G. Then, when the lagging power-factor current I passes through the armature winding, its mmf produces armature-reaction flux ϕ_{AR}, which, in turn, develops the armature-reaction voltage E_{AR}. Since E_{AR} directly opposes E, the terminal voltage V is their difference; observe also that the *difference* between ϕ_F and ϕ_{AR} is the resultant flux ϕ_R and that the latter is cut by the armature conductors to yield voltage V.

Next consider Fig. 160. For zero power factor leading, the voltage directions are given by Fig. 160a. The armature current directions are shown in Fig. 160b, where the poles are 90 electrical degrees *ahead* of their positions with respect to Fig. 160a because the current leads the voltage by 90 electrical degrees. Careful study of Fig. 160b indicates that the flux created by the armature mmf *adds* directly to the main flux. The latter may also be understood if the right-hand rule is used for the conductors between n_1 and n_2, assuming that these conductors form a spiral about point p. Thus the armature flux reacts with the main-pole flux in such a manner as to *increase* the latter in proportion to the value of the armature current *if the iron is unsaturated*.

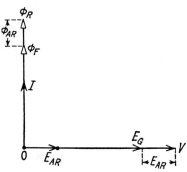

FIG. 161. Phasor diagram showing how the armature-reaction flux and voltage *add*, respectively, to the field flux and no-load voltage for a zero-leading power-factor load.

It follows, therefore, that for zero power factor leading, the voltage created by this armature flux can be assumed to add directly to the generated voltage E_G to produce a higher terminal voltage V.

A phasor diagram illustrating the relations for the zero-leading power-factor load is given in Fig. 161. Note that the main-field flux ϕ_F is responsible for the no-load generated voltage E_G. Then, when the leading power-factor current I passes through the armature winding, its mmf produces armature-reaction flux ϕ_{AR}, which, in turn, develops the armature-reaction voltage E_{AR}. Since E_{AR} directly aids E, the terminal voltage V is their sum; observe also that the sum of ϕ_F and ϕ_{AR} is the resultant flux ϕ_R and that the latter is cut by the armature conductors to yield V.

Alternator Phasor Diagram. The foregoing discussion leads to the complete phasor diagram of the alternator operating under load. Obviously, when the armature winding carries current, all three voltage drops—IR_A drop, IX_L drop, and IX_{AR} drop—must be added vectorially to the terminal voltage in order to determine the generated emf. Or to put it another way, if the load is removed from an alternator, the terminal voltage will change from V to E_G because at no load the generated voltage *is* the terminal voltage.

Figure 162 shows three complete phasor diagrams for the general types of load. For the unity power-factor load (Fig. 162a), the no-load voltage E_G is larger than the terminal voltage V. For the lagging power-factor load, the change from V to E_G is comparatively greater than it is for the

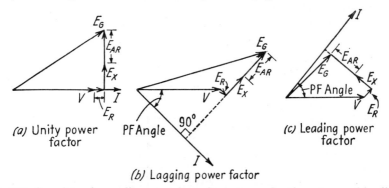

Fig. 162. Complete phasor diagrams of an alternator under three types of loading.

unity power-factor load. And note that the open-circuit voltage E_G is less than V when the power factor is leading.

In order to emphasize the practical importance of the phasor diagram in determining the percent regulation of an alternator, the following example is given.

EXAMPLE 8. A 500-kva 2,300-volt three-phase star-connected alternator has a full-load armature-resistance drop per phase of 50 volts and a combined armature-reactance (E_X) plus armature-reaction (E_{AR}) drop of 500 volts per phase. Calculate the per cent regulation of the alternator at (a) unity power factor; (b) 0.866 power factor lagging; (c) 0.8 power factor leading.

Solution

All calculations will be made on a "per phase" basis because its per cent voltage change is exactly the same as the per cent change in voltage between line terminals:

$$V \text{ (per phase)} = \frac{2{,}300}{\sqrt{3}} = 1{,}330$$

(a) Referring to Fig. 163a, $E_G = \sqrt{(1{,}330 + 50)^2 + (500)^2} = 1{,}470$

Per cent regulation (power factor = 1) = $\dfrac{1{,}470 - 1{,}330}{1{,}330} \times 100 = 10.5$

(b) Referring to Fig. 163b,

$$E_G = \sqrt{(1{,}150 + 50)^2 + (665 + 500)^2} = 1{,}680$$

Per cent regulation (power factor = 0.866 lag)

$$= \frac{1{,}680 - 1{,}330}{1{,}330} \times 100 = 26.3$$

(c) Referring to Fig. 163c,

$$E_G = \sqrt{(1{,}065 + 50)^2 + (800 - 500)^2} = 1{,}155$$

Per cent regulation (power factor = 0.8 lead)

$$= \frac{1{,}155 - 1{,}330}{1{,}330} \times 100 = -13.2$$

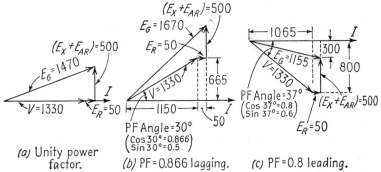

Fig. 163. Phasor diagrams illustrating Example 8.

Synchronous Reactance and Synchronous Impedance. When tests are performed upon an alternator to determine its regulation, it is customary to secure data that combine the value of E_X and E_{AR} into a single calculation. Both these voltage drops, the reactance voltage drop and a voltage drop resulting from the effect of armature reaction, bear the same 90° relationship to the current; moreover, they occur at full-load current when the alternator is operating at *synchronous speed*. It is proper, therefore, to designate this combined voltage drop, i.e., $E_X + E_{AR}$, as the *synchronous-reactance drop*. If this synchronous-reactance drop is divided by the full-load current I_L, the value in ohms (ohms = volts/amperes) is called the *synchronous reactance*. It is the latter term, symbolized by X_S, that is measured when the proper tests are performed upon an alternator.

Fig. 164. The synchronous impedance triangle of an alternator.

It is shown in a-c theory that resistance and reactance always bear the same relation with respect to each other as do the sides of a right-angle triangle. If this is done for the values of R_A and X_S in an alternator, the value of the impedance so determined is called the *synchronous impedance*, symbolized by Z_S. Z_S is equal, therefore, to $\sqrt{R_A{}^2 + X_S{}^2}$ (see Fig. 164).

To obtain data for the calculation of the regulation of an alternator, it is necessary to perform three simple tests. These are (1) the armature-resistance test, (2) the open-circuit test, and (3) the short-circuit test.

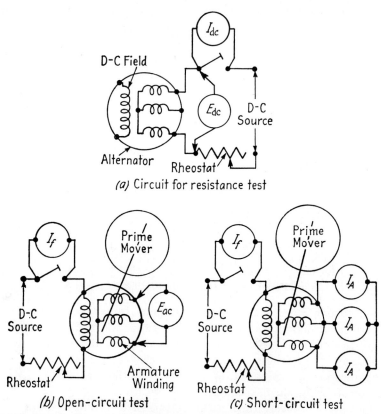

Fig. 165. Circuit diagrams for performing tests on an alternator to determine the regulation.

The Resistance Test. Assume that the alternator is Y-connected if it is three-phase. (Whether or not it is actually connected Y or Δ is immaterial to the final result; the same value will be obtained with either assumption, although the Y assumption leads to the simpler calculation.) With the d-c field winding *open*, measure the d-c resistance between each pair of terminals. The average of the three sets of resistance values is called R_t. Divide this value of R_t by 2 to obtain the *resistance per phase R_A* (see Fig. 165a). Since the resistance is very low,

ALTERNATING-CURRENT GENERATORS

it will be necessary to use a high-current rheostat to adjust the current to about rated value, a d-c ammeter of the proper range, and a low-reading voltmeter. The alternator should be at rest. In practice, it is customary to multiply the d-c resistance of R_A by a factor of 1.25 to 1.75 to obtain a value more nearly equal to the armature resistance when it carries alternating current.

The Open-circuit Test. With the armature-winding circuit open, operate the alternator at synchronous speed. Connect a d-c source to the

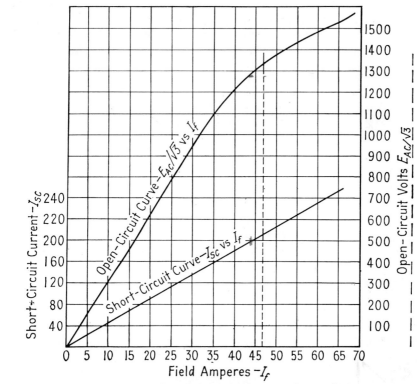

FIG. 166. Open-circuit and short-circuit curves for an alternator.

field, making provision to adjust the field current so that, starting at zero, it may be raised until the a-c voltage between any pair of terminals of the armature winding is somewhat above rated emf (see Fig. 165b). Record the data of I_f and E_{ac} for a sufficient number of points to plot the so-called *open-circuit saturation curve*. Before plotting the curve, divide E_{ac} by $\sqrt{3}$ to obtain the phase voltage (see Fig. 166), which gives the curve of open-circuit voltage $E_{ac}/\sqrt{3}$ vs. I_f.

The Short-circuit Test. Perform this test with great care. Referring to Fig. 165c insert a rheostat of sufficiently high ohmic value in the d-c field circuit to keep the current in that circuit very low. Connect three similar ammeters in Y to the armature-winding terminals; each ammeter should have a range high enough to register a current somewhat in excess of the rated full-load value. Operate the alternator at synchronous speed. Starting with a very low direct field current, progressively increase its value as the a-c ammeters increase their deflections to rated current and above; record I_f and the average of the three a-c ammeters. Plot the short-circuit characteristic I_{SC} vs. I_f, using Fig. 166 as a guide.

To determine the value of Z_S per phase, find the value of I_{SC} at the field current that gives the rated alternator terminal voltage per phase; Z_S will then be equal to the open-circuit voltage divided by the short-circuit current at that field current which gives the rated emf per phase.

EXAMPLE 9. The curves of Fig. 166 are for a 500-kva 2,300-volt three-phase alternator. If the average d-c resistance of the three armature-winding phases between pairs of terminals R_t is 0.8 ohm, calculate the values of Z_S and X_S. (Assume that the effective a-c resistance is 1.5 times the d-c resistance.)

Solution

From the curves at $E_{ac}/\sqrt{3} = 2,300/\sqrt{3} = 1,330$ volts,

$$I_{SC} = 204 \text{ amp}$$

$$Z_S = \frac{1,330}{204} = 6.52 \text{ ohms}$$

$$R_A = 1.5 \times \frac{0.8}{2} = 0.6 \text{ ohm}$$

$$X_S = \sqrt{(6.52)^2 - (0.6)^2} = 6.5 \text{ ohms}$$

EXAMPLE 10. For the data of Example 8, and using the values of R_A and X_S from Example 9, calculate the per cent regulation for a full-load power factor of 0.8 lagging.

Solution

$$I_L = \frac{500,000}{\sqrt{3} \times 2,300} = 126$$

$$E_R = 126 \times 0.6 = 75 \text{ volts}$$

$$I_L X_S = 126 \times 6.5 = 820$$

$$E_G = \sqrt{[(1,330 \times 0.8) + 75]^2 + [(1,330 \times 0.6) + 820]^2}$$

$$= \sqrt{(1,139)^2 + (1,618)^2} = 1,980 \text{ volts}$$

$$\text{Per cent regulation} = \frac{1,980 - 1,330}{1,330} \times 100 = 48.8$$

The phasor diagram for the solution of this problem is similar to that given in Fig. 163b.

Alternator Efficiency. The efficiency of an alternator is calculated in exactly the same manner as was the efficiency of d-c generators in Chap. 6. It will be recalled that it is first necessary to determine the various losses in the machine; with the alternator operating under load, these losses include:

1. Rotational losses
   a. Friction and windage
   b. Brush friction at the field collector rings
   c. Ventilation to cool the machine (if necessary)
   d. Hysteresis and eddy currents in the stator
2. Electrical losses
   a. Field winding
   b. Armature winding
   c. Brush contacts
3. Losses in the exciter used for field excitation
4. Stray-load loss

Concerning the power losses listed, the following comments may be made:

1a. The friction and windage may be determined by using a small auxiliary motor to drive the unexcited alternator at rated speed and measuring the power input to the former; next, with the auxiliary machine uncoupled and operating free, its power input is again measured. Then, when the power measurements are corrected for the losses in the auxiliary motor, the friction and windage loss is the difference of the two sets of readings. In high-speed alternators these losses may be extremely large, although the windage is greatly reduced by using hydrogen for cooling purposes.

1b. The brush friction at the collector rings, where power is delivered to the field for excitation purposes, is quite small and is often neglected in the efficiency calculation.

1c. In large alternators, which must be artifically cooled, power is required to circulate air or hydrogen; this power represents a loss chargeable to the alternator and is in addition to the windage loss indicated in 1a. The losses in the ducts from the blower to the alternator are, however, not included in this item but are charged to plant operation.

1d. The core loss involves hysteresis and eddy currents in the magnetized iron and results from the normal flux-density changes. It may be determined by measuring the power input to an auxiliary motor with and without the field excited; the difference between the two power measurements represents this loss. In accordance with the American

258   ELECTRICAL MACHINES—ALTERNATING CURRENT

Standards Association (ASA) rules, the field of the alternator should be excited to give a terminal voltage that is calculated by adding the full-load resistance drop to the rated voltage of the machine.

2a. The field copper loss is determined by multiplying the field-winding voltage by the field current, i.e., $E_F \times I_F$, or by using the equation $I_F{}^2 R_F$, where R_F is the field-winding resistance. This loss does *not* include the power required by the field rheostat, which is charged to plant operation.

2b. The armature-winding copper loss is calculated by the formula $nI_A{}^2 R_A$, where n, I_A, and R_A represent, respectively, the number of phases, the full-load stator-winding current, and the d-c resistance per phase at 75°C.

2c. The electrical loss at the brush contacts, between the brushes and the slip rings, is usually quite small and is often neglected in the efficiency calculation.

3. The losses in the exciter, the latter serving to excite the d-c field winding of the alternator, are generally omitted from the efficiency calculation and, as in the case of the field rheostat, are usually charged to plant operation.

4. The stray-load losses are generally classed as indeterminate because, with the alternator operating under load, they result from eddy currents in the armature copper conductors and because the distorted magnetic field increases the normal core losses as found in 1d. They may, however, be included in the efficiency calculation by using the *effective* value of the armature resistance in 2b instead of the d-c resistance.

After the total of the foregoing losses has been found, it may be used in the following equation to evaluate the alternator efficiency:

Per cent efficiency

$$= \left[1 - \frac{\text{kw losses}}{(\text{kva output} \times \text{PF}) + (\text{kw losses})} \right] \times 100 \quad (33)$$

EXAMPLE 11. A 2,000-kva 2,300-volt three-phase alternator operates at rated kilovolt-amperes at a power factor of 0.85. The d-c armature-winding resistance at 75°C between terminals is 0.08 ohm. The field takes 72 amp at 125 volts from exciter equipment. Friction and windage loss is 18.8 kw, iron losses are 37.6 kw, and stray-load losses are 2.2 kw. Calculate the efficiency of the alternator. (Assume that the *effective* armature-winding resistance is 1.3 times the d-c value.)

Solution

$$\text{Output} = 2{,}000 \times 0.85 = 1{,}700 \text{ kw}$$

$$I_L = \frac{2{,}000{,}000}{\sqrt{3} \times 2{,}300} = 503 \text{ amp}$$

$$R_A \text{ (per phase)} = \frac{0.08}{2} \times 1.3 = 0.052 \text{ ohm}$$

ALTERNATING-CURRENT GENERATORS

Losses	Kilowatts
Friction and windage	18.8
Iron	37.6
Field winding = (125 × 72) ÷ 1,000	9.0
Armature winding = 3 × (503)² × 0.052	39.4
Stray load	2.2
Total	107.0

$$\text{Per cent efficiency} = \left(1 - \frac{107}{1{,}700 + 107}\right) \times 100 = 94.1$$

Operation of Alternators in Parallel. Modern power plants usually have several alternators that are operated singly or in several parallel combinations to supply a common load. There are a number of good reasons for this practice, even though in some instances it might be cheaper to install only one or two very large alternators to take care of the plant load. (As a rule, large machines and their auxiliary equipment cost less per kilowatt than small machines.) In the first place, the efficiency of an electrical machine is a maximum at, or near, full load. Since the power-plant load fluctuates widely during each 24-hr period, it is most economical to use a small unit delivering approximately rated capacity when the demand is light, substituting larger alternators as the load increases, and employing several machines connected in parallel when heavy or peak loads occur. Obviously, changing from one unit to another or connecting an alternator in parallel with others must be carried out in such a manner that line disturbances do not occur, nor must users of electric energy be conscious of a power-plant change. Exactly how an alternator is *taken off* or *put on* a line without service interruption will be considered later; the latter process is called *synchronizing*.

Continuity of service is, of course, extremely important in the operation of electrical systems. It requires several alternators, or even power plants, connected in parallel, so that breakdown of an alternator, its prime mover, or an entire plant does not involve service interruption. Furthermore, it is considered good practice to give machines periodic inspections to forestall the possibility of failure. This can be done only when machines are at rest and implies that other alternators must be in service; thus, if there should be a breakdown, the repair work can be done carefully, since other units are available to maintain continuity of service.

Finally, additions are made from time to time as service is extended and increased. In fact, engineers usually make plans for future extensions when power plants are designed. Such extensions are made only as existing equipment begins to prove inadequate or when the installation of new units has been proved economically advisable.

260 ELECTRICAL MACHINES—ALTERNATING CURRENT

Before a three-phase alternator is connected to, or *put on*, a line that is already being served by one or more units, the following conditions must be fulfilled:

1. The effective voltage of the incoming machine must be approximately equal to the bus-bar voltage.

2. The frequency ($f = P \times \text{rpm}/120$) of the incoming machine must be exactly the same as that of the bus bars.

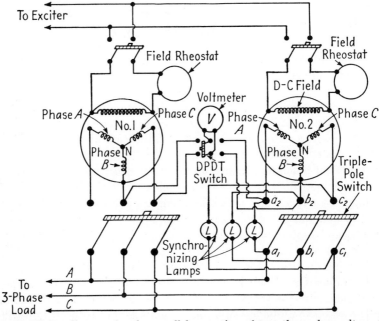

Fig. 167. Wiring diagram for the parallel operation of two three-phase alternators.

3. The phase sequence of the three phases of the incoming machine must be the same as that of the bus bars. This means that if the bus-bar voltage V_{AB} is 120° ahead of V_{BC}, and V_{BC} is 120° ahead of V_{CA}, then the incoming machine must have its three phases connected to bus bars A, B, and C, so that V_{AB} is 120° ahead of V_{BC} and V_{BC} is 120° ahead of V_{CA}.

4. With condition 3 fulfilled, it is necessary that *at the instant when the paralleling switch is closed*, voltage V_{AB} of the incoming machine must be *in phase opposition* to the bus-bar voltage V_{AB}; under this condition, the voltages V_{BC} and V_{CA} of the incoming machine and the bus bars will also be in phase opposition. This implies that there will be no circulating current between the windings of the alternators already in operation (the bus bars) and the incoming machine.

Consider Fig. 167, which represents a system of two three-phase alternators, where unit 1 is already connected to the bus bars A, B, and C.

It is desired to *synchronize* unit 2 with the bus bars, to transfer the entire load of unit 1 to unit 2, and to remove unit 1 from the line. Note that the three phases A, B, and C of both alternators may be traced through the triple-pole switches to the respective buses marked A, B, and C; that a voltmeter is connected to a DPDT switch so that the voltage of machines 1 and 2 may be measured; and that three synchronizing lamps are connected to the *open* triple-pole switch of machine 2.

Alternator 2 is brought up to synchronous speed by its prime mover, after which the field switch is closed. With the voltmeter switch closed to the right, the field rheostat is adjusted until the voltmeter registers the

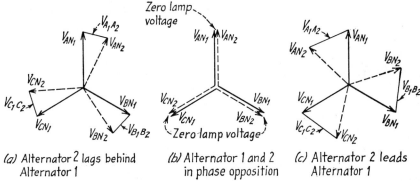

(a) Alternator 2 lags behind Alternator 1 (b) Alternator 1 and 2 in phase opposition (c) Alternator 2 leads Alternator 1

FIG. 168. Phasor diagrams illustrating why light flickering of synchronizing lamps occurs during the synchronizing process of two alternators.

same voltage as when closed to the left. At this point it will be noticed that the three synchronizing lamps will flicker *in unison*, the rapidity of the flicker depending upon the difference between the line frequency and the incoming machine frequency. Even a very slight frequency difference will cause the lamps to flicker. In order to synchronize alternator 2 with the bus, that is, fulfill condition 4, either increase or decrease the speed of the prime mover and its driven alternator until the flickering subsides and the lamps are *completely extinguished*. The operator may observe that a slight increase in speed will increase the flicker frequency, in which case the speed should be reduced. The lamps may not flicker in unison; if this happens condition 3 has not been fulfilled. This may be rectified very simply by interchanging any two leads from the alternator-winding terminals to the triple-pole switch. After careful speed adjustment, the extinguished lamps will indicate that the machines are in synchronism with each other; at this instant, *and no other*, the triple-pole switch should be closed.

The reasons for the flickering may be explained by referring to a set of diagrams (Fig. 168) in conjunction with the wiring diagram. The lamp voltages are always equal to the *potential differences* across the open triple-pole switch; these potential differences are $V_{A_1A_2}$, $V_{B_1B_2}$, and $V_{C_1C_2}$. Let V_{AN_1}, V_{BN_1}, and V_{CN_1} be the bus voltages, and let V_{AN_2}, V_{BN_2}, and V_{CN_2} be the voltages of machine 2. If machine 2 is not exactly in phase opposition with the bus, such as in Fig. 168a, the lamp voltages will be $V_{A_1A_2}$, $V_{B_1B_2}$, and $V_{C_1C_2}$; machine 2 is then behind machine 1. If machine 2 is speeded up, the gaps between the arrow tips will close. When the tips coincide, as in Fig. 168b, alternator 2 is synchronized with the bus. However, the period of lamp darkness may not last sufficiently long and machine 2 may swing far ahead of machine 1 (Fig. 168c), indicating one position when the lamps will burn very brightly. Obviously, when the *relative speed* of the two machines increases, the flickering becomes more rapid, and vice versa.

The maximum voltage across the individual lamps of Fig. 167 will occur when, during the synchronizing process, alternator 2 is *in phase* with alternator 1; this is, of course, the worst instant to close the synchronizing switch, since the lamp voltages are at their maximum values. Referring to Fig. 168, this condition would be represented by pairs of phasors V_{AN_1} and V_{AN_2}, V_{BN_1} and V_{BN_2}, and V_{CN_1} and V_{CN_2} that are oppositely directed with respect to one another. Remembering that the magnitudes of the individual voltage vectors in Fig. 168 are line-to-neutral quantities, the greatest lamp voltage is $2 \times V_N$ or $2 \times V_L/\sqrt{3}$, where V_N and V_L are, respectively, neutral and line voltages.

EXAMPLE 12. What maximum voltage can exist across a lamp in Fig. 167 when two 230-volt alternators are synchronized? How many 115-volt lamps, connected in series, should be used in place of the single units shown?

Solution

$$V_{\text{lamp}} = 2 \times 230/\sqrt{3} = 266 \text{ volts}$$

Use three 115-volt lamps in series for each phase.

After alternator 2 has been synchronized and is operating in parallel with alternator 1, it may be made to deliver part or all of the load, assuming, of course, that its rating is sufficient to do so. To shift the load from 1 to 2, *increase the mechanical power input to the prime mover of alternator 2 and simultaneously reduce the mechanical power input to the prime mover of alternator* 1. This may readily be done by admitting more steam (opening the steam valve) to the prime mover 2 (if it is a steam machine) and simultaneously closing the steam valve of prime mover 1. When machine 2 carries the entire load, as indicated by ammeters in the line circuits, the field switch of alternator 1 is opened; the latter unit is then shut down if desired.

One point which should be emphasized here is that load cannot be shifted from one alternator to another by field-current adjustment. This adjustment changes the power factor of the alternator load.

Two alternators operating in parallel and delivering a common load are in stable equilibrium because neither machine can pull out of synchronism. Should one machine attempt to speed up for some reason, this tendency would immediately be accompanied by a flow of current between the alternators that would circulate in the armature windings only. Since this circulating current does not pass into the load, it represents, in part, generator action for the faster alternator and motor action for the second machine. The result is that the increased electrical load on the fast alternator causes it to slow down, while the torque gained by the second alternator increases its speed. Thus it is seen that the circulating current serves to prevent two alternators from departing from exact synchronism because the energy represented by such current retards the fast one and accelerates the slow one. These actions are of short duration, but while they are taking place, the rotors of the two alternators swing forward and backward about the stable rotating positions; this latter phenomenon is called *hunting*.

It is also possible for one alternator to attempt to generate momentarily a slightly higher voltage than the other machine. When this happens, a circulating current immediately flows in the two armature windings, which has the effect of demagnetizing, by the proper amount, the field of the higher-voltage alternator and increasing the magnetism of the other alternator. Thus it is seen that the circulating current again serves to prevent one machine from assuming more than its proper share of the load.

Questions

1. In what respect are d-c and a-c generators similar?
2. How is it possible to convert a d-c generator into an alternator?
3. What is meant by *synchronous speed?* How is it calculated?
4. Give five reasons for the preference of revolving-field alternators over the revolving-armature type of machine.
5. Why is it possible to construct alternators of much larger capacities than generators for d-c service?
6. What is an exciter for an alternator? What type of machine is it? Where is it located?
7. Describe the construction of the stator for an alternator.
8. Describe the construction of the two general types of alternator field.
9. Why is it necessary that the speed of an alternator be maintained at a constant value at all times?
10. What three frequencies are generally used in the United States?
11. What voltage is generally used for field excitation?
12. What are collector rings, and what purpose do they serve?

13. Why is it possible to design alternators to generate much higher voltages than generators for d-c service?
14. Why is it frequently necessary to ventilate alternators by blowing air through them?
15. What special kind of insulating material is generally used in large alternators? Describe how this material is made.
16. Assuming a sine wave, upon what three factors does the generated voltage of an alternator depend?
17. Distinguish between *half-coiled* and *whole-coiled* armature windings in alternators.
18. Assuming the same voltage for two alternators that are similar in every other respect, how many turns per coil will there be in the whole-coiled winding with respect to the number in a half-coiled winding?
19. What is meant by a *group* of armature coils? How are the coils in such a group always connected?
20. What calculation should be made to determine the number of coils in a group for the whole-coiled winding? the half-coiled winding?
21. How are successive groups of coils connected in the whole-coiled winding? the half-coiled winding?
22. In what two ways can the three phases of a three-phase alternator be connected?
23. What is meant by a *full-pitch* winding? a *fractional-pitch* winding?
24. Why are fractional-pitch windings generally used in alternators?
25. What is meant by *pitch factor*?
26. What effect does the pitch factor have upon the generated voltage of an alternator?
27. Why are distributed windings generally used in alternators?
28. What is meant by *distribution factor*?
29. What effect does the distribution factor have upon the generated voltage of an alternator?
30. Define *alternator regulation*.
31. In a given alternator, upon what does the regulation depend?
32. What general limiting values can be assigned to the regulation of alternators delivering unity power-factor loads? lagging power-factor loads?
33. What three voltage drops occur in the alternator?
34. In making calculations for the regulation, how must the three voltage drops be used?
35. Why does the armature winding possess reactance? Upon what two factors does it depend?
36. Under what condition of loading is it possible for the generated emf in an alternator to be greater at full load than at no load?
37. What angular relation exists between resistance drop and reactance drop? between resistance drop and armature-reaction drop?
38. Why is it more important that automatic voltage regulators be used with alternators than with d-c generators?
39. Why does the load voltage tend to change more in a-c systems than in d-c systems?
40. Describe the general principle of operation of a *voltage regulator*.
41. Under what condition of loading does a voltage drop cause a drop in terminal voltage? a rise in terminal voltage?
42. Using the current as a reference phasor, how is the armature-resistance voltage drop related to it? the armature-reactance voltage drop? the armature-reaction drop?
43. Explain carefully, using Fig. 158, how a zero-lagging power-factor load current tends to demagnetize the main field.

ALTERNATING-CURRENT GENERATORS

44. Referring to the vector diagram of Fig. 159, explain how the armature-reaction voltage subtracts directly from the no-load voltage when the load power factor is zero lagging.
45. Explain carefully, using Fig. 160, how a zero-leading power-factor current tends to aid the main-field magnetization.
46. Referring to the vector diagram of Fig. 161, explain why the armature-reaction voltage is added directly to the no-load voltage when the load power factor is zero leading.
47. Define *synchronous reactance; synchronous-reactance drop.*
48. Define *synchronous impedance; synchronous-impedance drop.*
49. Explain carefully how the *resistance test* is performed on an alternator.
50. What is meant by the *effective* resistance of the armature winding? Why is it larger than the d-c resistance?
51. What relation exists between the resistance *between terminals* and the resistance *per phase* in a Y-connected winding? in a delta-connected winding?
52. Describe carefully how the *open-circuit test* is performed.
53. What useful information is obtained from the open-circuit test?
54. Describe carefully how the *short-circuit test* is performed.
55. What useful information is obtained from the short-circuit test?
56. Referring to Fig. 166, explain how the synchronous impedance may be determined.
57. List the various losses in an alternator.
58. Which of the losses listed in Question 57 do not usually occur in conventional d-c generators?
59. Explain how the friction and windage loss in an alternator may be determined.
60. What effect does hydrogen cooling have in the operation of large alternators?
61. Explain how the core loss in an alternator may be determined.
62. How is the copper loss in the field winding of an alternator determined?
63. How is the copper loss in the armature winding of an alternator determined?
64. What is meant by the *stray-load losses?* Where do they occur? Why are they indeterminate?
65. How can the stray-load losses be included in the efficiency calculation of an alternator? Why are these losses only approximate?
66. What would be the advantage of using a single large alternator in a power plant rather than several smaller units operating in parallel?
67. What are the important advantages of using several smaller alternators operating in parallel, rather than one large unit?
68. List the four important conditions that must be fulfilled before an alternator can be connected in parallel with the bus already supplying a load.
69. Explain carefully, using Fig. 167, how alternator 1 should be paralleled with alternator 2, assuming that the latter is already delivering load.
70. When lamps are used to synchronize one alternator with another, why do the lamps flicker in unison?
71. Under what condition will the lamps *not* flicker in unison?
72. Under what condition will the flicker disappear with the lamps completely extinguished? with the lamps burning brightly?
73. When two alternators are synchronized with lamps, what maximum voltage can exist across the latter? When does this occur? What provision must be made to insure against burning out the lamps?
74. Why will two alternators operating in parallel be in stable equilibrium even though the speed of one of them tends to increase? even though the voltage of one of them tends to decrease?

Problems

1. A 36-pole alternator is operated at 200 rpm. What frequency is generated?
2. At what speed should a six-pole alternator be driven to develop 40 cycles?
3. An alternator operates at a speed of 176.5 rpm. How many poles does it have if the voltage is generated at 50 cycles?
4. Calculate the *average* voltage generated in a six-turn full-pitch coil of a 25-cycle alternator if the flux per pole is 7.2×10^5 maxwells.
5. In Prob. 4, what is the effective voltage if the wave form is sinusoidal?
6. If the alternator of Prob. 4 has a total of 240 coils and the winding is Y-connected, what is the effective voltage between terminals? (Neglect the phase displacement of series-connected coils in the same pole-phase group.)
7. A 72-slot stator has a half-coiled four-pole three-phase winding. How many coils are there (a) per phase; (b) per group?
8. A 144-slot stator has a whole-coiled 12-pole three-phase winding. Calculate the number of coils: (a) per phase; (b) per group.
9. A 90-slot stator has a three-phase six-pole whole-coiled winding, each coil of which spans from slot 1 to slot 14. Calculate: (a) the pitch factor; (b) the distribution factor.
10. The following information is given in connection with an alternator: slots = 144; poles = 8; rpm = 900; turns per coil = 6; $\phi = 1.8 \times 10^6$; coil span = slots 1 to 16; winding = whole-coiled three-phase; winding connections = star. Calculate: (a) the voltage per phase; (b) the voltage between terminals.
11. What voltage would be generated per phase in the generator of Prob. 10 if the winding is connected for two-phase operation?
12. What voltage would be generated between terminals in the alternator of Prob. 10 if the winding is connected for six-pole operation?
13. A three-phase alternator has a rating of 5,000 kva at 13,200 volts. Calculate the full-load line current.
14. If the alternator of Prob. 13 delivers a load of 3,600 kw at a power factor of 0.82, calculate the line current.
15. The voltage of an alternator rises from 460 volts at full load to 535 volts at no load. Calculate the per cent regulation.
16. A three-phase Y-connected alternator delivers a unity power-factor load at 230 volts. If the synchronous-reactance voltage drop is 60 volts per phase, calculate the per cent regulation, neglecting the resistance voltage drop.
17. If the resistance voltage drop per phase in Prob. 16 is 5 volts, calculate the per cent regulation.
18. Assuming that the curves of Fig. 166 are for a 400-kva 2,200-volt three-phase alternator, calculate the synchronous impedance Z_S and the synchronous reactance X_S. The d-c resistance between terminals is 0.9 ohm.
19. Calculate the per cent regulation for Prob. 18 when the alternator delivers its full-load kilovolt-amperes at a power factor of 0.6 lagging.
20. Repeat the solution of Prob. 19 for a power factor of 0.8 leading.
21. A three-phase 1,000-kva 2,300-volt alternator is short-circuited and is operated at rated speed to produce rated current. With the short circuit removed and the same excitation, the voltage between stator terminals is 1,300. The effective (corrected) resistance between stator terminals is 2 ohms. Determine the per cent regulation of the alternator at a power factor of 0.8 lagging.
22. Repeat Prob. 21 for a power factor of 0.8 leading.
23. A 1,000-kva 4,600-volt three-phase alternator is short-circuited, and, operating

ALTERNATING-CURRENT GENERATORS 267

at rated speed, the field excitation is adjusted to give rated armature-winding current. The short circuit is then removed, and, with the same field current and speed, the open-circuit voltage between stator terminals is found to be 1,744. The *effective* (a-c) resistance per phase of the armature winding is 1.2 ohms. Calculate the per cent regulation of the alternator at a power factor of 0.8 lagging.

24. A 25-kva alternator has a total loss of 2,000 watts when it delivers rated kilovolt-amperes to a load at a power factor of 0.76. Calculate its per cent efficiency.

25. A 500-kva alternator operates at full-load kilovolt-ampere output at an efficiency of 93.5 per cent. If the load power factor is 0.83, calculate the total loss.

26. The d-c armature winding resistance between terminals of a 750-kva 4,400-volt three-phase alternator is 0.9 ohm. Calculate the copper loss in the winding at full load.

27. A 25-kva 220-volt three-phase alternator delivers rated kilovolt-amperes at a power factor of 0.84. The *effective* a-c resistance between armature-winding terminals is 0.18 ohm. The field takes 9.3 amp at 115 volts. If friction and windage loss is 460 watts and the core loss is 610 watts, calculate the per cent efficiency.

28. Each of two single-phase alternators has an armature winding whose resistance and synchronous reactance are, respectively, 0.025 and 0.06 ohm. If the machines are operating without load, calculate the circulating current in the windings if the alternators are paralleled: (*a*) when the emfs are equal at 230 volts, but are displaced 30° from a position of phase opposition; (*b*) when the voltages are in phase opposition, but one voltage is 230 and the other is 200 volts; (*c*) when the voltages are 230 and 200, and displaced 30° from a position of phase opposition.

29. Using the given data of Prob. 28, calculate the angle between two *equal* voltages of 230 if the circulating current is 1,000 amp when the alternators are paralleled.

CHAPTER 8

TRANSFORMERS

Transformer Action. When two coils of wire are inductively coupled, the flux that passes *through* one of them also passes entirely or in part *through* the other. This fact implies that the coils have a magnetic circuit that is common to both. Now then, if this flux created by a varying current through one coil changes, the mutual flux will change; under this condition, there will be created an *induced voltage* in the second coil. Note that the *secondary induced voltage* results because the flux changes through the coil, although this flux change is occasioned in the first place by the current change in the first, or *primary*, coil. The induced voltage of mutual induction in the secondary coil is called a *transformer voltage*, and the action that creates the emf is known as *transformer action*.

Transformer action takes place in any d-c system of coupled circuits when a switch is opened or closed, but its more important practical application is in connection with the operation of a-c apparatus such as transformers and induction motors. In such electrical equipment, one coil or set of coils is connected directly to an a-c source so that the current and the resulting flux automatically change periodically in both magnitude and direction. Since the linking flux in the coupled coil or set of coils changes, a voltage is induced in the latter by transformer action. Furthermore, if there is no relative motion between the coils, the frequency of the induced voltage in the second coil is *exactly* the same as the frequency of the primary impressed voltage. If the second coil is now connected to an electrical load such as incandescent lamps, electric motors, or heating devices, current will flow. Thus electrical energy can be transferred from one electric circuit to another electric circuit by transformer action, even though there is absolutely no metallic connection between the two. The whole remarkable process of energy transfer takes place by the *principle of electromagnetic induction*, first discovered by Michael Faraday in 1831.

The device that most commonly utilizes the principle of transformer action is the *static transformer*.

The foregoing may be summarized as follows: a transformer is a device that (1) transfers electrical energy from one electric circuit to another, (2) does so without a change of frequency, (3) does so by the principle of electromagnetic induction, and (4) has electric circuits that are linked by a common magnetic circuit. The energy transfer usually takes place with a change of voltage, although this is not always necessary.

Figure 169 illustrates a very simple transformer in which two coils are linked by a laminated magnetic steel core. The coil that is connected to the primary source of supply is called the *primary*, while the coil in which the voltage of mutual induction is induced and which "feeds" energy to the load is called the *secondary*. The primary coil takes electrical energy

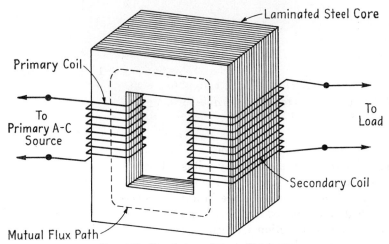

FIG. 169. Simple transformer diagram.

from the a-c source of supply, while the secondary, receiving this energy by electromagnetic induction, delivers it to the useful electrical units connected to its terminals. The static transformer has no moving parts and therefore requires little attention.

Since the losses in transformers are low, consisting only of copper and iron losses, the efficiency is extremely high compared with rotating electrical machines. Furthermore, the coils can be immersed in a tank of high-grade insulating oil for cool operation and at very high voltage, if desired.

Transformer Construction. There are two general types of transformer, distinguished from each other by the manner in which the primary and secondary coils are placed around the laminated steel core. In one type, the so-called *core type*, the coils surround a considerable part of the magnetic circuit. Figure 170 is a photograph of an assembled core type of transformer. Note that the primary and secondary coils are wrapped around the core sides, with the low-voltage coil leads at the top and the high-voltage leads at the bottom. In practice, the primary winding is

270   ELECTRICAL MACHINES—ALTERNATING CURRENT

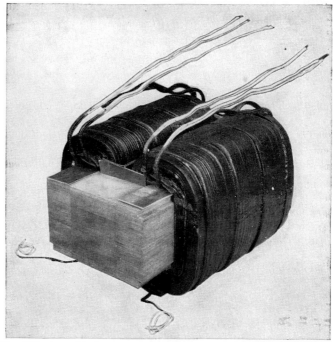

Fig. 170. Assembled core-type transformer.   (*Wagner Electric Corp.*)

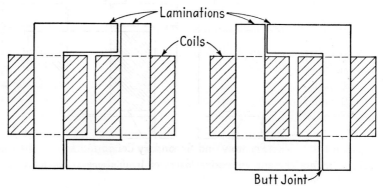

Fig. 171. Sketches showing successive layers of laminations assembled in a core type of transformer.

divided into an even number of separate coils, with half of them placed around one "leg" and the other half around the other "leg"; the same arrangement is used for the secondary winding. In some constructions, as in Fig. 170, the primary and secondary coils for each leg are assembled together to form a single unit, after which the assembly is dipped in an

Fig. 172. Partially assembled shell-type transformer. (*Wagner Electric Corp.*)

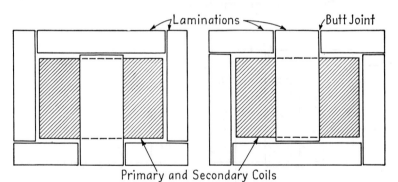

Fig. 173. Sketches showing successive layers of laminations assembled in a shell type of transformer.

insulating varnish and baked. Properly cut laminations are then pushed through the winding openings so that the butt joints of successive layers are interleaved between flat surfaces on both sides. One method of cutting the laminations for such an assembly is shown in Fig. 171. All pieces are exactly similar—die-cut.

In a second type of transformer construction, the so-called *shell type*, the magnetic circuit surrounds a considerable portion of the windings.

Figure 172 is a photograph of a partially assembled shell type of transformer. Note that all the primary and secondary coils are assembled (insulated from each other, of course), after which the entire coil assembly is dipped in an insulating varnish and baked. The properly cut laminations are then pushed through the coil opening and are butted to pieces surrounding the entire coil. As in the core type, the butt joints of successive layers are interleaved between flat surfaces on both sides. This may be clearly seen in the photograph, while Fig. 173 is a sketch showing the manner of assembly. It will be observed that in this construction the laminations are straight pieces, although there are four die-cut sizes.

Laminated cores are built up of good quality magnetic steel containing about 4 per cent silicon, an element that reduces the reluctance of the magnetic circuit as well as the core loss at high flux densities. For 60-cycle service, the individual laminations are usually 0.014 in. in thickness, coated with a varnish insulation of about 0.001-in. thickness; this insulation serves to lessen the effect of eddy currents. For lower frequencies, the laminations may be somewhat thicker.

Another construction uses a ribbon of steel strip for the core. In its manufacture, the steel strip of proper width is coiled around preformed coils. This is done by a special machine that winds the steel spirally through the winding openings and the outside of the coils. The advantages of this construction are (1) a more rigid core, (2) smaller size for a given kilovolt-ampere rating, (3) reduction in strains in the iron, normally set up by clamps, (4) lower iron losses at higher flux densities, and (5) reduction in the cost of manufacture. Transformers of this type are made in the smaller sizes.

The primary and secondary windings of practically all transformers, except those in very small sizes, are subdivided into several coils. These are then assembled around the core in such a manner as to reduce the flux that does not link both primary and secondary. Remembering that transformer action can only exist when flux—*mutual flux*—couples both primary and secondary, it should be clear that any flux that does not do so is, in effect, *leakage* flux. Interleaving primary and secondary coils has the effect of minimizing leakage flux. Another advantage resulting from winding subdivision is that it reduces the voltage per coil. This is particularly important in high-voltage transformers, in which insulation thicknesses make up a considerable part of the construction. In practice, it is customary to subdivide a winding so that the voltage across each coil does not exceed about 5,000 volts.

Transformers are frequently subjected to abnormally high voltage stresses caused by abnormal operating conditions, such as lightning and switching. This is especially true of very high-voltage transformers because a transformer possesses considerable self-inductance and mutual inductance, as well as capacitance between the turns in the coils. Although the capacitance effect is negligible at low-power frequencies, it is considerable during a high-frequency surge. Studies have shown that the first few turns—the end turns—are the ones that are subjected to voltages high enough to break down the insulation. It is for this reason

FIG. 174. Assembled high-voltage core-type transformer before being placed in a steel tank. This transformer has a rating of 6,667 kva at 138,000 volts. (*Wagner Electric Corp.*)

that the end turns are more widely separated and more strongly insulated than are the other turns. This construction is clearly visible in the assembled core-type transformer of Fig. 174.

Rigid bracing is always important in transformer assemblies. Unless this is properly done, the laminations and coils will vibrate with the current changes and thus cause objectionable noise—a humming sound—

and even insulation failure. Figure 175 shows how substantial bracing and rigid construction are carried out in a small shell-type transformer.

Transformers are generally placed inside tightly fitted sheet-metal cases or tanks filled with special insulating oil. This oil has been highly developed so that it serves both to insulate the windings and, through circulation, to keep the windings reasonably cool. When a smooth tank surface does not provide sufficient cooling area, the sides are corrugated or provided with radiators mounted on the sides. While modern transformer oils have, in general, excellent properties, they nevertheless have the disadvantage of tending to absorb moisture. The importance of sealing a transformer case tightly against this tendency to absorb moisture should be emphasized because the addition of only 8 parts of water in 1,000,000 reduces the insulating quality of the oil to a value generally recognized as below standard. Another thing to watch in oil is *sludging*, which is simply the decomposition of oil with continued use. It is caused principally by exposure to oxygen during heating and results in the formation of heavy deposits of dark, heavy matter that will eventually clog the cooling ducts in the transformer. Figure 176 shows a phantom view of a core-type transformer inside its case. This construction is for rural-line service and is equipped with a secondary circuit breaker that protects the transformer from short circuits and overloads, a special deion gap for surge protection, and a fuse link in the high-voltage bushing, the function of which is to disconnect the transformer from the line when there is a short circuit in the winding.

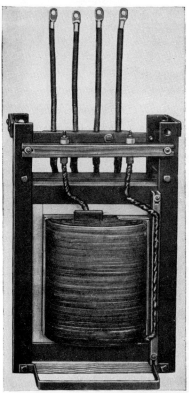

FIG. 175. Small assembled shell-type transformer. (*Wagner Electric Corp.*)

No feature in the construction of a transformer is given more attention and care than the insulating materials because the life of the unit depends, to a very large extent, upon the quality, durability, and handling of these materials. All insulating materials, such as paper, pressboard, cloth, mica, asbestos, and impregnating compounds, are selected on the basis of high quality and their ability to preserve this quality after many years of normal service. The coils are generally wound on forms and afterward dipped in good insulating varnishes and baked in an oven. The insulating compound is capable of providing a solid moisture-resisting surface after proper treatment.

All leads are brought out from the cases through porcelain bushings. There are many designs of these, their size and construction depending

FIG. 176. Phantom view of core-type transformer in case, for rural-line service. Note the many protective features against short circuit, overloads, and winding failure. (*Wagner Electric Corp.*)

upon the voltage at which the transformer winding is operated. In general, they look much like the insulators found on transmission lines, but they are designed primarily to protect the windings from grounds as the leads pass through the case.

Transformer Voltages and the General Transformer Equation. The impedance of the primary winding of a transformer is comparatively high.

If the secondary winding is open-circuited, that is, if there is no energy transfer, the impressed voltage V_P causes a very small current to flow in the primary winding. This no-load current I_N has two functions: (1) it must produce the mutual flux that varies between zero and $\pm \phi_m$, and (2) it must have a component that takes care of the hysteresis and eddy-current power losses—the iron losses—in the iron core. Since the mutual flux changes in magnitude and direction, there will be created an induced emf in the secondary winding, designated by E_S, and an induced emf in

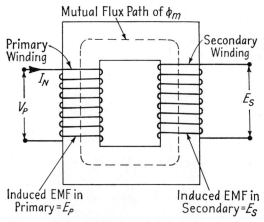

Fig. 177. Elementary diagram of a simple transformer with an open secondary winding.

the primary winding, designated by E_P; note that both induced emfs are created by the *same* mutual flux. Figure 177 is a simple diagram illustrating the elements discussed above. Since the no-load primary current I_N is extremely low, about 2 to 10 per cent of the normal current, there will be practically no voltage drop in the primary winding; under this condition V_P is very nearly equal to E_P. Also, no current flows in the secondary winding; therefore, the secondary terminal voltage is E_S at no load.

If it is assumed that the impressed emf V_P is a sine wave, the mutual flux will also vary sinusoidally. Under this condition, the induced voltages E_P and E_S will also vary as a sine function. Since the average induced voltage is equal to

$$E_{av} = N \times \frac{\phi_m}{t} \times 10^{-8} \quad \text{volts} \tag{34}$$

where E_{av} = average induced emf in coil
N = number of turns in coil
t = time for flux to change by ϕ_m maxwells

it follows that

$$E_{av} = N \times \frac{\phi_m}{1/4f} \times 10^{-8} = 4fN\phi_m \times 10^{-8} \quad \text{volts} \quad (35)$$

where f = frequency, cps

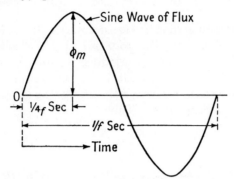

FIG. 178. Sine wave of flux of f cps. Note that the flux changes from zero to ϕ_m in $1/4f$ sec.

In Eq. (35), $1/4f$ is substituted for t because the flux changes from zero to ϕ_m in $1/4f$ sec, as indicated by Fig. 178. Now then, since for a sine wave the effective voltage E is equal to 1.11 times the average voltage E_{av},

$$E = 1.11 \times E_{av} = 4.44fN\phi_m \times 10^{-8} \quad \text{volts} \quad (36)$$

This is the *general transformer equation* and applies equally to the primary and secondary *induced* voltages. Thus

$$E_P = 4.44fN_P\phi_m \times 10^{-8} \quad \text{volts} \quad (36a)$$
and
$$E_S = 4.44fN_S\phi_m \times 10^{-8} \quad \text{volts} \quad (36b)$$

where N_P = number of primary turns
N_S = number of secondary turns

EXAMPLE 1. The 2,300-volt primary winding of a 60-cycle transformer has 4,800 turns. Calculate: (a) the mutual flux ϕ_m; (b) the number of turns in the 230-volt secondary winding.

Solution

(a) $$\phi_m = \frac{2{,}300 \times 10^8}{4.44 \times 60 \times 4{,}800} = 1.8 \times 10^5 \text{ maxwells}$$

(b) $$N_S = \frac{230 \times 10^8}{4.44 \times 60 \times 1.8 \times 10^5} = 480 \text{ turns}$$

EXAMPLE 2. The maximum flux in the core of a 60-cycle transformer that has 1,320 primary turns and 46 secondary turns is 3.76×10^6 maxwells. Calculate the primary and secondary induced voltages.

Solution

$$E_P = 4.44 \times 60 \times 1{,}320 \times 3.76 \times 10^6 \times 10^{-8} = 13{,}200 \text{ volts}$$

$$E_S = 4.44 \times 60 \times 46 \times 3.76 \times 10^6 \times 10^{-8} = 460 \text{ volts}$$

Voltage and Current Ratios in Transformers. Referring to Eqs. (36a) and (36b), it is clear that the *volts per turn* is exactly the same for both the primary and secondary windings because each equals $4.44 \times f \times \phi_m \times 10^{-8}$. This indicates that, in any transformer, *the primary and secondary induced voltages are related to each other by the ratio of the number of primary and secondary turns.* Thus

$$\frac{E_P}{E_S} = \frac{N_P}{N_S} \tag{37}$$

Equation (37) can also be derived mathematically by dividing Eq. (36a) by Eq. (36b).

EXAMPLE 3. The secondary winding of a 4,600/230-volt transformer has 36 turns. How many turns are there in the primary winding?

Solution

$$N_P = \frac{E_P}{E_S} \times N_S = \frac{4{,}600}{230} \times 36 = 20 \times 36 = 720$$

EXAMPLE 4. The volts per turn in a 25-cycle 2,400/230-volt transformer is 8. Calculate: (a) the primary and secondary turns; (b) the maximum flux in the core.

Solution

(a)
$$N_P = \frac{2{,}400}{8} = 300$$

$$N_S = \frac{230}{8} = 29$$

(b)
$$\phi_m = \frac{2{,}400 \times 10^8}{4.44 \times 25 \times 300} = 7.21 \times 10^6 \text{ maxwells}$$

Static transformers are extremely efficient because the only losses are those that occur in the copper windings (I^2R losses) and in the iron (hysteresis and eddy-current losses); there are no losses resulting from rotation, such as are present in rotating machines. If the input to a transformer is *assumed* to equal the output of a transformer (efficiency = 100 per cent) and the voltage drops are considered negligible, then

$$E_P \times I_P \times PF_P = E_S \times I_S \times PF_S$$

TRANSFORMERS

It is generally true that the secondary load power factor PF_S is practically equal to the primary input power factor PF_P. Therefore

$$E_P \times I_P = E_S \times I_S$$

This shows that
$$\frac{E_P}{E_S} = \frac{I_S}{I_P} \qquad (38)$$

and
$$\frac{N_P}{N_S} = \frac{I_S}{I_P} \qquad (39)$$

That is, the *voltage ratio* $E_P:E_S$ and the *turn ratio* $N_P:N_S$ are both proportional to the inverse current ratio $I_S:I_P$.

EXAMPLE 5. The secondary load current of a 2,300/115-volt transformer is 46 amp. Calculate the primary current.

Solution

$$I_P = \frac{E_S}{E_P} \times I_S = \frac{115}{2,300} \times 46 = 2.3 \text{ amp}$$

EXAMPLE 6. The primary and secondary currents of a transformer were measured and found to be 3.8 and 152 amp, respectively. If the secondary load voltage is 116 volts, what is the primary emf?

Solution

$$E_P = \frac{I_S}{I_P} \times E_S = \frac{152}{3.8} \times 116 = 4,640 \text{ volts}$$

Ratio of Transformation. The ratio of primary to secondary turns $N_P:N_S$, which equals the ratio of primary to secondary induced voltages $E_P:E_S$ [Eq. (37)], indicates how much the primary voltage is lowered or raised. The *turn ratio*, or the *induced-voltage ratio*, is called the *ratio of transformation*, and is represented by the symbol a. Thus

$$a = \frac{N_P}{N_S} = \frac{E_P}{E_S} \qquad (40)$$

Because the primary input voltage V_P and the secondary load voltage V_S are very nearly equal to the respective induced voltages, the terminal-voltage ratio $V_P:V_S$ is frequently called the *ratio of transformation*. The true ratio of transformation, Eq. (40), is constant, while the $V_P:V_S$ ratio varies about 1 to 8 per cent, depending upon the load and its power factor.

When the primary impressed voltage V_P is reduced to a lower secondary voltage V_S, the transformer is said to be a *step-down* transformer; conversely, if the voltage is raised, it is called a *step-up* transformer. In a step-down transformer, the ratio of transformation is greater than unity, while in a step-up transformer, a is less than unity. In practice, however, it is customary to specify the ratio of transformation a as a number *greater* than unity; this is done for convenience, but to eliminate the possibility of misunderstanding, it is well to add the term "step-down" or "step-up." Thus, a 2,300/230-volt transformer would be said to have a ratio of transformation of 10:1, step-down; a 13,200/66,000-volt transformer would be said to have a ratio of 5:1, step-up.

To determine the ratio of transformation of a transformer, it is necessary merely to measure the primary and secondary voltages at no load, that is, with the secondary open. For high-voltage transformers, this may involve the use of potential transformers (to be discussed latter), which

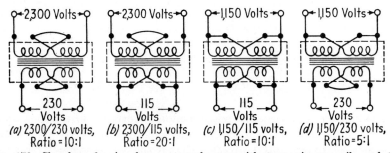

Fig. 179. Sketches showing how a transformer with two primary coils and two secondary coils, having a rating of 2,300–1,150/230–115 volts, may be connected for four voltage combinations.

are used in conjunction with standard low-range voltmeters to measure high values of emf. Another scheme is to impress a comparatively low voltage across the high side and then measure the voltages on both sides in the usual way. For example, in a 2,300/230-volt transformer, connect a 230-volt source to the 2,300-volt side, measuring this potential accurately with a standard 300-volt voltmeter; measure the secondary voltage with a voltmeter having a 30- or 75-volt range. The readings will, of course, be one-tenth of the operating voltages, but the ratio of primary to secondary voltage will be correct.

A transformer is frequently constructed with both the primary and secondary divided into halves, with four leads brought out from each side. The purpose of this construction is twofold: (1) it makes it possible to connect the primary and secondary coils either in series or in parallel, thereby extending its usefulness to four available voltage combinations, and (2) it makes it possible to connect the secondary coils to provide three-wire service. Figure 179 illustrates the four possible voltage combinations of a transformer having a rating of 2,300–1,150/230–115 volts. Such a rating means that the primary coils may be connected in series for 2,300-volt operation or in parallel when the source voltage is 1,150; the secondary coils may be connected in series for 230-volt service or in parallel for 115-volt service. Figure 180 illustrates how the same transformer may be connected for three-wire service at 230 and 115 volts when the source voltage is 2,300 or 1,150. In all the illustrations, it should be noted that the center leads for each side are crossed over before they are brought out through the transformer bushings. This is done to facilitate the parallel connection; the wires can be joined on the outside of the transformer without having a crossover.

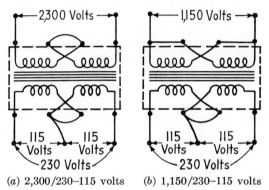

(a) 2,300/230–115 volts   (b) 1,150/230–115 volts

FIG. 180. Sketches showing how a transformer with two primary and two secondary coils, having a rating of 2,300–1,150/230–115 volts, may be connected for three-wire service at 230 and 115 volts.

Tapping a Transformer. When a transformer is used for distribution service, that is, when the secondary is connected directly to the "customer" load, it is called a *distribution transformer*. Transformers for such service are distinguished from so-called *power transformers*, which are employed in high-voltage transmission systems by public service companies for the transmission and distribution of comparatively large amounts of power. Power transformers do not come in contact directly with the actual "customer" load.

In relatively long distribution circuits, the voltage drop in the line wires increases with the distance from the primary source of voltage. This means that a distribution transformer connected at the beginning of a line will "receive" a higher voltage than one many miles away. Assuming that two similar distribution transformers are connected to the same line, but some distance from each other, the secondary voltages in the two cases will not be the same; the load voltage may be too high at the beginning of the line and too low at the far end. To overcome this obvious disadvantage, so that the proper voltage service may be provided for all loads, transformers are *tapped* on the primary side, the high side, in such a way that the ratio of transformation may be changed to suit the actual primary voltage at the point where the transformer is installed. Taps in the

282   ELECTRICAL MACHINES—ALTERNATING CURRENT

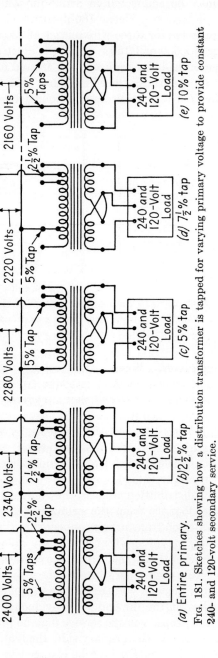

FIG. 181. Sketches showing how a distribution transformer is tapped for varying primary voltage to provide constant 240- and 120-volt secondary service.

primary are commonly made at convenient points from the ends of the winding and are specified in per cent of the entire winding. A common arrangement is to provide 2½-, 5-, 7½-, and 10-per cent taps. For a 2,400-volt primary, this would mean that the *same* secondary voltage could be obtained if the line potentials are 2,340, 2,280, 2,220, and 2,160 volts. Figure 181 shows how the same three-wire 240- and 120-volt service can be provided for variations in primary voltages up to 10 per cent.

EXAMPLE 7. A 4,800/240-volt transformer is provided with the following taps on the primary side: a 1¼ and a 5 per cent tap from one end, and a 2½ and a 5 per cent tap from the other. Calculate the primary voltages that may be used for constant 240-volt service and the ratio of transformation in each case. NOTE: The taps are made in the same way as in Fig. 181 with the exception that an additional 1¼ per cent tap is taken from the *left* side of the primary.

Solution

4,800 − 0 = 4,800 volts $a = 20{:}1$
4,800 − (0.0125 × 4,800) = 4,800 − 60 = 4,740 volts $a = 19.75{:}1$
4,800 − (0.025 × 4,800) = 4,800 − 120 = 4,680 volts $a = 19.5{:}1$
4,800 − (0.0375 × 4,800) = 4,800 − 180 = 4,620 volts $a = 19.25{:}1$
4,800 − (0.05 × 4,800) = 4,800 − 240 = 4,560 volts $a = 19{:}1$
4,800 − (0.0625 × 4,800) = 4,800 − 300 = 4,500 volts $a = 18.75{:}1$
4,800 − (0.075 × 4,800) = 4,800 − 360 = 4,440 volts $a = 18.5{:}1$
4,800 − (0.10 × 4,800) = 4,800 − 480 = 4,320 volts $a = 18{:}1$

Figure 182 shows a cutaway view of a typical pole-type distribution transformer in which tap changing is accomplished by turning a rotary ratio adjuster, visible at the rear left directly behind the high-voltage bushing. A pointer on the adjuster clearly indicates the ratio of the transformer.

Loading a Transformer. With the secondary of a transformer open-circuited, the primary is a simple high impedance taking a very low current that has two purposes: (1) to create the proper mutual flux ϕ_m and (2) to take care of the hysteresis and eddy-current power losses in the iron core. Since the secondary voltage depends upon the mutual flux, it should be clear that ϕ_m must not change appreciably if E_s is to remain substantially constant under normal conditions of loading. Furthermore, every value of power delivered by the secondary $V_S I_S \times PF_S$ must be matched by a corresponding power input to the primary $V_P I_P \times PF_P$ (assuming negligible power loss) if the transformer is to perform its function of transferring electrical energy from one electric circuit—the primary—to another electric circuit—the secondary. In other words, a transformer must perform under load, so that two conditions are fulfilled:

1. The mutual flux ϕ_m must remain practically constant.

2. $V_P I_P$ must equal $V_S I_S$, assuming that the primary and secondary power factors are equal.

FIG. 182. Cutaway view of a typical pole-type distribution transformer. Clearly visible are the four low-voltage leads (*in front*) and the rotary-type tap changer (*rear left*). A pointer on the tap changer clearly indicates the ratio connected. (*Wagner Electric Corp.*)

The following discussion should clarify the foregoing reasoning. Consider Fig. 183, which represents a simple transformer. With the secondary switch S open, the primary takes a very small current I_N because the induced counter voltage E_P is only slightly less than the impressed emf V_P. A major component of the current I_N creates the mutual flux ϕ; note the direction of the increasing flux in the core for the indicated current direction. If the switch S is now closed, a current I_S will flow to the load, the value and power-factor angle of which will depend upon the character of the load. However, the *direction* in which the current I_S will flow in the secondary winding at every instant will, by Lenz's law, be such as to *oppose any change in the flux*. Since the flux ϕ is assumed to be increasing, the current I_S must have the direction shown *if it is to oppose an increase in flux*. Thus the mmf represented by $N_S I_S$ actually tends to reduce the flux ϕ. Now then, if the flux ϕ is reduced, the induced emf in the primary is also reduced; the result is that the net voltage in the primary, represented by the *difference* between the impressed voltage V_P and

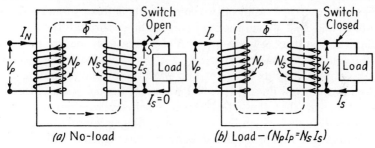

Fig. 183. Sketches illustrating a transformer operating without and with load.

the induced counter voltage E_P, is increased and with it the value of the primary current I_P. The fact that the primary current increases means that the two conditions, summarized above, are fulfilled because (1) the power input increases to match the power output and (2) the primary ampere-turns increases to offset the tendency on the part of the secondary ampere-turns to reduce the flux.

The above reactions occur instantaneously upon the application of load on a transformer. Moreover, the mutual flux ϕ_m drops very slightly between no load and full load (about 1 to 3 per cent), a necessary requirement if E_P is to drop sufficiently to permit an increase in I_P.

Since the no-load current is a negligible part of the primary current I_P when the transformer is loaded, it is correct to assume that the primary ampere-turns equal the secondary ampere-turns because it is only under this condition that the mutual flux can remain practically constant. Thus $N_P I_P = N_S I_S$, which leads to Eq. (39). In other words, when the transformer steps the voltage down, it steps the current up in exactly the same proportion. A welding transformer, for example, is a voltage step-down transformer, because the secondary must deliver a high load current. Or to put it another way, the primary and secondary currents are *inversely proportional* to the primary and secondary voltages as well as to the primary and secondary turns.

Regulation Calculations Using Voltage Values. The transformer serves as a source of supply when its secondary terminals are connected to a load; in this respect, it functions exactly like an alternator delivering a load. The load current I_S and its counterpart in the primary I_P [which equals $(N_S/N_P) \times I_S$] involve two distinct kinds of voltage drop within the transformer, as a result of which the load voltage V_S changes when the load changes. The voltage drops are quite similar in some respects to those discussed in Chap. 7 on alternators, although their magnitudes are, in general, much less. The first of these is the voltage drop caused by the resistance of the primary and secondary windings; they are simple to understand and calculate. The second voltage drop is somewhat more complex because it is a result of two components of flux caused by the load currents I_S and I_P. When the load current flows through the secondary winding, the resulting ampere-turns $(N_S I_S)$ create a separate flux, apart from the mutual flux ϕ_m produced by I_N, that links with the secondary winding only. This flux does not link with the primary winding and is therefore not mutual flux; it is called *secondary leakage flux*. Moreover, the load current that flows through the primary winding creates still another separate flux which links with the primary winding *only*. Since this flux does not link with the secondary winding, it is called *primary leakage flux*. The secondary leakage flux gives rise to an induced voltage that is not counterbalanced by an equivalent induced voltage in the primary; also, the primary leakage flux gives rise to an induced voltage that is not counterbalanced by an equivalent induced voltage in the secondary. As a result of these actions, the two induced voltages, created by the leakage fluxes, behave exactly like voltage drops. It follows, therefore, that the impressed voltage V_P suffers two kinds of voltage drop before the secondary winding delivers its load voltage V_S to the load. These are (1) the resistance drops in primary and secondary and (2) the voltage drops in primary and secondary caused by leakage fluxes. The latter are called *leakage-reactance voltage drops*.

Since the voltage drops are all directly proportional to the load current, I_S in the secondary and I_P in the primary, it should be clear that at no load there will be no voltage drop in either winding. Therefore, if a transformer delivers rated load at rated secondary terminal voltage, that voltage will change if the load is removed. And by definition, this voltage *change* between full load and no load divided by the full-load voltage is called the *regulation*. As a per cent, it may be written

$$\text{Per cent regulation} = \frac{V_{NL} - V_{FL}}{V_{FL}} \times 100 \qquad (41)$$

where the no-load and full-load voltages V_{NL} and V_{FL} are those measured at the secondary terminals.

EXAMPLE 8. Calculate the per cent regulation of a 2,300/115-volt transformer whose no-load voltage was measured and found to be 118 volts.

Solution

$$\text{Per cent regulation} = \frac{118 - 115}{115} \times 100 = \frac{300}{115} = 2.61$$

EXAMPLE 9. The per cent regulation of a 4,800/240-volt distribution transformer is 3.33 per cent. Calculate the voltage to which the secondary voltage will rise when full load is removed.

Solution

$$3.33 = \frac{V_{NL} - 240}{240} \times 100$$

$$V_{NL} = \frac{3.33 \times 240}{100} + 240$$

$$V_{NL} = 248 \text{ volts}$$

Leakage Reactance. Whenever a voltage is induced in the coil of a transformer winding by a changing flux through it, that induced voltage

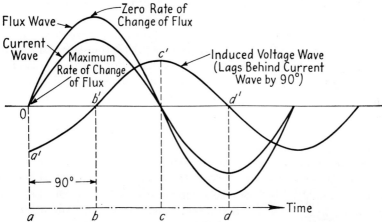

FIG. 184. Sine waves of current, flux, and the resulting induced voltage in a coil of wire.

lags behind the flux by exactly 90 electrical degrees. Also, since the flux and current are presumably in phase with each other, it follows that the induced voltage lags behind the current (that creates the flux) by 90 electrical degrees. The reason for this relationship may be understood by a study of Fig. 184.

Assume a sine wave of primary current that excites the transformer and, in phase with it, a sine wave of flux. At time a the flux is zero and is increasing at a *maximum* rate in a *positive* direction. Therefore the induced voltage in the coil will be a *maximum* and will, by Lenz's law, be in a *negative* direction, so that it will oppose an increase in flux; point a' on the voltage wave represents the induced emf in the coil at instant of time a. At time b the rate of change of flux is zero. Under this condition, the induced emf will be *zero;* point b' on the voltage wave represents the voltage at time b. At time c the flux is again zero and increasing at a maximum rate, but in a *negative* direction. At this instant, the induced voltage in the coil will be a *maximum*, but, also by *Lenz's* law, in a *positive* direction; point c' on the voltage wave represents the voltage at time c. Finally, at time d the rate of change of flux is again zero. Again the induced emf will be *zero;* point d' on the voltage wave represents the voltage at time d.

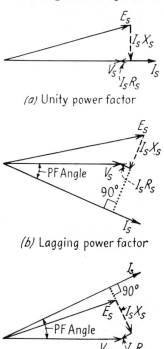

(a) Unity power factor

(b) Lagging power factor

(c) Leading power factor

FIG. 185. Phasor diagrams showing how the resistance and leakage-reactance drops are subtracted from the induced secondary voltage to yield the secondary terminal voltage.

Note particularly that the induced voltage wave lags behind the current wave by 90 electrical degrees.

Assume next that the secondary current creates a leakage flux, i.e., a flux that is in phase with the current and links with its own turns but *not* with the primary winding. Since this flux is varying sinusoidally, it will induce a sinusoidal voltage, which, as previously explained, will lag behind the leakage flux (or current) that creates it by 90 electrical degrees; its effective value will therefore react with the secondary induced emf E_S and attempt to change the terminal voltage V_S. In other words, the secondary winding, the seat of the transformer power output, must develop an induced emf that will not only supply a terminal voltage V_S for the load, but also take care of two internal voltage drops, namely, the voltage drop due to resistance—the $I_S R_S$ drop—and, as indicated above, the *leakage-reactance voltage drop*—the $I_S X_S$ drop—that results from the presence of leakage flux. The three components of secondary voltage V_S, $I_S R_S$, and $I_S X_S$ are, of course, added vectorially to yield the developed secondary emf, although, depending upon the magnitude and character

TRANSFORMERS

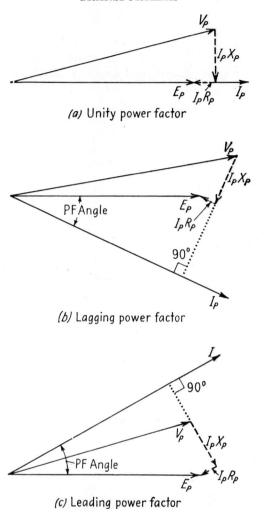

Fig. 186. Phasor diagrams showing how the resistance and leakage-reactance drops are subtracted from the impressed primary voltage to yield the primary induced voltage.

(lagging or leading) of the load power factor, the leakage-reactance drop may attempt to reduce or raise the terminal voltage.

The simple phasor diagrams of Fig. 185 show how the voltage V_S results after the leakage-reactance voltage drop $I_S X_S$ and the resistance voltage drop $I_S R_S$ are subtracted vectorially from the induced emf E_S for three general types of power-factor load.

If the same reasoning is applied to the primary side of the transformer, it should be clear that E_P results only after the resistance drop $I_P R_P$

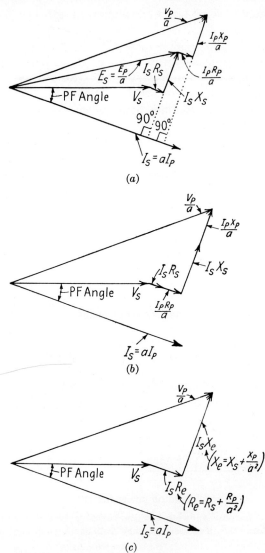

Fig. 187. Phasor diagrams for a lagging power-factor load, showing how the complex diagram can be greatly simplified by using *equivalent* resistance and reactance values.

and the leakage-reactance drop $I_P X_P$ are vectorially subtracted from V_P. This is shown in Fig. 186 for unity, lagging, and leading power factors.

Equivalent Resistance, Reactance, and Impedance. When regulation calculations are made for transformers, it is convenient to combine the resistance and reactance drops that actually occur on the primary and secondary sides into a single value of IR and a single value of IX. Obviously, this cannot be done by simple numerical additions for each component, because one set of voltage drops occurs where the voltage is high and the other set occurs where the voltage is low. One method of simplifying the calculations is to make use of the idea that a transformer having a ratio of transformation a can be converted into an *equivalent transformer* having a ratio of 1:1. When this conversion is made, primary and secondary IR drops can be added directly and primary and secondary IX drops can be added directly.

Consider Fig. 187a, which combines the phasor diagrams of Figs. 185b and 186b into a single diagram, but which also *reduces the primary voltages by dividing them by a*. (A step-down transformer is assumed for convenience.) Such a division by a converts the primary values into secondary terms and is equivalent to considering the transformer as having a ratio of transformation of 1:1. The next step is to alter the *arrangement* of the various voltages, but not the initial and final values. This is done in Fig. 187b. Now then, since $I_P = I_S/a$, it is possible to make the following substitutions:

$$(I_S R_S) + \left(\frac{I_P R_P}{a}\right) = (I_S R_S) + \left(\frac{I_S}{a} \times \frac{R_P}{a}\right) = I_S \left(R_S + \frac{R_P}{a^2}\right)$$

and

$$(I_S X_S) + \left(\frac{I_P X_P}{a}\right) = (I_S X_S) + \left(\frac{I_S}{a} \times \frac{X_P}{a}\right) = I_S \left(X_S + \frac{X_P}{a^2}\right)$$

The ohmic values in the final brackets of the foregoing equations are called the *equivalent resistance R_e* and *equivalent reactance X_e in terms of the secondary side*. Thus

$$R_e = R_S + \frac{R_P}{a^2} \quad \text{in secondary terms} \tag{42}$$

and

$$X_e = X_S + \frac{X_P}{a^2} \tag{43}$$

This analysis leads to the simplified phasor diagram of Fig. 187c. Note that it combines resistance and reactance drops into single phasors so that regulation calculations may be made more easily.

In the same manner, it is possible to show that the *equivalent resistance R_e* and the *equivalent reactance X_e* can be expressed *in terms of the primary side* as follows:

$$R_e = a^2 R_S + R_P \quad \text{in primary terms} \tag{44}$$

$$X_e = a^2 X_S + X_P \tag{45}$$

It will subsequently be shown how the equivalent values of R and X can be determined by the performance of simple tests.

Since resistance and reactance have properties that are represented 90° apart, their *vector summation* bears the same relation to the individual parts that the hypotenuse of a right-angle triangle bears to the two remaining sides. The term generally used for the vector sum of R and X is impedance Z. For the transformer, therefore,

$$Z_e = \sqrt{R_e^2 + X_e^2} \tag{46}$$

where Z_e = equivalent impedance of transformer

Equation (46) will be given in secondary terms if R_e and X_e are calculated by Eqs. (42) and (43); Eq. (46) will be given in primary terms if R_e and X_e are calculated by Eqs. (44) and (45).

EXAMPLE 10. A 25-kva 2,300/230-volt distribution transformer has the following resistance and leakage-reactance values: $R_P = 0.8$; $X_P = 3.2$; $R_S = 0.009$; $X_S = 0.03$. Calculate the equivalent values of resistance, reactance, and impedance: (*a*) in secondary terms; (*b*) in primary terms.

Solution

(*a*) Ratio of transformation $a = \dfrac{2{,}300}{230} = 10$

$$R_e = 0.009 + \frac{0.8}{100} = 0.017$$
$$X_e = 0.03 + \frac{3.2}{100} = 0.062 \quad \Bigg\} \text{ in secondary terms}$$
$$Z_e = \sqrt{(0.017)^2 + (0.062)^2} = 0.0642$$

(*b*)
$$R_e = (100 \times 0.009) + 0.8 = 1.7$$
$$X_e = (100 \times 0.03) + 3.2 = 6.2 \quad \Bigg\} \text{ in primary terms}$$
$$Z_e = \sqrt{(1.7)^2 + (6.2)^2} = 6.42$$

EXAMPLE 11. For the transformer of Example 10, calculate the equivalent resistance and reactance voltage drops for a secondary load current of 109 amp: (*a*) in secondary terms; (*b*) in primary terms.

Solution

(*a*) $I_S R_e = 109 \times 0.017 = 1.85$ volts $\Big\}$ in secondary terms
$I_S X_e = 109 \times 0.062 = 6.75$ volts

(*b*) $I_P R_e = \dfrac{109}{10} \times 1.7 = 18.5$ volts
$I_P X_e = \dfrac{109}{10} \times 6.2 = 67.5$ volts $\Bigg\}$ in primary terms

EXAMPLE 12. Using the data of Examples 10 and 11, calculate the per cent regulation: (a) for unity power factor; (b) for a lagging power factor of 0.8; (c) for a leading power factor of 0.866.

Solution

Using the same procedure as given for alternators in Example 8 of Chap. 7,

(a) $V_P = \sqrt{(2{,}300 + 18.5)^2 + (67.5)^2} = 2{,}320$

Per cent regulation (power factor = 1) = $\dfrac{2{,}320 - 2{,}300}{2{,}300} \times 100 = 0.87$

(b) $V_P = \sqrt{[(2{,}300 \times 0.8) + 18.5]^2 + [(2{,}300 \times 0.6) + 67.5]^2} = 2{,}360$

Per cent regulation (power factor = 0.8 lag) = $\dfrac{2{,}360 - 2{,}300}{2{,}300} \times 100$
$= 2.61$

(c) $V_P = \sqrt{[(2{,}300 \times 0.866) + 18.5]^2 + [(2{,}300 \times 0.5) - 67.5]^2}$
$= 2{,}280$

Per cent regulation (power factor = 0.866 lead) = $\dfrac{2{,}280 - 2{,}300}{2{,}300} \times 100$
$= -0.87$

Equivalent Circuit of a Transformer. The discussion of transformer performance under load has neglected the component of no-load current that flows in the primary winding only. This neglect is permissible in standard distribution and power transformers because I_N is, at most, but a small per cent of the load current; when included in the calculations, it adds very little to the accuracy of the final result. Neglecting it, however, simplifies the calculation procedure.

With this simplification, it is also possible to represent a transformer as an ordinary *series* electric circuit that has three elements: (1) the equivalent resistance, (2) the equivalent leakage reactance, and (3) the load. This is shown in Fig. 188, in which all quantities are represented in secondary terms. Note that the transformer, as an electric circuit, merely acts like an impedance voltage drop. That is, as the current passes through the transformer, there is a drop in voltage, so that the load voltage V_S is vectorially less than V_P/a, by the equivalent resistance and reactance voltage drops. To be sure, this drop depends not only upon the actual load current, but also upon the power factor of the load, as the previous example clearly demonstrates.

The Short-circuit Test. In order to determine experimentally the values of the equivalent resistance, impedance, and reactance, the so-called *short-circuit test* is generally performed. This test is an accurate attempt to make the windings carry rated currents without requiring that the transformer deliver a load; the power input to the transformer will then be extremely low. In this way, it is possible to simulate the pattern of leakage fluxes in the primary and secondary because the latter depend upon the load currents in the two windings. Since the short-circuit test is made with one of the windings short-circuited, the impressed voltage on the other winding must be very low, usually about 5 per cent of rated value. In the equivalent circuit diagram of Fig. 188, this implies that the load terminals are short-circuited, so that $V_S = 0$; under this condition the impressed emf V_P/a must merely overcome the full-load $I_S R_e$ and $I_S X_e$ voltage drops. Or, referring to the phasor diagram of Fig.

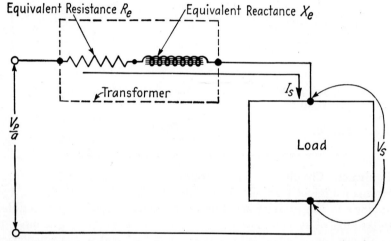

Fig. 188. Equivalent circuit of a transformer neglecting the no-load current.

187c, the voltage V_P/a must be the hypotenuse of the triangle formed by $I_S R_e$ and $I_S X_e$, since the phasor V_S is zero. Now then, when the voltage to which a transformer winding is connected is as low as 5 per cent of the rated voltage, the *mutual flux will be of the same order of magnitude*. And since the core loss is nearly proportional to the *square* of the mutual flux, it is seen that this loss will be practically zero. This means that a wattmeter, used to measure the power input under short circuit, will register only the copper loss because there is no power output or core loss.

Figure 189 represents the circuit diagram generally used to perform a short-circuit test upon a standard distribution transformer. Note that the low side is short-circuited while the instruments, i.e., voltmeter, ammeter, and wattmeter, are inserted in the high side. A rheostat in the input side is used to adjust the primary current to rated value; when rated current flows in the primary winding, rated current will also flow in the secondary winding by transformer action. Also, low-range instruments can be used because (1) the high side always carries the lower of the two currents, (2) the impressed voltage is about 5 per cent of rated value, and (3) the wattmeter registers only the copper loss in both windings—less than 3 per cent of the rated output.

From the data of watts, amperes, and volts obtained from this test, the values of equivalent resistance, equivalent impedance, and equivalent

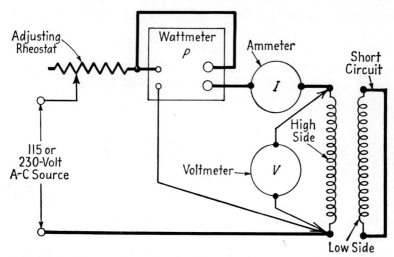

FIG. 189. Wiring diagram for performing short-circuit test upon a standard distribution transformer.

reactance can then be calculated, all in terms of the high side, using the following equations:

$$R_e = \frac{P_{SC}}{I_{SC}^2} \qquad (47)$$

$$Z_e = \frac{E_{SC}}{I_{SC}} \qquad (48)$$

$$X_e = \sqrt{Z_e^2 - R_e^2} \qquad (49)$$

where P_{SC}, I_{SC}, and E_{SC} are *short-circuit watts, amperes,* and *volts*, respectively.

The procedure for calculating the per cent regulation can then be carried on in the manner previously described.

EXAMPLE 13. The following data were obtained when a short-circuit test was performed upon a 100-kva 2,400/240-volt distribution transformer: $E_{SC} = 72$ volts; $I_{SC} = 41.6$ amp; $P_{SC} = 1,180$ watts. All instruments were placed on the high side, and the low side was short-circuited. Calculate: (*a*) the equivalent resistance, impedance, and reactance; (*b*) the per cent regulation at a power factor of 0.75 lagging.

Solution

(a)
$$R_e = \frac{1{,}180}{(41.6)^2} = 0.682 \text{ ohm}$$

$$Z_e = \frac{72}{41.6} = 1.73 \text{ ohms}$$

$$X_e = \sqrt{(1.73)^2 - (0.682)^2} = 1.59 \text{ ohms}$$

(b) Rated current $= \dfrac{100{,}000}{2{,}400} = 41.6$ amp

IR_e drop $= 41.6 \times 0.682 = 28.4$ volts

IX_e drop $= 41.6 \times 1.59 = 66.2$ volts

$$V_P = \sqrt{[(2{,}400 \times 0.75) + 28.4]^2 + [(2{,}400 \times 0.66) + 66.2]^2}$$

$= 2{,}460$ volts

Per cent regulation $= \dfrac{2{,}460 - 2{,}400}{2{,}400} \times 100 = 2.5$

As the foregoing discussion and the illustrative example showed, the full-load copper losses in the primary and secondary windings resulted from the circulation of *rated* current in these windings, and the losses indicated are obviously equal to the sum of $I_P{}^2R_P$ and $I_S{}^2R_S$. Note particularly that the copper losses vary as the *square* of the corresponding primary and secondary currents, and this means, of course, that they drop or rise sharply below or above the full-load value when the load is, respectively, higher or lower than rated kva. For example, when the secondary delivers a load equal to one-half of rated kva, the copper losses become one-fourth of the full-load value as found in the short-circuit test; at 1½ times rated kva they become 2¼ times as much as the full-load copper losses.

EXAMPLE 14. For the transformer of Example 13, calculate the copper losses when the load is (a) 125 kva, (b) 75 kva, (c) 85 kw at a power factor of 0.772.

Solution

(a) $\left(\dfrac{125}{100}\right)^2 \times 1{,}180 = 1{,}845$ watts

(b) $\left(\dfrac{75}{100}\right)^2 \times 1{,}180 = 663$ watts

(c) $\left(\dfrac{85/0.772}{100}\right)^2 \times 1{,}180 = 1{,}430$ watts

EXAMPLE 15. For what kilowatt load, at a power factor of 0.71, will the copper losses in the transformer of Example 13 be 1,500 watts?

Solution

$$1{,}500 = 1{,}180 \times \left(\frac{\text{kw}/0.71}{100}\right)^2$$

$$\text{kw} = 100 \times 0.71 \sqrt{\frac{1{,}500}{1{,}180}} = 80$$

The Open-circuit Test. When one side of a transformer is left open-circuited and the other side is connected to a source of alternating current whose voltage is the rated value, the current will be extremely low—about 2 to 10 per cent of rated load current. This no-load current I_N has two components, one producing the normal mutual flux ϕ_m and the other taking care of the hysteresis and eddy-current losses in the core.

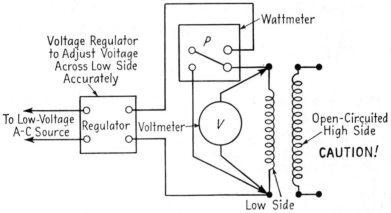

FIG. 190. Wiring diagram for performing open-circuit test upon a standard distribution transformer.

Since the current is so very low, the copper loss in the one winding in which I_N flows will be negligible. For all practical purposes, therefore, a wattmeter (used to measure the power input under the condition of open circuit) will register the *core loss*.

Actually, it does not matter whether the test is performed with instruments placed on the high side or on the low side, but it is obviously safer when it is done at the lower voltage. Figure 190 shows a wiring diagram of connections for the *no-load test* with instruments inserted on the low side. When the circuit is energized, great care should be taken not to come in contact with the high-voltage side. Before the wattmeter reading is recorded, it is essential that the impressed voltage be adjusted carefully to rated value because it is only at this value of **emf** that the proper core loss is measured.

As the magnetic flux alternates periodically, from a maximum value

$+\phi_m$ in one direction to a maximum value $-\phi_m$ in the other direction, two components of power loss are developed in the iron. They are (1) the *hysteresis loss*, which is purely magnetic, and results because the tiny magnetic particles produce a kind of molecular friction as they tend to change alignment with the rapid reversals of alternating current, and (2) the *eddy-current loss*, which is electromagnetic in character and is caused by the flow of currents in the iron in exactly the same way as in the transformer windings. Since these losses depend only upon the maximum value of the flux (or flux density), they are, for a given transformer, constant in magnitude and completely independent of the load. The hysteresis loss can, however, be minimized by employing high-quality magnetic steel, while the eddy-current loss is controlled by laminating, i.e., "slicing" the core.

Theory and experiment have shown that the hysteresis loss is

$$P_h = k_h f \mathcal{B}_m{}^{1.6}$$

while the eddy-current loss is

$$P_e = k_e f^2 \mathcal{B}_m{}^2$$

where k_h = proportionality constant depending upon volume and quality of steel

k_e = proportionality constant depending upon volume of the core, thickness of the laminations, and resistivity of steel

$\mathcal{B}_m$ = maximum flux density in core

f = frequency

But, by Eq. (36), $E = 4.44 f N \phi_m \times 10^{-8}$

Therefore, $$\phi_m = \mathcal{B}_m \times A = \frac{E \times 10^8}{4.44 f N}$$

and $$\mathcal{B}_m = \left(\frac{10^8}{4.44 N A}\right) \frac{E}{f}$$

where A = area of core

For a given transformer, N and A are constant; therefore the terms in the bracket may be set equal to some constant k. Thus

$$\mathcal{B}_m = k \frac{E}{f}$$

Substituting this value of flux density in the foregoing core-loss equations gives

$$P_h = k_h f k \left(\frac{E^{1.6}}{f^{1.6}}\right) = k_1 \left(\frac{E^{1.6}}{f^{0.6}}\right) \tag{50}$$

and $$P_e = k_e f^2 k \left(\frac{E^2}{f^2}\right) = k_2 E^2 \tag{51}$$

where k_1 and k_2 are simply proportionality constants. Since the total core loss P_c is the sum of the hysteresis and eddy-current losses, it follows that

$$P_c = k_1 \frac{E^{1.6}}{f^{0.6}} + k_2 E^2 \qquad (52)$$

It is especially important to note that, for a given impressed voltage E, the eddy-current loss is independent of the frequency; this loss is, however, directly proportional to the *square* of the impressed emf. The hysteresis loss does, on the other hand, depend upon both the applied voltage and its frequency.

EXAMPLE 16. A 4,400-volt 60-cycle transformer has core loss of 840 watts, of which one-third is eddy-current loss. Determine the core loss when the transformer is connected (a) to a 4,600-volt 60-cycle source, (b) to a 4,400-volt 50-cycle source, and (c) to a 4,600-volt 50-cycle source.

Solution

$$P_e = \frac{1}{3} \times 840 = 280 \text{ watts} \qquad P_h = 840 - 280 = 560 \text{ watts}$$

(a) $P_c = 560 \times \left(\dfrac{4{,}600}{4{,}400}\right)^{1.6} + 280 \times \left(\dfrac{4{,}600}{4{,}400}\right)^2$

$\qquad = (560 \times 1.073) + (280 \times 1.093) \quad = 601 + 306 = 907 \text{ watts}$

(b) $P_c = 560 \times \left(\dfrac{60}{50}\right)^{0.6} + 280 = (560 \times 1.121) + 280$

$\qquad\qquad\qquad\qquad\qquad\qquad\qquad = 624 + 280 = 904 \text{ watts}$

(c) $P_c = 560 \times \left[\left(\dfrac{4{,}600}{4{,}400}\right)^{1.6}\left(\dfrac{60}{50}\right)^{0.6}\right] + \left[280 \times \left(\dfrac{4{,}600}{4{,}400}\right)^2\right]$

$\qquad = (560 \times 1.073 \times 1.121) + (280 + 1.093)$

$\qquad\qquad\qquad\qquad\qquad\qquad = 674 + 306 = 980 \text{ watts}$

Regulation Calculations Using Short-circuit Data. When the short-circuit test is performed upon a transformer, it will be recalled that the impressed voltage E_{SC} is adjusted to permit *rated* current to circulate in the primary and secondary windings. Since there is no output voltage, it is obvious that E_{SC}, used completely to overcome the equivalent impedance of the transformer, is actually the impedance voltage drop at full load. As a per cent of the rated transformer voltage, in terms of the side on which the test is performed, it is called the *per cent IZ drop* and may be expressed as

$$\%IZ = \frac{E_{SC}}{E_{\text{rated}}} \times 100$$

300   ELECTRICAL MACHINES—ALTERNATING CURRENT

Also, the power input to the short-circuited transformer is used entirely to supply the copper losses, i.e., $P_{SC} = I_{FL}{}^2 R_e$. As a per cent of the volt-ampere rating of the transformer, this would be $[(I_{FL}{}^2 R_e)/(E_{\text{rated}} \times I_{FL})] \times 100$, which when properly reduced, becomes the *per cent* IR_e *drop*, or briefly, *per cent IR*; thus

$$\%IR = \frac{I_{FL}{}^2 R_e}{E_{\text{rated}} \times I_{FL}} \times 100 = \frac{I_{FL} R_e}{E_{\text{rated}}} \times 100 = \frac{P_{SC}}{(\text{volt-amp})_{\text{rated}}}$$

But the $\%IZ$ is related to the $\%IR$ and the $\%IX$ (the per cent reactance drop at rated load) as is the hypotenuse of a right triangle to the two sides; it follows, therefore, that

$$\%IX = \sqrt{(\%IZ)^2 - (\%IR)^2}$$

The values of $\%IR$ and $\%IX$ having been established from the short-circuit test, they may be substituted in an extremely convenient equation, not derived here but widely used, to determine the per cent regulation of a transformer. It is

$$\text{Per cent regulation} = \%IR \cos\theta + \%IX \sin\theta + \frac{(\%IX \cos\theta - \%IR \sin\theta)^2}{200} \quad (53)$$

In the derivation of Eq. (53) the angle θ, the power-factor angle, is arbitrarily made positive for a *lagging* power-factor load. For a *leading* power-factor load, with the angle θ *negative*, the signs of the two sine terms must therefore be changed so that $(+\%IX \sin\theta)$ becomes $(-\%IX \sin\theta)$ and $(-\%IR \sin\theta)$ becomes $(+\%IR \sin\theta)$. An example will now be given to illustrate the foregoing analysis.

EXAMPLE 17. Using the data given in Example 13, calculate the per cent regulation of the transformer (*a*) for a unity power-factor load, (*b*) for an 0.8 lagging power-factor load, (*c*) for an 0.866 leading power-factor load.

Solution

With the transformer rating $E = 2,400$ and volt-amp $= 100,000$, and $E_{SC} = 72$ volts, $I_{SC} = 41.6$ amp, and $P_{SC} = 1,180$ watts

$$\%IZ = \frac{72}{2,400} \times 100 = 3$$

$$\%IR = \frac{1,180}{100,000} \times 100 = 1.18$$

$$\%IX = \sqrt{3^2 - (1.8)^2} = 2.76$$

(a) For PF = 1, $\cos \theta = 1$ and $\sin \theta = 0$.

Per cent regulation
$$= (1.18 \times 1) + (2.76 \times 0) + \frac{[(2.76 \times 1.0) - (1.18 \times 0)]^2}{200}$$
$$= 1.18 + 0 + 0.38 = 1.218$$

(b) For PF = 0.8 lagging, $\cos \theta = 0.8$ and $\sin \theta = 0.6$.

Per cent regulation
$$= (1.18 \times 0.8) + (2.76 \times 0.6) + \frac{[(2.76 \times 0.8) - (1.18 \times 0.6)]^2}{200}$$
$$= 0.944 + 1.656 + 0.0075 = 2.51$$

(c) For PF = 0.866 leading, $\cos \theta = 0.866$ and $\sin \theta = -0.5$.

Per cent regulation
$$= (1.18 \times 0.866) - (2.76 \times 0.5) + \frac{[(2.76 \times 0.866) + (1.18 \times 0.5)]^2}{200}$$
$$= 1.022 - 1.38 + 0.0149 = -0.343$$

Negative regulation indicates that the voltage *drops* when the load is removed.

Efficiency Calculations Using Short-circuit and Open-circuit Data. When the secondary of a transformer delivers power to a load, an equivalent amount of power is supplied to the primary by the a-c source; the power output is generally delivered at a voltage that is different from that of the source. In addition to this useful power input that is transferred to the load by transformer action, the a-c source must supply the transformer with sufficient power to take care of the losses. Obviously, then, the power output is less than the power input by the amount of these losses; this implies that the efficiency of a transformer is always less than 100 per cent.

As previous discussions have shown, there are only two kinds of losses in a static transformer: (1) the copper losses in the primary and secondary windings and (2) the hysteresis and eddy-current losses in the laminated core. The sum of these two sets of losses is always an extremely small part of the total input in modern well-designed transformers, for which reason the efficiency is quite high. Transformer efficiencies greater than 95 per cent, especially in the larger units, are quite general.

It was also pointed out in connection with transformer operation that (1) the copper losses are measured by the wattmeter when the short-circuit test is performed, and (2) the core losses are measured by the wattmeter when the open-circuit test is performed. With these known losses it is possible, therefore, to determine the transformer efficiency by the equation

$$\text{Per cent efficiency} = \left(1 - \frac{\text{watts losses}}{\text{watts output} + \text{watts losses}}\right) \times 100 \quad (54)$$

The core losses in a transformer are practically constant for all conditions of loading as well as at no load. The fact that the core losses occur in the iron implies that they result because the flux varies periodically between $+\phi_m$ and $-\phi_m$. And since the maximum value of the flux is practically independent of the load, the core losses are unaffected by load changes, assuming, of course, that the frequency of the supply and the impressed voltage are substantially constant.

Remembering that the core losses in a transformer are practically constant for all conditions of load as well at no load and that the copper losses vary as the square of the kilovolt-ampere output, it is generally a simple matter to make calculations and plot an *efficiency-vs.-kva-output* curve for a transformer after the open-circuit and short-circuit tests have been performed. It will likewise be recalled that the tests indicated yield the core losses P_c and the full-load copper losses P_{SC}; also, in connection with the latter, the copper losses at $\frac{1}{4}$, $\frac{1}{2}$, $\frac{3}{4}$, $1\frac{1}{4}$, and $1\frac{1}{2}$ times rated kva are, respectively, $\frac{1}{16}P_{SC}$, $\frac{1}{4}P_{SC}$, $\frac{9}{16}P_{SC}$, $\frac{25}{16}P_{SC}$, and $\frac{9}{4}P_{SC}$, a convenient set of values when the results are tabulated as illustrated in the following example:

EXAMPLE 18. A 5-kva 2,300/230-volt 60-cycle standard distribution transformer was tested, with the following results: short-circuit test input = 112 watts; open-circuit test input = 40 watts. Calculate the efficiencies of the transformer for a power factor of 0.8 for the following fractions of rated kilovolt-ampere: $\frac{1}{4}$, $\frac{1}{2}$, $\frac{3}{4}$, 1, $1\frac{1}{4}$, and $1\frac{1}{2}$. Tabulate the results and plot an *efficiency vs. kilovolt-ampere output* curve.

Solution

Kva output	Losses			Watts		Per cent efficiency
	Core	Copper	Total	Output	Input	
0	40	0	40	0	40	0
$1\frac{1}{4}$	40	7	47	1,000	1,047	95.51
$2\frac{1}{2}$	40	28	68	2,000	2,068	96.71
$3\frac{3}{4}$	40	63	103	3,000	3,103	96.68
5	40	112	152	4,000	4,152	96.34
$6\frac{1}{4}$	40	175	215	5,000	5,215	95.87
$7\frac{1}{2}$	40	252	292	6,000	6,292	95.36

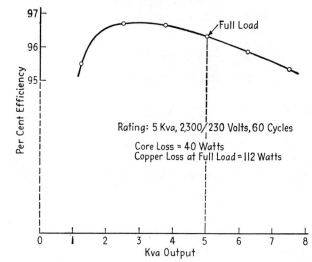

Fig. 191. *Efficiency vs. kilovolt-ampere output* curve for Example 14.

Figure 191 is an efficiency curve for this example.

Maximum Efficiency. In the analysis of the operation of a transformer as the load increases from zero to rated output and above, it will be noted (see Example 18 and Fig. 191) that the efficiency rises to a maximum and then proceeds to drop off; this is because efficiency calculations for progressively increasing values of load that rise linearly involve two kinds of losses, namely, iron losses that are substantially constant and copper losses that increase as the square of the load. It can, in fact, be shown that, because of these relationships, *the efficiency of a transformer is a maximum when the copper losses are equal to the iron losses.* This statement can therefore be written in equation form as

$$I_S^2 R_e = P_C$$

where I_S = secondary, or load, current
R_e = equivalent resistance in secondary terms
Rewriting the foregoing equation gives

$$I_S = \sqrt{\frac{P_c}{R_e}}$$

If both sides of this formula are then multiplied by $E_S/1{,}000$, the secondary kilovolts, and the right-side term is multiplied by I_{FL}/I_{FL}, it becomes

$$\frac{E_S I_S}{1{,}000} = \frac{E_S I_{FL}}{1{,}000 \times I_{FL}} \sqrt{\frac{P_c}{R_e}} = \frac{E_S I_{FL}}{1{,}000} \sqrt{\frac{P_c}{I_{FL}^2 R_e}}$$

304 ELECTRICAL MACHINES—ALTERNATING CURRENT

But $\dfrac{E_S I_S}{1{,}000}$ = kva load for maximum efficiency

$\dfrac{E_S I_{FL}}{1{,}000}$ = kva rating of transformer

and $I_{FL}^2 R_e$ = full-load copper losses P_{SC}

Hence $\text{kva}_{\text{max eff}} = \text{kva}_{\text{rated}} \sqrt{\dfrac{P_c}{P_{SC}}}$ \hfill (55)

Equation (55) is particularly interesting because it indicates that the iron and full-load copper losses, obtained from tests, may be used to determine both the kva load at which maximum efficiency occurs *and* the maximum efficiency. The following example illustrates how this is done:

EXAMPLE 19. Using the data of Example 17, calculate the kilovolt-ampere load (at a power factor of 0.8) when the efficiency is a maximum and the maximum efficiency.

Solution

$$\text{kva}_{\text{max eff}} = 5 \sqrt{\dfrac{40}{112}} = 2.99$$

$$\text{Per cent efficiency}_{\text{max}} = \left[1 - \dfrac{(40 + 40)}{(2{,}990 \times 0.8) + 80}\right] \times 100 = 96.76$$

All-day Efficiency. Distribution transformers must be ready at all times during the 24-hr day to provide electrical service for lighting and power. The primaries of such transformers are always energized, although the secondaries may supply practically little or no load much of the time, particularly where transformers serve residential lighting and power circuits. Since the core losses are supplied continuously and the load is very light during the greater part of the day, it is considered good practice to design such transformers so that the constant losses are very low. When this is done, the transformer will usually operate at maximum efficiency (core losses = copper losses) in a range of about one-half the rated kilovolt-amperes. This point is illustrated by Example 19.

On the basis of such operating conditions, it is obviously unfair to judge a transformer by its full-load efficiency; the latter might be much less than the maximum efficiency. A much more satisfactory method is to compare the so-called *all-day efficiencies* of transformers, since this rating takes into account operation over a 24-hr period. All-day efficiency is defined as *the ratio of the energy* (kilowatt-hours) *delivered by the transformer in a 24-hr period to the energy input in the same period of time.* In order to determine the all-day efficiency, it is necessary, of course, to know how the load varies from hour to hour during the day. An example should make this clear.

EXAMPLE 20. The transformer of Example 18 operates with the following loads during a 24-hr period: $1\frac{1}{2}$ times rated kva, power factor = 0.8, 1 hr; $1\frac{1}{4}$ times rated kva, power factor = 0.8, 2 hr; rated kva, power factor = 0.9, 3 hr; $\frac{1}{2}$ rated kva, power factor = 1.0, 6 hr; $\frac{1}{4}$ rated kva, power factor = 1.0, 8 hr; no load, 4 hr. Calculate the all-day efficiency.

Solution

Energy output, kw-hr

$W_1 = 1.5 \times 5 \times 0.8 \times 1 = 6.0$
$W_2 = 1.25 \times 5 \times 0.8 \times 2 = 10.0$
$W_3 = 1 \times 5 \times 0.9 \times 3 = 13.5$
$W_6 = 0.5 \times 5 \times 1.0 \times 6 = 15.0$
$W_8 = 0.25 \times 5 \times 1.0 \times 8 = 10.0$
Total 54.5

Energy losses, kw-hr

$(1\frac{1}{2})^2 \times 0.112 \times 1 = 0.252$
$(1\frac{1}{4})^2 \times 0.112 \times 2 = 0.350$
$1 \times 0.112 \times 3 = 0.336$
$(\frac{1}{2})^2 \times 0.112 \times 6 = 0.168$
$(\frac{1}{4})^2 \times 0.112 \times 8 = 0.056$
Iron = 0.04×24 = 0.960
Total 2.122

All-day efficiency = $\left(1 - \dfrac{2.122}{54.5 + 2.122}\right) \times 100 = 96.25$ per cent

Autotransformers. In principle and in general construction, the *autotransformer* does not differ from the conventional two-winding transformer thus far considered, but it does differ from it in the way in which the primary and secondary are interrelated. In the conventional transformer, the primary and secondary windings are completely insulated from each other but are magnetically linked by a common core. In the autotransformer, the two windings, primary and secondary, are *both* electrically and magnetically interconnected; in fact, a part of the single continuous winding is common to both primary and secondary.

The autotransformer may be constructed in either of two ways. In one arrangement, there is a single continuous winding with taps brought out at convenient points determined by the desired secondary voltages; in the other arrangement, there are two or more distinct coils which are electrically connected to form a continuous winding. In either case, the same laws governing conventional two-winding transformers apply equally well to autotransformers.

Consider Fig. 192a, which is a schematic diagram of an autotransformer. The input voltage V_1 is connected to the complete winding *ac* and the load is connected across a portion of the winding, that is, *bc*. The voltage V_2 will bear the same relation to V_1 as in the conventional two-winding transformer; that is, V_2 will equal V_1 multiplied by the ratio of the

ELECTRICAL MACHINES—ALTERNATING CURRENT

number of turns N_{bc} to the number of turns N_{ac}. Thus

$$V_2 = V_1 \times \frac{N_{bc}}{N_{ac}}$$

Assuming a unity power-factor load for convenience, the load current I_2 will equal V_2 divided by the load resistance R_L. If the transformer is assumed to be 100 per cent efficient, the power input $V_1 I_1$ must equal the load power $V_2 I_2$.

Note that I_1 flows in the portion of the winding ab, whereas the current $(I_2 - I_1)$ flows in the portion of the winding bc (see Fig. 192b). Under

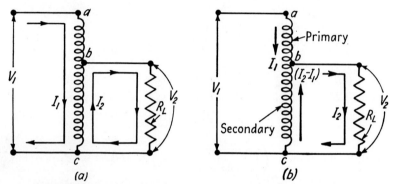

FIG. 192. Current and voltage relations in an autotransformer.

this condition, *the autotransformer acts exactly like a two-winding transformer if, from the standpoint of transformer action, it is considered that the portion of the winding ab is the primary and the portion of the winding bc is the secondary.* To prove that this is true, it must be shown that the ampere-turns $N_{ab}I_1$ in the primary equals the ampere-turns $N_{bc}(I_2 - I_1)$ in the secondary; it should be emphasized that a relationship of $(NI)_P = (NI)_S$ is fundamental to the operation of *all* transformers. This proof is now given.

$$N_{bc}(I_2 - I_1) = (N_{ac} - N_{ab})(I_2 - I_1)$$
$$= N_{ac}I_2 - N_{ac}I_1 - N_{ab}I_2 + N_{ab}I_1$$

But
$$\frac{N_{ac}}{N_{bc}} = \frac{I_2}{I_1}$$

Therefore
$$I_1 = \frac{N_{bc}}{N_{ac}} I_2$$

Substituting the value of I_1 in the second term gives

$$N_{bc}(I_2 - I_1) = N_{ac}I_2 - N_{ac} \times \frac{N_{bc}}{N_{ac}} I_2 - N_{ab}I_2 + N_{ab}I_1$$
$$= (N_{ac} - N_{bc})I_2 - N_{ab}I_2 + N_{ab}I_1$$
$$= N_{ab}I_2 - N_{ab}I_2 + N_{ab}I_1$$

which proves that $\quad N_{ac}(I_2 - I_1) = N_{ab}I_1$

The foregoing analysis indicates that *the power transformed* is

$$P_{\text{trans}} = (V_1 - V_2)I_1 = V_2(I_2 - I_1) \tag{56}$$

Since $a = V_1/V_2$ and $V_2 = V_1/a$,

$$P_{\text{trans}} = \left(V_1 - \frac{V_1}{a}\right)I_1 = V_1 I_1 \left(1 - \frac{1}{a}\right)$$

Hence $\quad P_{\text{trans}} = \text{power input} \times \left(1 - \frac{1}{a}\right) \tag{57}$

EXAMPLE 21. An autotransformer having a primary voltage of 116 and a secondary voltage of 80 delivers a load of 4 kw at unity power factor. Calculate the *transformed power* and the *power conducted* directly from the source to the load.

Solution

$$a = \frac{116}{80} = 1.45$$

$$P_{\text{trans}} = 4{,}000 \times \left(1 - \frac{1}{1.45}\right) = 1{,}240 \text{ watts}$$

$$P_{\text{conducted}} = 4{,}000 - 1{,}240 = 2{,}760 \text{ watts}$$

EXAMPLE 22. A conventional 3-kva 2,200/220-volt distribution transformer is to be connected as an autotransformer to step down the voltage from 2,420 to 2,200. (*a*) Make a wiring diagram showing how the transformer should be connected. (*b*) With the transformer used to *transform* rated power, calculate the total power input. Assume unity power factor.

Solution

(*a*) Figure 193 illustrates how the two-winding transformer is connected for autotransformer operation.

(*b*) $\quad a = \dfrac{2{,}420}{2{,}200} = 1.1$

$$\text{Power input} = \frac{P_{\text{trans}}}{1 - 1/a} = \frac{3{,}000}{1 - 1/1.1} = 33{,}000 \text{ watts}$$

The foregoing examples should make it especially clear that *an autotransformer of given physical dimensions can handle much more load power than an equivalent two-winding transformer*. In Example 21, 4 kw is handled with a transformer of about 30 per cent that rating, while in Example 18, a 3-kva transformer is capable of taking care of 11 times its rating. These great gains result only because an autotransformer transforms, by transformer action, a fraction of the total power; the power that

308 ELECTRICAL MACHINES—ALTERNATING CURRENT

is *not* transformed is *conducted directly* to the load without transformer action.

Autotransformers are cheaper in first cost than conventional two-winding transformers of similar rating. They also have better regulation, i.e., the voltage does not drop so much for the same load, and they operate at higher efficiencies. However, they are considered unsafe for use on ordinary distribution circuits because the high-voltage primary circuit is directly connected to the low-voltage secondary circuit; on the other hand, for connecting one high-voltage system, say, 22,000 volts, to

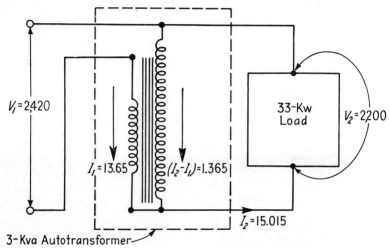

Fig. 193. Wiring diagram for solution to Example 22.

another, say, 13,800 volts, they are especially suitable, for the reasons given above. They are very frequently used in connection with the starting of certain types of a-c motors, so that lower than line voltage is applied during the starting period.

Autotransformers may be used to obtain the neutral of a three-wire 230/115-volt service in much the same manner as is done on the secondary side of a conventional two-winding transformer (see Fig. 180). The autotransformer in this case is merely a 230-volt reactance coil with a center tap, as shown in Fig. 194. The 230-volt load is "fed" by the a-c source and is unaffected by the presence of the autotransformer. Each of the 115-volt loads, however, is supplied with power from that portion of the autotransformer to which it is connected, the upper 115-volt load by winding *ab* and the lower 115-volt load by winding *bc*. When the two 115-volt loads are balanced, there will be no neutral current; the autotransformer merely acts as a simple high-impedance coil. If the load is unbalanced, the half of the winding that is connected to the heavier of the two loads is the secondary, while the other half of the winding is the primary; the autotransformer functions as a transformer only to the extent of the *unbalanced load*. Thus, if the load between *ab* is 4 kw and the load between *bc* is 10 kw, the autotransformer will transform 3 kw, with the upper half of the winding acting as the primary and the lower half acting as the secondary. (The autotransformer *transforms* one-half of the unbalanced load.)

Instrument Transformers. When it is necessary to measure comparatively high values of current or voltage, it is particularly desirable to use standard low-range instruments in conjunction with specially constructed *accurate-ratio* transformers. The latter are called *instrument transformers*

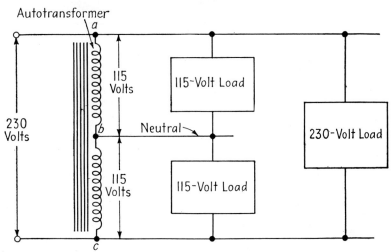

FIG. 194. Autotransformer used to provide three-wire service from a 230-volt a-c source.

and are of two kinds: *current transformers* and *potential transformers*. They are required to transform relatively small amounts of power because their only loads, called *burdens*, constitute the delicate moving elements of ammeters, voltmeters, and wattmeters.

A current transformer, as the name implies, is used with an ammeter to measure the *current* in an a-c circuit; a potential transformer, on the other hand, is used with a voltmeter to measure the *potential difference*, or voltage, in an a-c circuit. In practice, current transformers are usually connected to ordinary 5-amp ammeters, while potential transformers are generally employed with standard 150-volt voltmeters.

A current transformer has a primary coil of one or more turns of *heavy* wire, which is always connected in *series* in the circuit in which the current is to be measured. The *secondary* has a great many turns of comparatively *fine* wire, which *must always be* connected across the *ammeter termi-*

nals (5-amp range, as a rule); the latter *indirectly* indicates the current flowing in the primary.

In Fig. 195, it should be obvious that, in so far as voltages are concerned, the transformer is of the step-up type because the secondary has more turns than the primary. It follows, therefore, that current will be *stepped down* in the same ratio as the voltage is stepped up. Thus, if the current transformer has a ratio of 100:5, it means that the voltage is stepped up in the ratio 20:1, but more particularly that the current is stepped down by the same ratio. A designation like 100:5 is generally

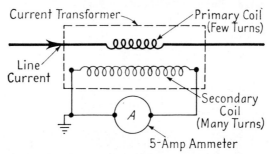

FIG. 195. Wiring connections for measuring a high current with a current transformer and a 5-amp ammeter.

used for current transformers because it furnishes information concerning the maximum allowable primary current and the proper ammeter range for the connected instrument.

EXAMPLE 23. An 80:5 current transformer is connected to a 5-amp ammeter. If the latter registers 3.65 amp, what is the line current?

Solution

$$I_{\text{line}} = \frac{80}{5} \times 3.65 = 16 \times 3.65 = 58.4 \text{ amp}$$

EXAMPLE 24. It is desired to measure a current of about 150 to 180 amp. (*a*) If a 5-amp ammeter is to be used in conjunction with a current transformer, what should be the ratio of the latter? (*b*) With this transformer, what should the instrument deflection be multiplied by to give the true line current?

Solution

(*a*) Use a 200:5 current transformer.
(*b*) Multiply the instrument deflection by 40.

An extremely practical design of current transformer is the *clamp-on* or *clip-on* type. This instrument has a laminated core so arranged that it may be opened out at a hinged section by pressing a trigger. When the core is opened, it permits the admission of the current-carrying conductor, whereupon the trigger is released and the core is closed tight by a spring. The current-carrying conductor thus acts as a single-turn primary, while the accurately wound secondary is permanently connected to the ammeter conveniently mounted in the handle. Figure 196 shows a

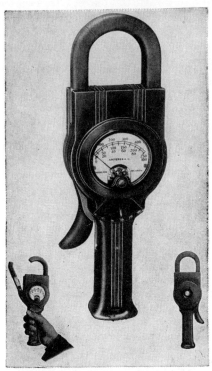

FIG. 196. A clamp-on ammeter. Note the open position in the lower left, ready for clamping over a current-carrying conductor, and the pointer and dial for changing the range. (*Weston Electrical Instrument Corp.*)

clamp-on current transformer and its ammeter. In this style, provision is made for changing the ratio to permit full-scale deflections of 10, 25, 50, 100, 250, and 500 amp by simply moving a pointer over a dial indicating the maximum current to be measured. The wiring connections for this multirange instrument are given in Fig. 197. Note that the current transformer secondary is connected to a d-c ammeter through a copper-oxide rectifier and a specially designed shunt for varying the ammeter range.

An important aspect of current-transformer operation is that its secondary must never be permitted to be open-circuited because, unlike distribution and power transformers, which are connected to constant-potential sources, the voltage across the primary winding varies over a wide range as the load changes; i.e., the voltage drop across the primary is a function of the current passing through it. This means, of course, that the flux in the core will change with varying values of impressed primary volts, since $E = k\phi$. For a high degree of accuracy it is therefore necessary to design the current transformer so that its core is never saturated when, in normal operation, secondary and primary ampere-turns can react; it is only under such conditions that the magnetizing component of the load current is reasonably low to limit the core density. However, when the secondary circuit is accidentally open-circuited, *its* ampere-turns are no longer available to react with the

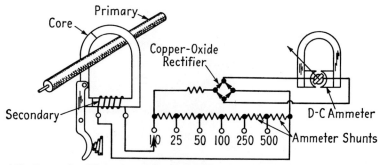

FIG. 197. Internal wiring connections of a clamp-on type of a-c ammeter. Note that the current transformer secondary is connected to a d-c ammeter through a copper-oxide rectifier and a properly designed shunt for varying the ammeter range.

primary ampere-turns, with the result that primary load current is *completely* magnetizing current; the core therefore saturates. Then, when the transformer is again put into operation, the hysteresis loop, instead of being symmetrical about the origin of coordinates, is displaced in the direction of the residual flux; this flux increases the exciting current and invalidates the original calibration of the instrument.

Potential transformers are carefully designed, extremely accurate-ratio step-down transformers. They are used with standard low-range voltmeters, the deflection of which, when multiplied by the ratio of transformation, gives the true voltage on the high side. In general, they differ very little from the ordinary two-winding transformers already discussed, except that they handle a very small amount of power. Since their secondaries are required to operate instruments and sometimes relays and pilot lights in electric circuits, they ordinarily have ratings of 40 to 100 volt-amp. Common ratios of transformation are 10:1, 20:1, 40:1, 80:1, 100:1, 120:1, and higher. As a rule, transformers for this service are of the shell type because this construction develops a higher degree of accuracy. Ordinary 150-volt voltmeters are generally used with such transformers to indicate *indirectly* the primary voltage. Figure 198 shows a potential transformer connected for the measurement of a high voltage with a 150-volt voltmeter. Note particularly that, for safety, the secondary circuit is extremely well insulated from the high-voltage primary and that, in addition, it is *grounded*. Such grounding serves to protect the

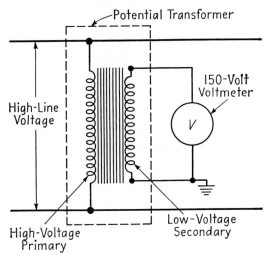

FIG. 198. Wiring connections for measuring a high voltage with a potential transformer and a 150-volt voltmeter.

operator from the high-voltage hazard should he come in contact with the wiring.

EXAMPLE 25. A 50:1 potential transformer and a 150-volt voltmeter are connected as in Fig. 198 for the measurement of the high-side voltage of a distribution transformer. If the voltmeter deflection is 133, what is the transmission-line voltage?

Solution

$$V_{\text{line}} = 50 \times 133 = 6{,}650 \text{ volts}$$

Figure 199 shows how measurements are made in a high-voltage circuit with the use of instrument transformers and a low-range voltmeter, ammeter, and wattmeter.

Transformer Polarity. Transformers are often connected in parallel to supply a common load, in much the same way as are alternators and d-c generators for the same purpose. More frequently, two or three transformers are connected together so that they may be used in polyphase systems, that is, where the service is two-, three-, or six-phase.

Before the connections are made, however, it is necessary that the *polarity* of the transformers be known. The *polarity of a transformer refers to the relative directions of the induced voltages in the primary and secondary windings with respect to the manner in which the terminal leads are brought out*

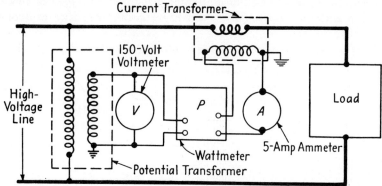

FIG. 199. Typical wiring connections showing instrument transformers and instruments inserted in a high-voltage circuit to measure volts, watts, and amperes.

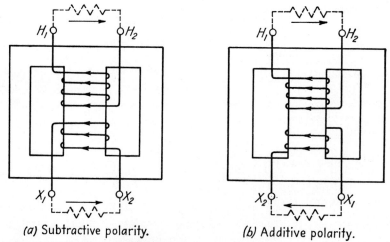

FIG. 200. Simplified sketches showing current directions due to induced voltages in high and low sides of a transformer.

and marked. Standard notations are *additive polarity* and *subtractive polarity*.

Consider Fig. 200, which represents the primary and secondary windings of a simple shell-type transformer. Since the *induced* voltages in both windings are produced by the *same* alternating mutual flux, the directions in both will always be the same. Thus current will *tend* to flow from H_1 to H_2 in an external circuit and through the high-voltage winding

TRANSFORMERS 315

at the same time that current will *tend* to flow from X_1 to X_2 in an external circuit and through the low-voltage winding. However, when the low-voltage leads are brought out as indicated in Fig. 200a, the direction from X_1 to X_2 is left to right, while the direction from X_1 to X_2 is right to left when the leads are brought out as shown in Fig. 200b. Note particularly that the subscripts 1 and 2 are attached to the symbols H and X on the basis of *induced emf*.

Next refer to Fig. 201, in which a voltage is *impressed* across the primary terminals H_1H_2. Since (1) this voltage is higher than, and opposite

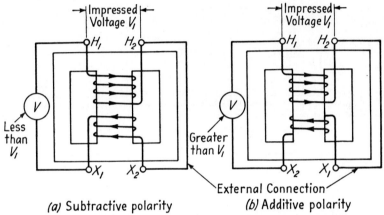

FIG. 201. Sketches illustrating why (a) a voltmeter registers less than V_1 for subtractive polarity, and (b) a voltmeter registers greater than V_1 for additive polarity. In (a) the two lower potential points H_2 and X_2 are joined together; in (b) the lower potential point H_2 is joined to the higher potential point X_1.

in direction to, *its* induced emf, and (2) the induced emf in primary and secondary are always similarly directed (see Fig. 200), it should be clear that the primary- and secondary-winding polarities are actually opposed. Thus, in Fig. 201, *arrows on the primary and secondary windings* are shown so that (1) the primary winding, acting as a load to the source, *takes* current from H_1 to H_2, while (2) the secondary winding, acting as a source, tends to *send* current to an external load from X_1 to X_2.

Referring now to Fig. 201a, if a wire is connected to the two terminals *on the same side*, that is, H_2 and X_2, a voltmeter across terminals H_1 and X_1 will register the *difference* between the induced voltages in the primary and secondary; the fact that the voltages *subtract* gives rise to the term *subtractive polarity* for this arrangement. Following Fig. 201b, it is seen that, if a wire is joined to H_2 and X_1 *on the same side*, a voltmeter connected across terminals H_1 and X_2 will register the *sum* of the induced voltages in the primary and secondary; the fact that the voltages *add* gives rise to the term *additive polarity* for this arrangement. The above statements can be verified by applying the right-hand rule to primary and secondary coils. Between H_1 and X_1 (subtractive) the voltmeter is connected to coils that are wound in opposition; between H_1 and X_2 (additive) the voltmeter is connected to coils that are wound in the same direction.

In summary, the polarity of a transformer can be determined in one of two ways: (1) noting the manner in which the terminals are marked

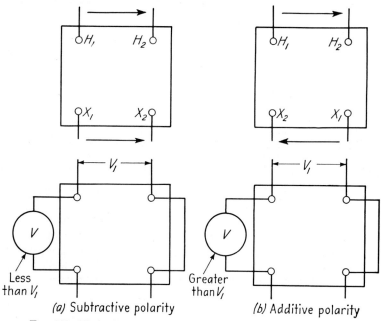

Fig. 202. Methods for determining the polarity of a transformer.

or (2) performing a simple voltmeter test, impressing a 115-volt source to the high side, and proceeding as in Fig. 202.

EXAMPLE 26. A 2,300/230-volt distribution transformer is tested for polarity in accordance with the standard method indicated by Fig. 202. If 118 volts is impressed across terminals H_1H_2, what will the voltmeter register if (a) the polarity is additive? (b) the polarity is subtractive?

Solution

(a) Voltmeter registers $118 + 11.8 = 128.8$ volts, if additive.
(b) Voltmeter registers $118 - 11.8 = 106.2$ volts, if subtractive.

Parallel Operation of Transformers. Several important conditions must be fullfilled if two or more transformers are to operate successfully in parallel to deliver a common load. These important conditions are:

1. The voltage ratings of both primaries and secondaries must be identical. This obviously implies that the transformation ratios are the same.

2. The transformers must be properly connected with regard to polarity.

3. The equivalent impedances should be *inversely proportional* to the respective kilovolt-ampere ratings.

4. The ratio of the equivalent resistance to the equivalent reactance ($R_e:X_e$) of all transformers should be the same.

The parallel operation of two or more transformers requires that the primaries be joined to the same source and that the secondaries be connected to the same load. Figure 203 shows how two transformers should

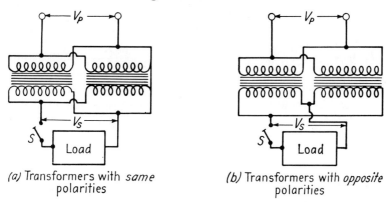

(a) Transformers with *same* polarities   (b) Transformers with *opposite* polarities

Fig. 203. Transformers connected for parallel operation.

be connected in parallel when they have the *same* polarity (Fig. 203a) and when they have *opposite* polarities (Fig. 203b).

No-load Operation. When the secondary load is removed, that is, with switch S open, the primaries will still be energized and the secondaries will remain connected in parallel. Since the latter are in phase opposition *with respect to each other*, no current can circulate in these windings if the induced voltages are exactly equal; this condition can exist only if the ratios of transformation of the two transformers are exactly equal. If transformer 1 has a ratio of transformation a_1 which is different from that of transformer 2, which has a ratio of transformation a_2, the circulating current in the secondary I_c will be*

$$I_c = \frac{(a_1 - a_2)V_S}{a_1 Z_{e_1} + a_2 Z_{e_2}} \tag{58}$$

* This equation is not strictly correct, because the impedance terms Z_{e_1} and Z_{e_2} are shown as *scalar* quantities; they are actually *phasor*, i.e., complex, quantities, and should be treated as such by applying the complex notation. However, since the latter is beyond the scope of this book, and the error is generally not significant in practice, it is omitted here to simplify the calculation procedure.

where Z_{e_1} = equivalent impedance of transformer 1 in secondary terms
Z_{e_2} = equivalent impedance of transformer 2 in secondary terms

EXAMPLE 27. The following information is given in connection with two transformers that are connected in parallel:

Transformer 1

Rating = 25 kva
2,360/230 volts
Z_e = 0.08, in secondary terms

Transformer 2

Rating = 35 kva
2,300/230 volts
Z_e = 0.06, in secondary terms

Calculate the secondary circulating current at no load.

Solution

$$a_1 = \frac{2{,}360}{230} = 10.26 \qquad a_2 = \frac{2{,}300}{230} = 10$$

$$I_c = \frac{(10.26 - 10)230}{(10.26 \times 0.08) + (10 \times 0.06)} = \frac{59.8}{0.821 + 0.6} = 42.1 \text{ amp}$$

It is interesting to note that the current circulating in the windings, without load, is about 28 and 38 per cent of the respective rated currents

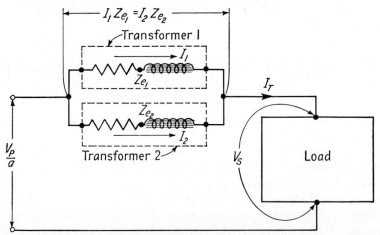

FIG. 204. Equivalent circuit of transformers, having *equal* ratios of transformation, connected in parallel and supplying power to a common load.

of the 25- and 35-kva transformers. This situation could easily exist if the improper tap were used on one of the transformers.

Load Operation—Equal Ratios of Transformation. When two transformers having *equal* ratios of transformation are connected in parallel, the total load current will divide between them inversely as their equivalent impedances. By making use of the equivalent circuit diagram of Fig. 188, it is possible to represent the system of parallel transformers

TRANSFORMERS

and the common load by the diagram of Fig. 204. From this electric circuit, two voltage equations may be written thus:

$$\frac{V_P}{a} = V_S + I_1 Z_{e_1} \qquad \frac{V_P}{a} = V_S + I_2 Z_{e_2}$$

Since $I_1 Z_{e_1}$ and $I_2 Z_{e_2}$ both equal $(V_P/a) - V_S$, it follows that

$$I_1 Z_{e_1} = I_2 Z_{e_2}$$

Therefore
$$\frac{I_1}{I_2} = \frac{Z_{e_2}}{Z_{e_1}} \qquad (59)$$

Equation (59) implies that when two transformers of different kilovolt-ampere ratings are connected in parallel, they divide the total load in proportion to their respective kilovolt-ampere ratings only when their equivalent impedances are *inversely* proportional to their respective ratings.

EXAMPLE 28. The following information is given for two transformers connected in parallel and delivering a total load of 300 kva:

Transformer 1
Rating = 150 kva
6,900/230 volts
Z_e = 9.4, in primary terms

Transformer 2
Rating = 250 kva
6,900/230 volts
Z_e = 5.8, in primary terms

Calculate the load current and kilovolt-amperes delivered by each transformer.

Solution

$$\text{Total current } I_T = \frac{300,000}{6,900} = 43.5 \text{ amp}$$

$$\frac{I_1}{I_2} = \frac{5.8}{9.4} \qquad I_1 = \frac{5.8}{9.4} \times I_2$$

Also,
$$I_T = I_1 + I_2$$

$$43.5 = \left(\frac{5.8}{9.4} \times I_2\right) + I_2 = 1.617 \, I_2$$

$$I_2 = 26.9 \text{ amp}$$
$$I_1 = 43.5 - 26.9 = 16.6 \text{ amp}$$

Therefore

$$\left.\begin{array}{l} \text{kva}_1 = 6.9 \times 16.6 = 114.4 \\ \text{kva}_2 = 6.9 \times 26.9 = 185.6 \end{array}\right\} \text{total} = 300 \text{ kva}$$

The foregoing analysis is not strictly correct unless the ratio $R_{e_1}:X_{e_1}$ is equal to that of $R_{e_2}:X_{e_2}$. If these ratios are not equal, the total current

320   ELECTRICAL MACHINES—ALTERNATING CURRENT

I_T will be *vectorially* equal to the sum of I_1 and I_2, and not arithmetically equal, as it was in Example 28. Practically, however, calculations by both methods generally yield results that do not differ sufficiently to warrant the more involved procedure by the vector solution (see footnote on p. 317).

It should also be pointed out that when $R_{e_1}:X_{e_1}$ is not equal to $R_{e_2}:X_{e_2}$, the power factor of the loads delivered by the two transformers will be unequal; they will both be slightly different from the common load power factor. The power factors of the loads carried by the individual transformers will be the same only when $R_{e_1}:X_{e_1} = R_{e_2}:X_{e_2}$, under which condition the transformers operate most satisfactorily.

Load Operation—Unequal Ratios of Transformation. When two transformers having unequal ratios of transformation are connected in parallel, the total load current will divide in accordance with the following equations:

$$I_1 = \frac{(a_2 - a_1)V_S + (a_2 Z_{e_2} I_T)}{(a_1 Z_{e_1}) + (a_2 Z_{e_2})} \tag{60}$$

$$I_2 = \frac{(a_1 - a_2)V_S + (a_1 Z_{e_1} I_T)}{(a_1 Z_{e_1}) + (a_2 Z_{e_2})} \tag{61}$$

EXAMPLE 29. The transformers of Example 27 deliver a total load of 46 kva. Calculate the secondary currents and the kilovolt-ampere load of each one.

Solution

$$a_1 = 10.26 \qquad a_2 = 10 \qquad Z_{e_1} = 0.08 \qquad Z_{e_2} = 0.06$$

$$I_T = \frac{46{,}000}{230} = 200 \text{ amp}$$

$$I_1 = \frac{(10 - 10.26)230 + (10 \times 0.06 \times 200)}{(10.26 \times 0.08) + (10 \times 0.06)}$$

$$= \frac{-59.8 + 120}{1.421} = \frac{60.2}{1.421} = 42.4$$

$$I_2 = \frac{(10.26 - 10)230 + (10.26 \times 0.08 \times 200)}{(10.26 \times 0.08) + (10 \times 0.06)}$$

$$= \frac{59.8 + 164}{1.421} = \frac{223.8}{1.421} = 157.6$$

$$\text{kva}_1 = 42.4 \times 0.23 = 9.75$$
$$\text{kva}_2 = 157.6 \times 0.23 = 36.25$$

The foregoing solution is not a theoretically correct one, since it neglects the fact that impedance and current values should be treated vectorially. However, the results obtained by this simplified solution are, in general, so close to those obtained by the more complex vector calculations that the procedure here presented may be considered entirely satisfactory (see footnote on p. 317).

Note particularly that the total load does *not* divide inversely as the transformer impedances when the transformation ratios are not the same. For distribution step-down transformers, the one having the lower ratio always tends to assume a larger share of the total load.

Three-phase Transformer Connections. Transformers that must handle a considerable amount of power are generally grouped together in *banks* for polyphase service. In three-phase systems, two or three

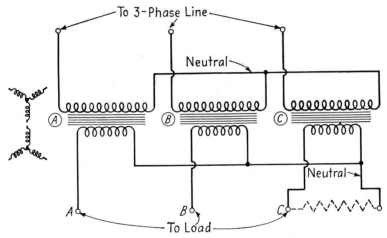

FIG. 205. The Y-Y connection of transformers.

identical transformers may be used in a bank for this purpose. There are four standard ways of connecting a three-transformer bank, namely Y-Y, Δ-Δ, Δ-Y, and Y-Δ. In a two transformer bank, the so-called *open-delta* (V-V) connection is most generally employed, while the T-T arrangement finds occasional use.

When polyphase banks are formed, it is customary to group together transformers that have similar kilovolt-ampere and voltage ratings and identical polarity markings. Not only are these conditions observed, but it is also considered good practice to use transformers manufactured by the same maker. In the wiring diagrams and discussions that follow, it is therefore assumed that all transformers are exactly alike.

The Y-Y Connection. Figure 205 shows a bank of three transformers connected in Y on both the primary and secondary sides. If the ratio of transformation of *each transformer* is a, the same ratio will exist between the *line* voltages on the primary and secondary sides. This connection will give satisfactory service only if the three-phase load is balanced; when the load is unbalanced, the electrical neutral will shift from its exact center to a point that will make the three line-to-neutral voltages unequal. In fact, a very low resistance approximating a short circuit may be connected across the secondary of one transformer (transformer C, for example), and only a low value of current will flow; under this condition, the neutral point will shift toward line C, which is equivalent to a reduction in voltage V_{CN} and a rise in voltages V_{AN} and V_{BN}. The so-called "floating neutral" is obviously objectionable, but the difficulty can be corrected by *grounding* the primary neutral, which automatically connects this point to the commonly grounded neutral of the Y-connected

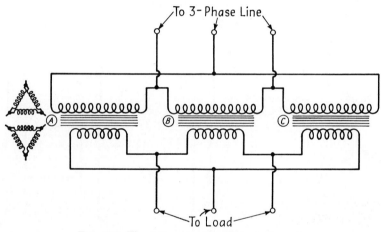

Fig. 206. The Δ-Δ connection of transformers.

alternator. The advantage of this system of connections is that the insulation is stressed only to the extent of the line-to-neutral voltage, which is 58 per cent of the line-to-line voltage.

The Δ-Δ Connection. Figure 206 shows a bank of transformers connected in Δ on both the primary and secondary sides. This arrangement is generally used in systems in which the voltages are not very high and especially when continuity of service must be maintained even though one of the transformers should fail. As will be pointed out subsequently, when one of the transformers is removed from a Δ-Δ bank, operation continues on what is known as *open delta*. The ratio of transformation existing between primary and secondary line voltage will be exactly the same as that of each transformer. No difficulty is experienced from unbalanced loading as in the Y-Y connection; the three-phase load voltages remain substantially equal, regardless of the degree of load unbalance.

The Δ-Y Connection. This scheme of connections, Fig. 207, is generally employed where it is necessary to step up the voltage, as, for example, at the beginning of a high-tension transmission system. On the high sides of the transformer, insulation is stressed only to the extent of 58 per cent of the line-to-line voltage. (In the Δ connection, the line voltage represents the transformer voltage.) Thus, on a 220,000-volt transmission system, the transformer insulation need be designed for an operating potential of about 127,500. Another advantage is that the neutral point is stable and will not "float" when the load is unbalanced. The voltage ratio of the line voltages will no longer be the same as the ratio of transformation a of each transformer. The ratio of the line-to-line voltages, high to low, will be 1.73a., i.e., $V_H:V_L = 1.73a$.

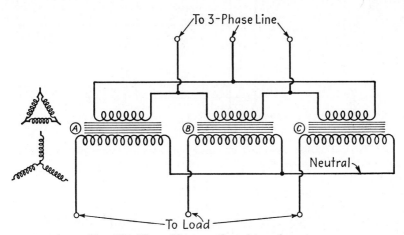

Fig. 207.—The Δ-Y connection of transformers.

In recent years, this connection has become popular in distribution systems in which a single bank of transformers is used to serve both the three-phase power equipment and the single-phase lighting circuits. In such cases the 2,400/120-volt transformer connections provide a four-wire secondary, with the neutral supplying the fourth wire. The special 208-volt three-phase power equipment is then connected to the line wires, while the lighting equipment is "fed" by the three 120-volt line-to-neutral circuits. Figure 208 shows how this is done.

The Y-Δ Connection. This connection is the reverse of the Δ-Y system and is represented by Fig. 209. It is used principally where the voltage is to be stepped down, as, for example, at the end of a transmission line. It is also employed in moderately low-voltage distribution circuits for stepping down from transmission voltages of 4,000 or 8,000 volts to 230 and 115 volts. The points made concerning Δ-Y connections apply equally well here.

EXAMPLE 30. Three 5:1 transformers are connected in Δ-Y (Fig. 207) to step up the voltage at the beginning of a 13,200-volt three-phase

324   ELECTRICAL MACHINES—ALTERNATING CURRENT

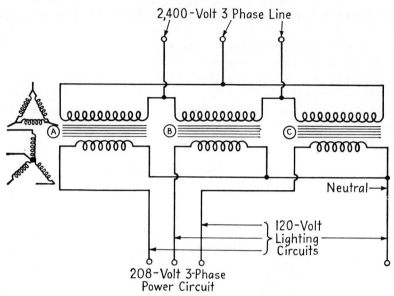

Fig. 208. The Δ-Y, 2,400/208-120-volt connection of transformers to serve power and lighting circuits simultaneously.

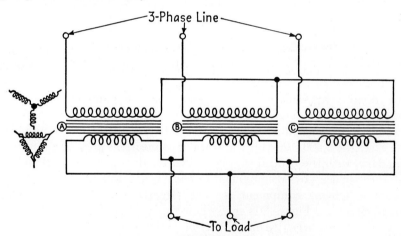

Fig. 209. The Y-Δ connection of transformers.

transmission line. Calculate the line voltage on the high side of the transformers.

Solution

$$\frac{V_H}{V_L} = 1.73a$$

$$V_H = 13{,}200 \times 1.73 \times 5 = 114{,}000 \text{ volts}$$

TRANSFORMERS

EXAMPLE 31. It is desired to step down the three-phase voltage from 6,600 to 460 volts using the Y-Δ transformer connection. What should be the ratio of transformation of each transformer?

Solution

$$\frac{V_H}{V_L} = 1.73a$$

$$a = \frac{6,600}{460 \times 1.73} = 8.3:1$$

EXAMPLE 32. Three 10:1 step-down transformers are connected Y on the primary side and Δ on the secondary side. If the primary line-to-line voltage is 3,980 and the secondaries deliver a rated balanced three-phase load of 180 kw at a power factor of 0.8, calculate: (a) the line voltage on the secondary side; (b) the current in each of the transformer windings on the secondary side; (c) the line current on the secondary side; (d) the current in each of the transformer windings on the primary side, i.e., the primary line current; (e) the kva rating of each transformer.

Solution

(a) $$V_S = \frac{3,980}{\sqrt{3} \times 10} = 230 \text{ volts}$$

(b) Since the load is balanced, each transformer delivers one-third of the total load, i.e., $180/3 = 60$ kw at a power factor of 0.8. Therefore

$$I_S \text{ (transformer winding)} = \frac{60,000}{230 \times 0.8} = 326 \text{ amp}$$

(c) I_S (line current) $= \sqrt{3} \times I_S$ (transformer winding)
$$= \sqrt{3} \times 326 = 565 \text{ amp}$$

Also $$I_S \text{ (line current)} = \frac{P_{\text{total}}}{\sqrt{3} \times V_S \times \text{PF}}$$
$$= \frac{180,000}{\sqrt{3} \times 230 \times 0.8} = 565 \text{ amp}$$

(d) $$I_P = \frac{60,000}{(3,980/\sqrt{3}) \times 0.8} = 32.6 \text{ amp}$$

Also $$I_P = \frac{P_{\text{total}}}{\sqrt{3} \times V_P \times \text{PF}} = \frac{180,000}{\sqrt{3} \times 3,980 \times 0.8} = 32.6 \text{ amp}$$

(e)   kva per transformer = 75

The V-V Connection. If one of the transformers of a Δ-Δ bank is removed and a three-phase source is connected to the primaries as shown in Fig. 210, three equal three-phase voltages will be measured at the secondary terminals at no load. This method of transforming three-phase power, using two transformers, is called the *open-delta*, or V-V, connection. It is employed (1) when the thee-phase load is comparatively small, so that the installation does not warrant a three-transformer bank; (2) when one of the transformers in a Δ-Δ bank fails, so that service may be continued until the faulty transformer is repaired or a good one is substituted; (3) when it is anticipated that the future load will increase to warrant the closing of the open Δ at some later time.

The total load that can be carried by a V-V transformer bank is *not* two-thirds of the capacity of a Δ-Δ bank, but 58 per cent of it. This is a

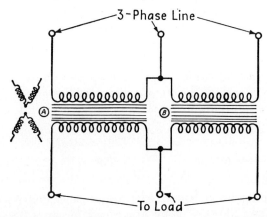

FIG. 210. The V-V (open-Δ) connection of transformers.

reduction of about 15 per cent from its normal rating. For example, when three 10-kva transformers are used in a Δ-Δ bank, the total rated load is 30 kva. However, if one of the transformers is removed, the resulting V-V connection has a rated capacity of $0.58 \times 30 = 17.4$ kva, and not 20 kva. The reduction in rating comes about because the average power factor at which the transformers operate is less than that of the load; this average power factor is, in fact, always 86.6 per cent of the balanced load power factor. A significant point in this connection is that, except for a balanced unity power-factor load, the two transformers in the V-V bank operate at different power factors. Another disadvantage of this connection is that the secondary terminal voltages tend to become unbalanced to a greater extent as the load is increased; this is true even when the three-phase load is balanced, a situation that does not exist when the load is supplied by a bank of three transformers. As in the Δ-Δ system, the ratio of the primary to the secondary line-to-line voltages is exactly the same as that of each transformer.

When an open delta (V-V) bank of two transformers delivers a balanced three-phase load whose power factor is $\cos \theta$, one of the units will operate

TRANSFORMERS   327

at a power factor of cos $(30° - \theta)$ while the power factor of the other unit will be cos $(30° + \theta)$. This means, for example, that:

1. When the load power factor is unity, the individual transformer power factors will be equal, but 0.866.
2. When the load power factor is 0.866, one transformer will operate at unity power factor while the power factor of the other will be 0.5.
3. When the load power factor is 0.5, one transformer will operate at zero power factor and deliver *no power*, while the other, operating at a power factor of 0.866, will handle the *entire load power*.

Writing these statements in general terms gives

$$\left. \begin{array}{l} P_1 = \text{kva}_t \times \cos(30° - \theta) \\ P_2 = \text{kva}_t \times \cos(30° + \theta) \end{array} \right\} \quad (62)$$

where P_1 = power delivered by transformer 1
P_2 = power delivered by transformer 2
kva_t = kilovolt-ampere load on each transformer
θ = power-factor angle of the load

Also
$$\text{kva}_t = \frac{VI}{1,000} \quad (63)$$

where V and I are, respectively, the voltage of, and the current in, the individual transformers, both on the same side, i.e., primary or secondary.

EXAMPLE 33. (a) What should be the kilovolt-ampere rating of each of two transformers in a V-V bank when the three-phase balanced load is 43.5 kva? (b) If a third transformer is added for Δ-Δ operation, what is the rated capacity of the bank? What per cent increase in load does this represent?

Solution

(a) Since the rated capacity of a V-V bank must be 15 per cent greater than the balanced three-phase load,

$$\text{kva}_t = 1.15 \times \frac{43.5}{2} = 25$$

(b) If a third transformer is added, the rating of the Δ-Δ bank will be

$$\text{Total kva} = 3 \times 25 = 75$$

The per cent increase in load will be

$$\frac{75 - 43.5}{43.5} \times 100 = 72.5$$

EXAMPLE 34. Two transformers are connected in *open delta* and deliver a balanced three-phase load of 245 kw at 460 volts and a power factor of 0.707. Calculate: (a) the secondary line (transformer) current; (b) the kva load on each transformer; (c) the power delivered by the individual transformers. (d) If a third transformer having the same rating as each of the other two is added to form a Δ bank, what total load can be handled?

Solution

(a) $$I_S = \frac{245{,}000}{\sqrt{3} \times 460 \times 0.707} = 435 \text{ amp}$$

(b) $$\text{kva}_t = \frac{460 \times 435}{1{,}000} = 200$$

(c) For a power factor of 0.707, $\theta = 45°$.

$$P_1 = 200 \times \cos(30 - 45°) = 193.2 \text{ kw}$$
$$P_2 = 200 \times \cos(30 + 45°) = 51.8 \text{ kw}$$
$$P_t = 193.2 + 51.8 \qquad = 245 \text{ kw} \qquad \text{(check)}$$

(d) $\quad$ kva (Δ bank) $= 3 \times 200 \quad = 600$

The T-T Connection. Another two-transformer method that can be used to transform three-phase power from one voltage to another is the T-T connection; it was first proposed by Charles F. Scott and is frequently called the *Scott* connection. Since it is also possible to use this scheme of connections to change from three-phase to two-phase, and vice versa, it has wide application when such a change is necessary. One of the two transformers, the so-called *main* transformer, must have at least two primary and two secondary coils so that a center tap may be brought out from each side. The other transformer, the so-called *teaser* transformer, must have primary and secondary windings the numbers of turns of which are 86.6 per cent of the respective turns of the main transformer. As a rule, it is more economical to manufacture both transformers alike with 86.6 per cent taps. Figure 211 is a wiring diagram showing how the connections are made.

To understand how the three-phase transformation results from this arrangement, it is desirable to think of the primary and secondary phasor voltages as forming geometrical Ts (from which the connection gets its name). Furthermore, assuming for convenience that the ratio of transformation is 10:1, step down, with a primary input of 1,000 volts, the various voltages and their relationships are represented in Fig. 212. The kilovolt-ampere ratings of the *main* and *teaser* transformers will be exactly the same, even though the voltage across the latter is only 86.6 per cent of that across the former. The reason for this is that the kilovolt-ampere loads carried by the *two halves of the main transformer* are out of phase by 60 electrical degrees; the result is that when these are vectorially added, their sum equals the kilovolt-ampere load on the teaser transformer.

It is also true that the total rating of the two Scott-connected transformers will be 58 per cent of the rating of a Δ-Δ bank, just as in the V-V

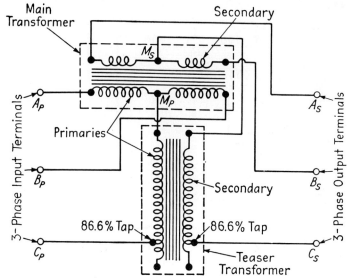

Fig. 211. The T-T (Scott) connection for transforming three-phase to three-phase.

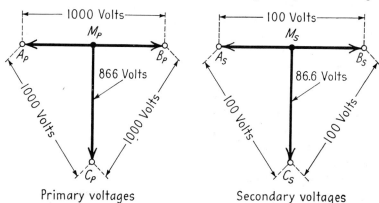

Fig. 212. Voltage relations in three-phase to three-phase Scott connection.

connection. This implies, therefore, that each transformer in a T-T connection must have a rating 15 per cent greater than each of the transformers in a three-transformer bank.

When it is necessary to change from three-phase to two-phase, or vice versa, the Scott transformation serves particularly well. This transformation is accomplished by connecting the three-phase side in T in the usual way; the two-phase side makes use of *all* the teaser windings (not the 86.6 per cent tap) as well as all the main winding. Figure 213 shows how this is done in transforming from three-phase to three-wire two-phase. Voltages are indicated for an input of 1,000 volts and a ratio of transformation of 10:1.

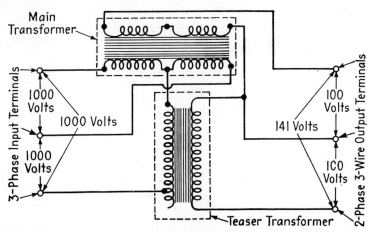

FIG. 213. Scott-connected transformers, changing from three-phase to three-wire two-phase. Assumed ratio of transformation is 10:1.

EXAMPLE 35. The Scott connection is used to transform 100 kva from 2,300 volts three-phase to 230 volts two-phase. Calculate: (a) the currents in the main and teaser primary windings; (b) the currents in the main and teaser secondary windings; (c) the kilovolt-ampere rating of the main transformer; (d) the kilovolt-ampere rating of the teaser transformer; (e) the ratio of the transformer capacity to the load kilovolt-amperes.

Solution

(a) For three-phase input, the current in each line is

$$I = \frac{\text{kva} \times 1{,}000}{\sqrt{3} \times 2{,}300}$$

Current in main and teaser primaries $= \dfrac{100 \times 1{,}000}{\sqrt{3} \times 2{,}300} = 25.1$ amp

(b) Since each of the two secondaries delivers 50 kva,

Current in main and teaser secondaries $= \dfrac{50{,}000}{230} = 217.5$ amp

(c)   kva rating of main transformer $= \dfrac{2{,}300 \times 25.1}{1{,}000} = 57.8$

(d)   kva rating of teaser transformer $= \dfrac{2{,}300 \times 25.1}{1{,}000} = 57.8$

(This rating is determined by considering that the entire 2,300-volt primary is available for use, although only 86.6 per cent of this winding is used.)

(e) Ratio of transformer capacity to load kva $= \dfrac{2 \times 57.8}{100} = 1.156$

Three-phase Transformers. When a considerable amount of power is involved in the transformation of three-phase power, it is more economical to use a three-phase transformer than, as previously discussed, a bank of

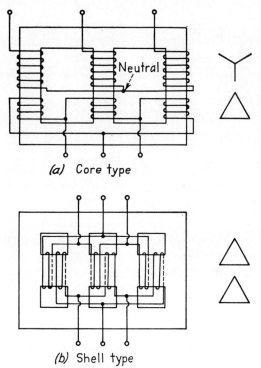

FIG. 214. Arrangements of core and windings in two types of three-phase transformer.

three single-phase transformers. The unique arrangement of the windings and the core makes it possible to save a great deal of iron by interlinking the magnetic circuits so that the same iron is used by the three phases simultaneously. Proper flux densities are maintained because the three-phase currents are displaced 120 electrical degrees with respect to each other. Moreover, since the entire three-phase assembly is reduced in size and is placed in a single tank, additional savings result because less transformer oil is used, fewer transformer bushings are needed (all polyphase connections are made inside the tank), and the complete transformer occupies less floor space than a bank of three transformers.

There are two general arrangements of the windings and the core. In one construction, the *core type*, the three primary and secondary windings surround a considerable part of the magnetic core, as shown in Fig. 214a. Note that the primary and the secondary of each phase is wound around its own "leg." In the *shell type* of construction (Fig. 214b), the magnetic circuits surround a considerable portion of the three-phase primary and secondary windings. One advantage of the shell type over the core type

Fig. 215. Three-phase core-type transformer before being placed in its tank. (*Wagner Electric Corp.*)

is that the former transformer can be operated in open Δ should one of the windings be damaged; operation can proceed at 58 per cent of the transformer rating if the primary and secondary of the damaged phase are disconnected from the circuit and are then short-circuited.

Figure 215 is a photograph of a three-phase core type of transformer before being placed in its tank. This particular unit has a rating of 450 kva for operation at 22,000 to 66,000 volts. The inside low-voltage winding is surrounded by a number of disk-type high-voltage coils, which are separated by insulating washers. The low-voltage winding is separated

TRANSFORMERS 333

from the circular-disk high-voltage winding by barriers and strips of pressboard, between which numerous large oil ducts permit free oil circulation.

A completely assembled three-phase transformer is shown in Fig. 216. It has a rating of 3,000 kva for operation on its high side at 161,000 volts. Note the three high-voltage bushings, taller than a man, the oil conservator, which eliminates the air in the transformer that would be present if

Fig. 216. Complete three-phase transformer having a rating of 3,000 kva at 161,000 volts. (*Wagner Electric Corp.*)

the oil level were established somewhere below the level of the transformer cover, and the oil-cooling radiators placed on opposite sides of the tank.

Three-phase transformers are much more difficult and costly to repair than are single-phase units. Moreover, when failure does occur and it becomes necessary to substitute a replacement unit to maintain service, the cost of the spare is much greater than it would be were a single-phase

334   ELECTRICAL MACHINES—ALTERNATING CURRENT

transformer to be used as a replacement in a three-transformer bank. Another point that must sometimes be considered is the difficulty of transporting a heavier three-phase transformer compared with the moving of each of the three single-phase transformers.

The Constant-current Transformer. Transformers of the type thus far discussed maintain substantially constant secondary voltage; they therefore serve electric circuits in which the current varies with the load. Not so common, however, is the circuit that requires a *constant current* whose

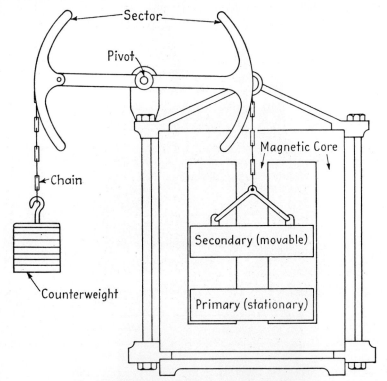

FIG. 217. Shell-type constant-current transformer.

voltage varies with changing load. The series street-lighting system is the best example of this practice. A specially designed type of transformer, the so-called *constant-current* transformer, is used for this purpose and functions so that the secondary voltage varies directly with the load impedance. That is, when the load consists of many lights in series, the impedance is high; in this case, the secondary voltage is high. On the other hand, when the number of series lights is reduced, the impedance drops, in which case the secondary voltage drops proportionately.

Figure 217 is a sketch of the constant-current transformer. The stationary primary coil and the movable secondary coil surround the center leg of the core; the core is quite long and operates at a rather high flux density. The secondary, free to move up or down, is suspended by a chain or cable from a sector that can turn about a pivot. A counterweight is attached to a similar sector and partially counterbalances the combined weight of the secondary and its suspension. At no load, that is, when the secondary circuit is open (infinite secondary impedance), the counterweight just permits the secondary to rest on the primary.

Referring to the electric circuit diagram of this system in Fig. 218, assume that the load consists of a great many lights in series. The secondary current I_S has its transformer counterpart in the primary I_P

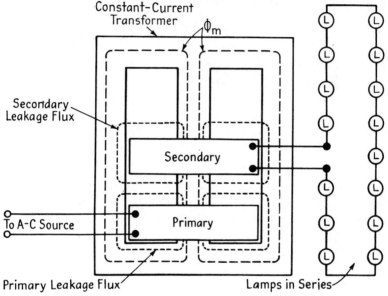

FIG. 218. Wiring diagram of constant-current transformer feeding a series-lamp circuit.

that is practically 180 electrical degrees out of phase with it. This implies that the two currents in the primary and secondary are, at every instant, in opposite directions in their respective coils. As a result, the mmfs created by the two coils cause them to repel each other; the secondary coil therefore moves upward until the force of repulsion between the coils exactly equals the unbalanced weight of the secondary coil and its suspension. Suppose next that the number of lamps is reduced. This is equivalent to a reduction in the load impedance, as a result of which a higher current flows in both windings and at once creates a greater force of repulsion between the coils, so that the secondary moves farther away from the primary. Motion continues until the increased leakage flux (the coils are farther apart) causes the secondary induced voltage to drop sufficiently so that the original value of secondary current flows again; at this point, the forces are once more in equilibrium. If the counterweight and the sector suspension are carefully adjusted, it is possible to make the transformer regulate for constant current over a wide range of load; in practice, this is usually for loads that vary between the rated number of lamps to about half as many.

The Induction Voltage Regulator. The fundamental principle of transformer action requires that the primary and secondary be coupled electromagnetically. If the flux created by the primary coil passes *through* the secondary coil, maximum voltage is induced in the latter. However, if the secondary is rotated into such a position that the flux

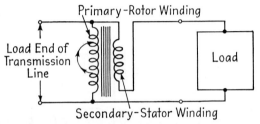

FIG. 219. Wiring diagram of induction-voltage regulator in a transmission line.

passes *across* (not through) it, no voltage will be induced in the secondary. This idea of rotating the primary with respect to the secondary is applied in the *induction voltage regulator*, a device frequently used to regulate the voltage at the load end of a long distribution system.

The induction voltage regulator is primarily a step-down transformer, but it is so constructed that the secondary voltage may be varied from zero to a certain maximum by merely altering the position of the primary-coil axis with respect to the secondary-coil axis. This is accomplished by placing the secondary in the slots of a stator similar to that of the alternator and winding the primary in a slotted rotor mounted concentrically inside. The primary is connected directly across the line wires, sometimes through slip rings and brushes, while the secondary is placed *in series* in one of the lines. A wiring diagram of the electrical connections is given in Fig. 219. Since the maximum voltage induced in the secondary can be made to add directly to, or subtract directly from, the line voltage by rotating the primary through an angle of 180°, the regulator is capable of "boosting" or "bucking" the line voltage; in the 90° position, no voltage is induced in the secondary.

A clearer picture of the construction and operation of the regulator is given in Fig. 220. Note that the secondary winding is placed in the stator slots, while the rotor slots carry the primary winding and a short-circuited winding. When the stator and rotor axes line up (Fig. 220a), maximum voltage is induced in the secondary; in one such position, the transformer is *additive,* while in a similar position, with the rotor rotated through 180°, the transformer is *subtractive.* In either of these two line-up positions, the short-circuited winding is not coupled to the primary (or secondary), because its axis is perpendicular to the flux axis; no voltage is therefore induced in it. If the rotor is turned through an angle of 90°, no voltage will be induced in the secondary. In this position, however, the secondary winding mmf will have no compensating primary component of mmf and will act as an impedance in the line. To correct this,

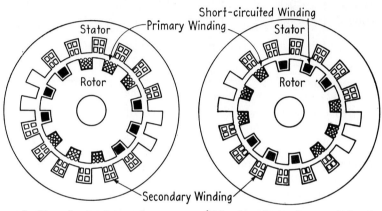

(a) Position of rotor for maximum induced voltage in secondary

(b) Position of rotor for zero induced voltage in secondary

Fig. 220. Induction-voltage regulator, with rotor shown in two positions.

the short-circuited winding, which is now lined up with the secondary, acts to produce an mmf that neutralizes the reactance of the secondary. The short-circuited winding obviously corrects for the proper amount of leakage reactance in any position of the rotor because it is placed on the rotor 90° displaced from the primary winding. In practice, the rotor is usually turned automatically by a small motor; the latter is properly connected to the line through a reversing switch and relay contacts. When the line voltage drops, the motor turns the rotor in one direction until the line voltage is raised to the correct value, whence the rotor stops; if the line voltage rises, the motor turns the rotor in the opposite direction until the line voltage is readjusted to the correct value, whence the rotor stops.

In electrical laboratories, hand-operated induction voltage regulators are frequently used in the performance of many experiments in which correct voltage adjustment is important. Figure 221 is a wiring diagram showing a very useful arrangement of the primary and secondary coils in connection with two DPDT switches. This sketch shows how the device can be used as a variable-ratio transformer and also as an induction voltage regulator. Provision is also made to connect the secondary coils in series or in parallel; when the coils are in parallel, the secondary voltage is one-half of that obtained for the series connection.

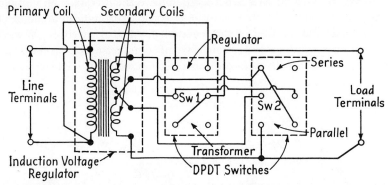

FIG. 221. Induction-voltage regulator wired to two DPDT switches for laboratory use. Switch No. 1 is used to operate device as a regulator or a variable-voltage transformer. Switch No. 2 is used to connect the secondary coils in series or parallel.

EXAMPLE 36. A single-phase induction regulator, used for laboratory purposes, has a 220-volt primary (shunt) winding and a secondary (series) winding with two 55-volt coils, each of which has a current rating of 20 amp. When used as a *two-circuit transformer* (see Fig. 221, for switch 1 closed down, i.e., on the transformer side) what maximum and minimum secondary voltages may be obtained, and what maximum kva rating does the regulator have (*a*) for the series connection? (*b*) for the parallel connection? When used as a *regulator* (see Fig. 221 for switch 1 closed up, i.e., on the regulator side), what maximum and minimum secondary voltages may be obtained and what maximum kva rating does the regulator have (*c*) for the series connection? (*d*) for the parallel connection?

Solution

(*a*) For the series connection:

$$V_S \text{ (maximum)} = (2 \times 55) = 110 \text{ volts}$$
$$V_S \text{ (minimum)} = 0$$
$$\text{kva (maximum)} = \frac{110 \times 20}{1,000} = 2.2$$

(*b*) For the parallel connection:

$$V_S \text{ (maximum)} = 55 \text{ volts}$$
$$V_S \text{ (minimum)} = 0$$
$$\text{kva (maximum)} = \frac{55 \times 40}{1,000} = 2.2$$

(c) For the series connection:
$$V_S \text{ (maximum)} = 220 + (2 \times 55) = 330 \text{ volts}$$
$$V_S \text{ (minimum)} = 220 - (2 \times 55) = 110 \text{ volts}$$
$$\text{kva (maximum)} = \frac{330 \times 20}{1{,}000} = 6.6$$

(d) For the parallel connection:
$$V_S \text{ (maximum)} = 220 + 55 = 275 \text{ volts}$$
$$V_S \text{ (minimum)} = 220 - 55 = 165 \text{ volts}$$
$$\text{kva (maximum)} = \frac{275 \times 40}{1{,}000} = 11$$

Questions

1. What is meant by *transformer action?* Under what conditions will it take place? Under what conditions will it not take place?
2. Explain why transformer action can take place in a d-c circuit.
3. If alternating current is impressed on one coil, what will be the frequency of the induced voltage in another coil with which it is coupled?
4. Define a *static transformer.*
5. Why are transformers more efficient than rotating electric machines?
6. Distinguish between *core-type* and *shell-type* transformers.
7. To what circuit is the primary of a transformer connected? the secondary connected?
8. Describe how the core and windings of a transformer are assembled.
9. What advantages are claimed for the transformer that is constructed by winding steel strips spirally through the openings and around the coils?
10. Why are transformer windings divided into several coils?
11. What is meant by *leakage flux?*
12. Why are the first few turns of high-voltage transformer coils especially well insulated?
13. Why do transformers hum? How can this hum be minimized?
14. What purposes are served by placing transformers in oil-filled tanks?
15. What properties should a good transformer oil possess?
16. List several common insulating materials for transformers.
17. Why are the tanks of some large transformers corrugated?
18. What is meant by oil *sludging?* How is it caused?
19. Describe how transformer coils are constructed.
20. What current flows in the primary of a transformer when the secondary is open-circuited? What function has this current?
21. Upon what factors does the induced voltage in the primary of a transformer depend? in the secondary?
22. What relation exists between primary and secondary voltages and turns?
23. What is meant by the *ratio of transformation?* How can this be determined experimentally?
24. What relation exists between the primary and secondary currents and turns?
25. What relation exists between the primary and secondary voltages and currents?
26. Distinguish between *step-up* and *step-down* transformers.
27. Why are some transformers constructed with the primary and secondary windings divided into two halves?

28. When a winding is divided into two halves, why are the center leads crossed before they are brought out?
29. Distinguish between *power* and *distribution* transformers.
30. Why are distribution transformers tapped?
31. Explain carefully how the primary current increases automatically in direct proportion to the current delivered by the secondary.
32. Why does the principle of transformer action require that the primary and secondary ampere-turns be equal?
33. What is meant by *voltage regulation* of a transformer?
34. What is the effect of the primary and secondary resistances upon the regulation of a transformer?
35. Distinguish between primary and secondary leakage fluxes.
36. What effect have the leakage fluxes upon the regulation of a transformer?
37. What is meant by the term *leakage-reactance drop?*
38. Upon what do the various voltage drops in a transformer depend?
39. Referring to Fig. 184, explain why the induced voltage lags behind the flux by 90°.
40. Why is it necessary to treat the various voltages in a transformer vectorially?
41. Assuming a constant impressed primary voltage, how does the secondary terminal voltage tend to change when the power factor is unity? is lagging? is leading at a very low value?
42. What is meant by the *equivalent resistance* of a transformer? How may it be calculated in primary terms? in secondary terms?
43. What is meant by the *equivalent reactance* of a transformer? How may it be calculated in primary terms? in secondary terms?
44. How can the equivalent impedance of a transformer be determined?
45. If the transformer is represented by an equivalent circuit like that shown in Fig. 188, what is neglected? Is this neglect permissible? Explain.
46. Outline carefully the procedure for performing the short-circuit test.
47. What useful information is obtained from the short-circuit test?
48. Outline carefully the procedure for performing the open-circuit test.
49. What useful information is obtained from the open-circuit test?
50. What are the two components of the core loss?
51. How is the hysteresis loss affected by a change in flux density?
52. Assuming a constant frequency, how does a voltage change affect the hysteresis loss? the eddy-current loss?
53. Assuming a constant impressed voltage, how does a frequency change affect the hysteresis loss? the eddy-current loss?
54. How do the copper losses vary with the load?
55. List the losses in a static transformer. How can these losses be determined experimentally?
56. Explain why the core losses are unaffected by the load.
57. Under what condition does the maximum efficiency occur in a transformer?
58. Why are *distribution* transformers frequently designed to develop maximum efficiency at loads that are somewhat lower than rated value?
59. Under what condition will the maximum efficiency occur at rated load?
60. Why is it desirable to have a *power* transformer operate at maximum efficiency when it is delivering rated load?
61. Define *all-day efficiency.*
62. Why is the all-day efficiency a more reasonable basis of comparison than the *conventional* efficiency?
63. What is an autotransformer?
64. What advantages are possessed by autotransformers?

TRANSFORMERS

65. List the disadvantages of autotransformers.
66. What is meant by the terms *transformed power* and *conducted power* when they refer to an autotransformer?
67. Under what assumption can an autotransformer be considered the equivalent of a two-winding transformer?
68. Make a sketch showing how a two-winding transformer may be connected to operate as: (a) a step-up autotransformer; (b) a step-down autotransformer.
69. How is it possible to use an autotransformer to obtain a neutral for a three-wire system?
70. What are current transformers? potential transformers?
71. What special precaution must be taken when a current transformer is used?
72. What instrument ranges are generally used in conjunction with instrument transformers?
73. How is the ratio of transformation of a current transformer specified? What is the significance of this notation?
74. Describe the *clamp-on* type of ammeter. What advantages does it possess?
75. Explain why the accuracy of a current transformer is impaired if, after its secondary is accidentally open-circuited, it is used again.
76. Referring to Figs. 195, 198, and 199, explain why secondary circuits should be grounded.
77. To what does the *polarity* of a transformer refer?
78. What names are given to standard polarities of transformers?
79. Explain how a test should be performed to determine the polarity of a transformer.
80. Under what conditions is it necessary to know the polarity of a transformer?
81. List the conditions that must be fulfilled before two transformers can be operated successfully in parallel.
82. Under what condition will there be no circulating current when two transformers are operated in parallel at no load?
83. Assuming that two transformers having equal ratios of transformation are connected in parallel, how is the total load divided between them?
84. Assuming that two transformers having unequal ratios of transformation are connected in parallel, indicate how the total load divides between them.
85. When two similar transformers have equal values of Z_e but different ratios of $R_e:X_e$, will the total load be divided equally between them? Will their power factors be equal?
86. List the four possible ways of connecting a bank of three transformers for three-phase service.
87. State the usual applications of each of the connections listed in Question 86.
88. Indicate which one of the connections listed above is unsatisfactory when the load is unbalanced. How is this usually corrected?
89. What are the advantages of the V-V connection?
90. What is the total load capacity of a V-V bank as compared with a Δ-Δ bank?
91. When two transformers are connected in open Δ, at what power factors will the individual transformers operate when a balanced three-phase load is delivered: (a) at unity power factor? (b) at a power factor of 0.866? (c) at a power factor of 0.5? (d) at a general power factor equal to cos θ?
92. What advantages are possessed by a three-phase connection in which the primaries are in Δ and the secondaries provide a neutral wire in addition to the three terminal leads?
93. Explain how the T-T connection is made for three-phase to three-phase operation.
94. Explain how the T-T connection is made for three-phase to two-phase operation.

95. List the advantages and disadvantages of three-phase transformers.
96. What are the two general types of three-phase transformer? Does either have any advantage over the other?
97. Describe the general constructional features of the constant-current transformer.
98. Explain the operation of the constant-current transformer in connection with its use in a series lighting circuit.
99. Describe the general constructional features of the induction voltage regulator.
100. Explain the principle of operation of the induction voltage regulator.
101. What is the function of the short-circuit winding in the induction regulator?
102. Referring to Fig. 221, what range of voltages would be possible if each secondary coil has 25 per cent as many turns as the primary and the primary input voltage is 120?

Problems

1. A 2,300/230-volt 60-cycle distribution transformer has 1,200 turns on the high side. If the net cross-sectional area of the iron is 9 sq in., calculate: (a) the total flux ϕ_m; (b) the maximum flux density in the core in lines per square inch; (c) the number of secondary turns.
2. A core-type transformer, Fig. 169, is constructed with 0.014-in. thick laminations having a uniform width of 2.75 in. If the maximum flux and flux density are, respectively, 6.2×10^5 maxwells and 8.2×10^3 maxwells per square inch and the air spaces between laminations occupy 8 per cent of the stacked core, calculate the number of laminations in the transformer.
3. A 25-cycle transformer has 2,250 primary turns and 150 secondary turns. If the maximum value of the mutual flux is 2.64×10^6 maxwells, calculate: (a) the primary and secondary induced voltages; (b) the ratio of transformation.
4. Calculate the full-load primary and secondary currents of a 5-kva 2,400/120-volt transformer.
5. The volts per turn of a 50-cycle 13,800/460-volt transformer is 6.68. Calculate: (a) the number of primary and secondary turns; (b) the flux in the core.
6. If the secondary load current of the transformer in Prob. 5 is 240 amp, calculate the primary current.
7. A transformer was tested under load and was found to deliver 62 amp at 228 volts when the primary current measured 3.2 amp. Calculate: (a) the primary input voltage; (b) the ratio of transformation.
8. A transformer has two 2,400-volt primary coils and two 240-volt secondary coils. Indicate by sketches the four possible ways to connect the transformer and, for each one, determine the primary-to-secondary voltages and the ratio of transformation.
9. A transformer has two 1,200-volt primary coils and two 120-volt secondary coils, and each of the primary coils has a 5 per cent tap. Indicate by sketches the five possible ways to connect the transformer for three-wire service, and, for each one, determine the primary-to-secondary voltages.
10. A 6,900/230-volt transformer is provided with 2½, 5, 7½, and 10 per cent taps on the primary side. Determine the primary voltages that may be used for a constant 230-volt service and the ratio of transformation in each case.
11. The no-load and full-load secondary voltages of a transformer were found to be 117 and 113 volts, respectively. Calculate the per cent regulation.
12. The per cent regulation of a 2,300/230-volt transformer is 2.61 per cent. Calculate: (a) the no-load secondary voltage; (b) the turns ratio.
13. A 100-kva 2,400/240-volt 60-cycle transformer has the following constants: $R_P = 0.42$; $X_P = 0.72$; $R_S = 0.0038$; $X_S = 0.0068$. Calculate the following values in primary and secondary terms: (a) R_e; (b) X_e; (c) Z_e.

TRANSFORMERS 343

14. Using the data of Prob. 13, calculate the following voltage drops in primary and secondary terms: (a) IR_e; (b) IX_e.
15. Using the data of Probs. 13 and 14, calculate the per cent regulation at unity power factor. See page 293
16. Repeat Prob. 15 for a power factor of 0.8 lagging.
17. Repeat Prob. 15 for a power factor of 0.8 leading.
18. A short-circuit test was performed upon a 10-kva 2,300/230-volt transformer with the following results: $E_{sc} = 137$ volts; $P_{sc} = 192$ watts; $I_{sc} = 4.34$ amp. Calculate in primary terms: (a) R_e, Z_e, and X_e; (b) the per cent regulation at a lagging power factor of 0.707.
19. The following data were obtained from a short-circuit test performed upon a 50-kva 2,300/115-volt 60-cycle transformer: $E_{SC} = 87$ volts; $I_{SC} = 21.75$ amp; $P_{SC} = 590$ watts. Calculate in primary terms: (a) R_e, Z_e, and X_e; (b) the per cent regulation at a power factor of 0.866 lagging.
20. For the transformer of Prob. 18, calculate the copper losses when the load is: (a) 8 kva; (b) 10 kw at a power factor of 0.85.
21. For the transformer of Prob. 19, calculate: (a) the kva load when the copper losses are 922 watts; (b) the kw load at a power factor of 0.8 when the copper losses are 714 watts.
22. The eddy-current loss in a 2,300-volt 60-cycle transformer is 280 watts. What will be this loss if the transformer is connected: (a) to a 2,300-volt 50-cycle source? (b) to a 2,400-volt 60-cycle source? (c) to a 2,200-volt 25-cycle source?
23. The hysteresis loss in a 6,600-volt 60-cycle transformer is 480 watts. What will be the loss when the transformer is connected: (a) to a 6,900-volt 60-cycle source? (b) to a 6,600-volt 40-cycle source? (c) to a 6,900-volt 40-cycle source?
24. A transformer has a core loss of 160 watts when its primary is connected to a 230-volt 60-cycle source. When the transformer is served by a 40-cycle supply, the impressed voltage is reduced to $^{40}\!/_{60}$ of 230 volts, or 153 volts, to maintain the same flux density as on 60 cycles; under this condition the core loss was found to be 96 watts. Calculate the 60-cycle hysteresis and eddy-current losses.
25. Using the data from the short-circuit test of Prob. 19, calculate: (a) the $\%IZ$, the $\%IR$, and the $\%IX$ drops; (b) the per cent regulation at a power factor of 0.707 lagging.
26. The following information is given in connection with tests performed on a 50-kva 4,600/230-volt 60-cycle transformer:

   Open-circuit test..... $E_{OC} = 230$ volts; $P_{OC} = 285$ watts
   Short-circuit test..... $E_{SC} = 150$ volts; $P_{SC} = 615$ watts; $I_{SC} = 10.87$ amp

   Calculate the per cent efficiency when the transformer is delivering rated kva at a power factor of 0.84.
27. For the transformer of Prob. 26, calculate the per cent efficiency when the load is 60 kva at a power factor of 0.86.
28. For the transformer of Prob. 26, calculate the per cent efficiency when the load is 48 kw at a power factor of 0.8.
29. For the transformer of Prob. 26, calculate the kva load, at a power factor of 0.84, when the efficiency is a maximum, and the maximum efficiency.
30. The following information is given in connection with tests performed on a 25-kva 2,400/240-volt 60-cycle transformer:

   Open-circuit test...... $E_{OC} = 240$ volts; $P_{OC} = 250$ watts
   Short-circuit test...... $E_{SC} = 72$ volts; $P_{SC} = 380$ watts; $I_{SC} = 10.4$ amp

Calculate: (a) the per cent regulation for a power factor of 0.866 lagging; (b) the per cent efficiency at 1½ times rated kva at a power factor of 0.8; (c) the kva load on the transformer when the efficiency is a maximum; (d) the maximum efficiency when the power factor is 0.8.

31. A 3-kva transformer has a core loss of 30 watts and a full-load copper loss of 75 watts. Calculate the all-day efficiency for the following loads: 1½ times the rated kilovolt-amperes, power factor = 0.85, 2 hr; rated kilovolt-amperes, power factor = 0.9, 5 hr; ¾ rated kilovolt-amperes, power factor = 0.95, 6 hr; ½ rated kilovolt-amperes, power factor = 1.0, 7 hr; no load, 4 hr.

32. An autotransformer, designed for 4,000- to 2,300-volt operation, supplies a load of 32 kw at a power factor of 0.8. Calculate: (a) the transformed power; (b) the conducted power.

33. For Prob. 32, calculate the current in : (a) the primary winding; (b) the secondary winding.

34. A 2,400/2,200-volt autotransformer delivers a load of 90.5 kw at a power factor of 0.75. Calculate the current in each winding section and the kva rating of the transformer.

35. A 5-kva, 2,300/460-volt distribution transformer is to be connected as an autotransformer to step up the voltage from 2,300 to 2,760 volts. (a) Make a wiring diagram showing how this should be done. (b) When used to *transform* 5 kva, calculate the kilovolt-ampere load output.

36. An autotransformer has a 5 and a 10 per cent tap from one end and a 20 per cent tap from the other. If 120 volts is impressed across the winding ends, calculate the values of all the voltages that may be obtained between any two taps and from either end and the taps.

37. An 80:5 current transformer is connected to a 5-amp ammeter that indicates a deflection of 4.62 amp. What is the line current?

38. (a) What should be the rating of a current transformer to measure a line current of about 75 amp? (b) With this transformer, what should the ammeter deflection be multiplied by to obtain the value of the line current?

39. A 20:1 potential transformer is used with a 150-volt voltmeter. If the instrument deflection is 118 volts, calculate the line voltage.

40. A polarity test is performed upon a 1,150/115-volt transformer. If the input voltage is 116, calculate the voltmeter reading if: (a) the polarity is additive; (b) the polarity is subtractive.

41. The following information is given for two transformers that are connected in parallel:

<table>
<tr><td>*Transformer 1*</td><td>*Transformer 2*</td></tr>
<tr><td>Rating = 10 kva</td><td>Rating = 7.5 kva</td></tr>
<tr><td>4,600/230 volts</td><td>4,485/230 volts</td></tr>
<tr><td>Z_e = 0.16 in secondary terms</td><td>Z_e = 0.22 ohm, in secondary terms</td></tr>
</table>

Calculate the secondary circulating current at no load.

42. If the tap of transformer 2 is shifted so that its voltage rating is the same as that of transformer 1, i.e., 4,600/230 volts, calculate the load assumed by each of the two transformers for a total load of 20 kva.

43. Using the data of Prob. 41 calculate the kilovolt-ampere load that will be carried by each transformer for a total load of 17.5 kva.

44. The following information is given for two transformers connected in parallel:

TRANSFORMERS 345

<table>
<tr><td>Transformer 1</td><td>Transformer 2</td></tr>
<tr><td>Rating = 75 kva</td><td>Rating = 50 kva</td></tr>
<tr><td>2,400/240 volts</td><td>2,400/240 volts</td></tr>
<tr><td>$Z_e = 2.22$</td><td>$Z_e = 4.15$</td></tr>
</table>

Calculate the kilovolt-ampere load carried by each transformer for a total load of 125 kva.

45. Three 10:1 transformers are connected Δ-Y for stepping up the 2,300-volt three-phase source. Calculate the secondary line voltage.

46. Three 30:1 step-down transformers are connected Y-Δ for stepping down the 132,000-volt three-phase transmission voltage. Calculate the secondary line voltage.

47. Three transformers, connected Δ on the primary side and Y on the secondary side, step the voltage down from 13,200 to 460 volts and deliver a 750-kva 0.8 power-factor load. Calculate: (a) the ratio of transformation of each transformer; (b) the kilovolt-ampere and kilowatt load on each transformer; (c) the current delivered to the load; (d) the current in the primary line wires; (e) the current in each secondary transformer winding; (f) the current in each primary transformer winding.

48. Three 25-kva transformers are connected Δ-Δ. (a) What total rated kilovolt-amperes can the bank deliver? (b) To what kilovolt-amperes should the load be reduced if one of the transformers is removed so that the bank operates V-V?

49. An open-Δ bank delivers a balanced three-phase load of 60 kva at 460 volts. (a) What current flows in the secondary of each transformer? (b) What kilovolt-ampere load does each transformer carry?

50. Two transformers are connected open-Δ and deliver a load of 86.7 kva at a power factor of 0.8. Calculate: (a) the kilovolt-ampere load on each transformer; (b) the power delivered by each of the two transformers; (c) the available transformer kilovolt-amperes if a third unit of the same rating as one of the other two is added to form a Δ bank.

51. The primaries of two transformers are connected to a 2,300-volt two-phase source. The secondaries are connected Scott, i.e., in T, and deliver a 460-volt 130-kw three-phase 0.866 lagging power-factor load. Calculate: (a) the current in each transformer winding on the primary side; (b) the current in each transformer winding on the secondary side; (c) the kilovolt-ampere rating of each of the two transformers.

52. The Scott connection is used to transform from 460 volts three-phase to 115 volts two-phase. If the load is 40 kva, calculate: (a) the current in the main and teaser primaries; (b) the current in the main and teaser secondaries; (c) the kilovolt-ampere rating of each of the transformers.

53. Referring to the induction regulator of Fig. 221, the input voltage is 220 and each of the coils in the secondary develops a maximum of 55 volts. Calculate the possible ranges of voltages obtainable.

CHAPTER 9

POLYPHASE INDUCTION MOTORS

Induction-motor Principle. In the electric motor, conversion of electrical power (or energy) to mechanical power (or energy) takes place in the *rotating* part of the machine. In the d-c motor and in one type of a-c motor, the electrical power is *conducted* directly to the rotor through brushes and a commutator; in this respect it is possible to designate such a machine as a *conduction* motor. In the most common type of a-c motor, electrical power is *not* conducted to the rotor directly; the rotor receives its power inductively in exactly the same way as the secondary of a transformer receives its power. It is for this reason that motors of this type are known as *induction motors*. In fact, it will become apparent, as the analysis proceeds, that it will be extremely useful to think of an induction motor as a sort of rotating transformer, i.e., one in which a stationary winding is connected to the a-c source, while the other winding, mounted on a structure that is free to turn, receives its power by transformer action *while it rotates*.

The principle of the induction motor was first discovered by Arago in 1824, when he observed the following interesting phenomenon: if a nonmagnetic disk and a compass are pivoted with their axes parallel, so that one or both of the compass poles are located near the edge of the disk, the compass will rotate if the disk is made to spin, or the disk will rotate if the compass is made to spin. The direction of the induced rotation in one element is always the same as that imparted to the other. Such an experiment can be readily performed if a simple copper or aluminum disk and a rather large compass are both mounted on the same vertical stem so that each may be rotated in its own bearing independently of the other. There is no more effective way to demonstrate the *principle* of the induction motor, of which there are several types. If the disk is rotated, the compass will follow at a speed always less than that of the disk; if the compass is rotated, the disk will follow the former at a lower speed.

Consider Fig. 222, which represents the edge of a *south* magnet pole that lies directly over, and close to, the edge of a nonmagnetic disk; both may be imagined as being supported on independent vertical pivots and free to rotate in a horizontal plane. Attention should be centered upon a small element of the disk that lies directly below the center of the pole. In (a) the flux passes vertically upward through the disk and is uniformly distributed. If the magnet is moved to the right, as in (b), the flux follows its usual practice of bending around the conductor element of the disk, so that a voltage is generated in the conductor away from the observer. The generated voltage causes a current to flow in the disk element, the path of which extends toward the pivot and then divides to the left and right, returning to the conductor element from both sides near the edge of the disk. (These current paths are like the rims of two coins that are both laid flat on the table and that touch each other.) The current in the disk then sets up a field of its own that encircles the conductor element in

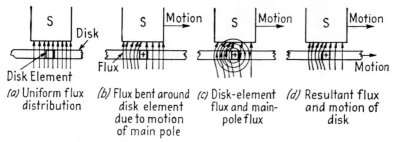

FIG. 222. Sketches illustrating principle of the induction motor.

a clockwise direction. This field and that of the south pole are shown in (c). The resultant field is next shown in (d), which indicates that the field is dense on the left side and weak on the right side of the disk element. Hence, *the conductor element and the disk move to the right, in the same direction as the motion of the pole.* It is never possible for the disk to move as rapidly as the pole. If it did, there would be no relative motion between the pole and the disk, there would be no generated voltage in the disk, and no flux would be created by the disk; this would immediately cause the disk to slow down until sufficient current flowed to develop the necessary power of rotation.

In this discussion it should be clearly understood that motor action (rotation of the disk) is developed by induction. The current in the rotor (disk) is the result of electromagnetic induction, which, it will be remembered, requires that there be relative motion between flux and conductors. Thus, if the mechanical load on the rotor increases, the rotor slows down; this slowing down means greater relative motion between flux and rotor, a greater voltage and current, and hence more power to take care of the added load. In other words, the power developed by the rotor automatically adjusts itself to the power required to drive the load.

In the actual motor, the rotor is obviously not a disk, but a well-designed structure consisting of a laminated core containing a winding;

nor is the main field a single concentrated pole moved by hand, but an even number of poles formed by a distributed winding in a slotted stator. The stator poles are formed by the interaction of the fields of two or three phases, the result of which is to create an *effect* that is equivalent to a set of revolving poles (see the Revolving Field, p. 354).

The Stator. The stationary part of the induction motor—the *stator*—consists of a cyclindrical laminated slotted core that is placed in a frame. In one manufacturing process, the armature winding is first wound in slots and properly connected. It is then dipped in an insulating varnish

FIG. 223. Wound stator and frame of an induction motor. (*The Louis Allis Co.*)

and baked, after which it is ready to be pressed into its frame. Such a stator and its frame are depicted in Fig. 223. In this particular construction it will be noted that the stator laminations are welded together at several places around the outer cylindrical surface. In another construction, the unwound stator is first pressed into its frame, as shown in Fig. 224. Note carefully that there are 36 slots for the armature winding, a significant point that will be considered later in connection with the study of motor windings. Also observe the shape of the teeth that partially close the slots. This design is quite general for induction motors because it tends to improve operation; it does, however, make it more difficult to place the winding into the slots because the individual wires of the coils must be fed into them one or two at a time. The completely wound stator, after it has been dipped and baked, is shown in Fig. 225. Note the sturdy construction and the ample provision for air circulation and cooling. The latter point is especially important in

POLYPHASE INDUCTION MOTORS 349

Fig. 224. Stator shown partially wound after the core is pressed into the frame. (*Reliance Electric & Engineering Co.*)

Fig. 225. Completely wound stator for induction motor. (*General Electric Co.*)

induction-motor operation because the temperature rise in the winding is a very definite limiting factor of motor output.

The Rotor. There are two general types of construction for the rotor of an induction motor: the *squirrel cage* and the *wound rotor*. Both designs employ a slotted laminated core tightly pressed on a shaft. A common practice in constructing the squirrel cage is to place the assembled core in a mold and then force the molten conducting material, often

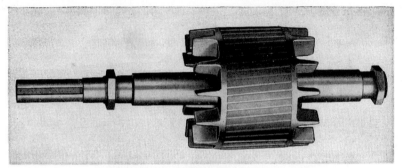

FIG. 226. Squirrel-cage type of rotor for induction motor. (*The Louis Allis Co.*)

FIG. 227. Cutaway view of squirrel-cage rotor for induction motor. (*General Electric Co.*)

aluminum, into the slots. End rings and projecting fins for cooling short-circuit the conductor bars at both ends. Figure 226 illustrates this one-piece construction, in which the conductors, end rings, and cooling fins, of aluminum, are clearly visible. Also to be noted is the fact that the slots are not parallel to the shaft axis; they are *skewed* because smoother, quieter operation results from this practice. In some cases the heavy conductor bars are driven into the slots with a tight fit and project a short distance from each end of the core. End rings, with holes lining up with the projecting conductors, are then forced over the latter, after which conductors and end rings are soldered or welded together. A good idea of the squirrel-cage rotor construction may be obtained from the cutaway view in Fig. 227. This photograph clearly shows the conductor material in the slot, the end-ring section, and the cooling fins.

The wound-rotor construction is employed when it is necessary to control the speed of the motor or to provide the machine with high starting torque. An insulated winding, similar to that used on the stator, is placed in the rotor slots. The ends of the winding are then connected to slip rings. As will be explained later, brushes riding on the rings are connected to a resistor controller which is used for speed and starting-torque control. Figure 228 depicts a completely wound rotor for an

Fig. 228. Complete wound rotor for a three-phase induction motor. Note the three slip rings and the cooling fans at both ends. (*The Louis Allis Co.*)

induction motor for three-phase operation. Two fans are provided in this particular design for cooling purposes.

The Stator Winding. The type of winding used in the stator (and wound rotor) of a polyphase induction motor is exactly the same as that discussed in Chap. 7 for alternators. This is the *whole-coiled lap winding* represented by Figs. 149 and 150, to which the student is referred. It will be recalled that in this type of winding (1) there are always as many coils in the entire winding as there are slots, (2) the number of coils per phase is equal to slots/phases, and (3) the number of coils in each pole group per phase equals slots/(poles × phases). Another important fact to remember is that the coils in each group must always be connected in series because the induced voltages in adjacent coils in each group are displaced from each other electrically by the angular distance between adjacent slots. If such coils were connected in parallel, the *resultant* voltage would circulate a comparatively large current in them to cause excessive heating and probable burnout of the winding.

An excellent view of the manner in which such a winding is placed in the stator core is given in Fig. 229. This shows a partially completed winding for a 36-slot four-pole three-phase stator. Plainly seen are the coils-per-pole-per-phase groups, separated by varnished cloth insulation.

This heavy insulation is necessary because the voltage between adjacent coils in *different* phases is the same as the voltage per phase. [There are three coils per pole per phase in this particular stator because $36/(4 \times 3) = 3$.]

Once the idea of the actual winding construction is grasped, it is possible to simplify its diagram greatly by illustrating coils in the same way as was done for transformers rather than by showing the separate coils

FIG. 229. Partially completed winding in the 36-slot stator of a four-pole three-phase induction motor. Note particularly the heavy insulation between phase groups. (*The Louis Allis Co.*)

properly located in the slots. Such a simplication will lead to a clearer understanding of the many combinations that are possible with each winding. For example, the complete winding diagram for the stator of Fig. 229 is drawn in this way in Fig. 230. This sketch shows at a glance that the stator has 36 slots and that the winding is for four poles and is connected in series-Y. The series-Δ connection is readily made by joining one line terminal to a junction of F_A and S_B, another to a junction of F_B and S_C, and the third to a junction of F_C and S_A.

The span of a coil is determined in much the same way as it was for alternator windings. It will be recalled that full-pitch windings are those in which all coils span exactly 180 electrical degrees, while fractional-pitch windings have coils that span less than 180°. Most induction-motor windings are fractional pitch for several reasons:

1. Leakage reactance is reduced a little, so that operation takes place at a somewhat higher power factor.
2. The winding is stiffer because the end connections are shorter.

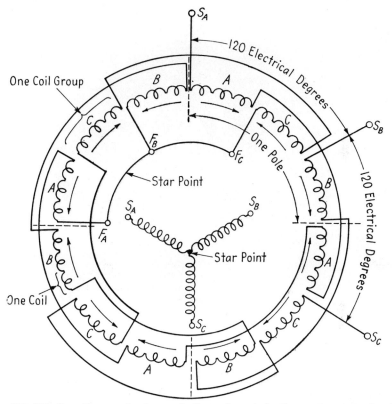

Fig. 230. Winding diagram for four-pole three-phase induction-motor stator having 36 slots. Total number of coils = 36; coils per pole per phase = 3; connection = series-Y.

3. The axial length of the machine is less, so that the distance between bearing centers is shortened.
4. Less copper is used in the winding so that the copper loss is a little smaller.
5. The designer finds it possible to adjust the number of turns in each coil to a desirable number for the purpose of effectively utilizing the whole slot area.

A completed stator ready to receive its rotor and end bells is shown in Fig. 231. Note the compact, trim appearance and the radial fins on the outside of the yoke frame for good motor cooling.

The Revolving Field. When a three-phase stator winding of an induction motor is connected to a three-phase source, three independent currents tend to flow, one in each of the winding phases. These currents will be displaced from each other by exactly 120 electrical degrees; that is, the current in phase A will reach its positive maximum 120° ahead

FIG. 231. Completely wound stator for polyphase induction motor. (*The Louis Allis Co.*)

of that in phase B, and the current in phase B will be a positive maximum 120° ahead of the current in phase C. The currents are therefore said to be *displaced in time* by 120°, with the phase sequence assumed to have the order A-B-C. The current waves are represented by Fig. 232a.

Next consider Fig. 232b, which schematically represents the three winding phases of a two-pole induction motor. The A phase is represented in a vertical axis, and the phase-A current can produce flux that acts only along a vertical axis; it will be assumed that when the current is *positive*, the mmf will produce flux directed vertically upward in phase A. The B phase is represented with a southeast-northwest axis, and the phase-B current can produce flux that acts only along this SE–NW axis; for phase B to be displaced in space from phase A by 120°, its mmf must

POLYPHASE INDUCTION MOTORS 355

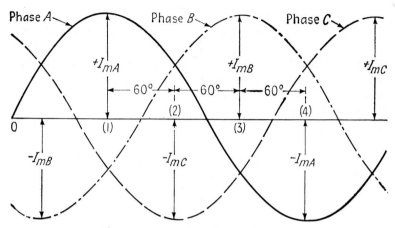

(a) Current waves in a three-phase system

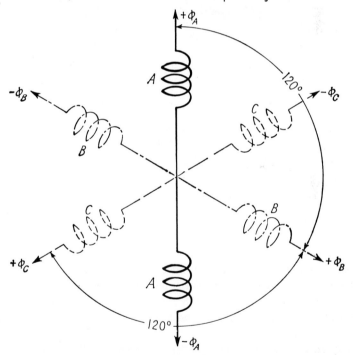

(b) Relative positions of the three phases in the stator of a two-pole, three-phase winding. The (+) and (−) flux directions correspond to the (+) and (−) current directions in *(a)*

FIG. 232. Relative directions of currents and fluxes in a two-pole three-phase winding of an induction motor.

produce flux in a SE direction when the current is *positive*. The C phase is represented with a southwest–northeast axis, and the phase-C current can produce flux that acts only along this SW–NE axis; for phase C to be displaced in *space* from phase B by 120°, its mmf must produce flux in a SW direction when the current is *positive*. With this arrangement, the three winding phases are said to be *displaced in space* by 120°.

It will now be shown that when the three phase currents I_A, I_B, and I_C are *displaced in time* by 120° and when the three phase windings are *displaced in space* by 120°, the *resulting magnetic field* (in this case, two

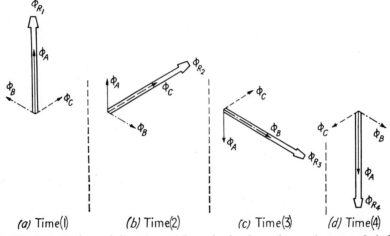

FIG. 233. Magnitudes and directions of fluxes in the three phases of a two-pole induction-motor winding, and the resultant flux, for four time instants. Note that the resultant field is constant in magnitude and rotates clockwise from a vertically upward direction to a vertically downward direction. (Refer to Fig. 232 when studying these diagrams.)

poles) *will rotate in space* as though physical poles were actually rotated mechanically. Furthermore, the field will rotate in the same direction as the space sequence of the windings A-B-C; i.e., for Fig. 232b it will be in a clockwise direction. And, as was previously explained, when poles revolve, the accompanying *revolving field* will cause a rotor to rotate in the same direction.

Referring to Figs. 232 and 233, four time instants will be selected and proof will be given that the *resultant field* is constant in magnitude and revolves the distance covered by one pole (180 electrical degrees) in one-half cycle.

Time 1. In Fig. 232a, I_A is a positive maximum; I_B and I_C are both negative and one-half their respective maximums. The fluxes will therefore have corresponding directions and proportional magnitudes. Refer

POLYPHASE INDUCTION MOTORS 357

to Fig. 232b for the directions and to Fig. 233a for the vector values and directions. When ϕ_A, ϕ_B, and ϕ_C, are combined, the resultant flux will be ϕ_{R_1}, ⅔ *the maximum flux per phase and directed vertically upward.*

Time 2. I_C is a negative maximum; I_A and I_B are both positive and one-half their respective maximums. The fluxes will therefore have corresponding directions and proportional magnitudes. When ϕ_C, ϕ_A, and ϕ_B are combined, the resultant flux (Fig. 233b) will be ϕ_{R_2}, ⅔ *the maximum flux per phase and directed 60° clockwise from* ϕ_{R_1}.

Note that time 2 is 60° later than time 1 and that the direction of the resultant field is rotated 60°.

Time 3. I_B is a positive maximum; I_A and I_C are both negative and one-half their respective maximums. The fluxes will therefore have corresponding directions and proportional magnitudes. When ϕ_B, ϕ_A, and ϕ_C are combined, the resultant flux (Fig. 233c) will be ϕ_{R_3}, ⅔ *the maximum flux per phase and directed 60° clockwise from* ϕ_{R_2}.

Note that time 3 is 60° later than time 2 and that the direction of the resultant field is rotated an additional 60°.

Time 4. I_A is a negative maximum; I_B and I_C are both positive and one-half their respective maximums. The fluxes will therefore have corresponding directions and proportional magnitudes. When ϕ_A, ϕ_B, and ϕ_C are combined, the resultant flux (Fig. 233d) will be ϕ_{R_4}, ⅔ *the maximum flux per phase and directed 60° clockwise from* ϕ_{R_3}.

Note that time 4 is 60° later than time 3 and that the direction of the resultant field is rotated an additional 60°.

Thus it is seen that the field revolves 180 electrical degrees, or one pole span, as the current I_A changes from a positive maximum to a negative maximum, that is, as it undergoes one-half cycle. Therefore *the field will revolve a distance covered by two poles for each cycle.* If there had been four poles, the revolving field would have rotated one-half a revolution (two poles) for each cycle; for six poles that revolving field would have rotated one-third of a revolution per cycle, etc. Hence the speed of the revolving field is *inversely proportional* to the number of *pairs* of poles. Also, with two poles the speed of the revolving field would be 1 revolution per second (rps) for a frequency of 1 cps, 2 rps for a frequency of 2 cps, or 60 rps for 60 cps. Therefore the speed of the revolving field is *directly proportional* to the frequency. If the foregoing two statements are combined for what is called the *synchronous speed*, the revolving-field speed in revolutions per second may be written

$$\text{rps} = \frac{f}{P/2} = \frac{2f}{P}$$

And since it is more convenient to specify speed in revolutions per minute (rpm), the foregoing equation becomes

$$\text{rpm}_{syn} = \frac{120 \times f}{P} \qquad (64)$$

Equation (64) gives the speed of the revolving field and *not* the speed of the rotor; the rotor always rotates at a lower speed, its value depending upon the mechanical load on the motor. Obviously, the revolving field speed is constant for a constant frequency.

EXAMPLE 1. Calculate the synchronous speeds of an eight-pole induction motor when supplied with power from: (*a*) a 60-cycle source; (*b*) a 50-cycle source; (*c*) a 25-cycle source.

Solution

(*a*) $\quad \text{rpm}_{syn} = \dfrac{120 \times 60}{8} = 900 \quad$ (for 60 cycles)

(*b*) $\quad \text{rpm}_{syn} = \dfrac{120 \times 50}{8} = 750 \quad$ (for 50 cycles)

(*c*) $\quad \text{rpm}_{syn} = \dfrac{120 \times 25}{8} = 375 \quad$ (for 25 cycles)

EXAMPLE 2. Calculate the synchronous speeds of 60-cycle induction motors having: (*a*) four poles; (*b*) six poles; (*c*) 10 poles.

Solution

(*a*) $\quad \text{rpm}_{syn} = \dfrac{120 \times 60}{4} = 1{,}800 \quad$ (for four poles)

(*b*) $\quad \text{rpm}_{syn} = \dfrac{120 \times 60}{6} = 1{,}200 \quad$ (for six poles)

(*c*) $\quad \text{rpm}_{syn} = \dfrac{120 \times 60}{10} = 720 \quad$ (for 10 poles)

Slip and Rotor Speed. Although the rotor of an induction motor must rotate in the same direction as the revolving field, it cannot do so at synchronous speed. In other words, there must be relative motion between revolving field and rotating rotor because rotor power can be developed only when the rotor conductors are cut by lines of force when the latter move faster than the former; under this condition the generated rotor voltage causes a current to flow in its low-resistance winding. The difference between the synchronous speed and the actual rotor speed is called the *slip speed;* it is the number of revolutions per minute by which a point on the rotor continues to slip behind a point on a fictitious revolving pole. Thus, in a four-pole 60-cycle motor, the slip would be 50 rpm if the actual rotor speed were 1,750 rpm. Slip is generally specified in terms of the synchronous speed and as a percentage becomes

$$\text{Per cent slip} = \frac{\text{rpm}_{syn} - \text{rpm}_{rotor}}{\text{rpm}_{syn}} \times 100 \tag{65}$$

When expressed as a decimal, the slip s may be written

$$s = \frac{\text{rpm}_{syn} - \text{rpm}_{rotor}}{\text{rpm}_{syn}} = \frac{(120f/P) - \text{rpm}_{rotor}}{(120f/P)}$$

from which

$$\text{rpm}_{rotor} = \frac{120f}{P}(1 - s) \tag{66}$$

EXAMPLE 3. The rotor speed of a six-pole 50-cycle induction motor is 960 rpm. Calculate the per cent slip.

Solution

$$\text{rpm}_{syn} = \frac{120 \times 50}{6} = 1{,}000$$

$$\text{Per cent slip} = \frac{1{,}000 - 960}{1{,}000} \times 100 = 4$$

EXAMPLE 4. Calculate the speed of a 60-cycle 14-pole motor if the slip s is 0.05.

Solution

$$\text{rpm}_{rotor} = \frac{120 \times 60}{14}(1 - 0.05) = 488$$

Generated Voltage and Frequency in a Rotor. The relative speed of the rotor with respect to the revolving field will affect both the generated voltage in the rotor and the frequency of the current in the rotor conductors. When the rotor is at rest, the field sweeps across its conductors at a maximum rate; under this condition, the voltage generated in each phase will be a maximum and will depend upon the number of stator turns, the stator voltage, and the number of rotor turns. As the field revolves, a back emf is generated in the stator winding which is very nearly equal to the impressed voltage; the same revolving field develops a voltage in the rotor. Since, at standstill, the flux is cutting each turn in the stator at exactly the same rate as in the rotor, it follows that the rotor voltage per phase is equal to the impressed stator voltage per phase multiplied by the ratio of rotor to stator turns per phase. Also, the frequency of the generated voltage in the rotor will be the same as that of the impressed emf.

At a slip of 50 per cent, the flux sweeps across the rotor conductors half as fast; under this condition, the *generated voltage* in each phase of the *rotor* and its *frequency* will be exactly one-half of the standstill values. At three-fourths of the synchronous speed, i.e., at 25 per cent slip, the *generated voltage* and *frequency* will be 25 per cent of the standstill values. In general, therefore,

$$E_R = s \times E_{BR} \tag{67a}$$

and
$$f_R = s \times f \tag{67b}$$

where E_R = generated voltage per phase in rotor at slip s
E_{BR} = blocked rotor generated voltage per phase
f_R = rotor frequency

EXAMPLE 5. A three-phase 60-cycle six-pole 220-volt wound-rotor induction motor has a stator that is connected in Δ and a rotor that is connected star. The rotor has half as many turns as the stator. For a rotor speed of 1,110 rpm, calculate: (*a*) the slip; (*b*) the blocked-rotor voltage per phase E_{BR}; (*c*) the rotor generated voltage per phase E_R; (*d*) the rotor voltage between terminals; (*e*) the rotor frequency.

Solution

(*a*) $s = \dfrac{1{,}200 - 1{,}110}{1{,}200} = 0.075$

(six-pole 60-cycle synchronous speed is 1,200 rpm)

(*b*) $\qquad E_{BR} = 220 \times \dfrac{1}{2} = 110$ volts

(*c*) $\qquad E_R = 0.075 \times 110 = 8.25$ volts

(*d*) Volts between rotor terminals $= \sqrt{3} \times 8.25 = 14.25$

(*e*) $\qquad f_R = 0.075 \times 60 = 4.5$ cycles

The foregoing example should emphasize the fact that the normal running-rotor voltage and frequency are extremely low.

Rotor Current and Power. The actual conversion of electrical power into mechanical power always takes place in the rotor of a motor. In the induction motor the power input to the rotor is not applied directly, i.e., conductively, but is transferred across an air gap inductively. Therefore, to understand clearly how induction motors operate, it is necessary to consider the electrical relations in the *rotor*.

It was shown in the previous section, Generated Voltage and Frequency in a Rotor, that the generated voltage in the rotor per phase is sE_{BR}. This voltage divided by the rotor impedance Z_R per phase will be the rotor current per phase. But the rotor impedance is made up of two components: (1) the rotor resistance R_R and (2) the leakage reactance sX_{BR}, where X_{BR} is the rotor reactance at standstill (note that the reactance is proportional to the slip s because it is a function of the rotor frequency). Thus

$$I_R = \frac{sE_{BR}}{\sqrt{R_R^2 + s^2 X_{BR}^2}} = \frac{E_{BR}}{\sqrt{(R_R/s)^2 + X_{BR}^2}} \tag{68}$$

POLYPHASE INDUCTION MOTORS 361

EXAMPLE 6. Using the data of Example 5, calculate the rotor current if $R_R = 0.1$ ohm and $X_{BR} = 0.5$ ohm.

Solution

$$I_R = \frac{110}{\sqrt{(0.1/0.075)^2 + (0.5)^2}} = \frac{110}{\sqrt{(1.333)^2 + (0.5)^2}} = \frac{110}{1.42} = 77.5 \text{ amp}$$

Equation (68) may be represented by a simple series electric circuit like that shown in Fig. 234. However, since it is convenient to have a

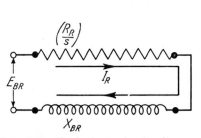
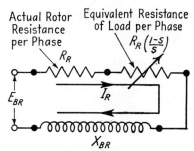

FIG. 234. Equivalent circuit diagram following Eq. (68).

FIG. 235. Modified equivalent circuit diagram of rotor, per phase. [See Eq. (69).]

circuit in which the actual rotor resistance R_R appears, R_R/s may be divided into two components whose sum is its equivalent. That is, let

$$\frac{R_R}{s} = R_R + R_R\left(\frac{1-s}{s}\right) \tag{69}$$

Figure 234 can now be modified so that the series rotor circuit may be represented by the more desirable form shown in Fig. 235. This circuit may be interpreted to mean that the total power delivered to the rotor per phase consists of two parts: (1) the power that causes a copper loss and (2) the electric power that is converted into mechanical power. Thus on a *per phase* basis

RPI (rotor power input) = RCL (rotor copper loss) + RPD (rotor power developed)

$$I_R^2 \times \frac{R_R}{s} = I_R^2 R_R + I_R^2 R_R\left(\frac{1-s}{s}\right) \tag{70}$$

where

$$RPI = I_R^2 \frac{R_R}{s} \tag{71}$$

$$RCL = I_R^2 R_R = RPI \times s \tag{72}$$

$$RPD = I_R^2 R_R\left(\frac{1-s}{s}\right) \tag{73}$$

It is significant that Eq. (73) states that *the mechanical load on an induction motor may be replaced by an electrical resistance whose value is* $R_R(1-s)/s$.

EXAMPLE 7. Using the data of Examples 5 and 6, calculate: (a) the rotor power input; (b) the rotor copper loss; (c) the rotor power developed by the rotor, in watts; (d) the rotor power developed by the rotor in horsepower (1 hp = 746 watts).

# of phases

Solution

(a) $\quad RPI = 3(77.5)^2 \times \dfrac{0.1}{0.075} = 24{,}000$ watts

(b) $\quad RCL = 24{,}000 \times 0.075 = 1{,}800$ watts

(c) $\quad RPD = 3(77.5)^2 \times 0.1 \times \dfrac{1 - 0.075}{0.075} = 22{,}200$ watts

(d) $\quad RPD = \dfrac{22{,}200}{746} = 29.8$ hp

It should be pointed out that the power developed by the rotor RPD is a little greater than the power required to drive the mechanical load because it includes the power that is necessary to overcome the friction and windage losses in the rotor.

Equation (73) may be written in another form. Thus

$$RPD = \frac{I_R^2 R_R}{s}(1-s) = RPI(1-s) \qquad (74)$$

In Example 7, $RPD = 24{,}000 (1 - 0.075) = 22{,}200$ watts.

Rotor Torque. It will be recalled that the *torque* developed by a motor is defined as *the force tending to produce rotation*. It is usually expressed in pound-feet, except in the case of very small motors, when it is given in terms of ounce-inches (oz-in.). Since

$$\text{hp} = \frac{RPD}{746} = \frac{2\pi \times \text{rpm}_{\text{rotor}}}{33{,}000} \times T$$

where T = torque in pound-feet,

$$T = \left(\frac{33{,}000}{746 \times 2\pi}\right) \times \frac{RPD}{\text{rpm}_{\text{rotor}}} = 7.04 \times \frac{RPD}{\text{rpm}_{\text{rotor}}}$$

But $\quad RPD = RPI(1-s) \quad$ from Eq. (74)

$$T = 7.04 \times \frac{RPI(1-s)}{\text{rpm}_{\text{rotor}}} = 7.04 \times \frac{RPI}{\text{rpm}_{\text{rotor}}/(1-s)}$$

Also $\quad \dfrac{\text{rpm}_{\text{rotor}}}{(1-s)} = \text{rpm}_{\text{syn}}$

Hence
$$T = 7.04 \times \frac{RPI}{\text{rpm}_{syn}} \quad (75)$$

EXAMPLE 8. Calculate the torque developed by the motor of Example 7.

Solution
$$T = 7.04 \times \frac{24{,}000}{1{,}200} = 140.8 \text{ lb-ft}$$

As the load on an induction motor increases, its speed decreases or, to put it another way, its slip increases. This increase in load can usually be carried somewhat beyond the rated value, although to do so subjects the motor to an excessive temperature rise if the overload is continued for some time. However, when the load reaches a definite maximum, determined by the manner in which the motor was designed and constructed, its developed torque is just sufficient to drive the mechanical machine to which it is applied. This torque is the *maximum torque* that can be developed and beyond which the motor will stall. It can be shown that maximum torque, sometimes called the *stalling torque*, will be developed when the slip will be very nearly equal to

$$s_{mt} = \frac{R_R}{X_{BR}} \quad (76)$$

Substituting this value of slip in Eq. (68) gives

$$I_{R_{mt}} = \frac{E_{BR}}{\sqrt{2}\, X_{BR}} \quad (77)$$

and substituting Eqs. (77) and (76) into Eq. (71) gives

$$RPI_{mt} = \frac{E_{BR}{}^2}{2 X_{BR}} \quad (78)$$

which can be used in Eq. (75) to determine the value of the maximum torque in pound-feet.

EXAMPLE 9. Calculate: (*a*) the maximum torque that can be developed by the motor of Example 6; (*b*) the speed at which this torque will occur.

Solution

(a) $$RPI_{mt} = 3\left[\frac{(110)^2}{2 \times 0.5}\right] = 36{,}300 \text{ watts}$$

$$T_{max} = 7.04 \times \frac{36{,}300}{1{,}200} = 213 \text{ lb-ft}$$

(b) $$\text{rpm}_{rotor} = 1{,}200\left(1 - \frac{0.1}{0.5}\right) = 1{,}200 \times 0.8 = 960$$

364 ELECTRICAL MACHINES—ALTERNATING CURRENT

Starting Torque. The torque developed by a motor at the instant it is started is called the *starting torque*. In some cases it is greater than the normal running torque, while in others it may be somewhat less than this. However, since the starting torque occurs when the slip s is unity,

$$I_{R_{st}} = \frac{E_{BR}}{\sqrt{R_R^2 + X_{BR}^2}} \quad \text{(at the instant of starting,)} \quad (79)$$

and

$$RPI_{st} = \frac{E_{BR}^2}{R_R^2 + X_{BR}^2} \times R_R \quad (80)$$

which can be used in Eq. (75) to determine the value of the starting torque in pound-feet.

EXAMPLE 10. Calculate the starting torque developed by the motor of Example 6.

Solution

$$RPI_{st} = 3 \left[\frac{(110)^2}{(0.1)^2 + (0.5)^2} \right] \times 0.1 = 14{,}000 \text{ watts}$$

$$T_{st} = 7.04 \times \frac{14{,}000}{1{,}200} = 82 \text{ lb-ft}$$

Induction-motor Efficiency. The calculation of the efficiency of an induction motor follows essentially the same procedure as it does for the d-c generator and motor, the alternator, and the transformer. It will be recalled that this involves first determining the various losses and then applying their sum in the equation for per cent efficiency, which is

$$\text{Per cent efficiency} = \left(1 - \frac{\text{watts losses}}{\text{watts output} + \text{watts losses}}\right) \times 100 \quad (81)$$

Three kinds of power loss occur in induction motors: (1) copper losses in the stator and rotor, (2) iron losses, i.e., hysteresis and eddy-current losses, in the stator, and (3) friction and windage losses in the rotor. Note that *no* iron losses are charged to the rotor, the reason being that the rotor frequency is extremely low under normal operating conditions (remember that iron losses are practically proportional to the *square* of the frequency).

The efficiency of an induction motor as well as other operating characteristics may be determined by performing three tests. These are (1) the *no-load test*, (2) *the load test*, and (3) the *stator-resistance test*. If the motor is a three-phase machine, it is customary to assume that the stator winding is Y-connected. (This may not be the actual case, but the result would be exactly the same if a Δ connection were assumed.) Power is applied to the motor operating free, i.e., without load, and measurements are taken of total power input, current, and voltage, the last being adjusted to a value that is close to the name-plate indication.

Omitting the starting equipment, Fig. 236 shows a wiring diagram of the connections to be made. If the motor is of the wound-rotor type, the external resistor should be cut out completely. When the two-wattmeter method is used to measure the power input, the total power is the *difference* between the wattmeter readings (not the sum) because the no-load power factor is considerably less than 0.5. The second test is performed upon the motor with load applied; the output need not be measured. Instruments must usually be changed to register the higher power and current indications; load adjustment should be made until the ammeters indicate

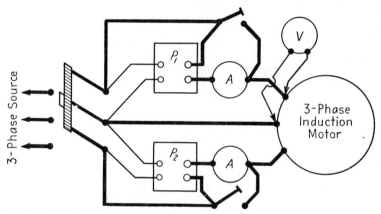

Fig. 236. Wiring diagram for performing no-load test on a three-phase induction motor.

name-plate value if the approximately full-load characteristics are desired. In this test, the wattmeter readings are additive because the power factor is much greater than 0.5. The third test is a simple d-c resistance measurement of the stator winding, per phase. It is customary to measure the resistance between three pairs of stator terminals; the average of these values is then divided by 2 to obtain the resistance per phase, a Y connection having been assumed. Since the effective a-c resistance is higher than the d-c resistance, the latter is usually multiplied by a factor of about 1.25 to obtain the former.

An example will now be given to illustrate the procedure outlined here.

EXAMPLE 11. A 5-hp 60-cycle 115-volt eight-pole three-phase induction motor was tested, and the following data were obtained:

No-load test: $V_{NL} = 115$; $P_1 = 725$; $P_2 = -425$; $I_{NL} = 10$

Load test: $V_L = 115$; $P_1 = 3{,}140$; $P_2 = 1{,}570$; $I_L = 27.3$; $\text{rpm}_{\text{rotor}} = 810$

D-c stator resistance between terminals = 0.128 ohm

Calculate: (a) the horsepower output; (b) the torque; (c) the per cent efficiency; (d) the power factor of the motor for the given load values.

Solution

$$\text{Effective a-c resistance of stator per phase} = \frac{0.128}{2} \times 1.25 = 0.08 \text{ ohm}$$

$$\text{No-load copper loss} = 3 \times (10)^2 \times 0.08 = 24 \text{ watts}$$

$$\text{Friction} + \text{windage} + \text{iron losses} = (725 \overset{P_{NWL}}{-} 425) - 24 = 276 \text{ watts}$$

(It is permissible to charge this loss *entirely* to the stator with no appreciable error, although the friction and windage portion is actually a part of the rotor loss.)

$$\text{Load copper loss in stator} = 3 \times (27.3)^2 \times 0.08 = 179 \text{ watts} \checkmark$$
$$\text{Total stator loss under load} = 276 + 179 = 455 \text{ watts} \checkmark$$
$$RPI \text{ under load} = (3{,}140 + 1{,}570) - 455 = 4{,}255 \text{ watts} \checkmark$$
$$s = \frac{900 - 810}{900} = 0.10$$
$$RCL \text{ [see Eq. (72)]} = 0.10 \times 4{,}255 = 426 \text{ watts} \checkmark$$
$$\text{Rotor power output} = 4{,}255 - 426 = 3{,}829 \text{ watts}$$

(a) $$\text{Horsepower output} = \frac{3{,}829}{746} = 5.13$$

(b) $$T = 7.04 \times \frac{4{,}255}{900} = 33.2 \text{ lb-ft}$$

(c) $$\text{Per cent efficiency} = \left(1 - \frac{455 + 426}{4{,}710}\right) \times 100 = 81.3$$

(d) $$\text{Power factor} = \frac{4{,}710}{\sqrt{3} \times 115 \times 27.3} = 0.866$$

The Blocked-rotor Test. The values of the rotor resistance and the rotor reactance, whether of the squirrel-cage or wound-rotor type, can be determined by performing the so-called *blocked-rotor test*. Great care should be taken when doing this because the motor temperature will rise very rapidly unless the current is not too high and the instrument readings are taken quickly. The same wiring diagram as that of Fig. 236 is used, except that some method must be employed to apply about 20 to 30 per cent of rated voltage to the stator. A good method is to insert three rheostats in the line wires, adjusting the three-phase motor voltage to a value at which approximately rated current flows *when the rotor is blocked and cannot rotate*. Readings are taken of power input, amperes, and volts. Since the blocked-rotor motor is, in fact, similar to a transformer with the secondary short-circuited, the principles discussed in Chap. 8 apply equally well here. Assuming a Y connection,

$$R_e = \frac{P_{BR}}{3 \times I_{BR}^2} \qquad (82)$$

$$Z_e = \frac{V_{BR}/\sqrt{3}}{I_{BR}} \qquad (83)$$

$$X_e = \sqrt{Z_e^2 - R_e^2} \qquad (84)$$

where R_e, Z_e, X_e = equivalent values of motor resistance, impedance, and reactance per phase, in stator terms, respectively
P_{BR} = total power input with rotor blocked
V_{BR} = voltage across stator terminals with rotor blocked
I_{BR} = stator current with rotor blocked

The rotor resistance per phase, in stator terms, can now be obtained by subtracting the effective stator resistance from the value obtained in Eq. (82). Thus

$$R_R = R_e - R_{\text{stator}} \qquad (85)$$

In practice it is generally assumed that the total equivalent reactance is divided equally between the stator and rotor. Therefore

$$X_{BR} = X_{\text{stator}} = \frac{X_e}{2} \qquad (86)$$

The following example illustrates the procedure discussed and the way in which the results can yield additional information concerning the characteristics of the motor:

EXAMPLE 12. A blocked-rotor test was performed on the motor of Example 11, and the following data were obtained: $V_{BR} = 26$; $I_{BR} = 32$; $P_1 = 1,430$; $P_2 = -860$. Calculate: (a) the equivalent resistance of the motor R_e; (b) the equivalent reactance of the motor X_e; (c) the rotor resistance and reactance R_R and X_{BR}; (d) the speed at which maximum torque will occur.

Solution

(a) $\quad R_e = \dfrac{570}{3 \times (32)^2} = 0.186$ ohm

(b) $\quad Z_e = \dfrac{26}{\sqrt{3} \times 32} = 0.47$ ohm

$\quad X_e = \sqrt{(0.47)^2 - (0.186)^2} = 0.432$ ohm

(c) $\quad R_R = 0.186 - 0.08 = 0.106$ ohm

$\quad X_{BR} = \dfrac{0.432}{2} = 0.216$ ohm

(d) $\quad \text{rpm} = 900\left(1 - \dfrac{0.106}{0.216}\right) = 458$

At the instant the motor is started, the rotor is not turning; the friction and windage loss is therefore zero. However, the iron loss at starting will be greater than at normal operating value because the rotor frequency is the same as the stator frequency. It is therefore fair to assume that the increased iron loss caused by the rotor exactly compensates for the friction and windage, under which condition the no-load loss may be considered constant. Furthermore, if the blocked-rotor test had been made at *rated* voltage, the stator current with the rotor blocked would have been equal to $(V_{rated}/V_{BR}) \times I_{BR}$, and the power input with the rotor blocked would have been $(V_{rated}/V_{BR})^2 \times P_{BR}$. With these points in mind, it is possible to calculate the starting torque.

EXAMPLE 13. Calculate the starting torque of the motor in Examples 11 and 12.

Solution

From Example 11, iron loss = 276 watts.

Stator current at 115 volts with rotor blocked = $\dfrac{115}{26} \times 32 = 142$ amp

Power input at 115 volts with rotor blocked = $\left(\dfrac{115}{26}\right)^2 \times 570$
$= 11,160$ watts

Stator copper loss = $3 \times (142)^2 \times 0.08 = 4,830$ watts

$RPI = 11,160 - 276 - 4,830 = 6,054$ watts

$$T_{st} = 7.04 \times \dfrac{6,054}{900} = 47.3 \text{ lb-ft}$$

In order to determine the maximum torque of the motor it is necessary that a load be applied until the speed drops to 458 rpm. Data should then be obtained and calculations made in the same manner as given in Example 11.

Starting Induction Motors. As a general rule, a motor is subjected to the most severe service during its starting and accelerating period; it is during this interval of time that an extremely high current tends to flow unless steps are taken to keep its value down to a reasonable level. In the case of an induction motor, the starting instant is equivalent to a transformer with a short-circuited secondary; this implies that a high rotor (secondary) current will, by transformer action, have to be matched by a correspondingly large stator (primary) current. Of course, as the rotor speeds up, its generated voltage E_R ($= sE_{BR}$) decreases, so that both the rotor and stator currents drop to values determined only by the mechanical load.

It is true that the starting of an induction motor does not impose electrical stresses that are as severe as those experienced by commutator-type motors; nevertheless, it is important that the starting current be minimized to such an extent that the line-voltage drop does not affect the operation of other equipment connected to the same line wires. Modern well-designed induction motors will usually take about 5 to 7 times the rated full-load current at the starting instant if rated voltage is connected to its terminals. Thus, for example, a 5-hp 220-volt 1,750-rpm motor, the rated full-load current of which is 13.6 amp, will draw about 90 amp at the starting instant if full voltage is impressed. One obvious way to reduce the starting current is to impress a lower starting voltage across the motor. And since the only limiting factor is the blocked-rotor equivalent impedance Z_e [see Eq. (83)], it follows that the starting current is *directly* proportional to the starting voltage. Therefore, in order to reduce the starting current by 50 per cent, it is necessary that the starting voltage be lowered by this same percentage. For the illustration given, then, this would mean that the starting current can be reduced to 45 amp if 110 volts is applied during the starting period.

One serious objection to the practice of using a reduced starting voltage is the large reduction in starting torque. Starting torque is proportional to the *square* of the impressed voltage because the former depends upon the flux as well as the starting current. Since both factors, ϕ and I, are equally affected by a reduction of the impressed voltage, the starting torque suffers on two counts. In the foregoing 5-hp-motor illustration, the starting torque will be only 7 lb-ft if 110 volts is used to start the motor, as compared with 28 lb-ft when the rated 220 volts is impressed. Since rated full-load torque is 15 lb-ft, the starting torque drops from about 187 per cent to 47 per cent of normal operating torque. Whether or not such a great reduction in starting torque is permissible will depend, of course, upon the application. But it is important to note that a desirable change in one factor (current) is accompanied by a more pronounced undesirable change in another factor (starting torque).

EXAMPLE 14. The following information is given in connection with a 50-hp 440-volt 1,160-rpm induction motor: full-load torque and current at 440 volts are 227 lb-ft and 63 amp, respectively; starting torque and current at 440 volts are 306 lb-ft and 362 amp, respectively. (a) Calculate the starting current and torque at 254 volts. (b) What percentages of the rated values are the values calculated in (a)?

Solution

(a) $$I_{254} = 362 \times \frac{254}{440} = 209 \text{ amp}$$

$$T_{254} = 306 \times \left(\frac{254}{440}\right)^2 = 102 \text{ lb-ft}$$

(b)
$$\text{Per cent } I = \frac{209}{63} \times 100 = 332$$

$$\text{Per cent } T = \frac{102}{227} \times 100 = 45$$

Induction-motor Starting Methods. Three methods generally used to start polyphase *squirrel-cage* motors may be classified as (1) full-voltage starting, (2) reduced-voltage starting, and (3) part-winding starting. *Wound-rotor* motors are started by inserting resistors in the leads connected to the brushes that ride on the slip rings (see Fig. 228).

1. *Full-voltage Method.* Motors that are started by being connected directly across the line are permitted to take extremely high values of

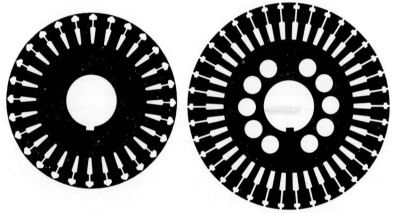

Fig. 237. Typical rotor laminations for double squirrel-cage induction motors.

starting current. Although there is no limitation on the size of the motor that may be started in this way, it should be understood that objectionable line-voltage dips will usually occur, especially if large motors are started frequently. Whether or not full-voltage starting is used will therefore depend upon such factors as (1) the size and design of motor, (2) the kind of application, (3) the location of the motor in the distribution system, and (4) the capacity of the power system and the rules governing such installations as established by public-service companies. Concerning the design of the motor, this generally refers to the *squirrel-cage rotor* construction and is usually one in which there are two squirrel cages, one above the other, or a design in which the rotor slots are deep and narrow to accommodate similarly shaped bars. Sketches of typical cross sections of double-cage rotors are shown in Fig. 237. Note particularly that there are two sets of slots separated by a narrow contracted section where the iron nearly meets. After the laminations are properly assembled and the molten metal, usually aluminum, is forced into the slots, two squirrel cages are formed, with the outer one having a higher resistance than the one on the inside. When a motor of this construction is started, the flux density in the rotor iron is high; this causes a considerable leakage flux to surround the inner cage, with the result that its reactance is comparatively greater—the rotor frequency is high when the motor is started—than the outer cage. The current is therefore forced to the high-resistance cage, and the rotor performs as though it has a single squirrel cage; the starting current is thereby limited to a reasonable

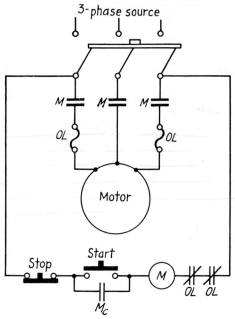

FIG. 238. Wiring diagram of a line-start automatic starter connected to a three-phase squirrel-cage induction motor.

value, and the motor develops good starting torque. As the motor speeds up, the rotor frequency diminishes—it is the slip frequency—and the reactance of the inner cage drops; the current now shifts mostly to the low-resistance inner cage. Operation then continues with two squirrel cages in parallel, a condition that permits the motor to perform very efficiently.

Line-start motors are generally used with full-voltage automatic starters, the latter being simple, inexpensive, and easy to install and maintain. A diagram of connections for such an installation is given in Fig. 238. To start the motor, it is merely necessary to press the START button. This energizes the M contactor through the overload relay contacts OL (normally closed), which, in turn, close the main contacts

M; contacts M_c also close to seal the main contactor, so that the START button may be released. Overload protection is provided by two thermal elements OL placed in the motor leads; should the motor overheat, the thermal elements open the OL contacts in the control circuit to deenergize the main coil; this opens the M contacts and disconnects the motor from its source.

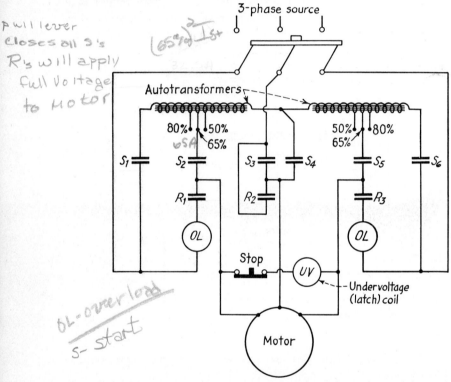

FIG. 239. Wiring diagram of a compensator connected to a three-phase squirrel-cage motor.

2a. Reduced-voltage—Compensator—Method. A diagram showing this type of starter is given in Fig. 239. It consists of two tapped autotransformers connected in open Δ, a set of six starting contacts, a set of three running contacts, two overload coils, an undervoltage coil-stop button control unit, and a handle (not shown) that is operated manually to actuate the *start* and *run* contacts. (Some starters contain three autotransformers connected in Y or Δ for slightly better current balance; the two-unit starter is, however, most common because a usual unbalance of about 15 per cent is not considered objectionable.) The autotransformers are generally tapped at the 50, 65, and 80 per cent points, so that adjustment at these voltages may be made for proper starting-torque requirements. Since the contacts frequently break large values of current, arcing is sometimes quenched effectively by having them assembled to operate in a bath of oil.

A spring-loaded lever, mounted on the outside of a steel cabinet to stand vertically in the OFF position, is pulled back to start the motor.

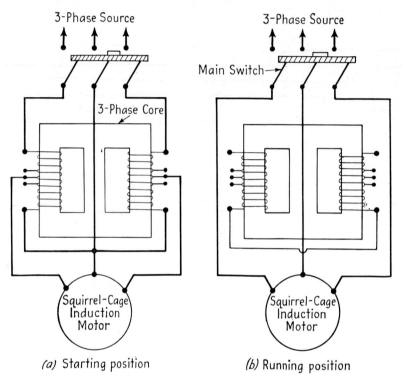

(a) Starting position   (b) Running position

FIG. 240. Wiring diagrams showing compensator, using an open-Δ connection, for starting a squirrel-cage induction motor.

This action closes the six start contacts, energizes the two autotransformers through contacts S_1, S_3, S_4, and S_6, and impresses reduced voltage (65 per cent in Fig. 239) across the motor through contacts S_2, S_4, and S_5. These connections are shown schematically in Fig. 240a. After the rotor has accelerated to about full speed, the handle is quickly pushed forward; this instantly opens the *start* contacts and *then*, a split second later, closes the *run* contacts R_1, R_2, and R_3. The motor is now connected across the full-voltage source as illustrated by the schematic sketch, Fig. 240b. A latch, fixed to the mechanism, is made to drop into a notch so that the operator is prevented from throwing the handle accidentally to the RUN position first; however, when the handle is quickly pushed from the START to the RUN position, the latch is kicked up to free the lever for its forward motion. An electromagnet, shown as the undervoltage (latch) coil in Fig. 239, continues to hold the handle in the RUN position until the STOP button is pressed; moreover, if the line voltage should fall to a value too low for efficient operation, the electromagnet will release and trip the holding mechanism. Another feature usually incorporated is a pair of overload coils OL that carry load currents and act to trip the lever in cases of sustained high currents.

It should be understood that the motor current, in starting, is directly proportional to the impressed voltage. Furthermore, neglecting the magnetizing currents taken by the autotransformers, the current on the primary side of the autotransformers is *less* than the motor current by the same ratio. Thus, if the motor starting current on full voltage is 6 times rated value, the use of autotransformers with 50 per cent taps will cause the *motor* to take 3 times rated current, and the *line* to carry 1½ times rated current, at the instant it is started.

EXAMPLE 15. The 65 per cent taps on a compensator are used in connection with a 10-hp 220-volt three-phase motor. Rated motor current is 26.4 amp, and blocked-rotor current at 220 volts is 150 amp. Calculate: (a) the starting current delivered to the motor; (b) the current on the line side of the compensator, neglecting the magnetizing currents in the autotransformers; (c) the per cent motor and line currents, in terms of rated amperes.

Solution

(a) $I_{motor} = 0.65 \times 150 = 97.5$ amp

(b) $I_{line} = 0.65 \times 97.5 = 63.4$ amp

(c) Per cent motor current $= \dfrac{97.5}{26.4} \times 100 = 369$

Per cent line current $= \dfrac{63.4}{26.4} \times 100 = 240$

2b. *Reduced-voltage—Line-resistance—Method.* The insertion of resistances in the line wires to start an induction motor is equivalent to the procedure followed in connection with the starting of a d-c motor. Figure 241 is a simple wiring diagram of a *manually* operated starter showing how this is done in a three-phase motor. The three line rheostats must obviously be identical, and the resistances must be gradually cut out *simultaneously* as the motor accelerates. It should be pointed out that each line resistance adds vectorially to the equivalent motor resistance and reactance per phase; therefore the addition of a given value of line resistance does not reduce the starting current to the same extent that it would in a d-c motor.

A wiring diagram showing how an *automatic* line-resistance starter is connected to a three-phase squirrel-cage induction motor is given in Fig. 242. It consists of three line resistors, two sets of three line contacts, two contactors, a timing relay, overload protection, a push-button station, and several auxiliary contacts. When the START button is pressed, the S contactor and the timing relay TR are energized. This causes the S contacts to close and start the motor with the three line resistors in series

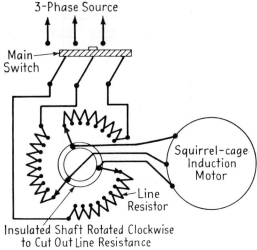

FIG. 241. Wiring diagram showing line-resistance method of starting a squirrel-cage induction motor.

with the former; also, the interlock contacts TR close to seal the S contactor and the TR relay. After a time delay, set by the timing relay TR, the latter times out to close the TR contacts, which, in turn, make the R contactor pick up. The R contacts then close, short-circuit the line resistors, and permit the motor to operate from its full-voltage source. As in Fig. 238, should there be a sustained overload, the OL contacts will open to stop the motor, as will the pressing of the STOP button.

2c. Reduced-voltage—Y-Δ—Method. If an induction motor is designed so that the stator winding is connected in Δ when in normal operation, the line voltage is also the voltage across each phase. However, when the same winding is connected in Y, the voltage per phase for the same source emf will be only $1/\sqrt{3} \times 100$, or 58 per cent of the line voltage. Advantage is taken of this fact in the so-called Y-Δ starting method, where, as in the *manually* operated starter of Fig. 243, a triple-pole double-pole (TPDT) switch connects the stator winding in Y when the motor is started and changes the connection to Δ after the machine has accelerated to normal speed. This method is sometimes favored because it is quite inexpensive, especially in installations in which the motors are not started frequently or where the starting requirements are not too severe. It

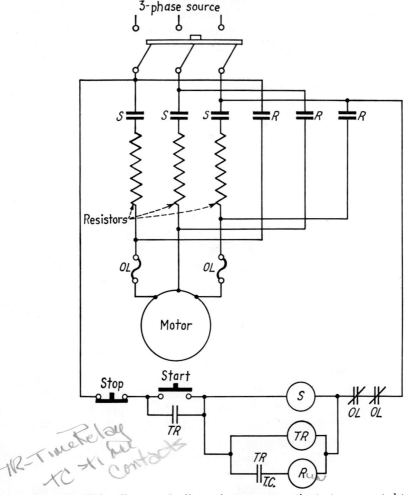

FIG. 242. Wiring diagram of a line-resistance automatic starter connected to a three-phase squirrel-cage induction motor.

does, however, lack flexibility in that only a single starting voltage is made available.

If it is desired to employ an *automatic* starter that will perform the same functions, the wiring connections shown in Fig. 244 may be used. When the START button is pressed, the M contacts close to connect winding terminals a, b, and c to the line, and the S contacts close to join winding terminals a', b', and c', to form the star point. After the timing relay TR times out, the S contacts open first, and this is followed by the closing of the R contacts; the latter connect the winding in Δ, with terminals a and b' forming one junction at R_3, b and c' forming a second junction at R_1, and c and a' forming a third junction at R_2.

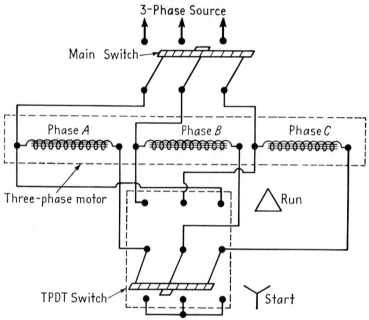

Fig. 243. Wiring diagram showing Y-Δ method of starting a squirrel-cage induction motor.

3. *Part-winding Method.* When a three-phase induction motor has a winding that is divided into two identical Y-connected sections and the latter are joined in *parallel* for *normal* operation, *part-winding starting* may be employed to limit inrush currents. The method is particularly adapted to dual-voltage motors, usually designed to operate at either of two voltages such as 220 or 440 volts; in such machines nine leads are brought out and labeled as indicated in Fig. 245a. For 440-volt operation, the series-Y connection is used with terminals T_4 and T_7 joined, T_5 and T_8 joined, T_6 and T_9 joined, and T_1, T_2, and T_3 serving as line terminals; for 220-volt operation, the two-parallel-Y connection is employed with terminals T_4, T_5, and T_6 connected to form a second Y point, and one line lead connected to a junction of T_1 and T_7, a second line lead connected to a junction of T_2 and T_8, and a third line lead connected to a junction of T_3 and T_9.

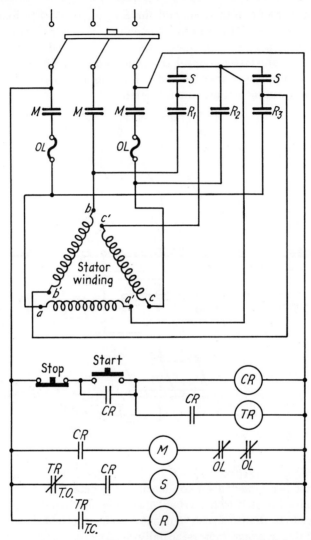

Fig. 244. Wiring diagram of a Y-Δ automatic starter connected to a three-phase squirrel-cage induction motor.

A motor that is to be prepared for part-winding starting, Fig. 245b, operates *normally* with its stator winding connected *two-parallel Y;* this means, therefore, that terminals T_4, T_5, and T_6 are permanently joined to form the second star point. Then, when the motor is started, only one section of the winding is used by connecting T_1, T_2, and T_3 to the three-phase source, for example, 220 volts; after the motor accelerates to about rated speed, the two sections are joined in parallel as previously described.

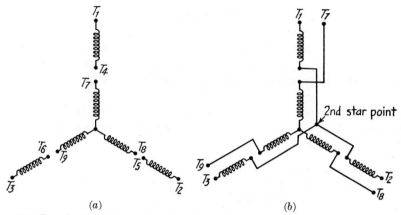

Fig. 245. Terminal markings and connections of a stator winding for (a) dual-voltage operations and (b) part-winding starting.

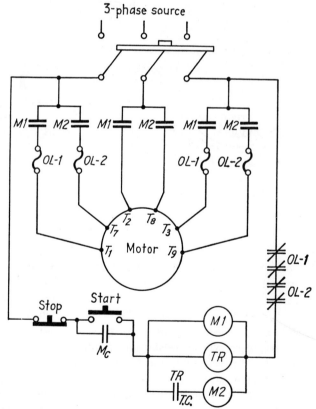

Fig. 246. Wiring diagram of a part-winding automatic starter connected to a three-phase squirrel-cage induction motor.

Such a starting procedure usually limits the inrush currents to about 60 to 70 per cent of the values resulting from full-voltage starting.

A wiring diagram showing a motor connected to an automatic starter for part-winding starting is given in Fig. 246. When the START button is pressed, the $M1$ contactor and the timing relay TR are energized; this closes the sealing contacts M_c and the three main contacts $M1$. The motor then starts with one-half of the stator winding connected across the line. After a given time delay, set by the timing relay TR, the

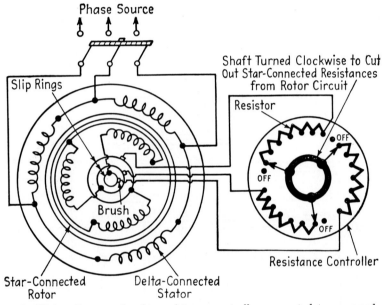

FIG. 247. Wiring diagram showing resistance controller connected to a wound-rotor motor for starting, speed control, and starting-torque-adjustment purposes.

latter times out; this causes the timed contacts $TR\text{-}T.C.$ to close, energize the $M2$ contactor, and close the second set of three main contacts $M2$. The two sections of the winding are now in parallel, and the motor proceeds to operate normally as a two-parallel-Y-connected machine.

The Wound-rotor Method. The wound-rotor induction motor (see Fig. 228) is frequently used when the starting requirements are particularly severe. Not only is it possible to limit the current to a reasonable value during the starting period, but also, as will be shown later, this rotor construction offers two additional advantages: (1) the starting torque can be adjusted to the motor's maximum torque and (2) the speed of the motor can be controlled over a wide range. Adjustments are readily made by simultaneously varying three resistors connected in the rotor circuit of the motor. This is accomplished (see Fig. 247) by using a Y-connected set of rheostats wired to the brushes that ride on the slip rings. In the OFF position the Y-connected rotor is open-circuited and will not rotate. When moved to the first set of studs, the resistance controller connects a proper value of resistance in each phase to limit the current in the rotor circuit [see Eq. (79)]. Since the stator current will, by transformer action, be matched by a correspondingly limited value of current, the motor will start with reduced input. As the motor speeds up, the three arms are rotated clockwise to cut out rotor resistance until, in the final position, the three brushes are short-circuited; the rotor winding is then electrically equivalent to a squirrel-cage rotor. It is customary to design the resistance controller so that the values of the starting resistors are such that the motor develops maximum starting torque. Speed control is accomplished by cutting in the proper values of resistance; the speed drops with increase in rotor resistance.

Operating Characteristics of Squirrel-cage Motors. The operating performance of squirrel-cage induction motors is largely determined by the *rotor resistance*, the *air gap* between stator and rotor, and the *shape of the teeth and slots* in the stator and rotor. These design factors affect such characteristics as starting current, starting torque, maximum torque, per cent regulation, and efficiency. And since there are a great many motor applications, differing in their requirements, manufacturers find it necessary to make machines that satisfy many specific needs.

One of the most important of the points noted above is rotor resistance. Its value is determined primarily by the material used in its construction and the area of cross section of the end rings. The two most widely used materials are copper and aluminum, the latter having a resistivity that is about 1.65 times as much as that of the former. In general, the following points can be made concerning the way in which an *increase in rotor resistance* affects the motor performance:

1. The starting torque increases until it reaches the maximum torque, after which it diminishes.
2. The starting current decreases.
3. The load efficiency decreases.
4. The per cent regulation increases.
5. The maximum torque is unaffected because it depends upon the rotor reactance only.

The shapes of the rotor teeth and slots affect the reactance, which, in turn, will have a bearing upon the starting current and the maximum torque. When the rotor slots are deep and narrow, as opposed to shallow and wide, the rotor reactance is high. The effect of the first condition is (1) to reduce the starting current and (2) to reduce the maximum torque.

Double-cage rotors combine two desirable slot designs so that (1) the starting current is comparatively low, (2) the starting torque is high, and (3) the load efficiency is high. As was previously discussed, motors using such rotors may be started directly from the line without the need for auxiliary starting equipment.

The air gap of an induction motor is always much less than that found in d-c machines. In fact, the clearance between stator and rotor is generally as small as mechanical tolerance will permit, owing to the component of stator current that must magnetize a circuit codsisting of the stator and rotor iron and the *air gap;* the latter requires a major part of the total mmf. Thus, if the air gap is large, the magnetizing current will be increased to such an extent that the motor power factor will drop. On the other hand, if the air gap is shortened to improve the motor power factor, it is usually necessary to use high-grade bearings, ball bearings being frequently employed in such cases.

Motor manufacturers have standardized on several general types of construction that seem to fulfill certain classes of application. Class A squirrel-cage motors are built for general-purpose duty requiring low starting torque. These develop starting torques of approximately 125 to 175 per cent of rated value and take about 5 to 7 times rated current when started at full voltage. Such motors are generally started at reduced voltage, by one of the methods previously described although they are suitable for line-voltage starting and many severe-service installations. They are used in many machine-tool applications, for ventilating equipment such as fans and blowers, for pumps of the centrifugal type, and to drive electric generators.

Class B motors are also built for general-purpose applications, and are usually started directly from the full line voltage. They have comparatively high reactance; their starting currents therefore are about $4\frac{1}{2}$ to 5 times rated at about 125 to 175 per cent starting torque when full voltage is impressed. The power factors of class B motors are consistently lower than motors of the class A type. The practical uses of these motors are similar to those given above.

Class C motors are the double-cage type and are generally started by applying full voltage. Starting currents are about $4\frac{1}{2}$ to 5 times normal, while starting torques reach values that are more than twice that at rated output. Frequent applications are compression pumps, large refrigerators, crushers, conveyor equipment, boring mills, woodworking equipment, and textile machinery.

Class D motors have rotors with comparatively high resistances and are used where the starting service is particularly heavy. Efficiencies of such machines are always lower than those listed above, and their regulations are considerably higher. These, too, are started by applying full voltage, under which condition the starting currents are about 4 to 5 times rated value. Their starting torques are about 2 to 3 times that developed at full load. Since their efficiencies are subnormal, they are used only where extremely high starting torque is an important requirement. The most usual applications are bulldozers, shearing machines, hoists, metal drawing equipment, stamping machines, laundry equipment, punch presses, foundry equipment, and stokers.

Table 7 lists squirrel-cage motors that fall in the *A* classification. The range covers outputs from 1 to 250 hp and includes the more important

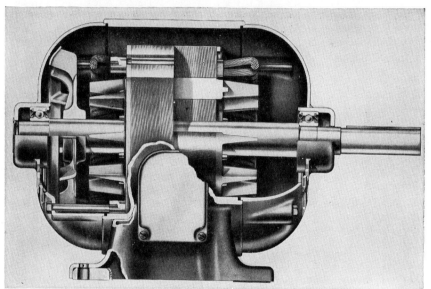

FIG. 248. Cutaway view of a completely assembled squirrel-cage induction motor. (*Westinghouse Electric Corp.*)

operating characteristics. A careful study of this table should prove extremely profitable.

A cutaway view of a completely assembled ball-bearing squirrel-cage induction motor is shown in Fig. 248. The rotor is of cast aluminum construction, with fins projecting from both end rings; good ventilation is provided by the latter.

In many machine-tool and woodworking applications, limited space is provided for the motor. In such cases it is particularly advantageous to build the motor as an integral part of the machine itself. The *shaftless motor* is well adapted to such installations because the machine itself provides the shaft for the rotor; a direct motor drive is therefore possible. This arrangement eliminates space-consuming, costly, and complicated mechanical transmissions that would otherwise be needed to transmit

TABLE 7. PERFORMANCE DATA FOR NORMAL TORQUE THREE-PHASE SQUIRREL-CAGE MOTORS

Hp	Poles	Full-load rpm	Full-load amperes			Torque, pound-feet			Full-load per cent efficiency	Power factor
			208 volts	220 volts	440 volts	Full-load	Max	Start		
1	2	3,450	3.2	2.9	1.45	1.52	4.5	2.6	82	0.82
	4	1,750	3.4	3.2	1.6	3.00	9.0	9.0	80	0.76
	6	1,160	3.8	3.6	1.8	4.53	12.4	9.0	80	0.68
	8	870	4.2	4.0	2.0	6.04	15.0	9.0	78	0.63
2	2	3,500	5.8	5.4	2.7	3.0	8.2	5.2	83	0.87
	4	1,750	6.2	5.8	2.9	6.0	18.0	15.0	83	0.81
	6	1,160	7.0	6.6	3.3	9.1	22.7	16.0	82	0.72
	8	870	8.0	7.6	3.8	12.1	27.4	18.0	79	0.65
3	2	3,500	8.2	7.8	3.9	4.5	11.2	7.8	85	0.89
	4	1,750	9.0	8.4	4.2	9.0	25.0	23.0	85	0.82
	6	1,160	10.2	9.6	4.8	13.6	34.0	24.0	84	0.73
	8	870	11.4	10.8	5.4	18.1	40.7	27.9	84	0.65
5	2	3,500	14.0	13.0	6.5	7.5	16.8	11.0	85	0.89
	4	1,750	14.4	13.6	6.8	15.0	33.7	28.0	85	0.85
	6	1,160	16.0	15.0	7.5	22.7	51.0	36.0	85	0.77
	8	870	17.4	16.4	8.2	30.2	68.0	39.0	84	0.71
7½	2	3,500	20.0	19.2	9.6	11.3	24.0	17.0	85	0.90
	4	1,750	21.0	20.0	10.0	22.5	48.0	39.0	85	0.87
	6	1,160	23.0	21.6	10.8	34.0	73.0	51.0	85	0.80
	8	870	24.0	23.0	11.5	45.3	97.0	56.0	84	0.76
10	2	3,500	27.0	25.6	12.8	15.0	30.0	22.0	85	0.90
	4	1,750	28.0	26.4	13.2	30.0	60.0	52.0	85	0.87
	6	1,160	30.0	28.0	14.0	45.3	90.0	68.0	85	0.82
	8	870	32.0	30.0	15.0	60.4	120.0	75.0	85	0.77
15	2	3,500	40.0	38.0	19.0	22.5	45.0	34.0	86	0.90
	4	1,750	41.0	39.0	19.5	45.0	90.0	74.0	86	0.88
	6	1,160	42.0	40.0	20.0	68.0	136.0	95.0	86	0.85
	8	870	46.0	43.0	21.5	90.6	181.0	113.0	86	0.79
20	2	3,500	53.0	50.0	25.0	30.0	60.0	45.0	87	0.90
	4	1,750	54.0	51.0	25.5	60.0	120.0	90.0	87	0.88
	6	1,160	56.0	53.0	26.5	90.6	181.0	122.0	87	0.85
	8	870	60.0	56.0	28.0	121.0	242.0	151.0	87	0.80
25	2	3,500	66.0	62.0	31.0	37.5	75.0	56.0	88	0.90
	4	1,750	66.0	62.0	31.0	75.0	150.0	112.0	88	0.89
	6	1,160	68.0	64.0	32.0	113.0	226.0	152.0	88	0.87
	8	870	72.0	69.0	34.5	151.0	302.0	188.0	88	0.81

Table 7. (Continued)

Hp	Poles	Full-load rpm	Full-load amperes			Torque, pound-feet			Full-load per cent efficiency	Power factor
			208 volts	220 volts	440 volts	Full-load	Max	Start		
30	2	3,500	78	74	37	45	90	67	89	0.90
	4	1,750	78	74	37	90	180	135	89	0.89
	6	1,160	80	76	38	136	272	183	89	0.87
	8	870	86	82	41	181	362	226	88	0.82
40	2	3,500	104	98	49	60	120	81	89	0.90
	4	1,750	104	98	49	120	240	180	89	0.90
	6	1,160	106	100	50	181	362	244	89	0.88
	8	870	112	106	53	242	484	302	88	0.84
50	2	3,500	130	122	61	75	150	93	89	0.90
	4	1,750	130	122	61	150	300	225	89	0.90
	6	1,160	132	126	63	227	454	306	89	0.88
	8	870	140	132	66	302	762	377	88	0.84
75	2	3,500	192	182	91	113	226	124	90	0.90
	4	1,750	192	182	91	225	450	337	90	0.90
	6	1,160	196	188	94	340	680	458	90	0.88
	8	870	204	192	96	453	906	565	89	0.86
100	2	3,500	256	242	121	150	300	165	90	0.90
	4	1,750	256	242	121	300	600	375	90	0.90
	6	1,160	262	248	124	453	906	565	90	0.88
	8	870	270	256	128	604	1,208	755	89	0.86
125	4	1,750	318	300	150	375	750	410	90	0.90
	6	1,160	324	306	153	566	1,132	705	90	0.90
	8	870	334	316	158	755	1,510	940	89	0.89
	10	690	342	324	162	951	1,902	1,140	89	0.89
150	4	1,750	374	354	177	450	900	495	91	0.91
	6	1,160	380	360	180	680	1,360	850	91	0.91
	8	810	388	366	183	906	1,812	1,130	90	0.90
	10	690	400	380	190	1,141	2,282	1,370	89	0.89
200	4	1,750	500	474	237	600	1,200	600	91	0.91
	6	1,160	508	480	240	906	1,812	1,130	91	0.91
	8	870	516	488	244	1,208	2,416	1,510	90	0.90
	10	690	536	506	253	1,522	3,044	1,825	89	0.89
250	4	1,750	624	590	295	750	1,500	750	91	0.91
	6	1,160	630	596	298	1,133	2,266	1,415	91	0.91
	8	870	640	604	302	1,510	3,020	1,885	90	0.90
	10	690	654	618	309	1,903	3,806	2,280	90	0.90

power from one part of the machine to another. A motor of this type is shown in Fig. 249. Practical applications of the shaftless motor are sensitive drills, vertical surface grinders, double-spindle shapers, precision boring machines, turret lathes, rod machines, cutoff saws, drill heads, and bed joiners. Added advantages of shaftless motor installations are unified and pleasing design, more efficient use of power because of locating the motor close to the point where power is used, reduction in weight, fewer wearing parts, centralized control, and the use of several motors to drive several parts of a complex machine, allowing a selection of the motor characteristics best suited to particular operations.

FIG. 249. Housing type of squirrel-cage induction motor for built-in applications. (*The Louis Allis Co.*)

Operating Characteristics of Wound-rotor Motors. The wound-rotor type of induction motor (see Fig. 228) is adapted to installations in which the loads must be started smoothly or in which extremely high starting torque is required at comparatively low values of starting current. The principal difference between squirrel-cage and wound-rotor motors is in the rotor winding. As was pointed out in connection with the squirrel cage, the rotor is a self-contained unit with a series of straight bars around the periphery that are permanently short-circuited at both ends by rings. It has no electrical or mechanical connection with the source of supply or the control circuit. Its resistance is fixed, thus providing fixed full-load operating speed, starting and maximum torques, and rate-of-acceleration characteristics. The wound-rotor construction differs from the squirrel cage in that it has a definite winding similar to the one used in the stator. The ends of this winding are connected to collector rings—slip rings—over which ride brushes. The latter are connected to a resistance controller, as shown in Fig. 247. Varying the values of resistance in the rotor circuit produces variation in starting torque and variation in operating speed with its possible smooth acceleration, both of which are definitely controlled at will. These motors permit the variation of the speed by as much as 50 to 75 per cent; the greater the resistance inserted in the rotor circuit, the lower is the speed *below* synchronous speed. When the resistance controller is cut completely out, the speed is somewhat less than that obtained with squirrel-cage rotors. These motors are ideally suited for the following applications: small electric excavators, printing presses, elevators, cranes, displacement pumps, compressors, hoists, bridges, turntables, strokers, large ventilating fans, crushers, and coal and ore loading. In all these it is essential that the motor start the high-inertia load smoothly and effectively.

Speed Control of Induction Motors. One of the most serious limitations of the induction motor is that its speed cannot be controlled so easily or so efficiently as that of the d-c motor. It will be recalled that the latter lends itself readily to control by the use of field or armature rheostats or by armature-voltage adjustment. Many methods have been devised to control the speed of the induction motor, but in all cases efficiency is generally low or the cost of the equipment high. It is for this reason, undoubtedly, that d-c motors are replacing a-c types where efficient speed control is essential to the application. The following methods are given to illustrate partially some of the methods commonly employed to control the speed of induction motors.

Wound-rotor Method. This is a widely used method, particularly for machines of up to about 500 hp. An important fact to recognize is that *the per cent efficiency* of a wound-rotor motor *is always less than* $(1 - s) \times 100$. Thus, if resistance is inserted in the rotor circuit so that the slip is 0.4, the motor efficiency is *less* than 60 per cent. For a slip of 0.6, the efficiency would be less than 40 per cent. In other words, the insertion of resistance in the rotor circuit to reduce the speed always involves a power loss in that external resistor. The general statement is proved as follows:

$$RPD = I_R^2 R_R \times \frac{1-s}{s} \qquad (73)$$

$$RPI = I_R^2 \frac{R_R}{s} \qquad (71)$$

Dividing Eq. (73) by Eq. (71) gives

$$\text{Per cent rotor efficiency} = \frac{RPD}{RPI} \times 100 = (1 - s) \times 100 \qquad (87)$$

Since the *motor efficiency* is obviously *less* than the *rotor efficiency* because there are stator iron and copper losses, it should be clear that speed control by this method is extremely inefficient.

Concatenation Method. Two motors may be coupled together to drive a common load. If the stator winding of one of them, a wound-rotor motor, is connected to the source, and its rotor winding is then joined to the stator winding of the second motor, the speed of the combination will be determined by the sum or difference of the numbers of poles in the two

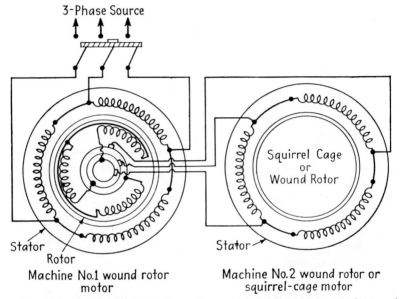

Fig. 250. Concatenation method of speed control. Both motors must be coupled together to drive the common load.

machines. In practice, it is customary to connect the rotor output of the first machine to the stator of the second machine so that the revolving fields of both are in the same direction; under this condition, the resulting speed will be determined by the *sum* of the poles in both machines. Figure 250 shows the method of connection. If further speed control is desired, the second machine may also be of the wound-rotor type. Thus, if the stator of machine 1 is wound for four poles and the stator of machine 2 is wound for six poles, the 60-cycle speeds will be (1) $1,800(1 - s)$, (2) $1,200(1 - s)$, and (3) $720(1 - s)$. As pointed out above, in-between speeds are obtainable by varying the rotor resistance in the second motor. The following proof is given for the foregoing statement. Since both

POLYPHASE INDUCTION MOTORS 389

machines *must* operate at the same speed, being coupled together,

$$\frac{120f_1}{P_1} \times (1 - s_1) = \frac{120f_1 s_1}{P_2} (1 - s_2)$$

or

$$\frac{1 - s_1}{P_1} = \frac{s_1 - s_1 s_2}{P_2}$$

But the product of s_1 and s_2 is usually very small compared with the other terms. Therefore

$$\frac{1 - s_1}{P_1} = \frac{s_1}{P_2}$$

and

$$s_1 = \frac{P_2}{P_1 + P_2}$$

Since

$$\text{rpm} = \frac{120f_1}{P_1}(1 - s_1) = \frac{120f_1}{P_1}\left(1 - \frac{P_2}{P_1 + P_2}\right)$$

it follows that

$$\text{rpm} = \frac{120f_1}{P_1 + P_2} \qquad (88)$$

Equation (88) gives the synchronous speed of the motors connected in *concatenation;* the actual speed will be less than this by the slip factor.

EXAMPLE 16. Two 60-cycle motors are connected in concatenation to drive a load. If machine 1 has six poles and machine 2 has eight poles, calculate the speed of the combination if the slip is 0.075.

Solution

$$\text{rpm} = \frac{120 \times 60}{6 + 8} \times (1 - 0.075) = 475$$

EXAMPLE 17. In Example 16, at what other speeds can the load be driven if each motor is operated separately, assuming the same value of slip?

Solution

$$\text{rpm}_6 = \frac{120 \times 60}{6} \times (1 - 0.075) = 1{,}110$$

$$\text{rpm}_8 = \frac{120 \times 60}{8} \times (1 - 0.075) = 832$$

Consequent-pole Method. In the usual four-pole motor, *successive* pole groups of *each phase* are connected so that they always produce poles of opposite polarity. Referring to Fig. 230, note that successive arrows of each phase are shown reversed with respect to each other; this connection is the usual one. However, if all the pole groups of each phase are connected so that their polarities are always the *same* at any instant, then, by *consequence,* four additional poles of the opposite polarity will be produced at the same instant. That is, by connecting all the coil groups for the same polarity (in a whole-coiled winding), the stator acts as though it has twice as many poles as pole groups. The additional poles so created are called *consequent poles,* and the winding is known as a *consequent-pole winding.* Figure 251 illustrates the difference between the standard connection and the consequent-pole connection.

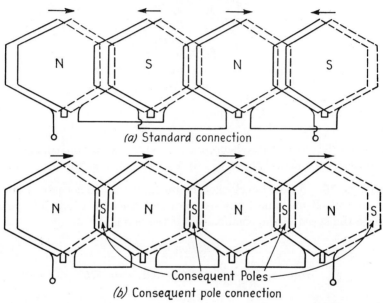

FIG. 251. Sketches showing the connections for one phase of a whole-coiled winding. The standard connection is for four poles and the consequent-pole connection is for eight poles.

By a unique switching arrangement, it is possible to have the standard connection on one side of a double-throw switch and the consequent-pole connection on the other side of the switch. Thus two speeds can be obtained, the higher speed with the standard connection and half speed with the consequent-pole connection. Several schemes for doing this are used in practice, one of which is shown in Fig. 252. In this wiring diagram, the stator is connected *two-parallel-star* for the high speed (four poles) and *series-star* for the low consequent-pole speed (eight poles). It should be stated that a two-parallel connection has two parallel paths in each phase, while a series connection has a single path for each phase. If the six-blade switch is closed to the right, the three-point side, each of the three phases A, B, and C can be traced to follow the dashed arrows for the eight-pole consequent-pole connection. When the switch is closed to the left, the six-point side, each of the three phases can be traced to follow the continuous arrows for the four-pole standard connection.

It should be understood that this type of two-speed motor winding always provides one speed that is half as much as the other. When three or four definite speeds are desired, it is necessary to have two separate windings, only one of which is used at a time. A popular arrangement is

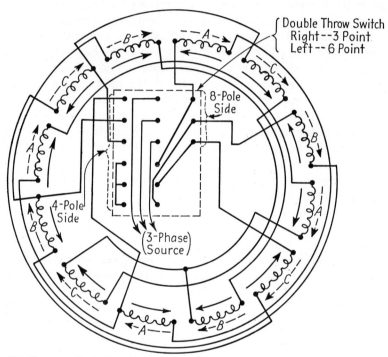

Fig. 252. Two-speed winding diagram for stator of an induction motor. With switch closed to right, follow dashed (→) arrows for eight-pole consequent-pole series-star connection. With switch to left closed, follow continuous (→) arrows for four-pole two-parallel star connection.

to have one winding provide a four- and an eight-pole combination and the other winding provide a six- and a twelve-pole combination. In this way, four synchronous speeds are possible: 600, 900, 1,200, and 1,800 rpm.

Multispeed motors have been used satisfactorily for machine-tool operations, for the movement of large volumes of air, in the printing industry, for woodworking machinery, in textile manufacturing operations, for foundry equipment, and for other applications.

Frequency-change Method. In some special installations in which the alternator supplies power to one or more independent motors, the speed of the latter can be changed by varying the speed of the alternator; this alters the frequency and therefore the motor speed [see Eq. (66)]. Such systems have been employed to a limited extent for ship propulsion and in the operation of some railways in Europe. Since a frequency change requires a change in the prime-mover speed as well as in that of the alternator, the method is not adapted to installations in which the speed must be varied widely. The reason for this is that the efficiency of large prime movers, such as turbines, decreases considerably when they are operated at speeds other than those for which they were designed.

Special Machine Methods. Other methods for varying the speed of an induction motor have been devised. Since these are beyond the scope of this book, they are commented upon briefly. The *Schrage* motor is a special commutator-type machine in which the rotor has two windings, one connected to the slip rings and the other connected to a commutator. A stator winding is connected to brushes riding on the commutator, while power is fed to the other rotor winding through brushes on the slip rings. Speed control is accomplished by a unique brush-shifting mechanism. The motor has excellent operating characteristics but is rather expensive. The *frequency-converter* system employs a wound-rotor motor coupled to a special type of machine called a *frequency converter*. The rotor of the induction motor is connected to brushes riding on the commutator of the frequency converter, while the slip rings of the latter are connected through regulating transformers to the same source that feeds the stator of the induction motor. Speed control is accomplished by varying the voltage of the regulating transformers that supply power to the frequency converter. The method is an excellent one, but it involves considerable expense in auxiliary equipment. Still another method is the *Scherbius* system, used to a limited extent in the operation of large steel-mill rolls. Its principle of operation is similar to that of the frequency-converter system because it introduces a voltage into the rotor of the main induction motor for speed reduction. In both these methods, speed change is accomplished without the great power loss in a resistance controller that is ordinarily involved in the operation of wound-rotor motors.

Figure 253 is a photograph of a completely assembled squirrel-cage motor used for general-purpose services. Note its trim, substantial appearance and the absence of the complex commutator and brush mechanism. It is the ideal type of machine for innumerable applications, giving excellent, trouble-free service for long periods of time. Such motors are particularly suitable for hazardous locations and for places where there may be considerable moisture and dirt.

Electric Braking of Alternating-current Motors. It is generally desirable to employ an electric brake in preference to one that operates mechanically when it is necessary to bring an a-c motor to rest quickly because electrical braking is smooth, provides accurately timed stops, and is not subject to the service problems of rubbing surfaces. As with d-c motors, the two well-known methods of *dynamic braking* and *plugging*, previously discussed (p. 207 and Figs. 134 and 136), are widely used on induction-type machines. Briefly, a motor is brought to a stop by dynamic braking when the a-c source is disconnected, and direct current is made to energize the winding. Also, when a three-phase motor is *plugged*, one pair of line wires that energize the winding are interchanged,

FIG. 253. Completely assembled squirrel-cage motor for three-phase operation. External connections are arranged for 220- or 440-volt service. (*The Louis Allis Co.*)

under which condition the revolving field reverses, with the result that the direction of rotation of the rotor attempts to reverse; in doing so, it must first come to a stop, and it is at this instant that a plugging switch (Fig. 134) disconnects the power supply.

Dynamic Braking. A complete wiring diagram of an automatic controller connected to a three-phase motor and arranged for starting and dynamic-braking service is shown in Fig. 254. Contactor M is energized when the START button is pressed; this causes the M_c contacts to close for sealing purposes and closes the main contacts M to start the motor in the usual way. To permit the motor to *coast* to a stop, the STOP button is pressed lightly so that contacts a and b are *not* bridged; under this condition the main contactor is deenergized, and the machine is disconnected from the source.

However, if the STOP button is pressed down hard, so that the ab contacts *are* bridged, a timing relay TR and a dynamic-braking relay DB

394   ELECTRICAL MACHINES—ALTERNATING CURRENT

are energized through normally closed contacts TR-$T.O.$; this occurs *after* the stop bar leaves the upper set of contacts and the M contactor is deenergized. The operation of the DB relay opens normally closed contacts $DB1$ to protect the motor against the possibility that alternating current will pass into the winding during the braking period and closes

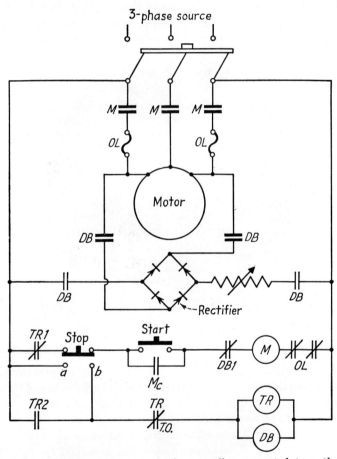

FIG. 254. Wiring diagram of an automatic controller connected to a three-phase squirrel-cage motor and arranged for starting and dynamic-braking service.

two sets of contacts DB to permit rectified direct current to energize the motor winding. Another precaution is a normally closed interlock on timing relay TR, which further opens the main-contactor circuit at $TR1$; this makes the START button ineffective while the motor is slowing down. At the same instant, contacts $TR2$ close to seal the timing relay when the STOP button is released.

Finally, the timing relay is set to time out at the instant the motor comes to rest. When this happens, the contacts *TR-T.O.* open; relays *TR* and *DB* then drop out, and the various relay contacts return to normal in preparation for another motor start.

Plugging. The principle of plugging as applied to induction motors depends upon the fact that the direction of rotation is the same as the

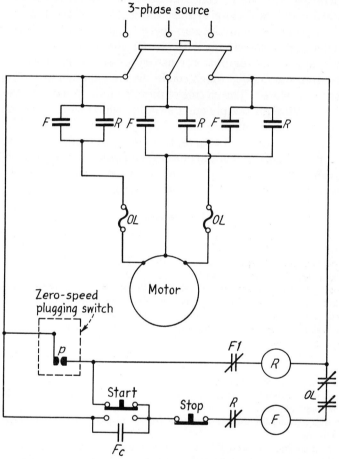

Fig. 255. Wiring diagram of an automatic controller connected to a three-phase squirrel-cage motor and arranged for starting and plugging service.

direction of the revolving field; the latter does, in turn, depend upon the phase sequence of the source with respect to the space sequence of the winding phases. This means, therefore, that the direction of rotation of a three-phase motor may be reversed by merely interchanging *one pair* of leads at the motor *or* line terminals because such a change reverses the direction of the revolving field.

Figure 255 shows a complete wiring diagram of an automatic controller connected to a three-phase motor and arranged for starting and plugging service. The machine is started in the usual way by pressing the START button. This causes the forward contactor F to pick up, whereupon (1) the normally closed contacts $F1$ open to prevent the reversing contactor from operating, (2) the F_c contacts close for sealing purposes, and (3) the forward contacts F close to start the motor. When the rotor starts turning, the contacts p at the zero-speed plugging switch close, but this has no effect upon the circuit with the $F1$ contacts open. To plug the motor to a stop, the STOP button must be pressed; this causes the F contactor to drop out, under which condition the main contacts F open and the $F1$ contacts close. The closing of the $F1$ contacts energizes the reverse contactor R (through the plugging switch), which now closes contacts R to reverse the stator connections. The motor thus develops a countertorque, opposed to the load torque; this results in rapid retardation of the application. Finally, when the motor speed falls to a very low value, the switch opens its contacts at p, the R contactor is deenergized, and the R contacts open to disconnect the motor from the source.

Questions

1. Distinguish between an *induction* motor and what might be termed a *conduction* motor.
2. Describe a simple experiment that might be performed to demonstrate the discovery made by Arago in 1824.
3. Referring to Fig. 222, explain why a disk will rotate if a pole is physically moved.
4. Explain why the disk of Fig. 222 cannot move as rapidly as the pole.
5. Describe the construction of the stator of an induction motor.
6. Describe the construction of the squirrel-cage type of rotor.
7. Describe the construction of the wound-rotor type of rotor.
8. What advantages are possessed by the cast-aluminum type of squirrel-cage rotor?
9. Why are the rotor bars skewed in a squirrel-cage rotor?
10. When are wound-rotor motors used?
11. Why is it desirable to use a number of stator slots that is divisible by the product of poles and phases?
12. Give five reasons for the preference of fractional-pitch stator windings in induction motors.
13. Referring to Figs. 232 and 233, explain carefully how a revolving field is produced in a three-phase induction motor.
14. What is the maximum value of the flux in a revolving field of a three-phase motor with relation to the flux produced by one phase?
15. How can the direction of the revolving field be reversed?
16. At what speed does the revolving field rotate? How can this be determined?
17. Why is it *not* possible for the rotor speed to be equal to that of the revolving-field speed?
18. What is meant by the *slip*? In what two ways may it be designated?
19. What voltage would be generated in the rotor if it could theoretically rotate at synchronous speed?

POLYPHASE INDUCTION MOTORS 397

20. What determines the generated voltage in the rotor when the latter is locked so that it cannot rotate?
21. What is the rotor frequency when the rotor is (a) locked? (b) revolving at normal speed?
22. What determines the rotor current in a given machine?
23. How is it possible to represent electrically the mechanical load of an induction motor? What is the advantage of doing this?
24. Distinguish between rotor power input *RPI* and rotor power developed *RPD*.
25. Why is the rotor power developed greater than the actual power applied to the mechanical load, i.e., the shaft power?
26. Define *normal torque, starting torque,* and *maximum torque.* Which of these is the largest?
27. How does the blocked-rotor reactance affect the maximum torque that can be developed by an induction motor?
28. What constructional practice is generally followed to minimize the rotor reactance?
29. List the various losses in an induction motor.
30. When measuring the no-load input to an induction motor, why is it necessary to subtract wattmeter readings when the two-wattmeter method of measuring power is used?
31. What losses are measured by the no-load test?
32. Following Example 11, outline the procedure to be followed to calculate the four important characteristics of an induction motor, namely, horsepower output, torque, per cent efficiency, and power factor.
33. Explain carefully how the blocked-rotor test is performed.
34. What important constants may be obtained from the blocked-rotor test?
35. At what speed will the maximum torque occur?
36. Why is the starting period the most severe period in the operation of a motor? Why is this period more serious in d-c motors than in induction motors?
37. What practice may be employed to reduce the starting current in squirrel-cage motors? in wound-rotor motors?
38. What is the objection to starting a squirrel-cage induction motor on reduced voltage?
39. List three methods generally used to start polyphase squirrel-cage motors.
40. Describe the construction of the double-cage rotor.
41. Explain why the starting current is minimized and the starting torque is comparatively high when a double-cage rotor is used in an induction motor.
42. Why does a double-cage-rotor motor have good efficiency under normal operating conditions?
43. Referring to Fig. 238, explain the operation of a full-voltage automatic starter when connected to a squirrel-cage motor.
44. Referring to Fig. 239, explain the operation of the compensator method of starting an induction motor.
45. Referring to Fig. 242, explain the operation of the line-resistance method of starting an induction motor.
46. Why does the line-resistance method of starting an induction motor reduce the current to a lesser extent than it does in the case of a d-c motor?
47. When an induction motor is started by the Y-Δ method, what voltage is impressed on the stator winding per phase at the starting instant?
48. Referring to Fig. 244, explain the operation of the Y-Δ method of starting an induction motor.
49. What is meant by *part-winding* starting of an induction motor?

398 ELECTRICAL MACHINES—ALTERNATING CURRENT

50. Referring to Fig. 246, explain the operation of the part-winding method of starting an induction motor.
51. Describe the wound-rotor method of starting an induction motor.
52. What other function can be performed by the wound-rotor controller besides starting when it is used with a wound-rotor motor?
53. To what value of resistance should the controller be adjusted when starting a wound-rotor induction motor?
54. What three important factors greatly affect the operating characteristics of induction motors?
55. What is the effect upon the starting torque and the efficiency of increasing the rotor resistance?
56. What is the effect upon the starting current and the maximum torque of increasing the rotor reactance?
57. Will deep, narrow rotor slots increase the rotor reactance? Give reasons for your answer.
58. Why is a short air gap so important to the operation of an induction motor? How short can the air gap be made?
59. Make a list of the four different classes of squirrel-cage motors and their starting currents and torques when full voltage is impressed.
60. List several practical applications for each of the four classes of squirrel-cage motors.
61. What is a shaftless motor? What advantages are possessed by such motors?
62. List several practical applications of shaftless motors.
63. Under what conditions is it advantageous to use a wound-rotor type of induction motor?
64. What happens to the power consumed by the resistance controller?
65. List several practical applications of the wound-rotor type of induction motor.
66. What is the effect upon the over-all wound-rotor motor efficiency when the speed is greatly reduced? What is this efficiency less than?
67. Describe the wound-rotor method of speed control.
68. Describe the concatenation method of speed control. What are its limitations?
69. What is meant by *consequent poles*? How are these produced?
70. What is the ratio of the two speeds obtainable in a consequent-pole motor?
71. How is it possible to obtain four speeds in an induction motor?
72. What are the objections to the frequency-change method of control for induction motors?
73. What advantages are possessed by electric brakes as compared with mechanical brakes?
74. Explain the principle of operation of *dynamic braking* as applied to the stopping of an induction motor.
75. Referring to Fig. 254, explain the operation of the automatic controller that is designed for starting and dynamic-braking service.
76. Explain the principle of operation of *plugging* as applied to the stopping of an induction motor.
77. Referring to Fig. 255, explain the operation of the automatic controller that is designed for starting and plugging service.

Problems

1. At what speed will a 14-pole 60-cycle induction motor operate if the slip is 0.09?
2. The name-plate speed of a 25-cycle induction motor is 720 rpm. If the speed at no load is 745 rpm, calculate: (*a*) the slip; (*b*) the per cent regulation.

POLYPHASE INDUCTION MOTORS 399

3. For how many poles is a 50-cycle induction motor wound if the name-plate speed is 460 rpm?

4. A three-phase 60-cycle 10-pole induction motor has a full-load slip of 0.075. What is the speed of (a) the rotor relative to the revolving field? (b) the revolving field relative to the stator?

5. A three-phase six-pole 60-cycle 230-volt wound-rotor motor has its stator connected in delta and its rotor in star. There are 75 per cent as many rotor conductors as stator conductors. Calculate the voltage and frequency between slip rings if normal voltage is applied to the stator when (a) the rotor is at rest; (b) the rotor slip is 0.04; (c) the rotor is driven by another machine at 800 rpm in a direction opposite from that of the revolving field.

6. The name plate of a squirrel-cage induction motor has the following information: 25 hp, 220 volts, three-phase, 60 cycles, 830 rpm, 64 amp per line. If the motor takes 20,800 watts when operating at full load, calculate: (a) slip; (b) per cent regulation if the no-load speed is 895 rpm; (c) power factor; (d) torque; (e) efficiency.

7. A four-pole wound-rotor induction motor has a Y-connected stator and a Δ-connected rotor, with the same number of conductors on each. The rotor resistance and reactance at standstill are 0.18 and 0.75 ohm per phase, respectively. If the voltage impressed on the stator is 110 and the frequency is 60 cycles, calculate the rotor current: (a) at starting; (b) when the speed is 1,720 rpm.

8. In Prob. 7, calculate the equivalent resistance per phase of the load.

9. A 60-cycle 230-volt induction motor has a 90-slot stator that contains a six-pole Y-connected double-layer winding. Calculate: (a) the voltage across one coil group of the winding (see Fig. 230); (b) the rotor speed and frequency.

10. A 5-hp 220-volt four-pole 60-cycle three-phase induction motor was tested and the following data were obtained:

   No-load test: $V_{nl} = 220$; $P_t = 310$; $I = 6.2$
   Load test: $V_l = 220$; $P_t = 3,650$; $I = 11.3$; rpm $= 1,710$
   Effective a-c resistance of stator per phase $= 0.3$ ohm

   Calculate: (a) friction, windage, and iron losses; (b) stator copper loss under load; (c) rotor power input; (d) rotor copper loss under load; (e) rotor output in watts; (f) horsepower output; (g) torque; (h) per cent efficiency under load; (i) load power factor.

11. A blocked-rotor test was performed upon the motor of Prob. 10, and the following data were obtained: $V_{BR} = 48$; $I = 18$; $P_t = 610$. Calculate: (a) the equivalent resistance of the motor per phase R_e; (b) the equivalent resistance of the rotor per phase R_R; (c) the equivalent blocked-rotor reactance per phase X_{BR}.

12. Using the results of Prob. 11, calculate the speed at which the torque will be a maximum.

13. Calculate the starting torque of the motor of Prob. 10 when it is started (a) at 220 volts; (b) at 110 volts.

14. A four-pole 230-volt 60-cycle induction motor has a rotor resistance of 0.15 ohm per phase and a rotor reactance of 0.5 ohm per phase. Calculate the horsepower output and the torque for a slip of 5 per cent.

15. The following data were obtained from tests performed upon a 10-hp 230-volt six-pole induction motor:

   No-load test: $V_{nl} = 230$; $P_t = 410$; $I = 8.5$
   Load test: $V_l = 230$; $P_t = 8,350$; $I = 23$; rpm $= 1,140$
   Effective resistance of stator per phase $= 0.14$ ohm
   Friction $+$ windage loss $= 210$ watts

Calculate: (a) stator iron loss; (b) stator copper loss under load; (c) rotor power input under load; (d) rotor copper loss under load; (e) horsepower output; (f) load torque; (g) load efficiency; (h) power factor.

16. If the starting torque of a 7.5-hp 440-volt 1,730-rpm motor is $2\frac{1}{2}$ times its rated full-load torque when rated voltage is impressed, calculate the starting torque when 230 volts is applied at the instant the motor is started.

17. The starting current of a 15-hp 440-volt three-phase induction motor is 132 amp when rated voltage is impressed. What voltage should be applied at the starting instant if the current is not to exceed 60 amp?

18. A 3-hp 208-volt three-phase motor is started by the Y-Δ method. If the starting current is 54 amp when rated voltage is applied, what current will flow when the Y-Δ switch is used?

19. When an induction motor is started directly from the line, the starting torque is 1.75 times full-load torque and the starting current is 8 times rated value. (a) What will be the motor starting torque and starting current, in terms of full-load values, when the machine is started by the Y-Δ method? (b) What will be the starting current on the line side under this condition?

20. Make a Y-connected winding diagram, similar to Fig. 230 for a six-pole three-phase stator in which there are 54 slots.

21. Make a Δ-connected winding diagram for a two-pole three-phase stator in which there are 36 slots.

22. Two 60-cycle motors having 10 and 14 poles are connected in concatenation. What synchronous speeds are possible?

23. The consequent-pole method is used to obtain the following 60-cycle synchronous speeds: 1,200 rpm, 900 rpm, 600 rpm, and 450 rpm. (a) What is the standard number of poles in each winding? (b) How many poles does each consequent-pole connection have?

24. Using Fig. 238 as a guide, show how a second push-button station should be added so that the motor may be started and stopped from two positions.

25. It is desired to operate a three-phase 440-volt 60-cycle motor from a 25-cycle source. What voltage should be applied to the machine if the air-gap flux density is to be maintained at its normal value?

26. A four-pole 50-cycle three-phase induction motor has a blocked-rotor reactance per phase that is four times the rotor resistance per phase. At what speed will the motor develop maximum torque?

27. A squirrel-cage motor develops 80 lb-ft of starting torque when the 50 per cent tap on a compensator is used to start the machine by the reduced-voltage method. What starting torque will be developed if: (a) the 65 per cent tap is used? (b) the 80 per cent tap is used?

CHAPTER 10

SINGLE-PHASE MOTORS

Types of Single-phase Motor. Great numbers of motors of comparatively small horsepower rating are designed to operate when connected to a single-phase source. Most of them, built in fractional-horsepower sizes, are technically termed *small motors*, a *small motor* being defined as "*a motor built in a frame smaller than that having a continuous rating of 1 hp, open type, at 1,700 to 1,800 rpm.*"* Accordingly, a ⅔-hp 900-rpm motor would *not* be classified as a fractional-horsepower machine, because a similar frame, designed for a speed of 1,800 rpm, would have a rating greater than 1 hp. Moreover, a 1½-hp 3,600-rpm motor would be classified as a fractional-horsepower machine because a 1,800-rpm motor in a similar frame would have a rating less than 1 hp. Single-phase motors perform a great variety of useful services in the home, the office, the factory, in business establishments, on the farm, and many other places where electricity is available. Since the requirements of the numerous applications differ so widely, the motor-manufacturing industry has developed several types of such machines, each type having operating characteristics that meet definite demands. For example, one type operates satisfactorily on direct current or any frequency up to 60 cycles; another rotates at absolutely constant speed, regardless of the load; another develops considerable starting torque; and still another, although not capable of developing much starting torque, is nevertheless extremely cheap to make and very rugged.

The type of motor that performs with about equal satisfaction on direct current or alternating current up to 60 cycles is the familiar d-c *series motor*. Such motors are generally constructed in small sizes, operate at high speed, and include special design features, so that commutation and armature-reactance difficulties are minimized. Since they may be connected to any of the commonly available sources of supply, they are appropriately called *universal* motors.

The induction principle is applied to several types of single-phase motor. This principle, it will be recalled from the study of polyphase induction

* American Standards Association (ASA) and National Electrical Manufacturers Association (NEMA).

machines, involves the production of a revolving magnetic field, several methods having been developed for doing this in single-phase motors. One of these methods is employed in the *shaded-pole* motor, an extremely popular small motor used in low-starting-torque applications. The so-called *reluctance-start* motor is a second type of machine, made in rather limited numbers in small sizes, that utilizes still another method to create the effect of rotating poles.

The *split-phase* type is perhaps the most widely used of all motors connected to single-phase sources of supply. It is manufactured in a great many sizes and styles, offering the user a choice of a number of desirable operating characteristics. There are, for example, (1) standard split-phase motors, (2) motors that employ a capacitor during the starting period only, and (3) motors that make use of one or more capacitors for starting and running duty. Much excellent design and development work has been done in recent years on this type of machine because motor manufacturers have felt that the great volume of business justifies their constant improvement.

Repulsion, repulsion induction, and *repulsion-start* motors are other types of single-phase machines that were widely applied until recent years. They have been largely replaced by split-phase motors of the capacitor type because the latter can be designed to perform as well as the repulsion types, offering in addition such advantages as lower cost and trouble-free service.

Synchronous motors, as the name implies, operate at synchronous speed for all values of load. There are several constructions of such machines, although they are usually manufactured in very small ratings. Depending upon the way in which they are made or their principle of operation, they have special names such as *reluctance* motors, *subsynchronous-reluctance* motors, and *hysteresis* motors.

The Universal (Series) Motor. *Principle of Operation.* The d-c series motor will operate if connected to an a-c source of supply because the direction of the torque is determined by *both* the field polarity and the direction of the current through the armature. Since the *same* current passes through the field coils and the armature winding, it follows that the a-c reversals from positive to negative, and vice versa, will simultaneously affect the field polarity and the direction of the current in the armature conductors. Figure 256 represents an elementary two-pole series motor in which the torque produces counterclockwise rotation for both directions of current flow. In Fig. 256a the top and bottom terminals are *positive* and *negative* and the upper and lower poles are *north* and *south*, respectively. With the current through the armature from *right* to *left*, the resulting assumed *crosses* and *dots* are indicated. Applying the rules studied in Chap. 5, it is seen that the armature will rotate as shown, i.e., counterclockwise. When the polarities of the line terminals reverse, as they would in the second half of an a-c cycle, the upper and lower poles become *south* and *north*, respectively. The current through the armature will now be from *left* to *right,* and the resulting crosses and dots in the conductors will therefore change. Figure 256b represents these conditions, from which it can be seen that the armature will continue to rotate counterclockwise.

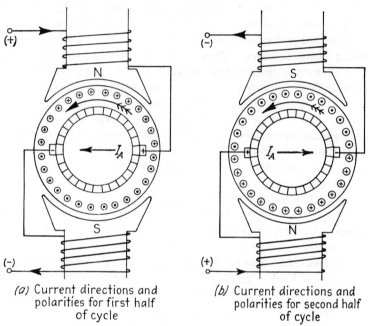

(a) Current directions and polarities for first half of cycle

(b) Current directions and polarities for second half of cycle

Fig. 256. Elementary series motor illustrating that the direction of rotation is unchanged when the line polarity is reversed.

Large Series Motors. When a series motor is comparatively large, the inductive effects of the field and armature tend to create serious commutation difficulties, which, if allowed to remain uncorrected, will cause the motor to operate unsatisfactorily when used on a-c circuits. Also, the iron losses, hysteresis, and eddy currents in the field, which are not present in d-c motors, tend to become abnormally high. To forestall such objectionable performance, large series motors are designed (1) with a well-laminated field, (2) for operation on the lower frequencies, such as 15 and 25 cycles, (3) with the number of series-field turns reduced to a minimum consistent with good operation, (4) so that the field iron is worked at low flux densities, (5) with the relative number of armature-winding turns with respect to the field turns much greater than in corresponding d-c series motors, (6) with special compensating and interpole windings, (7) with armatures having a comparatively large number of commutator segments, and (8) with special, added resistance between armature-coil ends and the commutator segments to which they are soldered. The most common application of large series motors is in connection with interurban and urban traction service.

Small Series Motors. Few of the foregoing design features are usually necessary for series motors of low horsepower ratings. Usually, they will have characteristics that are quite similar, whether connected to a d-c source or an a-c supply up to 60 cycles. However, if a particularly high grade of performance is desired, universal motors are designed with special compensating windings. The American Standards Association (ASA) has formally designated a *universal* motor to be a "series motor that may be operated either on direct current or single-phase alternating current at approximately the same speed and output." It further specifies that "... these conditions must be met when direct-current and alternating-current voltages are approximately the same and the frequency of the alternating current is not greater than 60 cycles per second."

The term "small motors" is, of course, relative, but it generally refers to those two-pole fractional-horsepower machines that usually operate at high speed and that have ratings from perhaps $\frac{1}{500}$ to $\frac{2}{3}$ hp. They are used in an unusually large number of applications, performing a wide range of services and having power consumptions of about the same order of magnitude as incandescent lamps, flatirons, space heaters, and the like. Their widest fields of application are under the following conditions:

1. When it is desired that a motor perform with complete satisfaction when connected to standard d-c or a-c sources of supply.

2. When it is important that a motor operate at a very high rate of speed—a high-speed motor can develop much more power per unit of weight than a low-speed motor.

3. When it is desired that the speed of a motor be capable of automatically adjusting itself to the magnitude of the load—the speed is high at light load and low at heavy load.

The field structure of the universal type of motor is always a stack of laminations bolted together, fitted with field windings, and usually placed inside a yoke ring. The armature is similar to that of large d-c motor armatures, except that the stack of laminations is pressed on a shaft so that the slots are usually *skewed*, i.e., set at an angle with respect to the shaft axis. This latter construction helps to keep the motor quiet and eliminates the tendency on the part of the armature to lock when started. Figure 257 illustrates several sets of field and armature laminations.

Speed-Load Characteristics of Universal Motors. The speed of series motors varies considerably with changes in load; when the load is heavy, the speed is low, while at light loads the speed rises to a very high value. In the case of small universal motors whose armature diameters are about 1 to 2 in., speeds as high as 20,000 rpm are not uncommon. Even with the external mechanical load completely disconnected, the maximum speed of such a machine is limited by its own friction and windage load, so that the so-called "runaway speed" (discussed in Chap. 5) is still a safe one. Figure 258 represents a typical *speed vs. load* curve for a universal motor. The significant point to be noted is the extremely wide speed

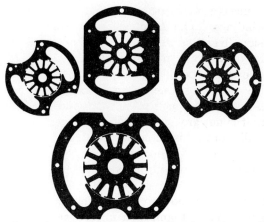

FIG. 257. Field and armature laminations for small series motors. (*Carnegie-Illinois Steel Corp.*)

variation; in some machines the no-load speed may be 5 or more times as great as the full-load speed.

If tests are performed on a universal motor, it will usually be found that the d-c and a-c (60-cycle) *speed-load* curves do not coincide. In some motors, the a-c speeds will always be higher than the d-c speeds, while in others the reverse is true; in still other designs, the curves cross between no load and full load, so that the a-c speeds are higher than the d-c speeds in one load range and lower in another load range. To understand why this is true, it should first be emphasized that several factors are responsible for the change in speed with a change in load. On direct current these are (1) *armature and field-resistance drops*, which tend to reduce the speed when load is applied, and (2) the effect of *armature reaction*, which tends to raise the speed when load is applied. When alternating current is used for the power source, a third factor is present in addition to the foregoing, namely, *reactance voltage drop*, which tends to lower the speed with increased loading. Note particularly that the effect of armature reaction is to reduce the air-gap flux, which, in turn, causes a rise in speed, while reactance voltage drop, appearing only when alternating current is used, acts to lower the effective motor voltage and therefore the speed. Whether or not the actual speed for a given load is higher on alternating than on direct current depends upon which of the above two factors is changed more. If the reactance voltage drop increases more than the air-gap flux decreases, the motor will tend to operate at a lower speed when alternating current is used; otherwise, the reverse is true. Also, at the higher loads the iron is saturated, so that the effective flux per a-c ampere is less than that produced per d-c ampere. The result of this condition

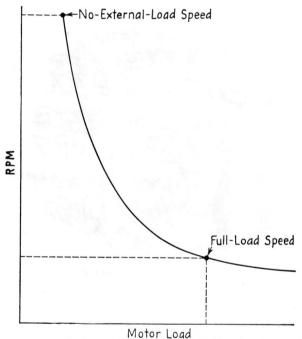

FIG. 258. Typical *speed vs. load* curve for a small universal motor.

is a tendency on the part of the motor to run faster with alternating current. (Always remember that speed is inversely proportional to flux.) Thus, if it were not for the fact that the reactance voltage drop acts in a direction opposite to that of the armature-reaction effect, the a-c speed would actually be greater under load than the d-c speed. The factor that predominates determines, in general, whether or not the a-c speed will exceed the d-c speed. At light loads, however, the reactance voltage drop is negligible, so that the motor tends to operate at a higher speed on alternating current than on direct current. In a good design, the reactance voltage drop is kept fairly low, and the difference in flux produced by direct and alternating current is reduced to a minimum by operating at low flux densities. The latter condition, i.e., a weak field motor, is gained at the expense of poorer commutation, correction for which must be made by choosing a proper grade of brush.

Armature Speed, Gears, and Gear Ratios. It is well known that motor output is a function of the armature speed because, other things being equal, a high peripheral velocity will permit a machine to cool better than one that rotates slowly. Series motors for small power applications, such as portable hand drills, food and drink mixers, and compressors, operate at very high speeds, but are geared to their loads so that the latter operate at conveniently low values. When such motors are well designed

Fig. 259. Cutaway view of a gear train on a small motor. Considerable speed reduction is accomplished in this way, permitting the motor to operate at an efficient high speed and the load at a desirable low speed. (*Bodine Electric Co.*)

and the gear losses have been kept down to a minimum, they are capable of doing jobs normally expected of those having larger physical dimensions. With regard to the gear mechanism, the practice seems to be to cut a single or double spiral thread on the extension of the armature shaft, which then engages one or more gears. Good lubrication is, of course, necessary in the gear box, one of which is shown in a cutaway view in Fig. 259. Gear ratios vary and depend on the applications; with motor speeds as high as 20,000 rpm, the actual load speed may sometimes be reduced to a few revolutions per minute.

Applications of Universal Motors. These motors are used in a considerable number of applications requiring all sorts of operating characteristics. In some, the actual motor speed is the load speed, a good example being a vacuum cleaner. In others, the load speed, through a gear train, is reduced considerably; good examples of this practice are drink mixers and portable drills. Another useful component of some

408   ELECTRICAL MACHINES—ALTERNATING CURRENT

installations is the so-called *constant-speed governor*, which permits the motor to be adjusted to a given speed that will remain practically constant, regardless of the load. Governors vary in construction and principle, but they all depend upon the opening and closing of electrical contacts mounted on the rapidly moving armature. Tension is applied to a set of spring-controlled weights, so that a dial may be used to adjust the mechanism to open and close the contacts within a very narrow preset speed band. The best example of this practice is the food mixer. Since all commutator-type motors, especially those that operate at high speeds, interfere with radio reception, universal motors are usually provided with interference eliminators. These are simply small capacitors connected directly across the line terminals, inside the motor.

In order to illustrate the wide range of applications of universal motors, a list of those more frequently used is given in Table 8.

TABLE 8. COMMON APPLICATIONS OF UNIVERSAL MOTORS

Advertising devices	Hammers	Sanders
Calculating machines	Hedge trimmers	Saws
Cast cutters	Hones	Screw drivers
Clippers	Lock mortisers	Sewing machines
Compressors	Motion-picture outfits	Sirens
Die sinkers	Nibblers	Small fans
Dishwashers	Nut setters	Small grinders
Drink mixers	Pipe threaders	Surgical instruments
Electric shavers	Planers	Vacuum machines
Files	Polishers	Valve grinders
Food mixers	Portable drills	Ventilating equipment
Grease guns	Pumps	Vibrators
Hair driers	Routers	Wood shapers

A disassembled small universal motor is shown in Fig. 260. Note particularly how one end of each field coil is connected to a brush holder.

The Shaded-pole Motor. *Construction and Principle of Operation.* One of the simplest and cheapest of manufactured motors is the so-called *shaded-pole* motor. It is essentially an induction machine, since its squirrel-cage rotor receives power in much the same way as does the rotor of the polyphase induction motor. There is, however, one extremely important difference between the two types, and this concerns their magnetic fields. Whereas the polyphase induction motor creates a true revolving field, in the sense that it is constant in magnitude and rotates at synchronous speed *completely around the entire core*, the field of the shaded-pole motor is not constant in magnitude and merely *shifts from one side of the pole to the other*. Because the shaded-pole motor does not create a true revolving field, the torque is not uniform but varies from instant to instant.

Figure 261 illustrates the construction of one pole of a shaded-pole

SINGLE-PHASE MOTORS

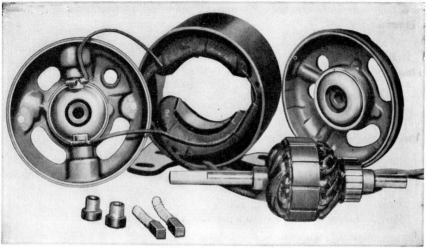

Fig. 260. A disassembled small universal motor. (*Robbins & Myers, Inc.*)

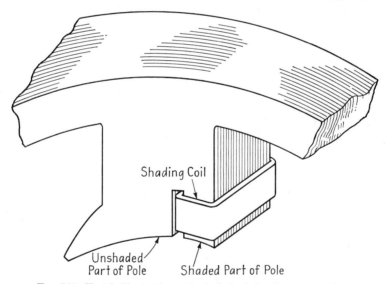

Fig. 261. Sketch illustrating a typical shaded-pole construction.

motor. Each of the laminated poles of the stator has a slot cut *across* the laminations about one-third the distance from one edge. Around the smaller of the two areas formed by this slot is placed a heavy copper short-circuited coil, called a *shading* coil; the iron around which the shading coil is placed is called the *shaded* part of the pole, while the free portion of the pole is the *unshaded* part. The exciting coil, not shown in the sketch, surrounds the entire pole.

When the exciting winding is connected to an a-c source of supply, *the magnetic axis will shift from the unshaded part of the pole to the shaded part of the pole.* This shift in the magnetic axis is, in effect, equivalent to an actual physical motion of the pole; the result is that the squirrel-cage rotor will rotate in a direction from the unshaded part to the shaded part. In order to understand how this comes about, it will be desirable to determine the positions of the magnetic pole centers for several instantaneous values of current in the exciting winding as the alternating current changes in magnitude and direction during a cycle. In this discussion it will be assumed that the current in the exciting winding follows a

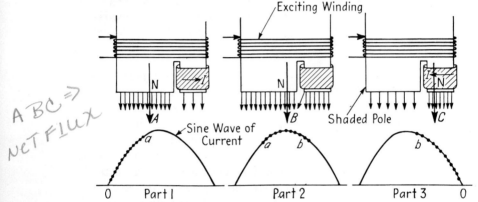

FIG. 262. Sketch illustrating how the magnetic axis shifts across the pole from the unshaded to the shaded parts of a shaded pole in a shaded-pole motor.

sine wave. In Fig. 262, the analysis is divided into three parts for current changes: (1) from zero to a *nearly* positive maximum, (2) in the region of maximum current, and (3) from a *nearly* positive maximum to zero.

Analysis of Current Changes

PART 1. Current change from O to a: The current rises very rapidly along a sine curve; under this condition, a voltage will be induced in the shading coil by transformer action. The current in the low-resistance turn will be comparatively high and will be in such a direction that it will oppose the rapidly changing flux that creates it. (This is in accordance with *Lenz's* law, which states that *the current set up by an induced voltage always tends to create flux in such a direction as to oppose any change in the exciting flux.*) As a result, the field flux is shifted mostly to the unshaded part of the pole; the magnetic axis will then be along the large arrow near the center of the unshaded part of the pole—arrow A.

PART 2. Current change from a to b: The current changes very little along the top of the sine wave; under this condition, practically no voltage is induced in the shading coil. This means that no current will flow in the shading coil, so that the latter produces no flux. And since the main exciting coil current is nearly always a maximum, the flux is also a maximum and is uniformly distributed over the entire pole face: the magnetic axis will now be moved to the center of the entire pole—arrow B.

PART 3. Current change from b to O: The current drops as rapidly along the sine curve as it rose from O to a; under this condition, a voltage will be induced in the shading coil that will send a comparatively large current through the low-resistance turn. This current will be in such a direction that flux will be created in the shaded part of the pole *to prevent the main-field flux from dying out;* i.e., it will create a strong north pole in the shaded part of the pole. The magnetic axis will thus be shifted to the center of the shaded part of the pole—arrow C.

Note particularly that the magnetic axis shifts across the pole from A to B to C, from the unshaded part to the shaded part of the pole. This shift actually occurs gradually, not in steps, from one side to the other. And as the negative half of the current wave flows in the exciting coil, a south pole trails along. The effect is *as though* a set of real poles were sweeping across the space from left to right. Simple motors of this type cannot be reversed, but must be assembled so that the rotor shaft extends from the proper end of the motor to drive the application in the correct direction. Specially designed shaded-pole motors have been constructed for reversing service, but are too complicated for general use. In one construction, two sets of open-circuited shading coils are used, one set occupying positions on one side of the poles and the second set being placed on the other side of the poles. A short-circuiting switch is provided to short-circuit either shading-coil winding (see Fig. 263a). In another construction, a slotted core is used, in which are placed two distributed windings and one set of shading coils. The arrangement of the three windings is such that clockwise rotation results when one main winding is used, while the rotor turns counterclockwise when the other main winding is used (see Fig. 263b).

A completely disassembled shaded-pole motor is shown in Fig. 264. Clearly seen are the stack of stator laminations in the upper left, the four exciting coils connected in series in the lower left, the squirrel cage and its shaft, the four shading coils in the center, bearings, cooling fan, housing, and miscellaneous parts.

Operating Characteristics and Applications. Offsetting the simple construction and the low cost of the shaded-pole motor are such disadvantages as low starting torque, very little overload capacity, and low efficiency. These motors are built commercially in sizes from less than

412 ELECTRICAL MACHINES—ALTERNATING CURRENT

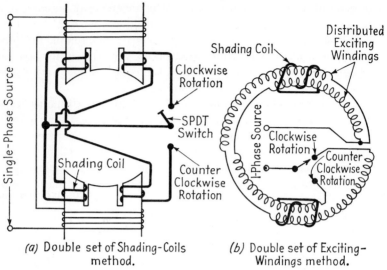

(a) Double set of Shading-Coils method. (b) Double set of Exciting-Windings method.

Fig. 263. Methods for reversing shaded-pole motors—two-pole construction.

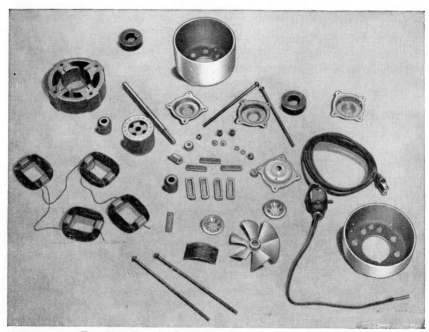

Fig. 264. Completely disassembled shaded-pole motor.

1/250 hp to about 1/6 hp for 25- to 60-cycle service. Full-load torques are developed at comparatively high values of slip, about 10 to 25 per cent, and maximum torques are of the order 1.25 times rated values. Efficiencies for the tiny sizes are as little as 5 per cent and may reach as much as 35 per cent in the higher ratings. When shaded-pole motors are used in applications requiring the movement of air, such as fans, blowers, ventilators, and circulators, no internal cooling fan is required, because a considerable amount of air moves over the motor to cool it. However, when applied to such devices as churns, animated signs, portable ironers, phonograph turntables, and advertising displays, they are usually provided with small fans so that the temperature rise may be kept down to reasonable values.

The impressed voltage across the winding of a shaded-pole motor greatly affects its operating speed under load; in other words, the slip of the rotor increases as the motor voltage is lowered. Advantage is taken of this fact to control the speed of such motors, particularly when they drive fans. Three fundamental schemes are generally employed to vary the winding voltage: (1) the use of an autotransformer that is tapped for several voltages; (2) the use of a tapped reactance coil (sometimes a resistance) that incurs a line-voltage drop; (3) the use of a tapped exciting winding, so that the constant-voltage source may be impressed across the entire winding or some part of it. Methods (1) and (2) are illustrated by Figs. 265a and b and need no explanation. By method (3),

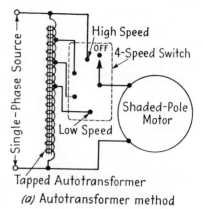

(a) Autotransformer method

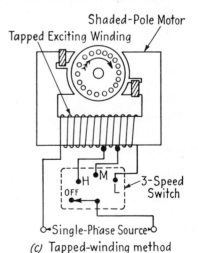

(b) Reactance-coil method

(c) Tapped-winding method

Fig. 265. Speed-control methods for shaded-pole motors.

Fig. 265c, the lowest speed occurs when the entire winding is used and the highest speed results when the smallest portion of the winding is used. The reason for this relationship is that the *volts per turn* increases as the number of turns decreases. Since the flux produced by the exciting winding is directly proportional to the volts per turn, it follows that a decrease in the number of turns is automatically accompanied by an increase in flux, or vice versa. Any flux change therefore immediately results in a corresponding variation in torque, which, in turn, affects the speed at which the motor rotates.

The torque of a small motor is generally given in terms of *pound-inches* (lb-in.) or *ounce-inches* (oz-in.). These are more convenient units than *pound-feet* (lb-ft) for motors of low horsepower ratings because their values are greater than unity. Since

$$\text{hp} = \frac{2\pi \times \text{rpm}}{33{,}000} \times T$$

where T is in pound-feet, therefore,

$$T = \frac{33{,}000}{2\pi} \times \frac{\text{hp}}{\text{rpm}} = \frac{5{,}250 \times \text{hp}}{\text{rpm}} \quad \text{lb-ft} \tag{89}$$

Since T (ounce-inches) $= 12 \times 16 T$ (pound-feet), it follows that

$$T = \frac{5{,}250 \times 12 \times 16 \times \text{hp}}{\text{rpm}} = \frac{\text{hp}}{\text{rpm}} \times 10^6 \quad \text{oz-in.} \tag{90}$$

EXAMPLE 1. Calculate the full-load torque, in pound-feet and ounce-inches, of a 1/50-hp 1,500-rpm shaded-pole motor.

Solution

$$T = \frac{5{,}250 \times 0.02}{1{,}500} = 0.07 \text{ lb-ft}$$

$$T = \frac{0.02}{1{,}500} \times 10^6 = 13.4 \text{ oz-in.}$$

EXAMPLE 2. A 1/20-hp 1,550-rpm shaded-pole motor has a maximum torque of 38 oz-in. and a starting torque of 12 oz-in. What percentages are these torques with respect to the full-load torque?

Solution

$$T = \frac{0.05}{1{,}550} \times 10^6 = 32.3 \text{ oz-in.} \quad \text{(full-load torque)}$$

$$\text{Per cent maximum torque} = \frac{38}{32.3} \times 100 = 118$$

$$\text{Per cent starting torque} = \frac{12}{32.3} \times 100 = 37$$

A cutaway view of a shaded-pole motor is shown in Fig. 266. Clearly visible are the laminated stator and the exciting winding, the squirrel-cage rotor with the skewed bars, the cooling fan at the left end, and the cooling fins in the outside shell.

The Reluctance-start Motor. Another method sometimes used to produce the effect of a shifting magnetic field similar to that described for the shaded-pole motor is to employ the construction illustrated in Fig. 267. Note that the air gap is not uniform under the pole face, but is

Fig. 266. Cutaway view of a shaded-pole motor. (*A. G. Redmond Company, Inc.*)

wider under one portion of the pole than under the other. This type of motor, which is called a *reluctance-start* motor, is also an induction machine, since the rotor power is received inductively. Its operating characteristics are about the same as those of the shaded-pole type; i.e., the starting torque, overload capacity, and efficiency are all comparatively low.

The direction of rotation of the reluctance-start motor is always from the wide air gap to the narrow air gap. When the field coils are excited from an a-c source, starting torque is developed because the pole flux has two out-of-phase components, one under the wide air gap and one under the narrow air gap. As a result of the two out-of-phase flux components, the motor acts as though there were two windings that are displaced in *space* and that carry currents which are out of phase in *time*. The latter two elements, as was learned in Chap. 9, are essential to the production of a revolving magnetic field and motor action.

To understand why two components of flux are created by a single a-c exciting winding, it is necessary to remember the following elementary facts:

1. When a coil of wire produces flux in a magnetic circuit containing no iron, the flux will be in phase with the exciting current.

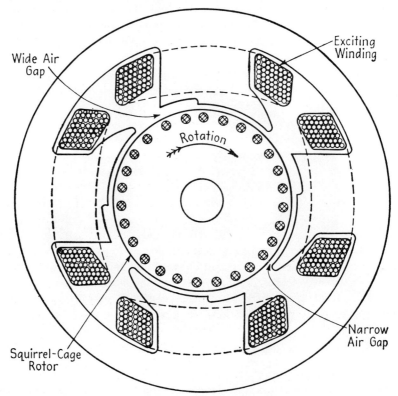

FIG. 267. Typical construction details of reluctance-start motors. Note that the magnetic reluctance in the wide air-gap space is greater than the magnetic reluctance in the narrow air-gap space.

2. When the mmf of a coil produces flux in a magnetic circuit that is made up entirely of iron, the flux that is created will lag behind the exciting current in time.

3. When two coils of wire, excited from the same source, produce fluxes that act in magnetic circuits, one of which has less iron, i.e., more reluctance, than the other, the fluxes will be out of phase in time, the one acting where there is less iron leading the one acting where there is more iron.

In the reluctance-start motor, the third condition prevails; furthermore, the single exciting coil can be properly considered as broken into two exciting coils, displaced in space, one acting on the wide-air-gap magnetic circuit, and the other acting on the narrow-air-gap magnetic circuit. The analysis therefore leads to the conclusion that time and space displacements exist.

Reluctance-start motors are not used so widely as shaded-pole motors. They are made in small sizes only and are applied where the starting and overload torque requirements are low. They cannot be reversed, but must be assembled so that the rotor shaft extends from the motor to drive the load in the proper direction. Speed control is accomplished by varying the voltage across the motor, using a tapped autotransformer or a tapped reactance coil for this purpose. Distributed windings can obviously not be employed.

The Split-phase Motor. One of the most widely used types of single-phase motor, made in larger ratings than are the shaded-pole and reluctance-start constructions, is the so-called *split-phase* motor. It is fundamentally an induction machine, since the rotor power is received inductively. Unlike the shaded-pole and reluctance-start motors, however, its field is almost constant in magnitude and revolves completely around the entire stator; it resembles the two-phase motor in this respect. Indeed, when split-phase motors are designed, an attempt is made to approximate the performance characteristics of two-phase motors, and this generally involves the creation of two independent sets of magnetic fluxes that are out of phase in *time* and *space* by 90 electrical degrees. In order to approximate the latter, it is necessary to employ two separate windings, connected in parallel, to magnetize the laminated iron core. One winding, called the *main* winding, has a comparatively low resistance and a high inductance; the second winding, called the *auxiliary* winding, has a comparatively high resistance and a much lower inductance than the other. When both windings are excited, the current in the auxiliary winding is more nearly in phase with the voltage than is the main winding current. The reason for this lies in the relative magnitudes of the resistance and the inductive reactance of the two windings; the ratio of $X:R$ for the auxiliary winding is less than it is for the main winding. The result is that the single-phase line current is *split* into two parts in such a manner that the main-winding current lags behind the auxiliary-winding current. It is from this *current-splitting* action that the name *split-phase motor* is derived. The second requirement is that the two windings, main and auxiliary, be placed on the laminated iron core in such a manner that their magnetic axes are not coincident. For the best results, the magnetic axes should be displaced with respect to each other by the greatest number of degrees; in a two-pole motor this would be 90°; in a four-pole motor, 45°; and in a six-pole motor, 30°.

It should be clear, therefore, that two conditions must be fulfilled if motor operation is to result by the split-phase principle. These are (1) time displacement between the two currents, resulting from the comparatively different values of resistance and reactance in the main and auxiliary windings, and (2) space displacement between the positions of the main and auxiliary windings.

The shift of the magnetic axes in a split-phase motor comes about in much the same way as it does in a shaded-pole motor because the same essentials are present in both. The shaded-pole motor has two windings: one of them is connected directly to the supply source, while the other, the

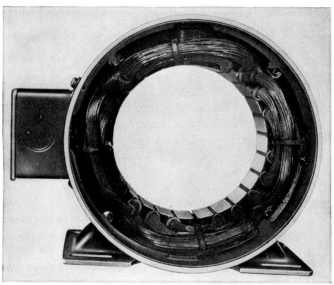

FIG. 268. Stator showing the main winding of a split-phase motor. (*Century Electric Co.*)

shading coils, receives its power by induction. The split-phase motor has two windings, both of which, main and auxiliary, receive their power from the same supply source. In both motors the two windings are displaced in space with respect to each other; in the shaded-pole motor, one winding surrounds the entire pole while the shading coil surrounds about one-third of the pole; in the split-phase four-pole motor, the two windings are 45° apart. In both motors the currents are out of phase with each other; i.e., there is a time displacement between the two currents.

From the standpoint of construction and operation, however, there is a considerable difference between the two types of motor. In the shaded-pole motor, the windings are generally placed around salient (shaped) poles (Fig. 264); in the split-phase motor, the windings are distributed completely around the core in slots. Figure 268 shows the manner in which the *main* winding is placed in a 24-slot stator of a split-phase motor. In the shaded-pole motor, the magnetic axis shifts from one side of a pole to the other; practically no action takes place in the interpolar space. In the split-phase motor, the magnetic axes revolve completely around the cylindrical core; the magnetic field is, in this respect, a true revolving field. And whereas the shaded-pole motor is incapable of developing much starting torque and overload power, the split-phase motor can be designed and constructed to perform admirably on the basis of starting and running characteristics. Furthermore, the efficiency of a split-phase

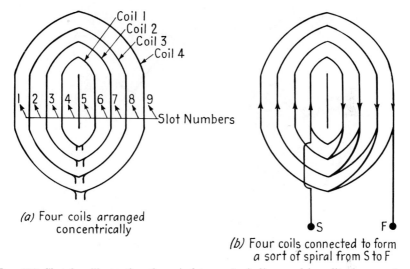

FIG. 269. Sketches illustrating the spiral type of winding used in split-phase motors.

motor is generally two or more times as much as that of the shaded-pole motor; the efficiency of a split-phase motor may be as high as 60 to 70 per cent, while that of the shaded-pole motor rarely exceeds 25 to 30 per cent.

The three general constructional variations of the split-phase type of machine are (1) the *standard* split-phase motor, (2) the *capacitor-start* split-phase motor, and (3) the *capacitor* split-phase motor, which uses one or two values of capacitance. As will be pointed out subsequently, the use of capacitors improves the starting torque characteristic, in addition to aiding other operating features when construction (3) is employed.

The Standard Split-phase Motor. All split-phase motors have slotted laminated stator cores like that of Fig. 224 and squirrel-cage rotors similar to that shown in Fig. 226. The winding construction usually differs from the whole-coiled lap type previously discussed in connection with polyphase motors. The so-called *spiral* winding and modifications of it are generally employed. As the name implies, the coils for each pole are formed and connected so that a sort of spiral is traced from the beginning to the end of the group. Figure 269a illustrates a group of four coils placed in slots and arranged concentrically, and Fig. 269b shows how the four coils are joined together in series. It will be noted that a sort of spiral will be traced in passing from start S to finish F. Both main and

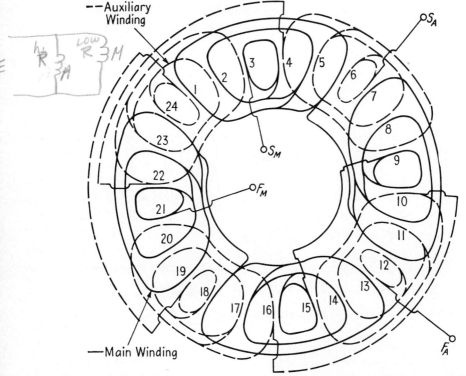

Fig. 270. Main and auxiliary winding diagrams for a 24-slot four-pole split-phase motor.

auxiliary windings are similar, although it is customary to use more turns and larger wire for the former. It should also be stated that there are usually different numbers of turns in the several coils; also, the number of coils in a group will depend upon the number of slots, the number of poles for which the machine is wound, and certain design considerations.

The number of poles for which a split-phase motor is wound is governed, as in polyphase motors, by the frequency and the desired synchronous speed. Thus

$$P = \frac{120 \times f}{\text{rpm}_{\text{syn}}} \qquad (91)$$

EXAMPLE 3. For how many poles is a split-phase motor wound if it operates at 1,750 rpm at full load from a 60-cycle source?

Solution

This machine obviously has a synchronous speed of 1,800 rpm. Therefore

$$P = \frac{120 \times 60}{1,800} = 4$$

Figure 270 represents a winding diagram, main and auxiliary, for a 24-slot four-pole stator. Note the arrangement of the two windings with

FIG. 271. Cutaway view of a split-phase fractional-horsepower motor. Note the stator windings, the squirrel-cage rotor, and the centrifugal switch. (*Century Electric Co.*)

respect to each other (90 electrical degrees apart) and the way in which successive groups are connected together to form poles of opposite polarities.

In operation, the two windings would be connected in parallel so that both are served by the same single-phase source. *Starting torque is developed only when the two windings are energized*, but when the rotor speed reaches about 75 per cent of rated value, the main winding develops nearly as much torque as both windings acting together. And when the speed of the motor reaches its rated value, the torque developed by both windings in the circuit is less than when only the main winding is energized. It seems logical, therefore, to provide some sort of mechanism, actuated by a centrifugal device or an electromagnetic relay, that will cut out the auxiliary winding at the proper time. Many such mechanisms have been developed. In the usual applications of the motor, centrifugally operated governors are employed, in which a set of spring-loaded weights, mounted on the shaft, fly out at a preset speed; as they do so, a switch mechanism fastened to the end bell disconnects the auxiliary winding from the circuit. Figure 271 shows a cutaway view of a split-phase motor in which the windings, the squirrel-cage rotor, and the centrifugal switch at the right end just inside the bearing are plainly visible.

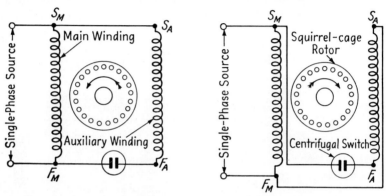

(a) Counterclockwise rotation   (b) Clockwise rotation

FIG. 272. Sketches showing how a split-phase motor is connected for either direction of rotation (see Fig. 270).

The direction of rotation of a split-phase motor is determined by the way the main and auxiliary windings are connected to the source and *with respect to each other*. Since the ratio of $X:R$ for the auxiliary winding is less than that of the main winding, its flux will always *lead* the main-winding flux. This means that *the field will revolve in a direction from a given auxiliary-winding pole to an adjacent main-winding pole of the same polarity*. Referring to Fig. 270, rotation will be counterclockwise if S_A is joined to S_M and F_A is joined to F_M. (The centrifugal switch would first, of course, be connected in series in the auxiliary winding.) Also, if S_A is joined to F_M and F_A is joined to S_M, rotation will be clockwise. Figure 272 illustrates schematically how this is done.

Since split-phase motors are usually started directly from the line with no current-limiting device, the initial current inrush may be as much as 4 to 6 times the rated value. This line current has two components, one of them (the larger of the two) passing through the main winding, while the other passes through the auxiliary winding. Advantage is taken of the high starting current in connection with the use of an electromagnetic type of relay that performs the same function as the centrifugal device.

This relay (Fig. 273) has a coil that is connected in series in the main-winding circuit, and a pair of contacts, normally open, that is in the auxiliary-winding circuit. During the starting period, when the main-winding current is high, the contacts close so that the auxiliary circuit is energized; after the motor reaches the proper speed, the main-winding current drops to a value that is low enough to cause the contacts to open. Such relays are commonly used on motors that are hermetically sealed in refrigeration units; they are located outside the motor, where they can be serviced or replaced, a situation that would be impossible if an internally mounted centrifugal switch were used. Figure 274 shows a cutaway view of one design of this type of relay used in hermetically sealed motors.

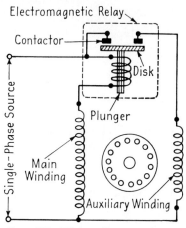

Fig. 273. Wiring diagram showing electromagnet relay in a split-phase motor.

Speed control of split-phase motors is accomplished with some difficulty. Special windings are generally necessary to change the speed, and even then the variation cannot be in small steps as in the case of d-c machines.

Fig. 274. Cutaway view of electromagnet type of relay used in connection with a hermetically sealed split-phase motor. (*Westinghouse Electric Corp.*)

The usual methods are to use two sets (main and auxiliary) of windings for two combinations of poles or to employ consequent-pole windings, discussed in connection with polyphase motors.

The speed regulation of split-phase motors is about the same as it is for those built for polyphase service; the speed varies about $2\frac{1}{2}$ to 5 per cent between no load and full load. For this reason, they are usually regarded as constant-speed motors, though not in the same sense as synchronous-type machines, the speeds of which are absolutely constant.

A well-designed split-phase motor will have a starting torque that is at least as much as the normal running torque. Starting torques as high as 150 per cent, i.e., where $T_{st} = 1.5 T_{FL}$, are rather common. When higher values of starting torque are required, it is necessary to use capacitor-type motors, which will be discussed in the next articles. In general, it is customary to employ standard split-phase motors in those applications that require comparatively low starting torques and have starting periods of brief duration. In connection with the latter point, a long period of starting may cause the auxiliary winding to burn out because that winding is designed for very short-duty service.

The Capacitor-start Motor. To overcome the lack of sufficient starting torque that is characteristic of the standard split-phase motor, these motors are provided with a capacitor in the auxiliary-winding circuit. The effect of this addition is to cause the auxiliary-winding and main-winding starting currents to be more nearly out of phase by 90 electrical degrees than was previously the case. Under this condition, the motor tends to approach two-phase operation. To illustrate the extent to which the capacitor alters the angular displacement between the two currents, tests indicate that the angles are about 82 and 25° for $\frac{1}{4}$-hp 60-cycle motors, with and without capacitors, respectively. Since the starting torque of a split-phase motor is proportional to the *sine of the angle* between the main and auxiliary starting currents, it should be clear that this change alone will increase the starting torque to values that are nearly $2\frac{1}{2}$ times that developed by the standard split-phase motor. In addition, other design improvements make it possible to increase the starting torque further so that starting torques with values as high as 350 to 450 per cent are readily achieved. Another important advantage that results from the addition of a capacitor in the auxiliary-winding circuit is that the initial inrush of current is reduced, though not to the same extent that the starting torque is increased.

A wiring diagram showing the connections for this type of machine, the so-called *capacitor-start split-phase* motor, is given in Fig. 275. Comparing that diagram with Figs. 272 and 273, note that it differs from the latter sketches by the simple addition of a capacitor. It is possible, in fact, to convert a standard split-phase motor into a capacitor-start type by adding a capacitor of the proper size in the auxiliary winding circuit. To do this, it is suggested that tests be carefully conducted to determine what its value should be with relation to the desired starting torque and

SINGLE-PHASE MOTORS 425

the resulting starting current and capacitor volts. As a guide, it might be indicated 200- to 250-μf (microfarad) capacitors are probable values for $\frac{1}{6}$- and $\frac{1}{4}$-hp 60-cycle motors, respectively.

In practice, capacitors are generally mounted on the outside of the motor, as shown in Fig. 276. They are of the electrolytic type and are designed for extremely short-duty service. Standard recommendations are on the basis of twenty 3-sec periods per hour, or some equivalent thereof; forty $1\frac{1}{2}$-sec periods or sixty 1-sec periods would be equivalent duty cycles.

In order to indicate typical capacitance values of practical capacitor-start split-phase motors, Table 9 is given. These are average values and may vary somewhat among the many designs.

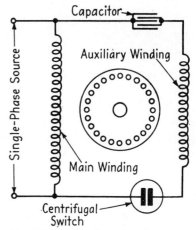

FIG. 275. Wiring diagram of a capacitor-start split-phase motor.

Also listed are approximate starting torques in percentages of full-load torques.

It should be clearly understood that the capacitor-start motor differs from the standard split-phase motor only during the starting period, for it

FIG. 276. Capacitor-start split-phase motor with an annular resilient-mounted base. Note the simple electrolytic-type capacitor on top. (*Wagner Electric Corp.*)

is then that it develops more starting torque with a lower inrush current. After the machine reaches normal speed and the auxiliary winding is cut out, its operating characteristics are identical to those of its counterpart. Such motors are therefore used in applications which require high starting torque, but have starting periods of short duration. The latter point is significant in view of the fact that electrolytic capacitors are rated for

TABLE 9. TYPICAL CAPACITANCE VALUES FOR 60-CYCLE CAPACITOR-START SPLIT-PHASE MOTORS

hp	Poles	rpm	Approx. capacitance values, µf	Approx. per cent starting torque	Dimensions of capacitors	
					Diameter, in.	Length, in.
1/8	2 4 6	3,450 1,725 1,140	80	350–400 400–450 275–400	1 3/8	2 3/4
1/6	2 4 6	3,450 1,725 1,140	100	350–400 400–450 275–400	1 3/8	3 1/4
1/4	2 4 6	3,450 1,725 1,140	135	350–400 400–450 275–400	1 3/8	3 3/4
1/3	2 4 6	3,450 1,725 1,140	175	350–400 400–450 275–400	1 3/8	4
1/2	2 4 6	3,450 1,725 1,140	250	350–400 400–450 275–400	2	3 1/8
3/4	2 4 6	3,450 1,725 1,140	350	350–400 400–450 275–400	2	4

short-duty service only and will be destroyed if the starting load does not permit the motor to come up to speed rapidly. Another point about electrolytic capacitors which should be mentioned is that they have maximum voltage ratings; if subjected to voltages that are more than 25 per cent greater than such ratings, they will be damaged. This voltage limitation means that a centrifugal switch must positively open at its pre-set speed; if the spring is not properly adjusted or if the load keeps the motor speed near the point at which the switch flutters, i.e., opens and closes rapidly, it is possible for the capacitor voltage to be twice rated value for an instant, in which event the capacitor will fail. This double voltage may result if the capacitor is fully charged at the instant the switch opens and then closes on the a-c voltage wave when it is a maximum in the opposite direction.

Because the capacitor-start split-phase motor develops an extremely high value of starting torque, it is particularly suitable for small portable

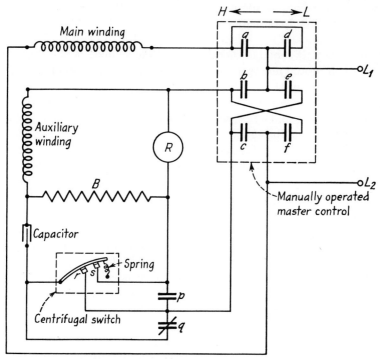

Fig. 277. Wiring diagram of a manually operated controller connected to a capacitor-start split-phase motor for a portable hoist.

hoist applications. For such service it must, however, be provided with a control circuit that will permit instant reversal of the direction of rotation. The arrangement frequently employed is a manually operated set of "up" and "down" cords, one of which is pulled for hoisting and the other for lowering. If, for example, the motor is hoisting and the centrifugal switch has operated to disconnect the auxiliary winding-capacitor circuit, pulling the "down" cord will instantly connect that circuit to the source, so that it is reversed with respect to the "up" position; this action therefore plugs the motor, with the result that it slows down quickly to a stop and speeds up in the opposite direction.

A wiring diagram of such an installation is given in Fig. 277. The control circuit consists of a special type of triple-pole double-throw (TPDT) switch that is actuated by pull cords, a relay R, with one normally open and one normally closed set of contacts p and q, and a protecting resistor B. For hoisting, a lever closes contacts a, b, and c. The closing of the a contacts connects the source L_1L_2 to the main winding, and current passes through the latter from right to left. Two other circuits are also established by the closing of contacts b and c. These are (1) through the auxiliary winding and its capacitor in a *downward* direction and normally closed contacts r and (2) through relay R and centrifugal-switch contacts s and r. When relay R operates to open contacts q and close contacts p, it remains energized but independent of the centrifugal switch. The motor now proceeds to accelerate and, at a pre-set speed, contacts r and s open.

To reverse the motor for lowering, the "down" cord is pulled. This open contacts a, b, and c *first*, and *then* closes contacts d, e, and f; relay R therefore drops out during the transition and causes contacts p to open and contacts q to close. Current now passes through the main winding, as before, from right to left. With centrifugal-switch contacts r and s still open, the auxiliary-winding circuit is energized through contacts q (the R relay is still dropped out); the motor is therefore plugged and slows down. Eventually contacts r and s close, relay R picks up, and the motor proceeds to stop and then accelerate normally in the opposite direction.

The function of the resistor B is important during a *momentary* power failure. Should this happen, relay R will drop out, contacts p will open, and contacts q will close. Then, when power comes *on* again, with the motor still running and contacts r and s open, the auxiliary-winding circuit is immediately energized. If resistor B were omitted, relay R could not pick up and the auxiliary winding and/or the capacitor would fail very quickly. However, with resistor B in the circuit as shown, there is sufficient voltage across the auxiliary winding to energize relay R through resistor B; it therefore picks up and instantly disconnects the auxiliary-winding circuit at contacts q.

The Two-value Capacitor Motor. In order to retain the advantages that result from the use of a capacitor during the normal running operation as well as the starting period, *two-value* capacitor motors have been developed. The reason two different capacitance values must be used is that the *starting* capacitor is much too large for good *running* performance; it is therefore necessary to employ a switching device that must change the capacitance in the auxiliary-winding circuit from a high to a low value when the motor attains rated speed. An additional requirement is that the *running* capacitor must have a continuous-duty rating, although as was pointed out in the previous article, *starting* capacitors are generally of the short-duty electrolytic construction.

Figure 278 is a wiring diagram for this type of motor in which an electrolytic (short-duty) and an oil-type (continuous-duty) capacitor are used in the auxiliary-winding circuit. When properly designed, such machines have operating characteristics that very closely correspond to those displayed by two-phase motors. Besides their ability to start heavy loads, they are extremely quiet in operation, have better efficiencies and power factors when loaded, and develop up to 25 per cent greater overload capacities. They are indeed splendid machines when the load requirements are

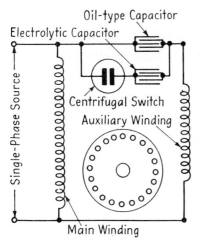 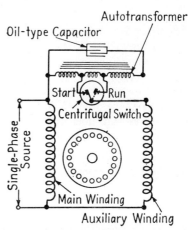

Fig. 278. Wiring diagram of two-value capacitor split-phase motor, using two capacitors.

Fig. 279. Wiring diagram of two-value capacitor split-phase motor, using an autotransformer and a capacitor.

severe. It will be noted in Fig. 278 that when the centrifugal switch is closed during the starting period, the two capacitors are in parallel, so that the total capacitance is the sum of their values; after the switch opens, only the continuous-duty oil-type capacitor remains in the auxiliary-winding circuit. In practice, the starting capacitor is about 10 to 15 times as large as the running capacitor.

Two-value capacitor motors are manufactured in which an autotransformer and a single capacitor unit are used. The idea is based on the principle that a capacitor of value C, connected to the secondary terminals of a step-up transformer, appears to the primary as though it had a value of a^2C, where a is the ratio of transformation. For example, if the step-up ratio is 6 and the actual capacitance connected across the high-voltage secondary is 5 μf, the low-voltage primary acts as though it had a 180-μf ($= 36 \times 5$) capacitor connected to its terminals. In other words, the effectiveness of a low-value capacitor can be increased greatly by the use of a step-up transformer. Figure 279 shows how this is done in one practical design in which the stepping up is accomplished by a comparatively inexpensive autotransformer. Note that the centrifugal switch shifts the capacitor from one voltage tap to another, so that the step-up ratio changes from one value on starting to another on running. In practice, the autotransformer and capacitor are placed in a can and mounted on top of the motor. The capacitor must be capable of withstanding a continuous a-c voltage of 330 and an intermittent voltage of 800; it is usually of the paper-tinfoil construction immersed in a high-grade insulation like wax or mineral oil, or in some other special compound such as Pyranol or Inerteen.

EXAMPLE 4. A 6-μf capacitor is connected to an autotransformer, as shown in Fig. 279. If the primary voltage is 110 on *starting* and 400 volts on *running*, what are the effective capacitances of the capacitor if the secondary voltage is 700 in either position?

Solution

On the starting position,
$$C = 6 \times \left(\frac{700}{110}\right)^2 = 243 \ \mu f$$

On the running position,
$$C = 6 \times \left(\frac{700}{400}\right)^2 = 18.4 \ \mu f$$

The Single-value Capacitor Motor. Split-phase motors, in which a single capacitor is *permanently* connected in the auxiliary-winding circuit, are frequently used where the required starting torque is low. They are commonly employed for air-moving equipment, such as fans and blowers, or, as in the case of oil burners, for applications in which quiet operation is particularly desirable. The starting torque of *single-value* capacitor motors, as they are called, is about 50 to 100 per cent of rated torque. No centrifugal switch is needed, since the motor is designed to run continuously with the auxiliary winding and its capacitor in series. The latter is of the oil-type construction because of its continuous-duty rating. A wiring diagram showing the connections for this motor is given in Fig. 280.

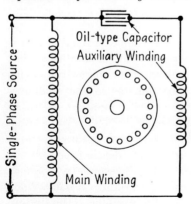

FIG. 280. Wiring diagram of a single-value capacitor split-phase motor.

A unique feature of this type of motor is that it may be very easily reversed *if the two windings are identical.* It was previously pointed out that the direction of rotation is always from a given auxiliary-winding pole to an adjacent main-winding pole of the same polarity. Therefore, if a SPDT switch is provided so that the capacitor may be connected in series with one or the other or the two *similar* windings, either one may be made to serve as the auxiliary or main winding; hence the direction of rotation will be determined by their relative positions. It is exactly as though the main and auxiliary windings of Fig. 270 interchanged positions. A wiring

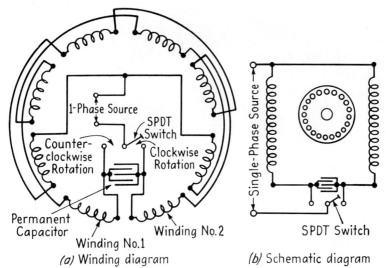

Fig. 281. Wiring diagrams showing how a single-value capacitor motor may be reversed by a SPDT switch.

diagram showing how this is accomplished in the single-value capacitor motors is given in Fig. 281.

Because of the simplicity of this reversing control scheme, the capacitor motor is often used to operate devices that must be moved back and forth frequently; these include rheostats, induction regulators, furnace controls, valves, and arc-welding controls.

The Repulsion-start Induction Motor. *General Constructional Details.* Before capacitor-start and two-value capacitor motors came into general use for applications requiring high starting torque, the *repulsion-start induction motor* was the accepted standard. Manufactured in integral- as well as fractional-horsepower sizes, this type of motor was used at one time in larger numbers than any of the others for single-phase service. Although it is the most costly of the various types of single-phase motor, it nevertheless possesses two extremely desirable characteristics: (1) high starting torque for periods that are of comparatively long duration and

(2) low starting current. In connection with the latter point, these motors draw starting currents that are about 60 to 70 per cent of that taken by corresponding sizes in the capacitor-start and two-value capacitor classifications. Its rotor construction is much more complicated than the squirrel cage of the split-phase motor; it has a standard d-c armature winding placed in skewed slots, a commutator of special design, and a centrifugal mechanism for short-circuiting the commutator when the

FIG. 282. Completely wound and assembled rotor of a repulsion-start induction motor. (*Wagner Electric Corp.*)

motor reaches its rated speed. Figure 282 shows such a rotor, in which the following points should be noted:

1. Skewed slots to reduce magnetic noise and to eliminate variation in starting torque at different positions of the rotor.

2. Governor weights at the right that are thrown outward by centrifugal force to move push rods which pass through the rotor and actuate the brush mechanism.

3. A cooling blower to provide ventilation for the rotor and stator.

4. A machine-wound armature winding securely braced and banded.

The brush mechanism is a very important part of this type of motor. It consists of a set of *short-circuited brushes* that ride on the vertical face commutator during the starting period. Figure 283 shows the construction of the brush rigging mounted on the inside of the end bell that fits on the commutator end of the machine. Note particularly the provision for shifting the brushes; this is done by loosening a set screw from the outside, shifting the brushes to a desired position, and then locking the rigging in place. (The reason for this will be explained later.) A short-circuiting necklace inside the hollow commutator is not in contact with the commutator bars until the governor weights fly outward. It is then that the push rods force a spring barrel forward until the short-circuiting necklace connects the commutator bars and the short-circuiting ring;

after that, the converted rotor, which is in effect a squirrel cage, permits the machine to operate as a squirrel-cage induction motor.

The stators of repulsion-start induction motors are identical with those found in split-phase machines, with the exception that only one winding is used; no auxiliary winding is needed for starting purposes, because the special rotor construction gives this motor its excellent starting characteristic. Figure 284 depicts a completely wound 36-slot stator in which the four-pole spiral type of winding, previously discussed, is clearly

FIG. 283. Inside view of an end bell for a repulsion-start induction motor. Note the brush rigging arrangement for use with a vertical commutator, and provision for shifting the brushes. (*Century Electric Co.*)

visible. Four winding leads are brought out in this design so that it may be connected for 115- or 230-volt service.

The Repulsion Principle. The repulsion principle was first employed in a practical motor by Dr. Engelbert Arnold in Germany and was commerically adopted in this country for some applications about 1896. Since that time it has been steadily improved both mechanically and electrically. It is still used in large numbers, particularly when *sustained* high starting torques at low values of starting currents are important; however, as previously pointed out, it has been replaced to a great extent by the more rugged, less costly capacitor-start or two-value capacitor types of motor. To understand how torque is developed by the repulsion principle, consider Fig. 285. Imagine a two-pole salient (shaped)-pole motor with the magnetic axis vertical. The armature is of standard d-c construction with commutator and brushes, but with the latter connected together, i.e., short-circuited. Assume that the *alternating current* flows in such a direction that it creates a *north* pole on top and a *south* pole on the bottom. Voltages will be induced in all the armature conductors by transformer action; by Lenz's law the direction of these induced voltages will be such that they will tend to create a flow of current in the conductors

FIG. 284. Completely wound 36-slot stator for a repulsion-start induction motor. (*Wagner Electric Corp.*)

so that the main field will be opposed. This means that the voltage in all the conductors to the left of the vertical line YY' will be directed toward ⊙ the observer and the voltages in all the conductors to the right of this line will be away ⊕ from the observer. This should be clear if the right-hand rule for determining the magnetic polarity is applied. However, it does *not* imply that the current directions in all the conductors will be the same as the voltage directions, because *the current directions in the various conductors will depend upon the position of the short-circuited brushes.* If the brushes are lined up with the vertical axis YY', as in Fig. 285, the current directions and the voltage directions in *all* the conductors will be the same. In other words, the brush axis and the main magnetic-field axis are collinear. As a result, the armature will become an electromagnet with a *north* pole on top, directly under the main *north* pole, and with a *south* pole at the bottom, directly over the main *south* pole. Under this condition, no torque will be developed because the two forces of repulsion, on top and on the bottom, occur along the YY' axis; there is no tangential force component.

If the short-circuited brushes are next moved over to position ZZ', as indicated by Fig. 286, torque *will* be developed. The reason is that the

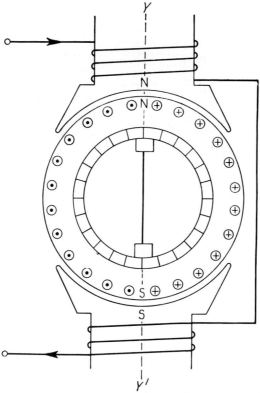

Fig. 285. Sketch representing a two-pole repulsion motor. Main magnetic axis and brush axis are collinear. The crosses and dots in the rotor conductors indicate induced voltages. *No rotation* occurs with the brushes in this position.

magnetic axis of the armature is no longer collinear with respect to the vertical axis of the main poles; it will now be along the axis ZZ' with the *north* and *south* poles shifted around by as many electrical degrees as the brushes are shifted. It is true that the induced voltages in conductors *a*, *b*, and *c* will oppose the other conductors above the brush axis and that the induced voltages in conductors *d*, *e*, and *f* will oppose the other conductors below the brush axis, but the net voltage acting at the brushes to produce a current flow will be sufficient to make the rotor core into a powerful magnet, *with a north pole at the upper brush position and a south pole at the lower brush position.* Obviously, the rotor *north* pole will now be *repelled* by the main *north* pole, and the rotor *south* pole will similarly be *repelled* by the main *south* pole, so that torque *will* be created. Note particularly that the main magnetic axis YY' and the rotor magnetic axis ZZ' are *not* collinear; there will now be tangential

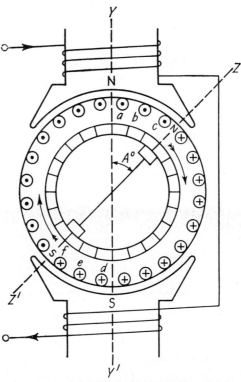

Fig. 286. Sketch representing a two-pole repulsion motor with brushes shifted clockwise by $A°$. The crosses and dots in the rotor conductors indicate current directions. Rotation will tend to be *clockwise*.

force components. Thus the rotor will rotate in a clockwise direction if it is free to do so. Also, since the forces are those of *repulsion (like poles repel)*, it is proper to designate this type of motor as a *repulsion* motor.

If the brushes are shifted in a counterclockwise direction, rotation will also be counterclockwise. This is shown in Fig. 287. It should be clear, therefore, that the direction of rotation is independent of the stator or rotor winding connections and depends only upon the brush position with respect to the main magnetic axis.

The discussion was based upon assumed salient poles because it was convenient to do so. However, the same theoretical analysis applies equally well to distributed windings, which are always used (see Fig. 284). Motors having stator cores with slots develop more uniform torque, are not likely to have dead spots when started, and, more particularly, can be wound for several different speed combinations. This last point is especially important in manufacture from the standpoint of cost, because a

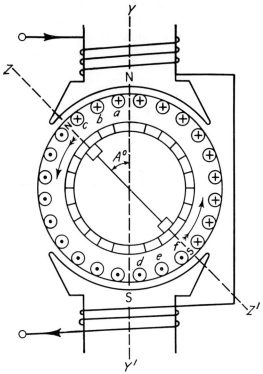

Fig. 287. Sketch representing a two-pole repulsion motor with brushes shifted counter-clockwise by $A°$. The crosses and dots in the rotor conductors indicate current directions. Rotation will tend to be *counterclockwise*.

single lamination can frequently be used in the design of two-, four-, six-, or eight-pole motors.

Operating Performance. In the single-phase repulsion-start motor, torque is developed by the principle of repulsion during the accelerating period only. At a predetermined speed, which is about 80 per cent of rated speed, the commutator is short-circuited by a mechanism similar to the one previously described; this action causes the rotor to become electrically equivalent to a squirrel cage, so that thereafter the machine has operating characteristics that are exactly like those of the split-phase motor. The brushes are also lifted from the commutator so that they will not wear unnecessarily or cause a friction loss.

The starting torque of repulsion-start motors is about the same as that of capacitor-start motors, average values being in excess of 350 and 300 per cent for four- and six-pole 60-cycle machines, respectively. They are, however, the preferred type when the starting period is of comparatively long duration because of high-inertia loads. Starting currents are somewhat lower than capacitor-start or two-value capacitor motors—about 30 to 40 per cent less—and this results in less line-voltage disturbance. Speed variation, i.e., change in speed between full load and no load, is about $2\frac{1}{2}$ to 5 per cent, which puts this type of motor in the constant-speed classification.

These motors are reversed by changing the position of brushes, which are usually located about 15 to 20 electrical degrees to either side of the main magnetic neutral. In most constructions, this is done by loosening a set screw and shifting the brush rigging until a pointer indicates the proper position for the desired direction of rotation. Obviously, the motor must first be permitted to come to rest before the brushes are shifted.

The Repulsion Motor. This motor is similar to the repulsion-start type, discussed in the previous article, with the exception that no short-circuiter or brush-lifting mechanism is provided; it operates continuously by the repulsion principle. It consists of a single-winding stator (Fig. 284) and a rotor very much like that found in d-c machines. The brushes are, of course, short-circuited and in contact with the commutator at all times. Unlike the repulsion-start type, the *repulsion* motor has a varying speed characteristic similar to that of the series motor, the operating speed of which depends upon the load; the speed is not a function of the line frequency and the number of poles, as it is in the case of induction motors, in which the revolving field determines the motor speed. Figure 288 represents a typical *speed vs. torque* curve for a repulsion motor and indicates how greatly its speed is affected by the load.

The direction of rotation is determined by the position of the brushes with respect to the main magnetic neutral. In practice, this is about 20 to 30 electrical degrees to either side of the YY' axis (Fig. 285). Sparking at the brushes is generally negligible at rated load, which usually occurs near synchronous speed, but becomes excessive when the speed is much above or below rated value.

Starting torques and starting-current values are about the same as those for repulsion-start motors, i.e., about 350 per cent starting torque and about 3 to 4 times rated current. Speed control is effected by varying the impressed voltage or by changing the position of the brushes. This latter point is an important reason for the use of these motors in preference to the repulsion-start type when it is necessary to adjust the speed of some application. A common use is the coil winder, in which the operator adjusts the speed by shifting the brushes; the motor is equipped with a special lever mechanism that shifts the brushes when a foot treadle is pressed.

The Repulsion Induction Motor. Another form of repulsion motor is a type that has a squirrel-cage winding in the rotor in addition to the usual

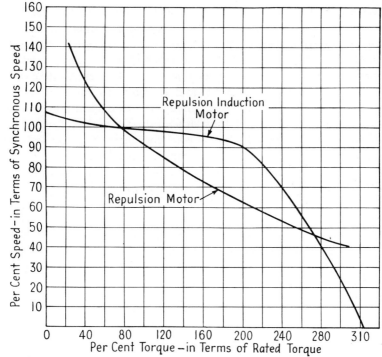

FIG. 288. Typical *speed vs. torque* curve for repulsion-type motors.

d-c winding connected to the commutator. The former is placed in a separate set of slots directly below the outside rotor slots. The brushes are short-circuited and ride on the commutator continuously. Starting torque is generally a little less than that developed by the repulsion type, about 300 per cent or more, while the starting current remains about the same. Used when the starting duty is heavy, they will carry any load which they can start. A common application is the air compressor that is subject to low temperatures.

Its *speed vs. torque* characteristic differs considerably from that of the repulsion motor (see Fig. 288) in that it is nearly flat between full load and no load; in this respect it may be classified as a constant-speed motor. It does, however, lose speed rapidly at torques that exceed about 200 per cent of normal and rises slightly above synchronous speed at loads somewhat below 100 per cent torque. Such a variation is desirable in applications in which the speed must remain substantially constant for most conditions of loading but must not deliver excessive horsepower load when the torque is high. In other words, the speed adjusts itself for heavy torque loads so that the horsepower output remains approximately constant. (It will be recalled that horsepower is a function of both the torque and the speed.) Typical applications are machine tools such as lathes and boring mills.

Like other types of repulsion motor, repulsion induction motors are reversed by shifting the brushes.

The Reluctance Motor. When the rotor of a split-phase motor has properly designed salient (shaped) poles, it will start as an induction motor, attain synchronous speed, and continue to run at a constant synchronous speed. The physical dimensions of such a machine must, however, be somewhat larger than other types of motor of equal horsepower rating. The usual method of constructing a rotor is to assemble it from standard squirrel-cage parts, except that some of the teeth are cut away. For example, in a 48-tooth four-pole rotor, the following teeth would be cut out:

1, 2, 3, 4, 5—13, 14, 15, 16, 17—25, 26, 27, 28, 29—37, 38, 39, 40, 41

This would leave four projecting poles consisting of the following teeth: 6 to 12, 18 to 24, 30 to 36, and 42 to 48; a rotor lamination for such a motor is shown in Fig. 311a. The squirrel cage would be installed in the usual manner.

In operation, the rotor comes up to nearly synchronous speed by induction-motor action. Since the mechanical load is *comparatively* light, the slip is negligibly small, in which case the revolving field permanently magnetizes the projecting rotor poles. The rotor poles then "lock in step" with the revolving poles of opposite polarities and continue to rotate at synchronous speed, the speed of the revolving field. The machine is, in effect, a synchronous motor without d-c excitation (discussed in the next chapter); it is called a *reluctance* motor, getting its name from the variable magnetic reluctance of the air gap. It must not be confused with the reluctance-start motor.

When the number of salient poles on the rotor is greater, by some multiple, than the number of electrical poles on the stator, the motor will operate at a constant average speed that is a submultiple of the apparent synchronous speed; it is then called a *subsynchronous* reluctance motor.

Reluctance motors are generally made in the fractional-horsepower sizes and use the conventional split-phase stator and the centrifugal switch to open the auxiliary winding. When they are built for heavier loads, sometimes in integral-horsepower ratings, the permanent-split capacitor construction with no centrifugal switch is preferred. Its constant-speed characteristic makes it particularly suitable for such applications as signaling devices, recording instruments, regulating and control equipment, many kinds of timers, and phonographs. The rotor of a small reluctance motor is shown in Fig. 289. Note how a portion of some of the teeth is cut away in the otherwise conventional squirrel-cage rotor.

The Hysteresis Motor. When the rotor of an induction motor is built up of a group of specially hardened steel rings instead of the usual thin silicon steel laminations, the effect of hysteresis is greatly magnified. As a result, the rotor will operate at synchronous speed because the hysteresis property of the rotor steel strongly opposes any change in the magnetic

Fig. 289. Rotor of a reluctance type of motor. (*Robbins & Myers, Inc.*)

Fig. 290. Rotor of a hysteresis type of motor. (*Robbins & Myers, Inc.*)

polarities once they are established. Effective synchronous motor action is therefore produced because the rotor poles lock in step with the revolving poles of opposite polarities. (In the conventional laminated rotor, the poles change position from instant to instant as the latter slips behind the revolving poles.) The fact that the rotor has no teeth or winding of any sort makes the motor extremely quiet in operation; it is not subject to the mechanical and magnetic vibrations that must always be present when there are teeth and cutaway sections in the rotor. Resilient mountings further reduce other possible noises, so that, virtually noiseless, hysteresis motors are particularly useful for sound-equipment applications.

These motors are made in very small sizes only, about $\frac{1}{100}$ hp or less, because hardened magnet steel must be worked at rather low flux densities for reasonable values of stator power input. This means that the rotor bodies would have to be too heavy for higher ratings. Figure 290 depicts

442 ELECTRICAL MACHINES—ALTERNATING CURRENT

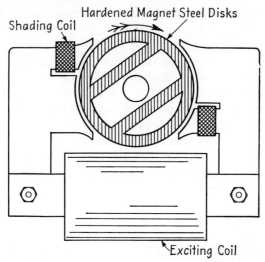

Fig. 291. Sketch illustrating Telechron type of hysteresis motor for operation of clocks. This motor is self-starting by the shaded-pole principle.

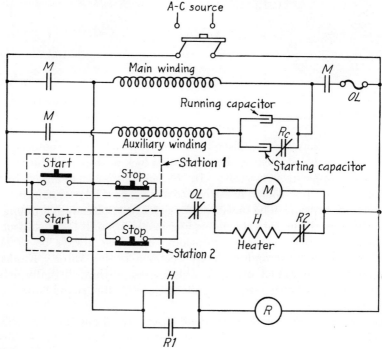

Fig. 292. Wiring diagram of an automatic starter connected to a two-value capacitor split-phase motor.

the rotor of a hysteresis motor of very low rating and illustrates the smooth cylindrical surface.

Electric-clock motors operate on the hysteresis-motor principle. In the *Telechron* design (Fig. 291), a two-pole revolving field is introduced into a sealed thin metal cylinder, in which a shaft, carrying one or more hardened magnet steel disks, drives a gear train. The latter requires about 2 to 4 watts input in the ordinary household clock.

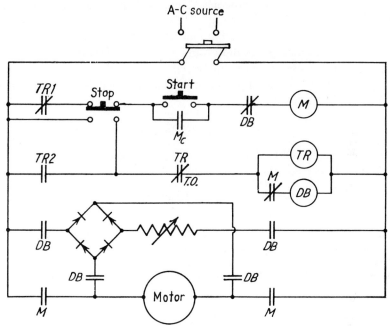

Fig. 293. Wiring diagram of an automatic starter connected to a split-phase motor and arranged for dynamic-braking service.

Automatic Starters for Split-phase Motors. Several types of automatic starters are used with split-phase motors, especially when the latter are large and must be started frequently; moreover, such equipment may be provided with more than one push-button station, sometimes remotely and conveniently located, as well as some form of electrical braking. Two typical arrangements are given in the following diagrams.

Starter for Two-value Capacitor Motor. The wiring connections for a large two-value capacitor motor (Fig. 278) and an automatic starter with two push-button stations are shown in Fig. 292. When either of the two parallel-connected START buttons is pressed, the M contactor is energized and a thermal timer H begins to carry current; the latter is made of a bimetallic strip around which is wound a heater coil, its action being similar to that of a thermostat. The operation of contactor M

causes the three M contacts to close and this starts the motor with two capacitors in parallel in the auxiliary-winding circuit. After the thermal timer times out, it closes the H contacts which, in turn, energizes the R relay; when the latter picks up, normally closed contacts R_c open to disconnect the electrolytic capacitor, the $R1$ contacts close to seal the R relay and the $R2$ contacts open to disconnect the heater; the H contacts then open. The motor may be stopped by pressing either of the two series-connected STOP buttons. Overload protection is provided as indicated.

Starter, with Dynamic Braking, for Split-phase Motor. The wiring connections for a split-phase motor and an automatic starter arranged for dynamic-braking service are illustrated by Fig. 293. Its operation is similar to that previously described on p. 393 for the polyphase motor and controller of Fig. 254.

Questions

1. Define a fractional-horsepower motor.
2. Explain why a motor having a rating greater or less than 1 hp is not necessarily classified as a fractional-horsepower motor.
3. Distinguish between a *conduction* motor and an *induction* motor.
4. Under what conditions must single-phase motors be used?
5. What is a *universal* motor? Indicate why it is so called.
6. Explain the principle of operation of the universal motor.
7. What factors must be specially considered in the design of large series motors so that operation will be satisfactory on alternating current?
8. What is meant when referring to "small motors"?
9. Why are the rotor slots of universal motors skewed?
10. Why is a high speed often desirable in the operation of a small motor such as a universal motor? What limits its no-load speed?
11. With alternating current, what factors are responsibile for the change in speed when the load changes?
12. Why does the effect of armature reaction tend to increase the speed of a series motor?
13. Explain how it is possible for a universal motor to operate at a higher speed on alternating current than on direct current for the same load?
14. Why is a gear box frequently used on a universal motor? List several practical applications when this is done.
15. What is a constant-speed governor? Describe the principle of operation of one such type.
16. Give several practical examples of the use of a constant-speed governor as applied to a universal motor.
17. Why does the operation of a universal motor interfere with good radio reception? How can such interference be eliminated?
18. Give several practical applications of universal motors, other than those listed in Table 8.
19. Why is a shaded-pole motor an induction machine?
20. Explain why the field produced by a shaded-pole motor is not a true revolving field in the same way as that created by a polyphase induction motor.

SINGLE-PHASE MOTORS

21. Describe the construction of the stator of a shaded-pole motor. Use a sketch in doing so.
22. In what direction will the rotor of a shaded-pole motor rotate?
23. Referring to Fig. 262a, explain why the pole is centered near the middle of the unshaded part of the pole.
24. Referring to Fig. 262b, explain why the pole is centered near the middle of the entire pole.
25. Referring to Fig. 262c, explain why the pole is centered near the middle of the shaded part of the pole.
26. Explain how a *simple* shaded-pole motor can be reversed.
27. Make a sketch illustrating how a shaded-pole motor with two sets of shading coils can be reversed.
28. Make a sketch illustrating how a shaded-pole motor with two sets of stator windings can be reversed.
29. List several practical applications of shaded-pole motors.
30. How is the speed of a shaded-pole motor controlled? Illustrate by wiring diagrams.
31. Describe the construction of the stator of a reluctance-start motor.
32. In what direction will the squirrel cage of a reluctance-start motor rotate?
33. Explain why the field shifts from one side to the other of the pole of a reluctance-start motor.
34. How is the speed of a reluctance-start motor controlled?
35. Describe the construction of the stator of a split-phase motor.
36. What is a centrifugal switch and why is it used in split-phase motors?
37. Explain carefully why the fluxes created by the main and auxiliary windings are out of phase in *time;* in *space.*
38. Discuss shaded-pole and split-phase motors with regard to their constructional similarities and differences.
39. How do shaded-pole and split-phase motors compare with regard to efficiency? starting current? overload capacity? starting torque? cost of manufacture? per cent regulation? per cent slip?
40. Why is it desirable to disconnect the auxiliary winding in a split-phase motor after the rotor reaches about 75 per cent of rated speed?
41. Explain carefully how the direction of rotation of a split-phase motor may be predetermined.
42. How is a split-phase motor reversed? Illustrate your answer by sketches.
43. Describe the construction and explain the operation of the electromagnetic relay used in a split-phase motor.
44. Why would it be undesirable to use a centrifugal switch in a hermetically sealed motor?
45. How is it possible to control the speed of a split-phase motor?
46. What is a *capacitor-start split-phase* motor?
47. What advantages are possessed by capacitor-start motors?
48. What type of capacitor is commonly used in capacitor-start motors?
49. How is it possible to convert a standard split-phase motor into a capacitor-start motor? Explain the exact procedure for doing this properly.
50. Why is a capacitor-start motor equivalent to a standard split-phase motor while operating normally under load?
51. Give several practical applications of capacitor-start motors.
52. What is meant by centrifugal-switch *flutter,* and why must it be avoided in capacitor-start motors?

53. Referring to Fig. 277, explain the operation of the manual hoist controller for a capacitor-start split-phase motor: (a) when hoisting; (b) when lowering; (c) when changing from lowering to hoisting.
54. What purpose is served by the resistor in the hoist controller circuit of Fig. 277?
55. What is the purpose of the two-value capacitor motor?
56. What types of capacitor are used in two-value capacitor motors? Explain.
57. What two methods are generally employed in the construction of two-value capacitor motors?
58. Explain how it is possible to magnify the effect of a capacitor by using an autotransformer.
59. Describe the single-value capacitor motor and indicate where it has its most important field of application.
60. What type of capacitor must be used in the single-value capacitor motor? Why?
61. Under what condition is it possible to reverse a single-value capacitor motor by using a simple SPDT switch? Illustrate by making a wiring diagram.
62. What is a repulsion-start motor?
63. What purposes are served by the centrifugal switch in a repulsion-start motor?
64. How do the stators of repulsion-start and split-phase motors differ? How do they resemble each other?
65. Describe the construction of the rotor of a repulsion-start motor.
66. How is a repulsion-start motor reversed?
67. Referring to Fig. 285, explain why torque is *not* produced when the short-circuited brushes line up with the main magnetic field axis.
68. Referring to Fig. 286, explain why torque is produced to rotate the rotor clockwise when the brushes are moved in a clockwise direction.
69. What would happen if the brush axis and the main magnetic-field axis in a repulsion-start motor were 90 electrical degrees with respect to each other?
70. Under what operating conditions are repulsion-start motors particularly desirable? List several practical applications.
71. Distinguish between a repulsion motor and a repulsion-start motor with regard to construction; operating performance.
72. What would happen if the load were removed from a repulsion motor?
73. How is the direction of rotation of a repulsion motor reversed?
74. What methods may be used to control the speed of a repulsion motor?
75. Distinguish between the rotor constructions of repulsion and repulsion induction motors.
76. What advantage is possessed by the repulsion induction motor when it is operating at overload torques?
77. Describe the reluctance motor.
78. What important characteristic is possessed by the reluctance motor that makes it suitable in certain applications?
79. Explain the principle of operation of the reluctance motor.
80. Give several practical applications of the reluctance motor.
81. When reluctance motors are built for the heavier loads, what stator construction is usually desirable?
82. What is meant by a *subsynchronous* reluctance motor? How does it differ in construction from the standard reluctance motor?
83. How is the hysteresis motor constructed?
84. How are hysteresis motors usually started?
85. What is the general field of application of the hysteresis motor?
86. Referring to Fig. 292, explain the operation of the automatic starter for a two-value capacitor motor.

87. Show by a simple sketch how a third push-button station should be connected into the circuit of Fig. 292.
88. Referring to Fig. 293, explain the operation of the automatic starter in which a split-phase motor may be started and stopped by dynamic braking.

Problems

1. Classify the following motors as either fractional or integral horsepower: (a) 1 hp, 3,600 rpm; (b) ½ hp, 1,800 rpm; (c) ¾ hp, 900 rpm; (d) ⅔ hp, 1,200 rpm.
2. Calculate the full-load power factor of a single-phase motor that has the following name-plate data: $E = 115$; $I = 4.6$; hp = ¼; rpm = 1,740. Assume an efficiency of 60 per cent.
3. What will a wattmeter register in the motor of Prob. 2?
4. Calculate the regulation and slip of a single-phase motor that has a no-load speed of 1,195 rpm and a full-load speed of 1,120 rpm when operated from a 60-cycle source.
5. Calculate the full-load torque of the motor in Prob. 2: (a) in pound-feet; (b) in ounce-inches.
6. A 1/20-hp 1,550-rpm shaded-pole motor takes 120 watts when operating at full load. What is its efficiency?
7. What per cent of the rated torque does a 1/25-hp 3,200-rpm shaded-pole motor develop at starting, if the average starting torque is 5.0 oz-in.?
8. Each coil of a four-pole shaded-pole motor has 185 turns for operation at 115 volts. All coils are connected in series. What change should be made in the winding if the motor is to be connected to a 220-volt source? (Assume that a voltage change must be accompanied by a corresponding change in the number of turns, so that volts per turn remains the same.)
9. Make a winding diagram for a six-pole split-phase motor having 36 stator slots. Use the diagram of Fig. 270 as a guide.
10. Connect the winding in the diagram of Prob. 9 so that it will rotate counterclockwise.
11. A ½-hp 1,750-rpm capacitor motor develops 350 per cent starting torque. What is the starting torque in pound-feet and ounce-inches?
12. For how many poles is a motor wound that operates at: (a) 960 rpm from a 50-cycle source? (b) 1,450 rpm from a 25-cycle source? (c) 1,130 rpm from a 60-cycle source?
13. Calculate the per cent slip for each of the motors of Prob. 12.
14. An autotransformer and a 4-μf capacitor are used in a two-value capacitor motor. If the step-up ratios on the starting and running positions are 6.8 and 1.5, respectively, calculate the effective values of the capacitance viewed from the primary side.
15. A two-pole subsynchronous 60-cycle hysteresis motor has a rotor with 32 teeth (poles). At what speed will it operate?
16. Make a sketch showing how a universal motor may be reversed by using a double-pole double-throw (DPDT) switch.
17. A ⅓-hp 115-volt 2,500-rpm repulsion motor yielded the following test data when started: $E = 115$ volts; $I = 11.7$ amp; $T = 4.2$ lb-ft; $P = 920$ watts. Calculate: (a) the ratio of the starting torque to the full-load torque; (b) the starting power factor.
18. Make a sketch showing how the stator winding for a 230-volt repulsion-start motor should be connected for 115-volt operation.
19. In performing a brake-load test upon a ¼-hp capacitor-start motor with its output

adjusted to rated value, the following data were obtained: $E = 115$ volts; $I = 3.8$ amp; $P = 310$ watts; rpm $= 1,725$. Calculate: (*a*) efficiency; (*b*) power factor; (*c*) torque in pound-inches.

20. A universal motor was operated under load from a 115-volt 60-cycle source, under which condition it took 284 watts, developed 3.2 oz-ft of torque, and ran at 6,000 rpm. When resistance was added in series to reduce the speed to 4,000 rpm while developing the same torque, the motor voltage was 90 volts and the power input 250 watts. Calculate the efficiency under each operating condition.

CHAPTER 11

SYNCHRONOUS MOTORS

General Facts Concerning Synchronous Motors. A synchronous motor, as the name implies, operates at an absolutely *average constant speed* regardless of the load; it departs from this average speed only instantaneously, during load changes. The regulation of this type of motor is therefore zero. If the countertorque exceeds the maximum torque that can be developed by the motor at the average constant speed, the motor will come to rest because its average torque then drops to zero. Only two factors determine the speed of a synchronous motor: (1) the frequency of the supply source and (2) the number of poles in the machine; the speed is directly proportional to the frequency and inversely proportional to the number of poles.

Unlike an induction motor, the excitation of which must come solely from the a-c supply lines to which it is connected, a synchronous motor receives its excitation from two sources of supply: (1) the a-c source through its stator winding and (2) a d-c source through its rotor field. And whereas the power factor of an induction motor is always lagging and fixed by the magnitude of the load, that of a synchronous motor is variable over very wide limits by changes in the d-c excitation. When the motor power factor is unity, the d-c excitation is said to be normal; overexcitation causes the motor to operate at a leading power factor, while underexcitation produces lagging power-factor operation.

The starting torque of a synchronous motor, as such, is practically zero. For this reason it is necessary to employ one of several methods to bring the machine up to synchronism (synchronous speed) before it is capable of assuming load. In one such method, widely used, the starting torque may be made sufficiently large to bring a rather heavy load up to synchronous speed.

Small synchronous motors, like those of the hysteresis and reluctance types, are used for timing devices in which it is essential that the speed be absolutely constant. Such machines, as was previously pointed out, have but one source of power; there is no d-c excitation. The motors that are considered in this chapter are the larger ones, which develop considerable power and which are served from both d-c and a-c sources. Except for certain special applications (for example, continuous-strip processes found in steel mills and paper manufacturing plants, where the many motors operating the rolls must be kept in a sort of synchronism), synchronous motors are not used because of their constant-speed characteristic. More particularly, they are preferred because they possess the following advantages:

1. They can be made to operate at a leading power factor and thereby improve the power factor of an industrial plant from one that is normally lagging to one that is close to unity.

2. They are less costly in certain horsepower and speed ranges (e.g., for 50- to 500-hp ratings at operating speeds less than 500 rpm and for horsepower ratings greater than 500 but operating at speeds that are more than 450 rpm).

3. They can be constructed with wider air gaps than induction motors, which make them better mechanically.

4. They usually operate at higher efficiencies, especially in the low-speed unity power-factor ranges.

Snychronous motors are frequently used for power-factor correction purposes only. Under such conditions, they operate without load and with greatly overexcited fields, so that they take currents that lead by nearly 90 electrical degrees; in this respect they act like capacitors. When operated in this way, they are called *synchronous condensers*. Such applications of the synchronous motor are common in large power systems in which considerable lagging power-factor equipment is installed. By operating in parallel with such lagging power-factor loads, the over-all power factor of a system can be greatly improved, and with it the voltage regulation and the transmission efficiency. In some recent developments, synchronous condensers have been enclosed in hermetically sealed casings and operated in an atmosphere of hydrogen, thereby reducing the effect of windage and increasing the cooling effect.

Synchronous Motor Construction. The stators of synchronous motors are identical with those of the alternators studied in Chap. 7. Whole-coiled lap windings are placed in the slots of the stationary laminated cores and are connected in the usual way for polyphase service. Figure 294 shows a completely wound stator having a rating of 7,350 kva at 360 rpm for operation from a 6,600-volt three-phase 60-cycle source. In accordance with alternator theory, the rotor for this machine would have 20 poles ($P = 120 \times 60/360$). Note the neat, compact arrangement of the coils and the end connections between the coils. Also clearly seen at the bottom are the leads that are later connected to the polyphase starting equipment. The form-wound coils are made up of multiple strands of copper strip in large machines such as this or of single strands in the small motors. Insulation may be cotton, asbestos, or glass, depending upon the size of the conductor, the motor voltage, and the class of insulation. The standard insulation is Class A, that is, fibrous materials such as cotton, paper, varnished cambric, etc., for use where ultimate temperatures do not exceed 105°C. Class B insulation includes mica, asbestos, and other materials capable of withstanding high temperatures.

FIG. 294. Completely wound stator of a synchronous motor having a rating of 7,350 kva, for operation from a 6,600-volt 60-cycle three-phase circuit. (*Allis-Chalmers Mfg. Co.*)

The coils are retained in the stator slots by means of fiber or leatheroid wedges treated with varnish to prevent moisture absorption. When starting service is particularly severe, the coils are securely braced to prevent movement.

The 20-pole rotor for the stator of Fig. 294 is shown in Fig. 295. The field poles are built up of laminations, i.e., steel punchings riveted together between cast-steel end plates. For the higher-peripheral-speed motors, as in this case, the poles have dovetail projections that fit into corresponding dovetail slots machined in the spider rim and secured by means of tapered steel keys driven in alongside the dovetail. Standard excitation for most field windings is 125 volts. The coil is usually made of strip copper wound on edge, which permits it to be easily ventilated. When 250-volt excitation is used, the fields generally have wire-wound coils. For high-speed machines, it is necessary to employ special bracing to prevent injury to the winding or insulation from centrifugal stresses. The field-winding ends are brought out to a pair of slip rings that are

FIG. 295. Completely assembled 20-pole salient-pole rotor for the stator of Fig. 294. (*Allis-Chalmers Mfg. Co.*)

fastened to the spider or mounted on the shaft extension. Current is led into the rings by means of carbon brushes; usually two or more brushes are provided for each ring so that one may be removed for inspection or replacement without interfering with the operation of the motor.

Figure 296 shows a 30-pole rotor, without shaft, placed inside its stator. This construction shows a plate-and-ring assembly instead of a spider. Note the rings fastened to the rotor plate and the brush holders and brushes secured to a tube that is bolted to the yoke frame. Clearly seen are the field-winding leads joined to the collector rings.

Exciters. Direct-current excitation may be obtained either from an exciting plant d-c system or, if no direct current is available, from separate d-c generators. Direct-connected exciters are frequently found on high-speed motors. These are overhung from specially constructed motor brackets with the exciter armature mounted on an extension of the motor shaft. On low-speed machines, belted exciters are sometimes used, notably on air compressor drives. Separate motor-generator sets are widely used for excitation of low-speed synchronous motors and less often on those that operate at high speed. A motor-generator set is shown in Fig. 297. The d-c generator (exciter) on the right is coupled to and driven by a standard polyphase induction motor. Where a group of

Fig. 296. Stator and rotor assembly for a 30-pole synchronous motor. (*Westinghouse Electric Corp.*)

synchronous motors is installed, it is sometimes economical to use a single motor-generator set for exciting the entire group, although greater flexibility is obtained when each motor is serviced by its own exciter.

Starting Synchronous Motors. It was previously pointed out that the synchronous motor, as such, has practically no starting torque; this situation exists only if there is no special winding in the pole faces of the rotor. The early machines were constructed in this simpler manner, for which reason it was necessary to start and bring them up to speed by an external motor. Several methods were used to accomplish this starting process, among which were the following:

1. The starting motor was a d-c machine coupled to the synchronous motor. In normal operation, the former was driven as a d-c generator and represented the synchronous-motor load. However, during the starting period the d-c machine acted as a motor to bring the synchronous motor up to speed, taking power from an available d-c source. The usual practice was to accelerate the synchronous motor, synchronize it

FIG. 297. Motor-generator exciter set for a synchronous motor. (*Westinghouse Electric Corp.*)

with the a-c supply, and then strengthen the field of the d-c machine until it began to function as a generator. After that the mechanical power output of the synchronous motor became the mechanical power input to the d-c generator.

2. The starting motor was the d-c exciter frequently mounted on an overhung synchronous-motor bracket and shaft extension (see Fig. 4). Here again, an available d-c source operated the exciter as a motor during the starting period; then, after the synchronous machine was brought up to speed and synchronized, the exciter assumed its normal function.

3. The starting motor was a small induction motor mounted directly on the synchronous-motor shaft. When this scheme was employed, the induction motor usually had two poles less than the synchronous motor, so that, with the usual induction-motor slip, the synchronous machine could be brought up to synchronous speed. After normal operation was established, the induction motor was sometimes uncoupled from the synchronous motor.

The method that is generally used at present is to incorporate a squirrel cage in the pole faces of the rotor so that the synchronous motor can be started as an induction motor. In manufacture, each pole, with its exciting winding, pole-face conductors, and short-circuiting bars, is assembled before it is placed on the rotor structure. Figure 298 illustrates one design in which five copper rods are brazed at both ends to heavy copper or copper-alloy segments. After the individual poles are properly fastened to the rotor rim, a complete squirrel cage is formed by joining together successive end segments. Figure 299 indicates how this is done by one manufacturer. The details shown in this rotor are for an 800-hp 2,200-volt three-phase 60-cycle 120-rpm synchronous motor. The

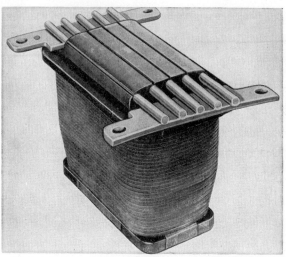

FIG. 298. Assembled pole for a synchronous motor, showing part of squirrel cage in the pole face. (*Westinghouse Electric Corp.*)

laminated core has five rods which, in different motors, vary in conductivity for different values of starting torque. The short-circuiting segments of each pole are seen connected to corresponding segments of the next pole by heavy locking nuts; they form a complete squirrel-cage winding. This construction is practically indestructible in ordinary operation.

When the synchronous motor is started by the induction-motor action of the squirrel cage, it is necessary to employ some method to reduce the current drawn from the line or to reduce the torque applied to the load. The former is done by one of the several methods described in Chap. 9, Polyphase Induction Motors. These are the line-resistance method (Fig. 242), the Y-Δ method (Fig. 244), the compensator method (Fig. 239), and the part-winding method (Fig. 246).

The general procedure that is followed in starting a synchronous motor is to short-circuit the field *before* the main three-phase switch is closed in the starting position in order to reduce the induced voltage in the field winding, which acts like the secondary of a step-up transformer; on starting, the revolving field cuts the field winding at synchronous speed. After the rotor reaches its approximate full speed, the short circuit is opened and the field is energized. This causes the rotor to "pull in step" with the revolving field; i.e., it is synchronized, with the least disturbance.

FIG. 299. Details of squirrel-cage construction in the rotor of a synchronous motor. Note the laminated core (1), the exciting winding (2), clamping bars (3), short-circuiting segments (4), locking nuts (5) and (6), and metal rods (7). (*Allis-Chalmers Mfg. Co.*)

At this point, full voltage is applied to the stator, or, if the part-winding method is used, the two halves of the stator winding are paralleled.

For automatic starting of a synchronous motor, the simple circuit of Fig. 300 may be used. Consisting of a main contactor M, a control relay CR, a relay D, a frequency relay FR, a reactor X, a resistor R, and several contacts, its operation is as follows. When the START button is pressed, the CR relay picks up and the two sets of CR contacts close. With the closing of the $CR2$ contacts, the M contactor is energized, and this, in turn, causes the M contacts to close *before* the M_a contacts close. The stator winding is thereby connected to the three-phase source and, acting as the primary of a transformer, induces an emf in the field-winding secondary, to which it is coupled electromagnetically. Current now flows in the field circuit at supply frequency (the rotor is still at rest), dividing between the *FR* relay and reactor *X*; however, since the reactance of *X* is much higher than the impedance of the series combination

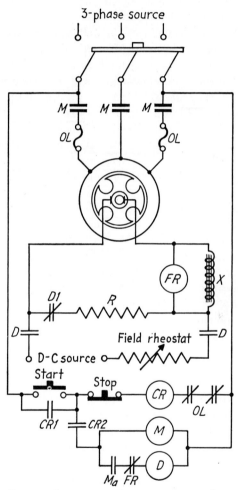

Fig. 300. Wiring diagram of an automatic starter connected to a synchronous motor.

of the *FR* relay and resistor *R*, most of the current passes through the latter. Relay *FR* then picks up and is fast enough to open normally closed contacts *FR before* the interlock contacts M_a close. The motor thus begins to accelerate because the field is closed through the normally closed contacts *D1*, resistor *R*, and the frequency relay *FR*. As the speed of machine increases, the frequency of the current in the field winding decreases (it is the slip frequency); this causes the reactance of X to diminish and a smaller part of the field-winding current passes through the FR relay. Finally, as the motor speed approaches synchronism, the current in the FR relay is reduced sufficiently so that it drops out. This closes the FR contacts, which, in turn, causes the D relay to pick up. The $D1$ contacts then open and, a fraction of a second later, the D contacts close to connect the field winding to the d-c source.

High-starting-torque Synchronous Motors. Synchronous motors with low-resistance squirrel cages for starting duty develop comparatively low values of starting torque. For many applications, such as blowers, air compressors, centrifugal and reciprocating pumps, pulp grinders, and the like, starting torques as low as 30 to 50 per cent are quite satisfactory. However, in certain other installations in which synchronous motors must develop starting torques of about 100 to 250 per cent, the design of the pole-face winding follows the practices used with induction motors. These are (1) the use of high-resistance squirrel cages and (2) the use of insulated windings in pole-face slots with connections made to slip rings (similar to the wound rotors of induction motors, Fig. 247).

Increasing the resistance of the squirrel cage of an induction motor has the effect of increasing the starting torque, increasing the full-load slip, and decreasing the efficiency. Since a synchronous motor behaves like an induction motor during the starting period, it is possible to improve the starting performance by constructing the squirrel cage with alloy materials that have resistivities greater than copper. Starting torques up to 200 per cent have been developed in this way for such applications as crushers and cement mills. A high-resistance squirrel cage does, however, prevent the rotor from coming up to a speed that is close to synchronism, particularly if the load is heavy; this means that when the d-c field is energized, the rotor must speed up rapidly to pull into step with the revolving field, with a consequent line disturbance. To lessen such disturbance, it is a good practice to short-circuit the field for an instant just before it is energized; this causes the rotor to pull ahead until it is rotating at nearly synchronous speed. In so far as low efficiency is concerned, a high-resistance squirrel cage has no effect upon the operating performance of a synchronous motor, because, unlike its continuous function in the induction motor, it acts only momentarily, i.e., during the starting period.

The wound-rotor idea is embodied in the so-called *simplex* type of rotor construction. An insulated three-phase winding is placed in the pole-face slots of the salient poles with the ends connected to three slip rings. Brushes riding on the slip rings are then connected to a resistance controller, which is used in exactly the same way as it is in the wound-rotor

SYNCHRONOUS MOTORS

induction motor; after the resistance is cut completely out, when the rotor reaches near synchronism, the phase winding becomes an electrical squirrel cage. The winding and controller resistances are adjusted to values that permit the synchronous motor to start extremely heavy loads; in some cases, such machines can develop torques that are $2\frac{1}{2}$ times as great as the normal full-load torques. A simplex rotor is shown in Fig. 301. Note particularly the half-coiled winding in the pole-face slots and the five slip rings: three rings are connected to the three-phase pole-face winding and two rings are joined to the d-c exciting winding. Such

FIG. 301. Simplex type of rotor for a synchronous motor. (*Westinghouse Electric Corp.*)

motors have many applications, among which are ball, rod, and tube mills, mining machinery, gyratory crushers, and line shafts in flour mills.

Principles of Operation. The operation of two a-c *generators* in parallel implies that both receive mechanical energy from prime movers. The latter supply the power for the common electrical load and the losses required by the machines themselves, except those necessary for excitation. If, by proper adjustment, the entire load is transferred to one of the alternators, the second machine will continue to take enough power from its prime mover and the a-c source for its own power losses, i.e., friction, windage, and iron losses. Next, when the prime mover is completely disconnected from the unloaded alternator, the latter will continue to operate at synchronous speed, now as a synchronous *motor;* under this condition, the other alternator will supply power to the synchronous motor to take care of the friction, windage, and iron losses. This is exactly the situation that exists when any synchronous motor is brought up to speed and is synchronized with the a-c source.

To understand why the rotor of a synchronous motor must always rotate at synchronous speed, it is necessary to recognize the fact that the stator field rotates continuously at synchronous speed and is equivalent to a set of physical poles that are locked in step with an equal set of strong rotor magnet poles; those of opposite polarities are always exposed to each other. Under no circumstances is it possible for a stator pole of one polarity to be opposite a salient pole of the same polarity. The lines of force that exist in the air gap between each stator and rotor pole may be considered as the "link" between the two; it is a flexible link in the sense that it can stretch when a heavy load tends to separate the two poles, but a link nevertheless that must never be broken if rotation is to continue. When the load, or countertorque, is light, the centers of the stator poles, as they rotate at synchronous speed, are nearly lined up with the centers of the rotor poles. As the load increases, the rotor poles slip a little behind so that the center lines of stator and rotor poles are displaced further. Still additional load widens the angular displacement until, at about 60 electrical degrees, the centers of stator and rotor poles are too far apart to remain locked in step. At this load, the motor stalls because the so-called "pull-out torque" is reached; under this condition, the motor will draw an extremely high value of current unless fuse or circuit-breaker protection is provided to open the line circuit.

As long as the mechanical load upon a synchronous motor is constant and there are no electrical disturbances, the rotor will revolve at an absolutely constant speed. When the load changes, however, there is a momentary change in the speed as the rotor-pole axes readjust themselves with respect to the stator-pole axes; if the load increases, the rotor poles slip backward, while a drop in load causes the rotor to speed up. During this adjustment period, which is of extremely short duration, the motor is not operating at exactly synchronous speed, but at slightly higher or lower speed. It is for this reason, therefore, that the term *average constant speed* should be used to designate the synchronous-motor speed accurately.

Loading a Synchronous Motor. All electric motors, whether d-c or a-c, act as generators at the same time that they are motoring. This is true because a counter emf is always generated when a motor is in operation. In a d-c machine, the rotating armature conductors cut lines of force under the stationary field poles; in an a-c machine, as in an induction or a synchronous motor, the rotating magnetic field cuts the stationary armature conductors of the stator. In fact, it is this very counter emf that limits the armature current to a value that is just enough to take care of the power requirements of the motor for a particular load.

The way in which the synchronous-motor input current adjusts itself to the given load can best be understood by first considering the reactions in a d-c shunt motor. It was shown in Chap. 5 that the armature current is $I_A = (V - E_C)/R_A$, where V is the impressed armature voltage, E_C is the counter emf, and R_A is the armature resistance. This is in accordance with Ohm's law, where $(V - E_C)$ is the *net* voltage acting across the armature resistance. Since, in a given machine, E_C depends upon the flux ϕ and the speed S, it follows that any change in either value will affect the input current I_A. Under no circumstances, however, can the value of E_C be equal to the impressed voltage V; it must always be *less* than V if the shunt machine is to behave like a motor. Thus, if the load increases, the speed drops; this drop in speed is immediately accompanied by a reduction in the counter emf E_C, as a result of which the net voltage $(V - E_C)$ rises. The higher net voltage then causes the armature current I_A to rise sufficiently so that the increased power requirements are met. Again, if the load decreases, the speed rises; this rise in speed is immediately accompanied by an increase in the counter emf E_C, as a result of which the net voltage $(V - E_C)$ drops. The lower net voltage then causes the armature current I_A to drop sufficiently so that the reduced power requirements are met. All these changes are assumed to take place without a change of flux.

This analysis is further clarified if Eq. (12) (Chap. 5, p. 138) is rewritten so that simple diagrams, *phasor* diagrams, can be drawn to illustrate how the current changes with changes in load. Similar diagrams will later be used to indicate how the same kind of reactions affect the operation of a synchronous motor. Thus $I_A = (V - E_C)/R_A$ may be written

$$I_A = \frac{V + (-E_C)}{R_A} \tag{92}$$

In this form, V and $-E_C$ can be plotted on a diagram in assumed positive and negative directions, as is done in Fig. 302, which illustrates how the armature current I_A is affected by changes in load. Note particularly that:

1. V is plotted to the right as a positive phasor and E_C is plotted to the left as a negative phasor.

2. Original load 1 develops a net voltage $(V - E_{C1})$ and an armature current I_{A1}.

3. Load 2 is *greater* than the original load, so that E_{C2} becomes less than E_{C1}, thus increasing both the net voltage $(V - E_{C2})$ and the armature current I_{A2}.

4. Load 3 is *less* than the original load, so that E_{C3} becomes greater than E_{C1}, thus decreasing both the net voltage $(V - E_{C3})$ and the armature current I_{A3}.

Before the reactions in a synchronous motor are analyzed, it is necessary to point out several important differences between its behavior and that of the shunt motor. These are:

1. The average speed of a synchronous motor is absolutely constant, regardless of the load, while that of the shunt motor changes as the load changes.

2. The counter emf E_C of a synchronous motor can have numerical values that are equal to, less than, or even greater than the impressed voltage V, while E_C must always be less than V in a shunt motor.

3. The vector position of E_C is never directly opposite to that of the impressed voltage V in a synchronous motor, while E_C and V are always diametrically opposed to each other in a shunt motor, as Fig. 302 shows.

4. The armature current I_A in a synchronous motor lags behind the *net* voltage by nearly 90 electrical degrees, while the armature current

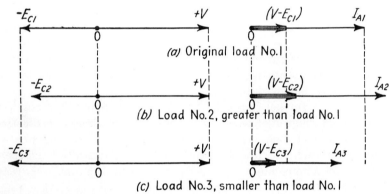

FIG. 302. Phasor diagrams illustrating how the armature current in a d-c shunt motor changes with changing load.

in a shunt motor is always in phase with the net voltage, as Fig. 302 shows.

Point 1 is obvious, of course, from what has previously been said. Point 2 states that the counter emf of a synchronous motor can have values that are equal to, less than, or more than the value of the impressed voltage; this is true because E_C depends upon the d-c excitation, which is *adjustable* over wide limits, and a *constant* revolving-field speed. Point 3 simply indicates the fact that the *position* of the counter emf with respect to the impressed voltage depends upon the angle between the centers of the stator and rotor poles; at light loads, E_C and V are almost directly opposite, while at heavy loads, the rotor poles may slip behind the direct opposite by angles that may be as much as 60°. Point 4 is true because the stator winding of an a-c armature is very highly inductive; in this case, the inductive reactance is very much larger than the resistance, so that the current will lag behind the net voltage by nearly 90 electrical degrees.

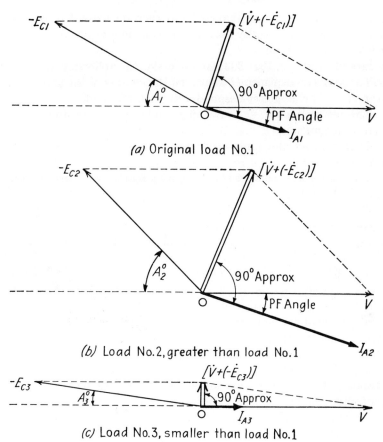

Fig. 303. Phasor diagrams illustrating how the armature current in a synchronous motor changes with changing load.

Equation (92), which refers to the shunt motor in the analysis, can now be written to symbolize the conditions that exist for *each phase* of a synchronous motor. Thus

$$I_A = \frac{V + (-E_C)}{Z_A} \tag{93}$$

where Z_A = impedance per phase of the stator winding, and the *bold-face italic symbols* mean that the various terms must be considered as *phasor* quantities.

In accordance with the above discussion, three phasor diagrams can now be drawn to illustrate qualitatively how the current adjusts itself for various conditions of loading. This is done in Fig. 303, in which an original load (a) is increased to (b) and then decreased to (c). In the sketches, it was conveniently assumed that the magnitude of E_C is exactly equal to the magnitude of V.

A careful study of Fig. 303 should make the following points clear:

1. The angular displacement between the centers of the stator and the rotor poles, $A°$, changes correspondingly with the load.

2. The net voltage $V + (-E_C)$ depends upon the position of E_C and not on its magnitude.

3. The value of the armature current I_A is proportional to the net voltage $V + (-E_C)$.

4. The armature current I_A is approximately 90° behind the net voltage $V + (-E_C)$.

5. The power-factor angle of the input (angle between V and I_A) changes slightly with changes in load.

Power-factor Adjustment of Synchronous Motors. One of the most important advantages of the synchronous motor is its ability to operate at a lagging or leading power factor that can be readily adjusted simply by changing the d-c excitation supplied to the rotating poles. As the d-c excitation is increased, the motor tends to take alternating current that is more leading, while the alternating current tends to become more lagging as the excitation is reduced.

Making use again of the phasor diagram analysis, it will be assumed that a synchronous motor delivers a *constant* load for several values of d-c excitation. Under these varying conditions, it will be shown that the *power input* remains essentially constant, although the current and power-factor angle *do* change with different degrees of d-c excitation. Consider Fig. 304a, which represents a phasor diagram for a given load, with the d-c excitation adjusted to such a value that the armature current I_{A1} is in phase with the impressed voltage V; under this condition, the value of E_{C1} is such that the power factor is unity.

Next, referring to Fig. 304b, assume that the d-c field-rheostat resistance is increased so that the d-c excitation is reduced. Since the speed of the synchronous motor cannot change, the value of E_C will drop; furthermore, the angle $A°$ between the centers of the stator and rotor poles will remain the same because the load is unchanged. The resulting net voltage $V + (-E_{C2})$ will therefore be shifted in a clockwise direction; under this condition, the armature current I_{A2}, lagging behind the net voltage by 90°, will also lag behind the impressed voltage V. Moreover, since the load remains unchanged, the current I_{A2} will have a projection upon V that is exactly equal to I_{A1} because the power input per phase must continue to be $V \times I_{A1}$. Thus a *reduction in the d-c excitation results in a lagging power factor with no appreciable change in power input.*

Referring next to Fig. 304c, assume that the d-c field-rheostat resistance is decreased so that the d-c excitation is raised. This time the value of E_C will rise, the angle $A°$ remaining the same as before. The resulting net voltage $V + (-E_{C3})$ will now be shifted in a counterclockwise direction; under this condition, the armature current I_{A3}, lagging behind the net voltage by 90°, will also shift counterclockwise so that it will lead the impressed voltage V. Furthermore, since there is no load change, the current I_{A3} will have a projection upon V that is exactly equal to I_{A1}, because, as before, the power input must continue to be $V \times I_{A1}$. Thus

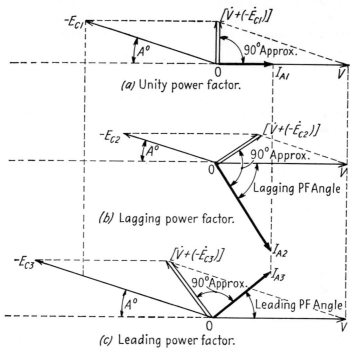

FIG. 304. Phasor diagrams illustrating how the power factors in a synchronous motor, operating under load, change as the d-c excitation is changed.

an increase in the d-c excitation results in a leading power factor with no appreciable change in power input.

Another explanation can be given of the effect upon the power factor of a synchronous motor when the d-c excitation is changed. Assume that an alternator supplies power to a synchronous motor whose load is fixed. Assume further that the alternator is equipped with a regulator so that the line voltage remains constant. Under such operating conditions, the synchronous motor receives two sources of excitation: (1) the d-c excitation for the rotor, from its own exciter, and (2) the a-c excitation for the revolving field in the stator, from the alternator. Moreover, the total d-c and a-c excitation required by the synchronous motor remains constant so long as the load and the line voltage do not change. This implies, for example, that a decrease in the d-c excitation of the synchronous motor will be immediately accompanied by a corresponding increase in its a-c excitation supplied by the alternator; the latter results from an automatic adjustment of the regulator that causes the alternator to receive more d-c excitation from its exciter. Conversely, an increase in the d-c excitation of the synchronous motor is immediately absorbed by the revolving field of the alternator, so that the d-c excitation of the latter is automatically reduced by regulator adjustment. Since an underexcited alternator operates at a leading power factor (see Chap. 7), the synchronous motor that it serves does so also.

With the foregoing statements in mind, assume that a synchronous motor operates at unity power factor (Fig. 304a); it therefore receives both d-c excitation from its exciter and a-c excitation from the alternator. If the d-c excitation is reduced, the alternator makes up the deficit by increasing the a-c revolving-field excitation. This additional alternating current is a component of the total current that lags behind the voltage by 90°. The synchronous motor therefore takes a current, larger than before, which lags behind the impressed voltage V; its in-phase component is the same, however. Figure 304b illustrates this operating condition.

Assume next that the d-c excitation is increased above the value needed for unity power factor. This excess is immediately transferred to the alternator, reducing its d-c excitation. And, since an underexcited alternator operates at a leading power factor, the synchronous motor which it serves does so also. Therefore, overexciting a synchronous motor causes it to take a larger current than before, leading the impressed voltage and having an in-phase component that remains unchanged. Figure 304c illustrates this operating condition.

The Synchronous Condenser. When a synchronous motor is operated without an external mechanical load, it still takes power for its own rotational losses, i.e., friction, windage, and iron losses. Although the in-phase component of the stator current is usually comparatively small under no-load operation, the total current may be rather large, since it depends upon the d-c excitation. Moreover, the power factor of the motor may be controlled over wide limits by simply adjusting the d-c field excitation; if the latter is increased, the power factor tends to become more leading. Figure 305 illustrates how the power factor of an unloaded synchronous motor is changed from unity to a value that is almost zero *leading*. Note particularly that this is done by merely increasing the value of E_C through d-c excitation adjustment.

The ability of a synchronous motor to take a very low leading power-factor current makes it behave like a capacitor. Advantage is taken of this fact in connection with its use in electrical systems to counteract part or all of the lagging component of the load. When this is done, the overall power factor of the system is improved and with it the transmission efficiency and regulation as well as the general operating performance of other apparatus because the current delivered by the generating equipment approaches a minimum for a given kilowatt load as the power factor approaches unity.

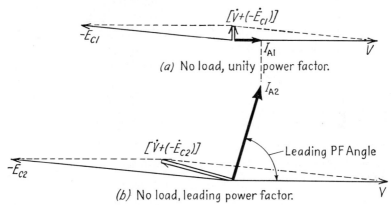

FIG. 305. Phasor diagrams illustrating how an unloaded synchronous motor can have its power factor changed from unity to a very low value leading.

A synchronous motor used in this way is said to "float" on the line because it has no mechanical output. And as such, it is generally called a *synchronous condenser;* it should, more properly, be referred to as a *synchronous capacitor.*

Several problems will now be solved to illustrate the use of a synchronous condenser in connection with power-factor correction.

EXAMPLE 1. An industrial plant has a load of 1,500 kva at an average power factor of 0.6 lagging. Neglecting all losses, calculate: (a) the kilovolt-ampere input to a synchronous condenser for an over-all power factor of unity; (b) the total kilowatt load.

Solution

Problems of this type are best solved by the use of simple kilovolt-ampere phasor diagrams, such as Fig. 306. An arbitrary reference voltage phasor is drawn horizontally first. The kilovolt-ampere load is then drawn in a lagging direction at the proper angle (in this case the angle whose cosine is 0.6 = 53°) to a convenient scale.

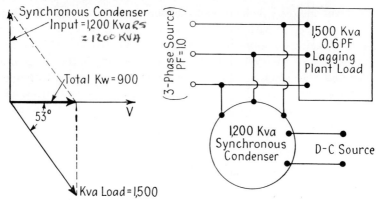

Fig. 306. Kva and wiring diagrams for solution of Example 1.

(a) For an over-all power factor of unity, the synchronous condenser will have to counteract the *vertical component* of 1,500 kva, which is

$$1{,}500 \times \sin 53° = 1{,}500 \times 0.8 = 1{,}200 \text{ kva}$$

(b) $\qquad$ Total load $= 1{,}500 \times 0.6 = 900$ kw

A wiring diagram for this problem is also shown.

EXAMPLE 2. A 750-kva synchronous condenser is available and is used to correct the lagging power factor of the plant in Example 1. Neglecting all losses, calculate: (a) the total kilovolt-amperes of the plant; (b) the over-all power factor.

Solution

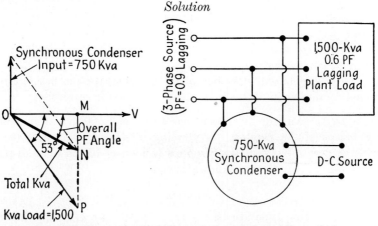

Fig. 307. Kva and wiring diagrams for solution of Example 2.

Referring to Fig. 307, $OP = 1{,}500$ kva; $OM = 900$ kw; $MP = 1{,}200$ kva. The synchronous condenser correction is $NP = 750$ kva,

$$MN = MP - NP = 1{,}200 - 750 = 450 \text{ kva}$$

(a) Total kva $= \sqrt{(OM)^2 + (NM)^2} = \sqrt{(900)^2 + (450)^2} = 1{,}006$

(b) Over-all power factor $= \dfrac{OM}{ON} = \dfrac{900}{1,006} = 0.895$ lagging

EXAMPLE 3 It is desired to purchase a synchronous condenser to correct the 2,400-kva 0.67-lagging-power-factor load in an industrial plant to 0.95 lagging. Neglecting losses, calculate: (a) the kilovolt-ampere input rating of the required synchronous condenser; (b) the total kilovolt-amperes of the plant.

Solution

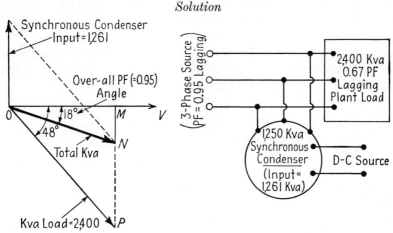

FIG. 308. Kva and wiring diagrams for solution of Example 3.

Referring to Fig. 308,

$$OP = 2,400 \text{ kva}$$
angle $POM = 48°$ (since cos 48° = 0.67)
angle $NOM = 18°$ (since cos 18° = 0.95)
$$MP = OP \times \sin 48° = 2,400 \times 0.743 = 1,783 \text{ kva}$$
$$OM = OP \times \cos 48° = 2,400 \times 0.67 = 1,608 \text{ kva}$$
$$\frac{MN}{OM} = \tan 18° = 0.325$$

Therefore $MN = OM \times 0.325 = 1,608 \times 0.325 = 522$ kva

(a) Synchronous condenser rating $= MP - MN = 1,783 - 522$
$= 1,261$ kva (Use a 1,250-kva machine.)

(b) Total kilovolt-amperes $= \sqrt{(OM)^2 + (MN)^2}$
$= \sqrt{(1,608)^2 + (522)^2} = 1,690$ kva

A synchronous condenser is, strictly speaking, a synchronous motor operating without mechanical load but with strong d-c excitation. For these reasons it is generally modified mechanically and electrically from the usual motor designs. Having no mechanical output, it has no shaft extension, nor is the shaft diameter as large as those required by loaded motors. And, since the d-c excitation must be very much stronger than

470   ELECTRICAL MACHINES—ALTERNATING CURRENT

that required for ordinary synchronous-motor performance, the construction of the rotating poles and windings is somewhat heavier. Moreover, the synchronous condenser must develop only sufficient starting torque to bring its own rotor up to speed.

The Dual-purpose Synchronous Motor. More usually, the synchronous motor is used for the dual purpose of delivering a mechanical load and correcting an otherwise low lagging power factor. This is particularly true in systems that must supply power to a considerable number of lightly loaded induction motors and other types of induction equipment. In such plants, it is frequently advantageous to install a synchronous motor to raise the power factor to a suitable value and at the same time to have it drive some constant-speed load such as a d-c generator, a large pump, or other more or less continuous-duty equipment.

To illustrate the use of the synchronous motor in this way, the following problems are given.

EXAMPLE 4. An industrial plant has an average load of 900 kw at a power factor of 0.6 lagging. A synchronous motor is to be installed to drive a d-c generator and to raise the over-all power factor to 0.92. If preliminary estimates indicate that the input to the synchronous motor will be about 250 kw, calculate: (*a*) its kilovolt-ampere input rating; (*b*) the power factor at which it will operate.

Solution

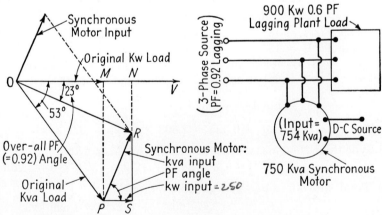

FIG. 309. Kva and wiring diagrams for solution of Example 4.

Referring to Fig. 309,

OM = 900 kw
angle POM = 53°   (since cos 53° = 0.6)
angle RON = 23°   (since cos 23° = 0.92)

$$\frac{OM}{OP} = \cos 53° = 0.6$$

Therefore
$$OP = \frac{900}{0.6} = 1{,}500 \text{ kva}$$

$$\frac{MP}{OP} = \sin 53° = 0.8$$

Therefore
$$MP (= NS) = 1{,}500 \times 0.8 = 1{,}200 \text{ kva}$$
$$PS = MN = 250 \text{ kw}$$
$$ON = OM + MN = 900 + 250 = 1{,}150 \text{ kw}$$
$$NR = ON \times \tan 23° = 1{,}150 \times 0.424 = 488 \text{ kva}$$
$$RS = NS - NR = 1{,}200 - 488 = 712 \text{ kva}$$

(a) PR = synchronous motor input
$$= \sqrt{(PS)^2 + (RS)^2} = \sqrt{(250)^2 + (712)^2} = 754 \text{ kva}$$

(b) Power factor of synchronous motor = $\cos$ angle $SPR = \dfrac{PS}{PR}$

$$= \frac{250}{754} = 0.332 \text{ leading}$$

EXAMPLE 5. The average input to a manufacturing plant is 3,000 kva at a power factor of 0.72 lagging. A synchronous motor having a rating of 1,300 kva is installed for the purpose of operating a new line shaft and improving the plant power factor. Assuming that the synchronous-motor load is about 600 hp at an efficiency of 89.5 per cent, and that it is operated at rated kilovolt-ampere input, calculate: (a) the over-all kilovolt-ampere load; (b) the over-all power factor.

Solution

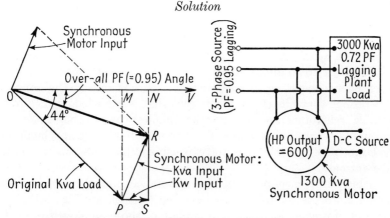

FIG. 310. Kva and wiring diagrams for solution of Example 5.

Referring to Fig. 310,
$$OP = 3{,}000 \text{ kva}$$
$$\text{angle } POM = 44° \quad (\text{since } \cos 44° = 0.72)$$

Since
$$\text{Kilowatts input} = \frac{\text{horsepower} \times 0.746}{\text{efficiency}}$$

and
$$PS = \frac{600 \times 0.746}{0.895} = 500 \text{ kw}$$

Synchronous motor input = PR = 1,300 kva

$RS = \sqrt{(PR)^2 - (PS)^2} = \sqrt{(1{,}300)^2 - (500)^2} = 1{,}200$ kva

$PM \; (= SN) = 3{,}000 \times \sin 44° = 3{,}000 \times 0.695 = 2{,}085$ kva

$NR = SN - RS = 2{,}085 - 1{,}200 = 885$ kva

$OM = 3{,}000 \cos 44° = 3{,}000 \times 0.72 = 2{,}160$ kw

$ON = OM + MN = 2{,}160 + 500 = 2{,}660$ kw

(a) Over-all kilovolt-ampere load = $\sqrt{(ON)^2 + (NR)^2}$
$$= \sqrt{(2{,}660)^2 + (885)^2} = 2{,}800 \text{ kva}$$

(b) Over-all power factor = $\dfrac{ON}{OR} = \dfrac{2{,}660}{2{,}800} = 0.95$

It is interesting to note in this example that the addition of the 1,300-kva synchronous motor *reduces* the over-all kilovolt-ampere load and *raises* the power factor.

The Synchronous-induction Motor. The synchronous reluctance motor, generally used on single-phase circuits and particularly in fractional-horsepower ratings, p. 440, has also been designed and constructed for polyphase service. Although built in oversize frames, they are nevertheless limited to low-power applications, operate at poor efficiency and power factor, and develop low values of starting torque. Their advantage is that they run at constant (synchronous) speed without the need for d-c excitation; they are, in this respect, "singly fed" and, in contrast to "doubly fed" motors, receive power only from an a-c source.

A recent design development that takes advantage of the reluctance principle is a *synchronous-induction motor*, also *singly fed*, that can be built in ratings up to about 40 hp (100 hp for special applications) and have excellent operating characteristics. This is indeed the range of output ratings that fills the gap between what is regarded as the largest acceptable reluctance motor and the smallest practicable *doubly fed* (d-c excited) synchronous motor; the former is too bulky and inefficient in the larger sizes and the latter too expensive to build and operate in the smaller ratings.

The chief difference between the reluctance and the synchronous-induction motors is the rotor *design and construction* (see Fig. 311); their stators are, however, identical and similar to those employed in *doubly fed* synchronous machines. Although they have, strictly speaking, squirrel-cage rotors, the latter are designed in such a manner that the fluxes entering from the stator are *guided* through low-reluctance paths with effective flux barriers to form distinct magnetic circuits. As Fig. 311b illustrates, the laminations have the same number of notched salient iron poles as the stator winding has wound poles and, in addition, have cutaway sections that tend to restrict the fluxes to definite paths. Thus, when the rotor is turning, the rotor poles lock in step with the poles of the rotating stator field in a position of minimum reluctance, and any tendency on the part of the rotor to slow down is opposed by the exposure of flux barriers which introduce high-reluctance air spaces. Moreover,

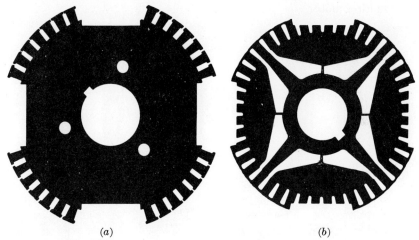

FIG. 311. Rotor laminations for reluctance and synchronous-induction motors. (a) Reluctance-motor lamination. (b) Synchronous-induction motor lamination.

unlike the standard squirrel-cage motor, whose operation depends upon the difference between the stator-field speed and the rotor speed (the slip), the synchronous-induction motor runs at synchronous speed to some point beyond full load when it pulls out of step; for loads greater than that represented by the pull-out torque, the motor rotates at less than synchronous speed until, at some maximum torque, it stalls.

In the horsepower ratings in which it is manufactured, the synchronous-induction motor has found many constant-speed applications, hitherto impractical with other types of standard motor. These include:

1. A six-pole motor that drives a 40-pole alternator for 400-cycle constant-frequency service.

2. A motor that is coupled to a constant-frequency alternator, with the latter supplying power to a group of induction motors that drive a constant-speed conveyor system.

3. High-frequency applications in which the constant-speed motor may be operated satisfactorily at speeds as high as 8,500 rpm.

4. Frequency-converter drives (Chap. 12) in which a precise frequency source can be furnished by a wound-rotor motor acting as an alternator.

5. Adjustable frequency installations (5 to 120 cps) in which a motor is made to operate over a wide range of speeds to drive glassmaking, plywood manufacturing, plastics processing, papermaking, and multicolor printing machines.

An unassembled synchronous-induction motor is illustrated by Fig. 312. Note particularly that it resembles a standard squirrel-cage type of induction motor.

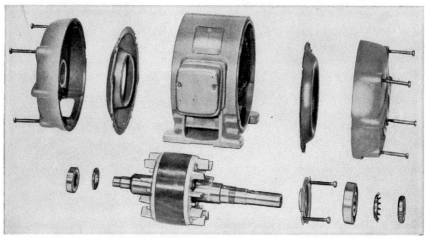

Fig. 312. Unassembled synchronous-induction motor. (*The Louis Allis Co.*)

Hunting and Damping of Synchronous Motors. It was previously shown that an increased synchronous-motor load causes the angle $A°$ (Fig. 303) between the centers of the stator and rotor poles to increase. This is called the *torque angle* because it depends upon the torque that is developed by the motor. When the mechanical load is constant, the rotor settles down to an absolutely constant speed with the torque angle fixed by the particular delivered horsepower. Should the load be changed, however, the rotor speed changes momentarily until the torque angle adjusts itself to the new horsepower; if the load increases, the rotor slips backward to an increased torque angle, while a load reduction causes the rotor to advance to a smaller torque angle. But because of the mechanical inertia of the rotating parts, the rotor overshoots the final position, slowing down or speeding up more than it should. In slowing down as a result of an increased load, for example, it passes the proper torque angle, giving up some of the kinetic energy; under this condition, the motor develops more torque than it requires and speeds up. Acceleration to locate the correct torque angle causes the rotor to move forward a little more than it should; this results in less than the required torque, so that the motor slows down again. Understand that this periodic speed change is only momentary while the rotor is attempting to settle down to a correct torque angle and that it goes on while the *average speed* is constant. This rapid forward and backward motion of the rotor as it revolves at the average constant speed is called *hunting;* i.e., the rotor may be said to be "hunting" for its correct position with respect to the absolutely constant speed of the revolving field. In some cases, when the mass of the rotating parts has an oscillating period that is the same as, or some multiple of, the hunting period, it is possible for the swings to grow progressively greater; under this condition, the torque angle might even exceed the pull-out value, causing the motor to drop out of synchronism and stall.

Hunting is, of course, an objectionable characteristic of all synchronous motors, since it produces severe mechanical stresses as well as great variations in current and power taken by the motor. Fortunately, the very same pole-face squirrel cage that provides the machine with its starting torque is instrumental in *damping* the oscillations. Its effectiveness in doing this, however, depends upon the resistance of the squirrel cage; the lower the resistance, the stronger is the damping action. But since a high squirrel-cage resistance is necessary if the synchronous motor is to have a good starting torque, it is customary to employ a compromise value for the cage. The term *amortisseur* is generally applied to the squirrel cage in connection with its damping action. This damping action results because, by Lenz's law, any change in the flux linking the amortisseur as it attempts to oscillate back and forth causes a current to flow in the cage; this current then flows in such a direction as to oppose a change in the flux that normally links the armature and the rotor.

The Supersynchronous Motor. Extremely high starting torque is sometimes a necessary requirement of a synchronous motor for certain applications. The simplex rotor construction has already been described (Fig. 301), its starting torque reaching values that may be as high as 250 per cent of rated torque. Another design, mistakenly designated as a *supersynchronous* motor, is capable of developing a starting torque that is equal to the pull-out, or stalling, value. It is a costly motor to build because the armature structure, which is stationary in the conventional machine, is mounted in a cradle supported by bearings; it is thus free to revolve just as is the rotor. A band brake is provided around the outside yoke frame to lock the armature at the proper time.

When the supersynchronous motor is started, the brake is released and power is applied to the armature winding. Since the inertia of the rotor with its heavy mechanical load is greater than the surrounding armature structure, the developed starting torque rotates the latter in its bearings in a direction *opposite* to that of the normal rotor direction. When the armature finally reaches synchronous speed, the field is stationary in space. If the band brake is next tightened, the stator slows down as the rotor with its load speeds up; the relative speed of the revolving field and the armature will always be synchronous speed. Finally, after the stator has been gradually brought to rest, the rotor with its load is functioning normally at synchronous speed. This design makes it possible to develop a starting torque that is equal to the pull-out torque and may reach values of 300 per cent or more.

Synchronous Motor Applications. The proper selection of a motor for a given application always involves a complete understanding of essential operating and physical requirements. Such information is gained only through first-hand experience in the installation and operation of actual industrial equipment. To be sure, the field of industrial practice is so wide that it is often necessary to consult specialists when expert advice is needed about motor installations in some highly developed industries. Recognizing certain well-established practices, however, manufacturers have developed standard types of motors having characteristics that meet the general and special needs of most kinds of mechanical equipment. Such motors are usually listed in catalogues and bulletins in which are indicated such important items as horsepower output, speed range, starting torque, pull-out torque, starting amperes, etc.

In so far as synchronous motors are concerned, they are generally made in the larger sizes for constant-duty service. Frequent starting of these motors is not particularly desirable. Loads may be directly coupled, in which case the application has the same speed as the motor; in other installations, a belt or gears may be necessary so that the loads operate at a lower speed than the motor. In still others, the load and the motor poles are mounted on *one continuous shaft* for direct connection. Motor speeds for 60-cycle service range from 1,800 rpm (four poles) down to about 120 rpm (60 poles).

Fans, blowers, compressors, and pumps are usually coupled loads, operating at high speeds; the latter may be 1,800 down to about 514 rpm, depending upon the size of the motor. For such applications, the horsepower rating may be 40 to 750. Pulp beaters and grinders, line shafts, tube and ball mills, steel- and metal-rolling mills, and rock and ore crushers are generally coupled or geared to the motors. Such applications may require machines that have horsepower ratings up to 4,000. Speeds as low as 120 rpm are used in the larger sizes, and when particularly high starting torques are required, the simplex type of rotor is used. For d-c generator operation, direct connection between synchronous motor and generator is a usual practice. Such motors are built in sizes from

SYNCHRONOUS MOTORS

about 40 up to 1,000 or more hp and operate at speeds of 450 rpm and less.

Questions

1. What is meant by *average constant speed* when it refers to a synchronous motor?
2. What factors determine the speed of a synchronous motor? Indicate whether the speed is directly or inversely proportional to these factors.
3. Explain how a synchronous motor differs from an induction motor with regard to its excitation and the power factor at which it operates.
4. Why is it necessary to employ special starting equipment or constructions for the purpose of starting a synchronous motor?
5. What two important characteristics are possessed only by the synchronous type of motor?
6. Explain briefly how the power factor of the input power to a synchronous motor may be adjusted.
7. What is meant by a *synchronous condenser?* What important function is performed by such a machine when it is installed in an electrical system?
8. Describe the construction of the stator core and winding of a synchronous motor.
9. What determines the number of poles for which the stator is wound?
10. What kinds of insulation are used in the stator windings? Indicate whether these materials are class A or class B insulations.
11. Describe the general constructional details of the rotor of a synchronous motor.
12. What voltages are common for the d-c excitation?
13. Why is special bracing necessary in the rotor construction of synchronous motors operated at high speed?
14. What is the advantage of using a spider in the construction of the rotor?
15. How is the d-c excitation supplied to the revolving poles?
16. What is an exciter?
17. What methods may be used to drive an exciter?
18. Explain how a synchronous motor is started when a d-c generator (the normal load) or an exciter is used for this purpose.
19. When a small induction motor is mounted on the shaft of a synchronous motor for starting purposes, why does it have two poles fewer than the synchronous machine?
20. Describe the construction of the squirrel cage, now generally used on the synchronous-motor rotor, for starting purposes.
21. Referring to Fig. 300, explain the operation of an automatic starter when used to start a synchronous motor.
22. How does the resistance of the squirrel cage affect the starting torque of a synchronous motor?
23. When a squirrel cage is used in the rotor of a synchronous motor, how is the starting current kept down to reasonably low values?
24. Why is it desirable to short-circuit the d-c field when a synchronous motor is started?
25. What methods are generally employed to develop high starting torque in synchronous motors?
26. Describe the construction of the *simplex* type of rotor of a synchronous motor.
27. Why is it not possible for the rotor of a synchronous motor to operate at a speed other than synchronous speed?
28. What determines the angular displacement between the centers of the stator revolving field and the rotating rotor poles?

478 ELECTRICAL MACHINES—ALTERNATING CURRENT

29. Under what condition will a synchronous motor stall?
30. Explain why it is proper to speak of the *average* constant speed, and not the constant speed, of a synchronous motor.
31. Is it possible for the counter emf of a shunt motor to be equal to, or greater than, the impressed armature voltage? Explain.
32. Is it possible for the counter emf of a synchronous motor to be equal to, or greater than, the impressed stator voltage per phase? Explain.
33. Referring to Fig. 303, explain why the armature current changes for different conditions of loading.
34. Referring to Fig. 304, explain why the armature current and the power factor change for different conditions of d-c excitation.
35. Referring to Fig. 305, explain why an overexcited synchronous motor operating without load acts like a capacitor, taking a current that leads the voltage by almost 90°.
36. What special constructional details may be observed in synchronous condensers? Give reasons for such modifications from the standard synchronous-motor constructions.
37. Under what conditions would it be desirable to install a dual-purpose synchronous motor?
38. Describe the constructional differences of rotors for reluctance and synchronous-induction motors.
39. Referring to Fig. 311b, give reasons for the special construction of the rotor lamination for a synchronous-induction motor.
40. Distinguish between a *singly fed* and a *doubly fed* synchronous motor.
41. What is meant by *pull-out torque* when referring to a synchronous-induction motor? At what approximate load does it occur?
42. Distinguish between *pull-out* torque and *maximum* torque when referring to a synchronous-induction motor.
43. List several applications of the synchronous-induction motor not previously mentioned.
44. Explain carefully the phenomenon of *hunting* in a synchronous motor.
45. Under what conditions may hunting become particularly serious?
46. What method is generally employed to dampen the oscillations that result from hunting?
47. Describe the *supersynchronous* motor and explain exactly how such a machine is started.
48. What important advantage is possessed by a supersynchronous motor? What limits its general use?
49. In selecting a motor for a given application, what factors must be taken into consideration?
50. List several synchronous-motor applications not previously mentioned.

Problems

1. At what speed will a synchronous motor operate if it has 10 poles and is connected to a 50-cycle source? a 25-cycle source? a 60-cycle source?
2. How many poles does a synchronous motor have if it operates at 200 rpm when connected to a 60-cycle source?
3. A 10-pole 25-cycle alternator is directly coupled to, and is driven by, a 60-cycle synchronous motor. How many poles are there in the latter?
4. A 60-cycle synchronous motor is coupled to, and drives, a 50-cycle alternator.

SYNCHRONOUS MOTORS

How many poles does each machine have, and at what speed does the motor generator set operate?

5. A 50-hp synchronous motor has a full-load efficiency of 91 per cent and operates at a power factor of 0.8 leading. Calculate: (a) the power input to the motor; (b) the kilovolt-ampere input to the motor.

6. The power input to the d-c field of a 15-kva (input) synchronous motor is 6 per cent of the rated a-c input. What is the direct current that excites the field at 120 volts?

7. The starting torque of a 25-hp 720-rpm synchronous motor is 150 per cent of its rated torque. Calculate the starting torque in pound-feet.

8. The inrush current to a synchronous motor is 235 amp when it is started from a rated-voltage source of 2,200. What per cent tap on a compensator should be used if the starting current on the line side is to be limited to 100 amp? What voltage will be impressed upon the motor at the starting instant?

9. When a synchronous motor is delivering rated load and the generated voltage E_c per phase is adjusted so that it is equal to the impressed voltage V per phase, the torque angle $A°$ is 30° (see Fig. 303). Calculate the impedance voltage drop in the armature winding as a percentage of the terminal voltage.

10. The full-load losses in a 5,000-kva three-phase 6,600-volt synchronous condenser are 160 kw. Calculate the full-load current and power factor.

11. It is desired to correct the 2,400-kva 0.65-lagging-power-factor load in a plant to unity by the installation of a synchronous condenser. Calculate the kilovolt-ampere rating of the latter.

12. What should be the rating of the synchronous motor in Prob. 11, if the over-all power factor is to be 0.92?

13. An electrical system has a load of 5,000 kw at a lagging power factor of 0.67. If a 3,000-kva synchronous condenser is installed for power factor correction purposes, calculate: (a) the over-all power factor; (b) the total kilovolt-amperes on the system.

14. An industrial plant has a load of 800 kw at a power factor of 0.8 lagging. It is desired to purchase a synchronous motor of sufficient capacity to *deliver* a load of 200 kw and also serve to correct the over-all plant power factor to 0.92. Assuming that the synchronous motor has an efficiency of 91 per cent, determine its kilovolt-ampere input rating and the power factor at which it will operate.

15. A factory takes a load of 2,400 kva at 0.6 lagging power factor. A synchronous motor having an input rating of 1,500 kva is to be installed to carry an additional load of 1,200 kw (output) and also to improve the power factor. Assuming a motor efficiency of 92.4 per cent, calculate: (a) the over-all kilovolt-ampere load; (b) the over-all power factor; (c) the synchronous motor power factor.

16. A factory load of 900kw at 0.6 power factor lagging is to be increased by the addition of a synchronous motor that takes 450 kw. At what power factor must this motor operate and what must be its kilovolt-ampere input if the over-all power factor is to be 0.9 lagging?

17. A manufacturing plant has a load of 3,600 kva at a lagging power factor of 0.707. A 500-hp synchronous motor having an efficiency of 90 per cent is installed and is used also to improve the over-all power factor to 0.937 lagging. Calculate the kilovolt-ampere input rating of the synchronous motor and the power factor at which it must operate.

18. The average input to a manufacturing plant is 3,000 kva at a power factor of 0.72 lagging. A synchronous motor having an input of 1,300 kva is then installed and delivers a load of 600 hp at an efficiency of 89.5 per cent. Calculate: (a) the over-

all kilowatt input; (b) the over-all kilovolt-ampere input; (c) the over-all plant power factor.

19. The input to an industrial plant is 1,440 kw at a power factor of 0.6 lagging. It is desired to connect a synchronous motor that operates at a leading power factor of 0.8 to the power mains and have it correct the over-all power factor to 0.9. What should be the power input to the synchronous motor?

20. For the industrial plant of Prob. 19, calculate the horsepower output of a synchronous motor operating at a leading power factor of 0.8 and an efficiency of 87.4 per cent that will correct the over-all power factor to unity.

CHAPTER 12

CONVERTERS AND RECTIFIERS

Converters—Types and Uses. A converter is a machine (or combination of machines) in which electrical energy of one form is changed to electrical energy of another form. The three most widely used types of converter are: (1) synchronous converters, in which alternating current is changed to direct current, or vice versa; (2) frequency converters, in which electrical energy at one a-c frequency is changed to electrical energy at another frequency; and (3) phase converters, in which a single-phase system is changed to a polyphase (two- or three-phase) system.

The usual sources of electrical energy are large power stations where polyphase alternating current is generated. Great networks of transmission systems then distribute the energy over wide areas, the voltages being raised or lowered by transformer equipment when it is desirable to do so. At the load centers, where the a-c voltages have been stepped down to suitable values, the electrical devices generally consume energy at the same frequency that is supplied; as a rule, the polyphase equipment is operated directly from the available two- or three-phase lines. There are many installations, however, that require direct current, alternating current at a different frequency, or even a different number of phases from what is available from the supply lines. Under such operating conditions, it is therefore necessary to employ the proper types of conversion equipment, previously mentioned.

The applications of d-c electricity are numerous in industry, on the farm, in railway transportation, in communication systems (telephones, telegraphs, and radio), and in the home. Traction systems, for example, are best operated by d-c series motors, the starting and overload torque characteristics of which are unexcelled for such service. And direct current is, of course, absolutely essential in such electrolytic processes as electroplating, electrorefining, electrotyping, and the production of aluminum and fertilizers. Also, many motor applications such as elevators, printing presses, and certain steel-mill equipment are generally superior when operated from d-c systems. In all these, and others, the incoming alternating current must be converted to direct current, and this is often accomplished by the synchronous converter.

Frequency conversion is not quite so common as conversion from alternating to direct current. In some cases, when electrical systems having different frequencies are to be interconnected so that power may be supplied from one to the other, it is necessary to link them together through a frequency converter. The latter is sometimes used to convert from 60 cycles to 120 or 180 cycles for the purpose of operating induction motors at very high speeds. This is particularly desirable in the operation of woodworking machinery, in which two-pole induction motors will have speeds that are two or three times the maximum obtainable at the lower frequency. Also, frequency converters are extremely useful in laboratories and testing departments of manufacturing plants, where a wide range of frequencies must be available for experimental and other needs.

When it is necessary to change from one polyphase system to another (for example, from three- to two-phase, and vice versa), transformers are generally used; this method for polyphase conversion is simple and efficient. However, when it is required to convert from single- to three- or two-phase, transformers do not function with satisfaction; this is true because of the pulsating nature of the single-phase power, which is actually negative during a portion of every cycle. But since a rotating machine always stores kinetic energy in its rotating portion, it can be used to convert from single-phase to polyphase because the rotor can slow down during each interval of negative power to give up some of its kinetic energy and thus maintain a constant flow of polyphase power. It is true that the speed of such a converter varies from instant to instant, but the departure from constant speed is so slight that it is not objectionable. Phase converters are sometimes used in railway electrification, the Virginian Railway, the Norfolk and Western Railway, and portions of the Pennsylvania Railroad being notable examples. In these systems, single phase at 11,000 volts is fed to the locomotive through a single overhead trolley and the tracks; the single phase is then transformed to a lower voltage and converted into three phase to drive the polyphase motors.

Synchronous Converter Construction. A synchronous converter is a machine that receives alternating current through a set of slip rings at one end of an armature that rotates at synchronous speed (rpm = $120f/P$) and delivers direct current from the opposite end through a commutator and a set of brushes. It is, in fact, a combination synchronous motor and d-c generator.

Its operation as a *synchronous motor* involves the rotation of an armature inside a set of stationary poles; this is just the reverse of the *conventional* synchronous motor (studied in Chap. 11), in which a set of poles revolves inside a stationary armature. Figure 313 illustrates the arrangement of the parts of a synchronous converter. Note that it has essentially the same construction as any d-c machine, with the exception that a set of slip rings is added at the end of the armature opposite the end of the commutator. As will be discussed later, the armature winding, which is essentially the type found in any d-c machine, is properly tapped and connected to the slip rings. Brushes riding on the latter are fed from an a-c source, which is usually three or six phase. Direct current is delivered to the load from the commutator and brushes as in any d-c generator. The d-c excitation for the field is provided by the pole windings connected across the d-c brushes; i.e., the machine is self-excited. It should be recognized at the outset that the ability of a single machine to perform

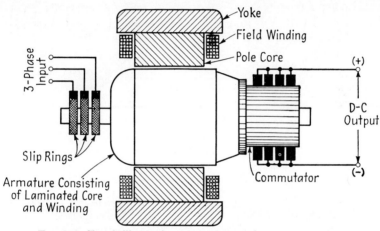

FIG. 313. Sketch illustrating parts of a synchronous converter.

the function of motor and generator simultaneously is not unusual, because every motor generates a counter emf while in operation; in the synchronous converter, the generated a-c counter emf is rectified by the commutator and brushes and is made available as a source of direct current at the brushes.

The Single-phase Converter. The single-phase synchronous converter is rarely used in practice because it is inefficient, heats excessively, and has a tendency to hunt; moreover, it is physically much larger per kilowatt than are polyphase converters. An understanding of its behavior does, however, serve to simplify the basic principles underlying the operation of three-, six-, and 12-phase converters.

Refer to Fig. 314, which represents a simple two-pole single-phase converter. The armature winding has 24 coils, and the commutator is tapped at segments 1 and 13, which are diametrically opposite and connected to the two slip rings. It is assumed that the flux-density distribution follows a sine wave, that the converter is supplied with power from a single-phase source, and that rotation is clockwise. Considered from the standpoint of a d-c generator, the voltage between positive and negative brushes will be constant and will be fixed by the flux per pole ϕ, the speed of rotation, and the number of turns of wire in each of the two armature-winding paths. On the other hand, the a-c voltage between

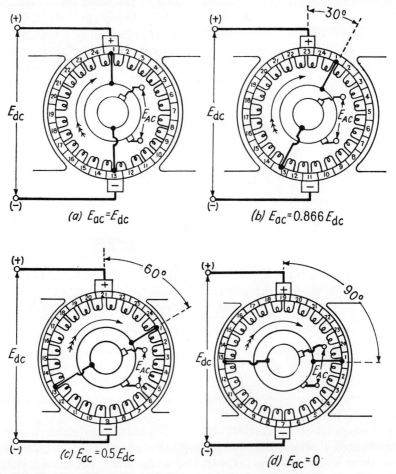

FIG. 314. Sketches illustrating how the a-c voltage between slip rings varies as the tap coils rotate with respect to the brush positions. A sine flux-density distribution is assumed for a two-pole single-phase synchronous converter.

slip rings will *not* be constant in magnitude or direction, but will vary sinusoidally. When the slip-ring taps are directly under the brushes (Fig. 314a), the a-c voltage will be exactly equal to the d-c voltage, because both voltages result from the generated emfs in the same coils; under this condition, the a-c voltage is the maximum value of the sine wave. When the armature is rotated 30° (Fig. 314b), the a-c voltage between slip rings will drop to 86.6 per cent of the maximum value; this should be clear by reference to Fig. 315, which shows part of a sine wave and the per cent voltages in terms of the maximum value. Continued rotation to a position of the tapped coil 60° from the vertical (Fig. 314c) will result in a slip-ring voltage of $0.5E_{dc}$. And when the tapped coils

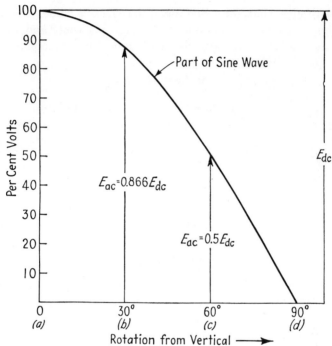

FIG. 315. Voltage relations in a single-phase converter for various positions of tapped coils with respect to the brush positions (see Fig. 314).

reach a position 90° from the vertical (Fig. 314d), the slip-ring voltage will be zero. A close study of Fig. 315 should emphasize the following:

1. The d-c voltage is independent of the position of the tapped coils, since the number of series coils between brushes does not change.

2. The a-c voltage between slip rings depends upon the net voltage existing between tapped coils *assuming that the impressed emf and the counter emf are always equal.*

3. In position (d) the net counter emf is zero because the voltage in one-half of the armature conductors per path opposes an equal voltage in the other half of the armature conductors.

4. For any intermediate position between (a) and (d) the net voltage results from a partial subtraction of some armature conductor voltages from those remaining.

486 ELECTRICAL MACHINES—ALTERNATING CURRENT

The foregoing analysis leads to the following relation between the a-c voltage E_{ac} between rings and the d-c voltage E_{dc} between brushes.

$$E_{ac} = \frac{E_{dc}}{\sqrt{2}} \quad \text{(single-phase)} \tag{94}$$

(It will be recalled that, for a sine wave, the effective value $= E_{max}/\sqrt{2}$.)

If all losses are neglected (efficiency = 100 per cent) and the power factor is assumed to be unity, the power input may be equated to the power output. Thus

$$E_{ac} \times I_{ac} = E_{dc} \times I_{dc} \tag{95}$$

EXAMPLE 1. A 1.5-kw single-phase converter operates at full load from a 230-volt a-c source. Neglecting losses, and assuming a power factor of unity, calculate: (*a*) the d-c voltage and current; (*b*) the a-c input current.

Solution

(a) $$E_{dc} = \sqrt{2}\, E_{ac} = \sqrt{2} \times 230 = 325 \text{ volts}$$

$$I_{dc} = \frac{P_{dc}}{E_{dc}} = \frac{1{,}500}{325} = 4.62 \text{ amp}$$

(b) $$I_{ac} = \frac{P_{ac}}{E_{ac}} = \frac{1{,}500}{230} = 6.52 \text{ amp}$$

EXAMPLE 2. A 2.5-kw single-phase converter is operated *inverted*, i.e., it converts direct to alternating current. If the d-c input voltage is 230, calculate: (*a*) the a-c voltage; (*b*) the alternating and direct currents at full load.

Solution

(a) $$E_{ac} = \frac{230}{\sqrt{2}} = 162.5 \text{ volts}$$

(b) $$I_{ac} = \frac{2{,}500}{162.5} = 15.4 \text{ amp}$$

$$I_{dc} = \frac{2{,}500}{230} = 10.9 \text{ amp}$$

The Polyphase Converter. *Three-phase.* In the three-phase synchronous converter, the armature winding is tapped at points 120 *electrical* degrees apart and connected to three slip rings. For a two-pole machine, this means only three tappings. However, when four or more poles are used, there must be *three taps for every pair of poles* because every pair of poles spans 360 electrical degrees. Figure 316 illustrates how the winding of a four-pole machine is tapped for three-phase operation. It should be pointed out that it is customary to tap the coils at the rear bend, i.e., at the end of the armature opposite the commutator, because the slip rings are located there. For convenience, however, tappings are shown in the illustrations at the commutator; this, of course, is electrically equivalent to the usual practice.

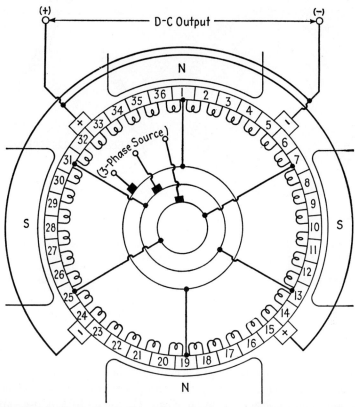

FIG. 316. Sketch illustrating a three-phase four-pole synchronous converter tapped and connected to three slip rings.

In order to determine the relations between the direct and alternating voltages and currents, consider the greatly simplified sketch of Fig. 317, which represents a two-pole three-phase converter. (The commutator has been omitted for convenience, and the brushes are located diametrically opposite.) With tappings m, n, and p brought out exactly 120° apart, this is actually the equivalent of a Δ connection when considered from the standpoint of a-c winding practice. Thus the a-c voltage E_{mn} is equivalent to 0.866 multiplied by the a-c voltage E_{mq}. Since the a-c voltage E_{mq} (= E_{ac}) equals $E_{dc}/\sqrt{2}$, it follows that

$$E_{ac} = \frac{E_{dc}}{\sqrt{2}} \times 0.866 = 0.612 E_{dc} \quad \text{(three-phase)} \quad (96)$$

Again assuming 100 per cent efficiency and unity power factor, the

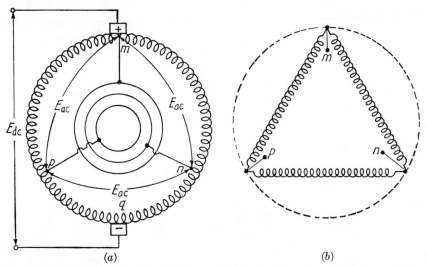

FIG. 317. Sketches illustrating (a) how a two-pole three-phase converter winding is tapped, and (b) the equivalent a-c Δ winding connection.

three-phase power input equals the d-c power output. Thus

$$\sqrt{3}\, E_{ac} \times I_{ac} = E_{dc} \times I_{dc} = \frac{E_{ac}}{0.612} \times I_{dc}$$

Therefore $\quad I_{ac} = \dfrac{I_{dc}}{\sqrt{3} \times 0.612} = 0.943 I_{dc} \quad \text{(three-phase)} \quad (97)$

EXAMPLE 3. A 25-kw three-phase converter has a d-c output voltage of 230 volts. Assuming negligible losses and unity power factor, calculate: (a) the a-c input voltage between rings; (b) the full-load d-c output; (c) the full-load a-c input per line.

Solution

(a) $\quad E_{ac} = 0.612 \times 230 = 141$ volts

(b) $\quad I_{dc} = \dfrac{P}{E_{dc}} = \dfrac{25{,}000}{230} = 108.7$ amp

(c) $\quad I_{ac} = 0.943 \times 108.7 = 102.3$ amp

Six-phase. In the six-phase synchronous converter, the armature winding is tapped at points 60 electrical degrees apart and connected to six slip rings. For a two-pole machine, this means only six tappings, but for four or more poles there must be six taps for every pair of poles because every pair of poles spans 360 electrical degrees. If a sketch like that of Fig. 316 is made for a four-pole six-phase converter, it will be necessary to show six rings with taps taken from the winding every 30 mechanical degrees. Each ring will have two connections made to it. (The student is urged to make such a diagram.)

To determine the relations between the direct and alternating voltages and currents, consider the simplified sketch of Fig. 318, which represents a

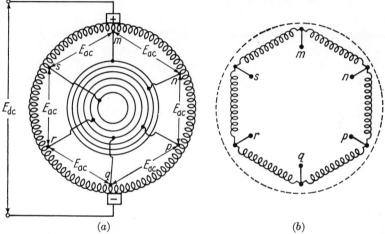

Fig. 318. Sketches illustrating (a) how a two-pole six-phase converter winding is tapped and (b) the equivalent a-c mesh winding connection.

two-pole six-phase converter. With tappings m, n, p, q, r, and s brought out, this is equivalent to the so-called six-phase *mesh* connection when considered from the standpoint of a-c winding practice. Thus the a-c voltage E_{mn} is equal to one-half the a-c voltage E_{mq}. Since the a-c voltage E_{mq} ($= E_{ac}$) equals $E_{dc}/\sqrt{2}$, it follows that

$$E_{ac} = \frac{E_{dc}}{\sqrt{2}} \times 0.5 = 0.354 E_{dc} \quad \text{(six-phase)} \tag{98}$$

As before, assuming 100 per cent efficiency and unity power factor, the six-phase power input equals the d-c power output. Thus

$$6 E_{ac} \times I_{ac} = E_{dc} \times I_{dc} = \frac{E_{ac}}{0.354} \times I_{dc}$$

Therefore $\quad I_{ac} = \dfrac{I_{dc}}{6 \times 0.354} = 0.472 I_{dc} \quad \text{(six-phase)} \tag{99}$

EXAMPLE 4. A 1,500-kw six-phase synchronous converter has a full-load d-c voltage of 600. Determine: (a) the a-c voltage between rings; (b) the d-c output; (c) the a-c input per line.

Solution

(a) $E_{ac} = 0.354 \times 600 = 212.4$ volts

(b) $I_{dc} = \dfrac{P}{E_{dc}} = \dfrac{1{,}500{,}000}{600} = 2{,}500$ amp

(c) $I_{ac} = 0.472 \times 2{,}500 = 1{,}180$ amp

Twelve-phase. The 12-phase synchronous converter is used in very limited numbers because of the complexity of the transformer connections when transforming from three- to 12-phase as well as the mechanical problems involved in bringing out 12 leads per pair of poles for connection to the slip rings. When constructed for large power requirements, however, a given size is capable of delivering about 11 per cent more power than a six-phase converter and about 16 per cent more power than a three-phase converter.

Without attempting to derive the relations between the direct and alternating voltages and currents, it can be shown that

$$E_{ac} = 0.182 E_{dc} \quad \text{(12-phase)} \tag{100}$$

and $I_{ac} = 0.236 I_{dc}$ (12-phase)   (101)

EXAMPLE 5. A 5,000-kw 1,200-volt 12-phase synchronous converter operates at full load at an efficiency of 96 per cent and a power factor of 0.95. Calculate: (a) the a-c voltage between rings; (b) the d-c output; (c) the alternating current in each of the 12 wires.

Solution

(a) $E_{ac} = 0.182 \times 1{,}200 = 218.4$ volts

(b) $I_{dc} = \dfrac{5{,}000{,}000}{1{,}200} = 4{,}167$ amp

(c) I_{ac} (efficiency = 100 per cent and power factor = 1)
$$= 0.236 \times 4{,}167 = 984 \text{ amp}$$

I_{ac} (efficiency = 96 per cent and power factor = 0.95)
$$= \dfrac{984}{0.96 \times 0.95} = 1{,}080 \text{ amp}$$

Table 10 summarizes the results concerning direct and alternating voltages and currents in synchronous converters.

Heating and Ratings of Synchronous Converters. All commutating-type machines have limited output for two reasons: (1) the temperature rise resulting from heating must not exceed certain well-established practices, and (2) sparking at the commutator becomes destructive at values of direct current that are too high. However, both these factors

CONVERTERS AND RECTIFIERS 491

tend to limit the output of d-c motors and generators to a greater extent than they do the output of polyphase synchronous converters. The latter machines are therefore physically smaller and have longer commutators than is the case with d-c motors and generators of equal ratings. These gains come about because the effective or net current in the armature winding of a *polyphase* synchronous converter is actually less than would flow if the machine were operated as a d-c motor or generator. Thus, under steady operating conditions the magnetic field is less distorted because armature reaction is not so strong; this means better commutation. Moreover, the armature copper loss ($I_A{}^2 R_A$) is not so much as

TABLE 10. CONVERSION RATIOS IN SYNCHRONOUS CONVERTERS

	Formula*	2-ring 1-phase	3-ring 3-phase	6-ring 6-phase	12-ring 12-phase
E_{ac}	$\dfrac{E_{dc}}{\sqrt{2}} \times \sin\dfrac{\pi}{n}$	$0.707 E_{dc}$	$0.612 E_{dc}$	$0.354 E_{dc}$	$0.182 E_{dc}$
I_{ac}	$\dfrac{2.83}{n} \times I_{dc}$	$1.414 I_{dc}$	$0.943 I_{dc}$	$0.472 I_{dc}$	$0.236 I_{dc}$

* n = number of rings; efficiency = 100 per cent; power factor = 1.0.

would result if the machine were operated as a d-c motor or generator; this means a lower temperature rise. Polyphase converters are therefore extremely efficient machines.

To understand the points made above, it must first be recognized that a synchronous converter is simultaneously a synchronous motor and a d-c generator. The armature winding receives alternating current through the slip rings; this is the *motor current*, that is, it causes motor action. When the armature revolves, generator action takes place, that is, a counter emf is generated. If the counter emf is then permitted to deliver a direct current to a load through the commutator and brushes, one set of armature-winding conductors must carry *both* the motor and generator currents. And, since these flow in opposite directions (counter emf and impressed emf are oppositely directed), the resulting current will be the difference between them.

In a single-purpose machine, i.e., a generator or motor, the current in every coil in a continuous-series path is the same at any instant. This is not the case in the dual-purpose synchronous converter, because the d-c and a-c components are out of phase with each other by different amounts in the various coils. For example, in a coil *midway* between two taps, the d-c and a-c components are out of phase with each other by 180 electrical degrees if the power factor is unity. This is represented in **Fig. 319a**,

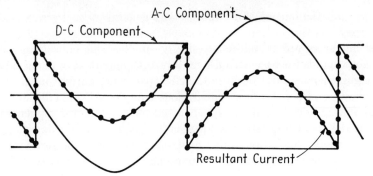

(a) Unity-power-factor current in a coil midway between coils tapped to slip rings. Note that the d-c and a-c components are 180 degrees out of phase

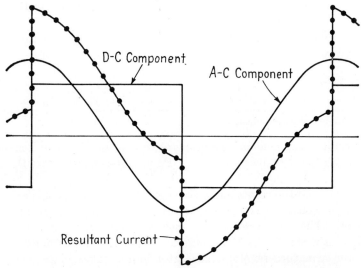

(b) Unity-power-factor current in a coil tapped to slip rings Note that the d-c and a-c components are 90 degrees out of phase

FIG. 319. Graphs showing the great difference between the currents in the mid-coil and tapped coil in a synchronous converter. Sketches illustrate why the tapped coil has a higher temperature than the mid-coil.

which also shows the resultant current. On the other hand, the d-c and a-c components are out of phase with each other by only 90 electrical degrees in a *tapped* coil, also at unity power factor, as is shown by Fig. 319b. Obviously, the heating differs in the various coils, the greater heating occurring in the tapped coil. Now then, since the rating of a converter is determined by an allowable safe temperature of the *hottest* coil, it is desirable, from a practical standpoint, that the temperature variation between hot and cool coils be as little as possible. This temperature variation is a maximum for a single-phase converter because the mid-coil is far removed from the tap coil; it becomes less as the number of tap coils increases because the distance between tap and mid-coils diminishes. It is for this reason that the rating of a converter increases with the number of phases.

As in any synchronous motor, the power factor at which a synchronous converter operates is determined by the d-c field excitation; the latter is readily controlled by properly adjusting the field rheostat. In practice, it is customary to operate at or near unity power factor because heating is lessened and output increased by doing so.

Table 11 shows how the *relative heating* and *relative output* vary with the number of phases and the power factor. These *relative* values merely indicate how the heating and output of the dual-purpose synchronous converters compare with those of the single-purpose d-c generators. All calculations were made assuming 100 per cent efficiency.

TABLE 11. EFFECT ON HEATING AND OUTPUT OF A SYNCHRONOUS CONVERTER IN COMPARISON WITH OPERATION AS A DIRECT-CURRENT GENERATOR

Power factor	Three rings		Six rings		Twelve rings		Direct current
	Relative heating	Relative output	Relative heating	Relative output	Relative heating	Relative output	
1.0	0.565	1.33	0.268	1.93	0.209	2.19	1.0
0.95	0.693	1.20	0.364	1.66	0.299	1.83	1.0
0.90	0.843	1.089	0.476	1.45	0.404	1.57	1.0
0.85	1.020	0.99	0.609	1.28	0.528	1.38	1.0
0.80	1.230	0.90	0.768	1.14	0.676	1.22	1.0

A study of Table 11 should emphasize several important facts:

1. The output of a synchronous converter (three-, six-, and 12-phase) is generally larger than the output of the same machine operating as a d-c generator.

2. The relative output increases with the number of rings.

3. The relative output diminishes as the power factor drops.

4. The temperature rise of a synchronous converter is less, in general, than the temperature rise of the same machine operated as a d-c generator.

Starting Synchronous Converters. Several methods may be used to start synchronous converters. These are: (1) operating the machine from the d-c side as a shunt motor; (2) using a small auxiliary motor mounted on an extension of the converter shaft; (3) operating the machine from the a-c side as a polyphase induction motor. Of these, the last is the most generally employed.

Shunt-motor Method. In many converter power stations, a number of machines are operated in parallel to supply d-c power to a common load. This makes sufficient direct current available so that a unit may be brought up to speed as a d-c motor and synchronized on the a-c side in the usual way (the procedure is quite similar to that described in Chap. 11, Synchronous Motors). If there is a series field in addition to the shunt field, the former must be short-circuited during the starting period, when the d-c end is functioning as a motor instead of as a generator. Neglecting to short-circuit the series field would cause the machine to become a differentially compounded motor because the shunt-field current does not reverse, while the series-field current does. Under this condition, the main field might be weakened to cause instability or even reversal of polarity.

Since all polyphase converters are supplied with a-c power through a set of step-down transformers, synchronizing can be carried out on either the high side or the low side. In practice it is customary, however, to do this on the high side.

Auxiliary-motor Method. When this method is employed, a small squirrel-cage induction motor having two poles fewer than the converter brings the latter up to speed. The slip of the induction motor is generally high enough so that it rotates the converter armature at a speed that is just about equal to its synchronous speed. Synchronizing is then carried out in the usual way, this being done, as a rule, on the high side of the step-down transformers.

Method in Which a Converter Serves as an Induction Motor. Modern synchronous converters are provided with squirrel-cage windings in the pole faces exactly as are synchronous motors. In the latter type of machine, it will be recalled, the squirrel cages perform two important functions: (1) they tend to prevent hunting, and (2) they permit the motor to be started by induction-motor action. The same two purposes are served by such squirrel cages in the operation of converters because these machines are essentially synchronous motors when viewed from the a-c input side. It should be noted, however, that whereas the squirrel cage is placed in the pole faces of the rotating field of the synchronous motor, it is a part of the stationary-field poles in the synchronous converter.

To start a converter by induction-motor action, it is necessary to impress a reduced voltage, about 50 per cent of rated, across the slip rings during the accelerating period. The polyphase currents in the armature then produce a revolving field that *attempts* to cause rotation of the squirrel cage and its field structure in the same direction. Failing to do this, since the poles cannot move, *the armature rotates in a direction opposite to that of the revolving field.* In other words, reaction between the revolving field in the armature at synchronous speed and the induced currents in the stationary squirrel cage causes the machine to develop torque. After the armature reaches a speed that is *nearly* synchronous speed (it can never reach exactly synchronous speed as a squirrel-cage motor), the revolving field will be *nearly* stationary in space and will rotate forward at a speed equal to the slip speed; this situation results because the fictitious revolving field always rotates forward at synchronous speed with respect to the stationary poles, while the armature actually rotates backward at *nearly* synchronous speed. Thus the speed of the armature poles will be

$$\left(\frac{120f}{P}\right) + \left(\frac{-120f}{P}\right)(1-s) = 120f\left(\frac{s}{P}\right)$$

Since the field is revolving in a direction opposite to that of the actual armature rotation, the flux in the stationary poles is reversing very rapidly; the result is that all the residual flux is removed. At this point, the armature attempts to approach synchronous speed so that a voltmeter connected to the d-c brushes may be observed to swing slowly up and down scale as the induced *north* and *south* polarities in the main poles change; this is caused by the slowly revolving armature poles as they sweep across the main poles. If the d-c field switch is then closed, the motor pulls into step to operate at synchronous speed; full voltage is then applied to the slip rings. Furthermore, the polarity at the d-c brushes will be determined by the polarity of the field poles (*north* and *south*) at the instant that the machine pulls into step. Thus, if a definite brush polarity is desired, the field switch should be closed as the d-c voltmeter needle starts to swing up scale, assuming that the correct polarity is indicated by such a deflection. However, in most cases the field switch is closed without observing the rapid swinging of the voltmeter needle, and the machine is arbitrarily permitted to come up with either of the two conditions of brush polarity. Then, if the latter is incorrect, giving a down-scale deflection on the voltmeter, the rotating armature must be made to slip a pole. This is done first by reversing the field connections from the brushes to the field windings (by using a reversing switch), and then, after the armature has slipped back a pole (it must do so because armature core and field poles always expose opposite polarities to one another), the reversing field switch is thrown back to its original position.

A wiring diagram showing how a six-phase converter is connected to a set of three transformers through a triple-pole double-throw (TPDT) switch for starting and running operation is given in Fig. 320. This illustrates the primaries connected in Δ and the secondaries connected diametrical (also see Fig. 325). Note particularly that with the switch closed *up* only one-half of each of the transformer secondaries is used to energize the converter; then, after the machine is up to speed, the switch is quickly closed *down*, in which position full secondary voltage is impressed across the converter.

When a converter is started by induction-motor action, the shunt field acts as the secondary of a transformer, the primary being the armature

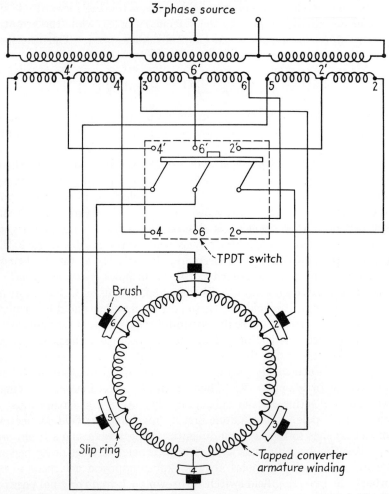

Fig. 320. Wiring connections for starting a six-phase rotary converter on half voltage by using a TPDT switch.

winding to which the a-c source is connected. Since the shunt field has many more turns of wire than the armature, a very high voltage is induced in the former because of the step-up ratio. To forestall the possibility of insulation failure caused by such a dangerously high emf, it is customary to split the field into a number of sections by using a sectionalizing switch; this practice reduces the insulation voltage to a safe value determined only by the voltage per section. Figure 321 illustrates how this is done in a four-pole converter in which the sectionalizing switch divides the entire field into four parts when it is in the open position. After the motor

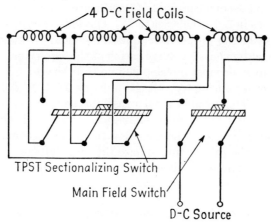

Fig. 321. Sketch illustrating how the d-c field of a four-pole converter is sectionalized during the starting period.

approaches synchronous speed, the sectionalizing switch may be closed because the induced emf in the field drops to a low value as a result of the reduced frequency.

Another point that must be recognized in starting a converter by this method, if there is a series field and a diverter, is that a switch must be provided to open this circuit. If this is not done, a very high current will flow in the series field and the diverter because of transformer action; in this case, the action is that of a step-down transformer. Here again the circuit may be closed after the machine approaches synchronism.

Transformers and Transformer Connections for Converters. In the operation of polyphase synchronous converters, it is necessary that transformer equipment be used between the a-c source and the slip rings. There are several reasons for this:

1. A definite voltage ratio exists between the a-c voltage input and the d-c voltage output of a converter (see Table 10).

2. The a-c source voltage is usually much higher than the desired d-c output voltage.

3. In the case of six- and 12-phase converters it is necessary to change the three-phase source to six- or 12-phase power, and this is best accomplished by the use of properly connected transformers.

4. When the d-c output must be a three-wire service, the neutral wire is provided by the neutral point of the Y-connected transformer secondaries.

Three-phase Converters for Two-wire Direct-current Service. A bank of three single-phase transformers is generally used for such installations, with the primaries and secondaries connected either in Y or in Δ. Four combinations are therefore possible: (1) primaries and secondaries in Y; (2) primaries and secondaries in Δ; (3) primaries in Y, secondaries in Δ; and (4) primaries in Δ, secondaries in Y. Obviously, the connection that is used must yield the proper three-phase secondary voltage, the latter being determined by the a-c voltage rating of the converter. Thus, if the three-phase converter must develop 230 volts at the d-c end, the a-c slip ring, or transformer secondary voltage, must be $0.612 \times 230 = 141$ volts (see Table 10). This usually means that the transformer equipment is specially constructed for a given converter, being used with the latter only and connected so that the incoming three-phase primary voltage is stepped down to the proper three-phase slip-ring voltage.

The following example is given to illustrate the application of transformers for three-phase converter service:

EXAMPLE 6. A 250-kw three-phase synchronous converter has a d-c output voltage of 240 volts. It receives its a-c power from a 2,300-volt source through a bank of three transformers connected in Δ on the primary side and Y on the secondary side. Assume that the machine operates at full load at a power factor of 0.92 and an efficiency of 93 per cent, and calculate: (*a*) the d-c output; (*b*) the three-phase a-c voltage at the slip rings; (*c*) the alternating current delivered by the transformer secondaries to the slip rings; (*d*) the ratio of transformation of each transformer; (*e*) the current in each of the transformer primary coils; (*f*) the current in each of the line wires on the primary side.

Solution

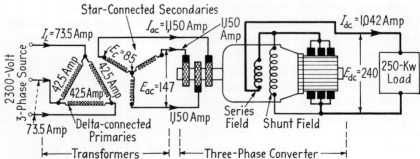

FIG. 322. Wiring diagram illustrating solution to Example 6 for 250-kw three-phase synchronous converter.

Refer to Fig. 322, which represents a wiring diagram for the problem and indicates the various current and voltage values calculated below.

(a) $$I_{dc} = \frac{250,000}{240} = 1,042 \text{ amp}$$

(b) $$E_{ac} = 0.612 \times 240 = 147 \text{ volts}$$

(c) $$I_{ac} = \frac{1,042 \times 0.943}{0.92 \times 0.93} = 1,150 \text{ amp}$$

The equation in Table 10 is for power factor = 1 and efficiency = 100 per cent. The transformer secondary current can also be calculated as follows:

$$I_{ac} = \frac{P}{\sqrt{3} \times E_{ac} \times \text{PF} \times \text{eff}} = \frac{250,000}{\sqrt{3} \times 147 \times 0.92 \times 0.93} = 1,150 \text{ amp}$$

(d) Volts per transformer coil secondary $= \dfrac{147}{\sqrt{3}} = 85$

$$a = \frac{2,300}{85} = 27.1:1$$

(e) Current per transformer coil primary $= \dfrac{1,150}{27.1} = 42.5$ amp

(f) Primary line current $= 42.5 \times \sqrt{3} = 73.5$ amp

The primary line current can also be calculated as follows:

$$\text{Primary } I_L = \frac{P}{\sqrt{3} \times E_L \times \text{PF} \times \text{eff}} = \frac{250,000}{\sqrt{3} \times 2,300 \times 0.92 \times 0.93} = 73.5 \text{ amp}$$

Three-phase Converters for Three-wire Direct-current Service. When the d-c output of a synchronous converter must be three-wire service, the neutral wire for the system may be obtained from the *star point* of the Y-connected transformer secondaries. In addition to their function as transformers, therefore, the latter act as reactance coils for the unbalanced direct current in the neutral wire; in this respect, the transformers behave exactly like the reactance coils of Fig. 125 in Chap. 6, used in conjunction with three-wire generators.

The neutral wire of the three-wire service, however, cannot be obtained from the Y point of a *simple* Y connection as shown in Fig. 322 because the direct neutral current will saturate the three individual transformer cores; when this happens, the transformers will cease to function properly as such. In order to make a neutral point available from the Y-connected secondaries, it is necessary that the latter be interconnected in such a way that the effect of any direct current in one-half of a secondary winding is neutralized by the effect of the same direct current in the other half of the secondary winding. This is accomplished by utilizing two identical secondary coils for each transformer secondary and using the so-called *zigzag* connection. Figure 323a illustrates the improper simple Y connection, in which the iron cores would be saturated, while Fig. 323b shows how the same transformers are connected zigzag so that no core saturation exists. Comparing the correct and incorrect connections, the following points should be noted:

1. In the incorrect connection, the core of transformer A will be saturated, because the direct current is in the *same* direction for halves

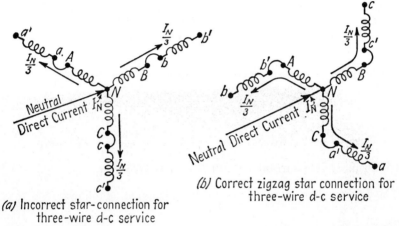

(a) Incorrect star-connection for three-wire d-c service

(b) Correct zigzag star connection for three-wire d-c service

FIG. 323. Wiring diagrams showing incorrect and correct Y connections of transformer secondaries for three-wire d-c service.

NA and aa'—in the correct connection, the core will not be saturated because the direct current will be from N to A in one-half of the secondary and in the *reverse* direction, from a' to a, in the other half of the winding.

2. In the incorrect connection of transformer B, the direct current will be in the *same* direction for halves NB and bb'—in the correct connection, the direct current will be from N to B in one-half of the secondary and in the *reverse* direction, from b' to b, in the other half.

3. In the incorrect connection of transformer C, the direct current will be in the same direction for halves NC and cc'—in the correct connection, the direct current will be from N to C in one-half of the secondary and in the reverse direction, from c' to c, in the other half.

Since the three-phase voltages between terminals a', b', c' for the simple Y connection are 15 per cent greater than those existing between terminals a, b, c for the zigzag connection (assuming the same total number of turns per transformer for both), it is always necessary to have about 15 per cent

CONVERTERS AND RECTIFIERS 501

more turns when the zigzag connection is used if the proper voltage is to be supplied to the slip rings.

Figure 324 illustrates the wiring connections for a synchronous converter and its transformers supplying 230/115-volt three-wire d-c service.

Six-phase Converters. It was shown in Chap. 7 that two transformers can be connected together by a Scott connection so that a three-phase system may be changed to a two-phase system, and vice versa. Transformers may also be employed to change the power of a three-phase system to that of a six-phase system, three single-phase units being necessary to accomplish this. For such service, the primaries may be connected either in star or in delta in the usual way for three-phase operation,

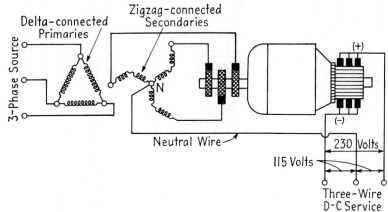

FIG. 324. Wiring diagram showing three-phase synchronous converter connected to provide three-wire d-c service.

while three general connection arrangements are possible for the six-phase secondaries. These are (1) the diametrical connection, (2) the *star* connection, and (3) the double-Δ connection.

The star connection (sometimes called the *double* Y) must be used when the d-c output is to be a three-wire service, the neutral point of the star system supplying the neutral wire; in this arrangement, each transformer secondary has two identical coils. The diametrical connection is generally employed when the d-c service is two wire, as in traction systems; in this arrangement, the secondary of each transformer is a single coil. The double-Δ connection is infrequently applied to the operation of synchronous converters.

Figure 325 shows a wiring diagram of a bank of transformers connected for three- to six-phase operation, with the primaries in Δ and the *secondaries in star*. Note that the Y point N provides the neutral for the three-wire d-c service, just as it does in the zigzag three-phase connection of

Fig. 324. As in the latter, the six-phase star permits the transformers to operate so that the core is not saturated by any d-c component.

Figure 326 shows a wiring diagram of a bank of transformers connected for three- to six-phase operation, with the primaries in Δ and the secondaries joined to diametrical points (180-electrical-degree points) on the

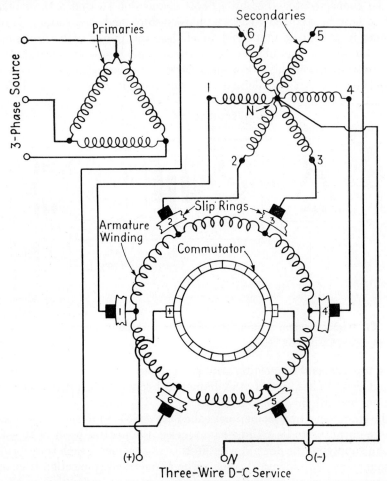

FIG. 325. Transformers connected for three-phase to six-phase operation. Primaries are in Δ and secondaries are in Y. The star point of the transformer secondaries serves as the neutral of the three-wire d-c system.

synchronous-converter armature winding. Note particularly that the general directions of the transformer windings have been made to correspond to points selected on the armature winding.

An example will now be given to illustrate the operation of a six-phase converter and its associated transformer bank.

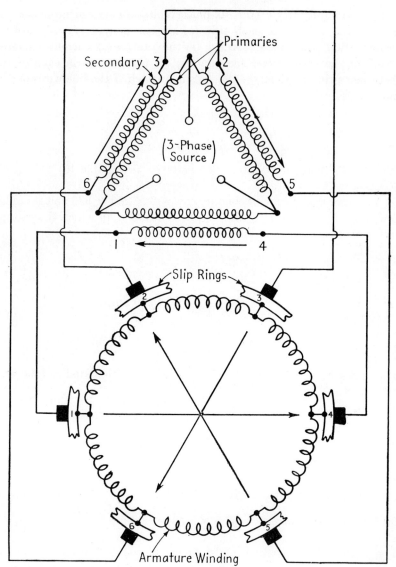

Fig. 326. Transformers connected for three-phase to six-phase operation. Primaries are in delta and secondaries are joined to diametrical points on the converter armature winding.

EXAMPLE 7. A 2,000-kw 750-volt d-c six-phase synchronous converter is fed by a bank of transformers connected in delta on the primary side and *diametrical* on the secondary side. The three-phase primary line voltage is 13,200 volts. Assume that the machine operates at full load at an efficiency of 96 per cent and a power factor of 0.95, and calculate: (*a*) the

504 ELECTRICAL MACHINES—ALTERNATING CURRENT

direct-current output; (*b*) the six-phase voltage between adjacent slip rings; (*c*) the voltage of each transformer secondary; (*d*) the ratio of transformation of each transformer; (*e*) the alternating current delivered by each of the transformer secondaries; (*f*) the current in each of the transformer primary coils; (*g*) the current in each of the line wires on the primary side.

Solution

(*a*) $\quad I_{dc} = \dfrac{2{,}000{,}000}{750} = 2{,}667 \text{ amp}$

(*b*) $\quad E_{ac} = 0.354 \times 750 = 265 \text{ volts} \quad$ (see Table 10)

(*c*) Volts per transformer secondary coil $= 2 \times 265 = 530$ volts. (This follows from the geometry of the six-phase hexagon, in which the voltage across a diagonal $= 2 \times$ volts per side.)

(*d*) $\quad a = \dfrac{13{,}200}{530} = 24.9:1$

(*e*) $\quad I_{ac} = \dfrac{2{,}667 \times 0.472}{0.96 \times 0.95} = 1{,}380 \text{ amp}$

(The equation in Table 10 is for power factor $= 1$ and efficiency $= 100$ per cent.)

The transformer secondary current can also be calculated as follows:

$$I_{ac} = \dfrac{\text{power per transformer}}{\text{volts per transformer} \times \text{PF} \times \text{eff}}$$

$$= \dfrac{(2{,}000{,}000/3)}{530 \times 0.96 \times 0.95} = 1{,}380 \text{ amp}$$

(*f*) Current in each transformer coil primary $= \dfrac{1{,}380}{24.9} = 55.5 \text{ amp}$

(*g*) Primary line current $= 55.5 \times \sqrt{3} = 96 \text{ amp}$

The primary line current can also be calculated as follows:

$$\text{Primary } I_L = \dfrac{P}{\sqrt{3} \times E_L \times \text{PF} \times \text{eff}}$$

$$= \dfrac{2{,}000{,}000}{\sqrt{3} \times 13{,}200 \times 0.96 \times 0.95} = 96 \text{ amp}$$

A wiring diagram illustrating the solution to this example, with all currents and voltages shown, is given in Fig. 327.

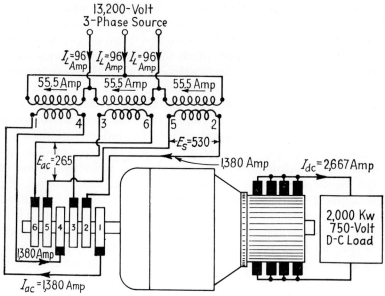

Fig. 327.—Wiring diagram representing the solution to Example 7.

Armature Reaction and Commutating Poles in Converters. When the armature of any commutator-type machine carries current, it creates a field of its own that tends to distort the main field. Unless such field distortion is counteracted in the neutral zone by the use of interpoles, serious sparking occurs between brushes and commutator (discussed in Chap. 4). Fortunately, the effect of armature reaction is usually much less in synchronous converters than in equivalent d-c machines, particularly when they operate at unity power factor. This is the result of two oppositely directed components of armature reaction, one of them resulting from the direct current delivered to the load and the other resulting from the alternating current taken by the armature so that it may operate as a synchronous motor. The difference between these two armature-reaction effects is generally so small that, in so far as sparking at the commutator is concerned, heavy overloads are permissible. Moreover, when commutating poles are employed further to improve the commutation process, they need not be so strong as those in corresponding d-c generators or motors, because their sole purpose is to aid current reversal and not to counteract a shift in the magnetic neutral. In fact, it is possible to design the commutating poles for converters so that they are 15 to 25 per cent weaker than those used in d-c generators of equal ratings.

Since the actual current in the armature of a synchronous converter is much less than in a d-c generator of similar capacity, the size of the wire in the armature winding is correspondingly smaller; this saving in copper thereby permits a further reduction in the size of the armature core. As a result of such economies, the size of the commutator, which carries the full direct load current, is usually much longer in proportion to the armature core than in single-purpose machines. This is an interesting constructional feature and is immediately apparent when converters are structurally compared with d-c generators (compare Figs. 3 and 5).

Controlling the Direct-current Voltage of a Converter. It is always desirable to have some degree of control over the terminal voltage of a generator, because fluctuations in load tend to produce variations in the armature and line drops. In d-c generators, this voltage control is readily accomplished by compounding, i.e., by using a series field whose flux aids the main-field flux in direct proportion to the load current. Since compounding of alternators is impossible, the load current being alternating, regulators are employed in such machines so that variations in the a-c voltage are accompanied by automatic adjustments in the rheostat resistance of the d-c field circuit. In both types of single-purpose machine, voltage control through flux adjustment is therefore simple and effective.

Field-rheostat Control. The situation is quite different, however, in dual-purpose machines such as synchronous converters. In these, the d-c voltage output is fixed by the slip-ring a-c voltage input, and is affected only slightly by variations in flux. Remembering that the converter is a synchronous motor when viewed from the a-c side, a change in the flux will result in a change in the power factor; strengthening the field will cause the machine to take a leading-power-factor current, while weakening the field will make the power factor lag. Since the kind of power factor (lagging or leading) determines whether the internal reactance drop will aid or buck the induced voltage, the terminal voltage may be adjusted to some extent in this way. Thus, if the field is strengthened above the value that produces unity power factor, the motor will operate at a leading power factor; under this condition, the internal reactance drop *aids* the induced a-c emf, so that the d-c terminal voltage rises. On the other hand, if the field is weakened below the value that produces unity power factor, the motor will operate at a lagging power factor; under this condition, the internal reactance drop *bucks* the induced a-c emf so that the d-c terminal voltage drops. Control of the d-c voltage by this method, i.e., by adjusting the rheostat resistance in the field circuit, is limited to about 5 per cent above or below the unity power-factor value. Moreover, since the voltage control and the power factor are interdependent, the efficiency and output of the converter will be reduced if it is necessary to operate at a leading or lagging power factor.

Series-reactance Control. If reactance coils are placed in the line wires between the transformers and the slip rings, the d-c voltage output may be controlled by adjusting the resistance in the field rheostat. This follows directly from what was said in the foregoing article. The addition of such reactors increases the reactance above that possessed by the armature winding alone, so that a greater degree of control is possible. Following the previous analysis, an increase in the d-c excitation causes the synchronous motor to operate at a leading power factor, under which condition the a-c slip-ring voltage rises; as a result, the d-c brush voltage rises, since it depends upon the former. And conversely, a decrease in the d-c excitation causes the synchronous motor to operate at a lagging power factor, under which condition the a-c slip-ring voltage drops, and with it the d-c brush voltage. Control of the d-c voltage by this method is limited to about 10 per cent above and below the unity power-factor condition. It is open to the objection, however, that the converter efficiency and the output are reduced as the power factor departs from unity. If a series field is used, the scheme becomes automatic, since an increase in the d-c load is met by an increase in flux, a leading power factor, and greater a-c slip-ring and d-c brush voltages; the reverse is true if the load decreases.

Induction-regulator Control. Since the d-c voltage depends directly upon the a-c voltage applied to the slip rings, a logical method of controlling the potential between d-c brushes is to employ an induction regulator on the input side of the converter. When placed between the slip rings and the transformer secondaries, it must obviously have the same number of phases as the converter. As is usually the case in practical installations, however, when it is placed on the primary side of the transformers, it is designed for the common three-phase high-voltage input, under which condition the regulator current is correspondingly reduced. This method of control has the advantage that the power factor is independent of the voltage change, although it does involve the added expense of another piece of equipment. A variation of about plus or minus 10 per cent is generally permissible with induction regulators.

Synchronous Booster Control. In some of the larger rotary converters, in which the control of the d-c voltage output is extremely important, it is customary to employ a separate alternator whose armature winding is connected *in series* with that of the converter winding. In the usual construction, this alternator is of the revolving armature type, with this armature placed on an extension of the converter armature. Connections between the alternator and the converter windings are made directly from one to the other, so that no extra slip rings are required; obviously, both must have the same number of poles and phases. The field of this alternator is excited by the same d-c output voltage that excites the converter field.

This additional component of the synchronous converter is called a *series booster,* and the entire unit is designated a *synchronous booster converter.* A 13,000-amp 270-volt six-phase 300-rpm machine is shown in Fig. 328. Clearly visible is the a-c end, with its six slip rings, and the series booster. In operation, the series-booster voltage is controlled by varying the d-c excitation through its field rheostat. Since the voltage of the latter unit may be made to add to, or subtract from, the main converter voltage, the slip-ring voltage is adjustable over a considerable range; in other words, the series booster can *boost* or *buck* the converter

FIG. 328. Synchronous booster converter having a rating of 3,500 kw at 270 volts direct current. The machine has 24 poles and operates at 300 rpm from a 60-cycle source. (*Westinghouse Electric Corp.*)

voltage by an amount readily controlled through a simple field-rheostat adjustment. It should therefore be clear that this is an excellent control method because the d-c output voltage is adjusted by changing the very voltage, the a-c slip-ring voltage, upon which it depends.

As pointed out in the foregoing paragraph, the series booster can be made to aid or oppose the converter voltage. In the boosting function, the unit absorbs mechanical power from the main synchronous motor because it acts like an alternator. When it performs as a bucking unit, however, it behaves like a motor, receiving electrical energy and delivering mechanical energy to the common shaft.

Parallel Operation of Converters. In systems in which a considerable amount of d-c power is required, it is customary to operate several converters in parallel to supply the common load; a good example of such practice is the converter station for traction service. No difficulty is experienced in doing this if certain precautions, mentioned below, are taken. Paralleling a converter on the d-c side is accomplished in a manner that is similar to the procedure for connecting a d-c generator in parallel with the bus bars; of course, this must be done after the converter is properly synchronized on the a-c side. When a compounding series field is used—as it is in most converters of substantial size—it is necessary that there be an equalizer connection.

In order to assure stable operation of converters in parallel, each unit must be serviced by its own bank of transformers. This service is particularly necessary if the latter are designed with considerable reactance; in some cases in which the transformer reactance is not sufficient, it is found desirable to insert line-reactance coils between the transformers and the slip rings. The practice also tends to prevent a flow of current from one converter to another.

One of the difficulties that might arise is a power failure on the a-c side while the d-c side is still connected to the bus bars or a storage battery. When this happens, the machine continues to rotate in the same direction, but as a d-c motor. Since armature reaction is now stronger than before, the field is greatly weakened; moreover, if the converter is compounded, the current through the series field is reversed, which further weakens the field. The result of these field-weakening effects is that the d-c motor speeds up, often to a very dangerous value. To forestall this runaway tendency, converters that are paralleled with others are equipped with speed-limiting devices mounted on the end of the shaft. These function to open the d-c circuit breakers as soon as the speed of the converter reaches a predetermined unsafe value. Often, reverse-current relays are also installed to open the d-c circuit breakers should the power flow be reversed.

Several methods may be employed to alter the division of load between converters connected in parallel. All are based on the principle that the d-c voltage may be raised to increase the load it delivers (or lowered to reduce the load it delivers) if the slip-ring voltage is properly adjusted. If the machine is a synchronous booster converter, the booster voltage is controlled to change the direct current delivered to the load. Another possibility is to increase the d-c excitation, which makes the converter take a leading power-factor current and causes the line- and transformer-reactance drops to aid the induced voltage and thus increase the slip-ring emf. Still another method is to use induction regulators on the primary side of the transformers; the boosting or bucking action of these is made available to raise or lower the slip-ring voltage.

The Inverted Converter. The usual operation of the synchronous converter is to change alternating to direct current. As previously explained, the a-c input causes the machine to function as a synchronous motor at *constant* (synchronous) *speed*, regardless of the magnitude of the load or field excitation. An increase in field current merely causes the input current to lead the voltage more, while a decrease in field current produces the reverse effect.

A converter can be used to change direct to alternating current, in which case the d-c input causes the machine to operate as a shunt or compound motor, while the a-c slip-ring voltage is applied to the load. When functioning in this way, the machine is called an *inverted converter*.

An inverted converter no longer operates at an absolutely average constant speed, as does a synchronous motor, because it runs as a d-c motor, and therefore its field strength and its load both determine what the speed will be. It will be recalled that, in general, an increase in load or flux tends to reduce the speed, while a decrease in load or flux has the opposite effect upon the speed. Assuming that the speed is adjusted to develop a given a-c frequency for a certain unity power-factor load, a higher speed and frequency result for the same kilovolt-ampere load if the power factor is lagging. This comes about because a lagging power-factor load current weakens the field strength. Moreover, since the frequency rises with increased speed, the inductive reactance of the load increases; the result is that the lagging power factor is lowered further and, in turn, causes the speed to rise still more. In fact, this tendency on the part of an inverted converter to speed up with lagging power-factor load is cumulative and will, if permitted to remain unchecked, cause the machine to race. It is for this reason that converters operating in this way are usually equipped with speed-limiting devices so that the runaway tendency may be avoided. Another scheme that has had some practical application is to mount an exciter on the converter shaft to provide the d-c field with current. When such an exciter is designed so that its voltage is sensitive to speed changes, any tendency on the part of the converter to speed up as a result of a weakened field caused by a lagging power-factor load is immediately counteracted by an increased excitation current from the exciter.

Since an inverted converter does not operate at an average constant speed, it cannot be properly called a *synchronous* machine. In practice, however, the speed is usually held within close limits in order to maintain as constant a frequency as possible. Heating, efficiency, and voltage ratios of these machines are about the same as those of straight synchronous converters. They are seldom used today, because large a-c power stations serve wide areas.

Frequency Converters. It is sometimes necessary to have a frequency that is different from the usual 25-, 50-, or 60-cycle source available from the supply lines. Also, in some special cases, two sources of supply of different frequencies must be interconnected so that there may be a flow of power from either one to the other. Under such circumstances, *frequency converters* must be used.

Synchronous-Synchronous Frequency Converter. One method is to couple two synchronous machines together so that both rotors, having the proper number of poles, operate at the same speed. Since rpm = $120f/P$,

$$\text{rpm} = \frac{120f_1}{P_1} = \frac{120f_2}{P_2}$$

so that
$$\frac{f_1}{f_2} = \frac{P_1}{P_2} \tag{102}$$

Thus it is seen that each machine must have a number of poles that is directly proportional to the frequency to which its stator winding is connected.

EXAMPLE 8. What are the fewest numbers of poles that can be used in two synchronous machines if it is desired to convert from (*a*) 60 to 25 cycles? (*b*) 40 to 60 cycles? Determine the speed of the set in each case.

Solution

(a) $P_1/P_2 = f_1/f_2 = {}^{60}\!/_{25} = 1\frac{2}{5}$. This means that if machine 1 has 12 poles, machine 2 should have five poles. Since a machine can obviously *not* have an *odd* number of poles (i.e., 5), it follows that machine 2 must have 10 poles, while machine 1 must have 24 poles. The speed of the set will therefore be

$$\text{rpm} = \frac{(120 \times 60)}{24} = \frac{(120 \times 25)}{10} = 300$$

(b) $P_1/P_2 = f_1/f_2 = {}^{40}\!/_{60} = \frac{4}{6}$. Therefore, machine 1 must have four poles while machine 2 must have six poles. The speed will be

$$\text{rpm} = \frac{(120 \times 40)}{4} = \frac{(120 \times 60)}{6} = 1,200$$

Induction Frequency Converter. It is possible to use a wound-rotor induction motor, whose rotor is driven at the proper speed and in the proper direction, to change the frequency of the source to another value. In the usual practical installation, the source frequency is changed to a higher value, as, for example, from 60 to approximately 100, 120, or 180 cycles. Such higher frequencies are then employed in connection with the operation of induction motors at much higher speeds than are obtainable when they are serviced by the lower frequency. Remembering that the speed of an induction motor is directly proportional to the frequency and inversely proportional to the number of poles [rpm = $(120f/P)(1-s)$], it should be clear that the maximum speed of an induction motor is about 3,600 rpm when a two-pole machine is connected to a 60-cycle source. For higher speeds, the frequency must obviously be greater than 60 cycles because there can be no fewer than two poles.

An *induction frequency converter* is essentially a wound-rotor motor with its stator winding connected to the available source of supply, usually 60 cycles, and its rotor winding connected to the motor, whose speed is to be different from what it would be if it were connected to the incoming power lines. To obtain current from the rotor of the frequency converter at a *higher frequency* than that of the source, it must be *driven* in a direction *opposite* to that in which it would normally run as a motor. Actually, therefore, a frequency converter is a special type of a-c generator.

The generated rotor frequency of this type of converter can be varied by varying the operating speed. This can be accomplished by using multispeed or adjustable-speed driving motors; in belt-driven sets, it can be done by changing the pulley ratio. In this way, several rotor frequencies may be produced from the same converter to vary the speed of motors that it operates. For example, if the converter is to change 60 to 120 cycles, the rotor of the converter must be driven at *synchronous speed* in a direction that is *opposite* that of the revolving field; under this condition the rotor conductors cut the revolving field at twice synchronous speed. Also, for 180 and 100 cycles, the rotor's opposite-direction speed must be, respectively, twice and two-thirds synchronous speed. The rotor of this type of frequency converter can, moreover, generate voltages whose frequencies are less than those in the stator winding if it is made to operate in the *same* direction as the revolving field; under such conditions the speed of the rotor conductors with respect to the revolving field is less than the speed of the latter. Thus, for output frequencies of 30 and 40 cycles, the rotor's same-direction speed must be, respectively, one-half and one-third synchronous speed.

Writing the foregoing statements in general terms gives

$$f_{conv} = f\left(1 \pm \frac{\text{rpm}_{conv}}{\text{rpm}_{syn}}\right) \qquad (103)$$

where the plus sign is used when the rotor is driven against the revolving field's direction and the negative sign when the rotor turns in the same direction as the stator field.

EXAMPLE 9. A four-pole 60-cycle wound-rotor motor is to be used as a frequency converter. At what speed must the rotor be driven, and in what direction must it rotate, if the output frequency is to be: (a) 90 cycles? (b) 150 cycles? (c) 50 cycles? What will be the rotor frequency if the speed is: (d) 2,200 rpm against the field's direction? (e) 400 rpm in the same direction as the field?

Solution

The four-pole 60-cycle synchronous speed of the revolving field is 1,800 rpm.

(a) $$90 = 60\left(1 + \frac{\text{rpm}_{conv}}{1,800}\right)$$

$$\text{rpm}_{conv} = \left(\frac{90}{60} - 1\right)1,800 = 900 \text{ opposite direction}$$

(b) $$\text{rpm}_{conv} = \left(\frac{150}{60} - 1\right)1,800 = 2,700 \text{ opposite direction}$$

(c) $$50 = 60\left(1 - \frac{\text{rpm}_{conv}}{1,800}\right)$$

$$\text{rpm}_{conv} = \left(1 - \frac{50}{60}\right)1,800 = 300 \text{ same direction}$$

(d) $$f_{conv} = 60\left(1 + \frac{2,200}{1,800}\right) = 133\tfrac{1}{3} \text{ cps}$$

(e) $$f_{conv} = 60\left(1 - \frac{400}{1,800}\right) = 46\tfrac{2}{3} \text{ cps}$$

The rotor voltage delivered by a frequency converter varies directly as the rotor frequency; this is because the rotor voltage, like the rotor frequency, depends upon the rate at which the rotor conductors cut the revolving field flux, and this, in turn, is a function of the *relative motion* of the rotor with respect to the field. For example, if the rotor voltage is 220 at 60 cycles, it would be 367 volts at 100 cycles ($220 \times 100/60$), 440 volts at 120 cycles ($220 \times 120/60$), etc. Similarly, a squirrel-cage induction motor rated for a given frequency and voltage, but designed for a range of frequencies, will operate satisfactorily at proportionately increased frequency and voltage. For example, a motor rated 60 cycles and 220 volts, if designed for a range of frequencies, may be operated at 100 cycles 367 volts or 120 cycles 440 volts with complete satisfaction.

A wiring diagram showing the usual connections and control for an induction frequency converter driven by an induction motor is shown in Fig. 329. General practice is to use separate switches to connect the driving motor and the stator winding of the frequency converter to the supply source. The motor starter may be of either the manual or the magnetic type. The switch of the converter stator winding should always be of the magnetic type, with the operating coil connected to the *motor side* of the starter for the driving motor. With this arrangement, there is no danger of damage to the frequency-converter windings should the converter windings be connected to the line while the driving motor is shut down.

Figure 330 shows a cutaway view of an induction frequency converter (on the left) coupled to a squirrel-cage motor (on the right).

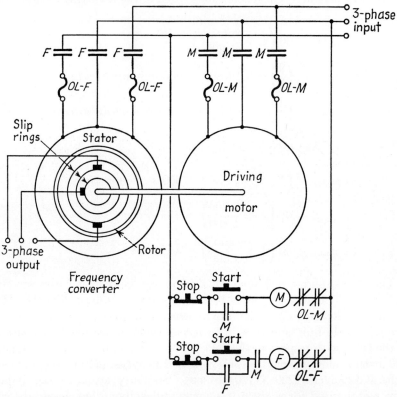

FIG. 329. Wiring connections for an induction frequency converter coupled to a squirrel-cage motor.

Induction frequency converters are widely used in the furniture and woodworking industries and in automotive and metal-working operations in which high-speed induction motors are essential.

Phase Converters. Transformers are generally used when it is necessary to change from one *polyphase* system to another. This was treated in Chap. 8, in which the Scott transformation, three- to two-phase and vice versa, was discussed. Also Figs. 325, 326, and 327 illustrate how transformers are connected to change a three-phase system to a six-phase system.

A transformation from single-phase to two- or three-phase cannot be made with transformers; for this purpose an *induction phase converter* must be used. Its principal application is in connection with the operation of electric locomotives, in which single-phase power is fed through one overhead trolley wire and the track to a phase converter; this machine then generates a three-phase system, which, in turn, can be used to operate the excellent three-phase traction motors. The latter arrangement eliminates the disadvantage of having two overhead trolleys, which would be necessary in a three-phase distribution-system.

An induction phase converter is actually a polyphase induction motor that is operated from a single-phase source. The stator winding is so

FIG. 330. Cutaway view of induction frequency converter coupled to a squirrel-cage induction motor. (*Westinghouse Electric Corp.*)

designed that one of its two phases is connected to the single-phase source, while the other phase is displaced 90 electrical degrees from the first and is connected in T with it. The result is that when a revolving field is created as the single-phase motor operates, two generated emfs are developed in the two windings that are 90° out of phase. By properly designing the phase-converter winding and by using a step-down transformer to lower the high trolley-to-track voltage, it is possible to obtain a substantially balanced three-phase system. Figure 331 shows a simple wiring diagram of an arrangement such as the one described.

In operation, the phase converter receives its single-phase power from the step-down transformer. Since this power is pulsating, the motor action is not steady, slowing down and speeding up from instant to instant. The speed variation is, however, only slight, so that the three-phase voltages remain fairly well balanced.

Types of Rectifier. As defined by the American Standards Association (ASA), "*a rectifier is a device which converts alternating current into* *unidirectional current by virtue of a characteristic permitting appreciable flow of current in only one direction."* There are many types of rectifier, but all exhibit the important and useful property indicated because the resistance to the flow of current in the transmitted direction is extremely low compared with that in the blocked direction.

Rectifiers are widely used as converters to link together existing a-c sources of supply and devices or power equipment that must operate from d-c sources. Where the power requirements are comparatively small, it is customary to employ solid material types such as copper-oxide or selenium rectifiers, or the glass-enclosed hot-cathode type such as the

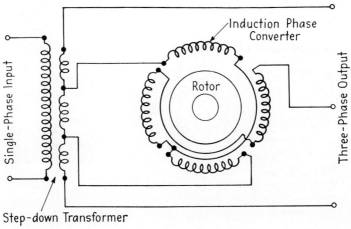

FIG. 331. Wiring connections for an induction phase converter, changing a high-voltage single-phase system to a low-voltage three-phase system.

Tungar or Rectigon; mechanical and electrolytic types of rectifier have been used on occasion, but they are generally subject to operating difficulties and are of low efficiency. For installations requiring large amounts of power such as traction motors and the electromechanical industries, the mercury-arc type of rectifier is the accepted standard; recently *silicon-type* rectifiers have been greatly improved and developed, and give promise of being widely used. The commutator and its brushes on a d-c machine are, of course, a mechanical rectifier that performs an important and reliable function, especially when used with interpoles and compensating windings, but the discussions that follow will not be concerned with this type of rectifier but with those, previously mentioned, that are employed primarily to service d-c devices or machines from an a-c scource.

Half-wave and Full-wave Rectification. Since the d-c voltage supplied to the load by the output terminals of one or more rectifiers depends upon the a-c voltage input to the rectifier, a transformer having the proper ratio of transformation is generally connected between the rectifier and the available a-c source so that the proper d-c voltage may be obtained at the output terminals. For *half-wave rectification* a single unit is used as illustrated by Fig. 332a, and current is transmitted to the load during all or part of one-half of the cycle and is blocked during the other half. Thus with a resistance load, Fig. 332b, current i_L flows as soon as the voltage v_L becomes positive, continues to flow, although it changes in magnitude, as the voltage varies from 0 to V_m to 0, and is prevented from flowing when v_L is negative. Note particularly that full voltage will appear across the rectifier during the nonconducting period because it is

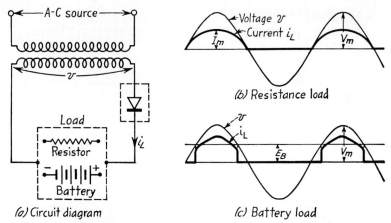

Fig. 332. Circuit and wave diagrams for a half-wave rectifier.

then that the device acts like an open circuit; this implies that the so-called *inverse voltage* should not be permitted to exceed the specified peak or inverse-voltage rating of the rectifier if breakdown of the latter is to be avoided.

When the load is a battery on charge, Fig. 332c, the latter introduces a countervoltage that prevents a current flow until the output voltage of the transformer is equal to the battery voltage E_B; charging current then flows, rising to a maximum at V_m, and then decreasing to zero at cutoff when v is again equal to the battery voltage E_B.

For *full-wave rectification*, i.e., for a flow of load current during both halves of the a-c cycle, two connection schemes are possible. In one of them, Fig. 333a, two rectifiers are used and the transformer secondary must have an available mid-tap; in the second scheme, Fig. 333b, four rectifiers are connected to form a so-called *bridge* circuit.

Referring to Fig. 333a, current flows through the load *from m to n* regardless of the polarity of the transformer. When terminal x is positive, the left half of the transformer nx serves as the source and establishes a current through rectifier 1 from x to m to n; when terminal y is positive, the right half of the transformer ny becomes the source and sends a current through rectifier 2 from y to m to n. Thus *current passes through the load on both halves of the cycle and in the same direction;* moreover, when rectifier 2 blocks a current flow, rectifier 1 is active, and vice versa.

In the bridge circuit of Fig. 333b, current also flows through the load from m to n on both halves of the cycle, but the entire transformer secondary and two diagonal rectifiers are active for each pulse. Thus,

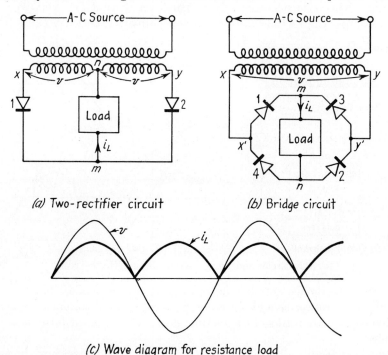

FIG. 333. Circuits and wave diagram for a full-wave rectifier.

when terminal x is positive, the current takes a path from x', through rectifier 1, through the load from m to n, through rectifier 2, and then to y' and y; when terminal y is positive, rectifiers 3 and 4 are active as the current passes through the load in the same direction (m to n) from y' to x' and x.

Another point that should be mentioned is that the waveshape of the load current will be similar to the voltage variation, assuming ideal rectifiers and a pure resistance load; as Fig. 333c illustrates, this means, of course, that the current wave will be sinusoidal if the voltage wave is.

The Copper-Oxide Rectifier. One of the most widely used rectifiers for small power application is the dry-disk (contact) type, in which a layer of cuprous oxide, formed on the surface of a sheet of copper, permits current to flow readily *from oxide to copper* but prevents such passage in the opposite direction. Called a *copper-oxide rectifier*, it usually consists of a number of units connected in one of several ways for half-wave or full-wave rectification. Figure 334a shows a cross section of such a rectifier that consists of two washers, one of which, made of soft metal, usually lead, is firmly pressed against the formed copper-oxide surface of another, made of copper. When four such units are assembled on an

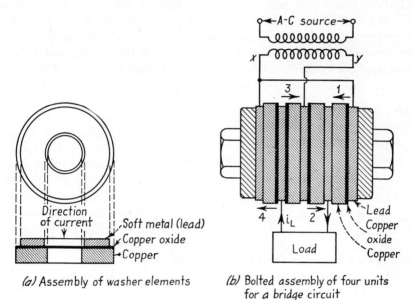

(a) Assembly of washer elements (b) Bolted assembly of four units for a bridge circuit

Fig. 334. Sketches illustrating one unit and a bridge circuit arrangement of a copper-oxide rectifier.

insulating rod, then bolted together and connected as shown in Fig. 334b, a bridge circuit is established to serve as a full-wave rectifier. Thus, when terminal x is positive, units 1 and 3 are active and current passes into the load as indicated; also, when terminal y is positive, units 2 and 4 become effective and permit the load current to flow in the same direction as before.

Since the voltage that may be impressed across a single unit is limited by a certain breakdown potential, several of them are frequently connected in series (as is done with dry cells or batteries) in each of the bridge arms if the circuit voltage is sufficiently high; in such cases, each of the elements of Fig. 334b would be replaced by a stack of series-connected units.

Applications of the copper-oxide rectifier are numerous, several of them being battery-charging circuits, many kinds of control circuits (see Figs. 254 and 293, where they are used to obtain direct current for dynamic braking), d-c instruments that are used in a-c circuits (the clamp-on ammeter), and many communication circuits.

The Selenium Rectifier. Another type of dry-disk rectifier consists of an iron plate that is coated with selenium and a thin layer of special alloy, the latter providing a contact surface and at the same time ensuring uniform current density in the active material. A very thin layer of selenium, about 0.05 mm thick, is deposited to one side of a roughened iron disk and, after a series of heat treatments which produce a crystalline structure, is covered by a soft metal layer, the counter electrode. An arrangement of materials such as this readily permits current to pass *from iron to selenium* but offers an extremely high resistance to a flow in the opposite direction. Figure 335 illustrates the construction of one unit made up of washers, with the thicknesses greatly exaggerated. Since the voltage across one unit is usually limited to 6 volts (the inverse peak voltage is 18 volts), it is customary to stack several units in series when such rectifiers are used in the higher voltage circuits. Moreover, the cross-sectional area of the disks is determined by the current they must pass, on the basis of 0.25 amp per sq in. for full-wave rectification.

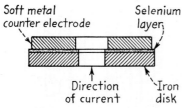

FIG. 335. Sketch of a selenium-rectifier unit.

Like the copper-oxide type, the selenium rectifier is commonly used in small power applications such as battery charging, signaling and control circuits, and certain types of measuring instruments. In recent years they have, however, been so highly developed that they have been used successfully in installations requiring considerable amounts of power; these include field excitation for synchronous motors, welders, crane and shop sources for d-c motors, magnetic brakes, chucks and clutches, arc lighting systems, power supplies for laboratories, elevator service, and electroplating tanks. When required to serve such large power equipment, six stacks of selenium rectifiers of the proper voltage and current rating are usually connected to a three-phase source through a set of three transformers as illustrated by Fig. 336. As shown, the equipment is put in operation by pressing the START button and, when the main a-c contacts close and the d-c switch is closed, the three phases "fire" in sequence through their respective rectifiers A, B, and C as each of the voltages successively becomes more positive than the others; when a rectifier such as A "fires," unidirectional current passes to the load and returns in two equal successive periods through the rectifiers shown below B and C.

The Hot-cathode Gaseous Rectifier. When an evacuated closed vessel (a glass tube) contains a heated filament that is separated from a positively charged plate, electrons that leave the *cathode* (the tungsten filament) are attracted to the *anode* (the graphite plate) and are transmitted through a circuit if the tube elements are connected externally to a load; such an electron flow is actually an electric current but opposite in

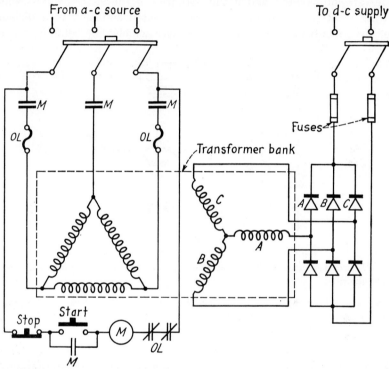

Fig. 336. Wiring diagram showing six stacks of selenium rectifiers connected to a three phase source through a bank of Δ-Y transformers to supply a large d-c load.

direction to the conventional direction as considered in this book. Moreover, if this so-called *diode* is filled with argon gas to a pressure of 5 cm, evaporation of the filament is inhibited and the voltage drop in the tube, i.e., between anode and cathode, may be kept reasonably low under load conditions and the device may be operated with good efficiency. Also, the fact that current can flow only in one direction gives this device rectification properties.

Tubes of the type described, together with auxiliaries such as a transformer and ballast resistor (and sometimes a reactor), are manufactured under the trade names of Rectigon (Westinghouse Electric Corporation) and Tungar (General Electric Company). A circuit diagram showing how this type of rectifier is connected to a battery for charging purposes is given in Fig. 337. Note that the filament is energized by a separate section of the transformer secondary, taking about 14 amp at 2.5 volts in the tube that delivers about 5 amp to the battery load. During the a-c period when m is positive, the negatively charged electrons that are emitted by the heated filament are attracted to the positively charged anode (graphite plate), and if the potential of the latter exceeds about 15 volts positive, the electrons are accelerated sufficiently to dislodge other electrons from the gas molecules with which they collide; the new

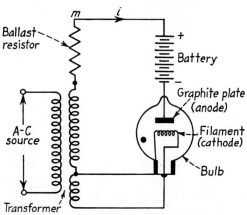

Fig. 337. Circuit diagram of a hot-cathode type (Tungar or Rectigon) of half-wave rectifier charging a battery.

electrons are then also attracted to the anode. The slow moving positively charged gas ions, on the other hand, are attracted to and surround the filament, where they completely neutralize the negative space charge to permit the cathode to function properly.

While the rectifier is in operation, there is a voltage drop within the tube which decreases slightly as the load current increases. This means, therefore, that with no ballast resistor in the circuit, the load current would tend to rise to an excessive value because any increase in load current will be accompanied by a reduction in the tube drop, a corresponding increase of the load voltage, and a further rise in the load current; the addition of the ballast resistor, however, prevents such unstable performance.

These rectifiers are designed primarily for battery chargers and are made in ratings of 2 to 15 amp. The tube drop varies between 6 to 40 volts, depending upon the size of the tube and the load, and the operating efficiency is about 35 per cent in the smaller sizes and as high as 75 per cent in the larger ratings. They can, moreover, be adapted for full-wave rectification when two tubes and a center-tapped transformer secondary

CONVERTERS AND RECTIFIERS 523

are used. The student is urged to draw a circuit diagram for such a full-wave rectifier.

EXAMPLE 10. A Rectigon battery charger delivers a d-c load of 10 amp at 12 volts. Assuming sinusoidal voltage and current variations for the half-wave rectifier and a tube drop of 10 volts, calculate the current and voltage ratings of the transformer secondary.

Solution

Since the unidirectional load current in a half-wave rectifier flows during one-half of each cycle, its *average value* is

$$I_{dc} = \frac{1}{2} \times \frac{2}{\pi} I_m = 0.318 I_m$$

where I_m is the maximum value of the sine wave (see Fig. 332a).

Also, the *effective value* of such a current in a half-wave rectifier is

$$I_{ac} = \frac{1}{\sqrt{2}} \times \frac{I_m}{\sqrt{2}} = \frac{I_m}{2}$$

It follows, therefore, that

$$I_{ac} = \frac{I_{dc}}{2 \times 0.318} = \frac{10}{2 \times 0.318} = 15.7 \text{ amp}$$

The total d-c output voltage, which is the sum of the d-c load volts and the tube drop, is also equal to

$$V_{dc} = 0.318 V_m$$

where V_m is the maximum value of the sine wave (see Fig. 332a).

But the *effective value* of the voltage across the transformer secondary, appearing during *both* halves of the cycle, is

$$V_{ac} = \frac{V_m}{\sqrt{2}}$$

If follows, therefore, that

$$V_{ac} = \frac{V_{dc}}{\sqrt{2} \times 0.318} = \frac{12 + 10}{\sqrt{2} \times 0.318} = 48.9 \text{ volts}$$

EXAMPLE 11. The transformer of a 5-amp Tungar battery charger has a 115-volt primary and a 52-volt secondary. If the d-c load voltage is 15 volts, calculate: (*a*) the tube drop, T.D.; (*b*) the effective transformer secondary current; (*c*) the transformer primary current, neglecting the exciting current.

Solution

(a) $$V_{ac} = \frac{V_{dc}}{\sqrt{2} \times 0.318}$$

$$52 = \frac{(15 + \text{T.D.})}{\sqrt{2} \times 0.318}$$

$$\text{T.D.} = (52 \times \sqrt{2} \times 0.318) - 15 = 8.4 \text{ volts}$$

(b) $I_{ac_s} = \dfrac{I_{dc}}{2 \times 0.318} = \dfrac{5}{2 \times 0.318} = 7.85$ amp   (secondary current)

(c) $I_{ac_p} = \dfrac{V_s}{V_p} \times I_{ac_s} = \dfrac{52}{115} \times 7.85 = 3.55$ amp   (primary current)

The Mercury-arc Rectifier. Although the hot-cathode type of rectifier (Rectigon and Tungar) performs well in the small-power applications for which it is designed, its rating and life are limited by the size of the cathode; the latter, the source of the electrons necessary to the proper functioning of the device, operates at a high temperature and therefore slowly volatilizes to become less effective with time. The limitation indicated has been largely overcome by the mercury-arc rectifier, in which a pool of mercury serves as the hot cathode and supplies mercury vapor to the evacuated enclosure from which negative and positive ions are produced. When in operation, this highly developed rectifier (placed in glass containers in the smaller sizes and in steel tanks in the large power units that often substitute for rotating converters) depends upon the valve action of a metallic (mercury) arc which burns between two electrodes. Current can pass in one direction only if the cathode, the negative electrode, is excited, i.e., made white-hot by electronic emission, and if the anode, the positive electrode, is maintained at a temperature at which no electrons are liberated. Mercury is used in commercial rectifiers because it is readily vaporized and can easily be condensed and returned to the cathode. Moreover, these rectifiers are usually provided with a suitable number of anodes, each of which is connected to a secondary phase of a transformer; the anodes in the same enclosure do not interfere with one another, and *each one carries current only during that part of the cycle when its voltage is more positive than that of the others.* The mercury pool serves as the cathode for all of the anodes and carries current at all times.

The Two-anode Single-phase Mercury-arc Rectifier. Unlike the hot-cathode type of rectifier in which a copious supply of electrons is constantly emitted by a heated filament, the mercury pool cathode of the mercury-arc rectifier must be maintained at a high temperature by the arc between cathode and anode; moreover, since the arc is, in fact, the actual load current, it must not be permitted to go out completely, because the deionization time of the vapor is only a few microseconds. A simple single-anode construction is, for this reason, not feasible because, as a half-wave rectifier, no arc current would flow during each half of every cycle. (In the larger, more complex ignition arrangements a special auxiliary igniter is used to strike the arc on each positive voltage pulse,

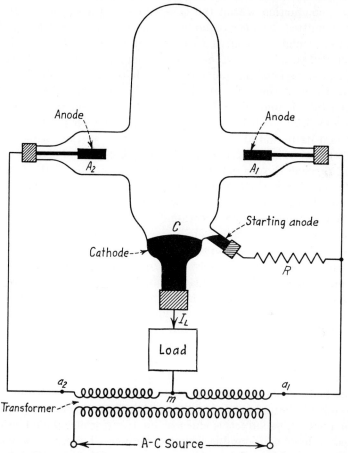

FIG. 338. Two-anode full-wave mercury-arc rectifier for single-phase service.

but this modification of the single-anode rectifier is generally employed, with others, in polyphase systems.) The two-anode full-wave mercury-arc rectifier is therefore the standard construction for single-phase service.

A sketch of a glass-enclosure type of mercury-arc rectifier is illustrated by Fig. 338 and is shown connected to a load and a center-tapped transformer secondary. To initiate operation, the tube is tilted slightly until mercury spills over to bridge the cathode pool and the starting anode;

this contact makes it possible for the right-hand half of the transformer secondary to send current through a circuit consisting of the load, the mercury in the cathode and auxiliary anode, and the current-limiting resistor R. When the tube is restored to an upright position again, the mercury bridge is broken, an arc is formed, and the resulting temperature rise at the surface of the cathode pool causes some of the mercury to vaporize; conduction within the tube can now take place because electrons are emitted from the cathode and are driven by the electric field to that anode which is positive with respect to the cathode. Moreover, collisions between the electrons and the mercury vapor ionize the latter, creating positive ions that move from anode to cathode. Thus a current I_L flows through the load, as indicated, in the conventional direction. Once the tube is ionized, full-wave rectification takes place because the two anodes fire repeatedly in succession, each one on the half of the a-c cycle when its potential is positive with respect to the cathode. Thus, when terminal a_1 is positive, transformer winding ma_1 delivers current to the load through the arc between A_1 and C; on the next half of the cycle, when terminal a_2 is positive, transformer winding ma_2 becomes effective, whereupon the arc instantly forms between A_2 and C to continue the flow of load current in the same direction.

Two important performance characteristics of the mercury-arc rectifier are responsible for the *sustained arc stream* and the tube's continuous operation; these are: (1) the formation of a glowing *cathode spot* and (2) the effect of inductance in the load circuit which causes a time lag of the current behind the voltage. Concerning the first of these, the positive ions, formed when the electrons collide with the gas molecules as they move rapidly to the anode, bombard the mercury pool; the energy thus developed by a current of about 4,000 amp per sq cm causes the temperature of a very small area to rise and produce an intensely hot spot on the surface of the mercury. This cathode spot dances about at random and does, in fact, vaporize sufficient mercury to replace the vapor that is condensed on the enlarged dome-shaped bulb to trickle back into the cathode pool. The inductive effect of the transformer, the latter usually designed to have considerable leakage reactance, is clearly shown in the wave diagrams of Fig. 339a. Note particularly that the anode currents i_1 and i_2 do *not* drop to zero when the voltage waves pass through zero, but continue to flow for short periods to sustain ionization; this "spillover" or residual current is extremely important because the arc is maintained while the positive potential rises sufficiently to permit a smooth transfer of the electron stream from one anode to another.

Since the cathode or *load current* results from the firing of either anode, and for a short overlapping period of both anodes, it should be clear that its waveshape will be a composite of i_1 and i_2; this is shown in Fig. 339b and indicates that the current varies only slightly and may, for practical reasons, be assumed to be fairly constant. This means, therefore, that the *average anode current*, on the basis of a rectangular wave form, is one-half of the uniform direct load current; i.e., $I_{A_{av}} = 0.5 I_{dc}$, and the *effective anode current* (the current in each half of the transformer secondary winding is $I_{A_{eff}} = 0.707 I_{dc}$, where I_{dc} is the direct current in the load circuit).

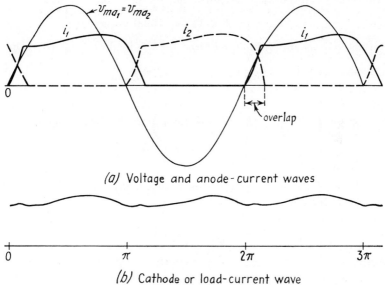

Fig. 339. Waveshapes for a two-anode mercury-arc rectifier.

The flow of current within the tube between anode and cathode involves a voltage drop, called *arc drop*, and is, for the moderately small rectifiers of the two-anode design, about 15 volts. (For the large high-voltage rectifiers that require a considerable separation between the anodes and cathode, the arc drop may be as high as 35 volts.) Moreover, the arc drop is substantially constant over a wide range of loads, and this means, of course, that the output voltage changes very little during normal operation; the voltage regulation of a mercury-arc rectifier is, in fact, much better than that of rotating converters.

EXAMPLE 12. A 10-kva 2,300/230-115-volt transformer is connected to a two-anode mercury-arc rectifier. Assuming a 15-volt arc drop, calculate: (*a*) the full-load effective anode current; (*b*) the direct current in the load; (*c*) the d-c load voltage; (*d*) the kw load output; (*e*) the full-load primary current.

Solution

(a) $$I_{A\text{eff}} = \frac{10{,}000/2}{115} = 43.5 \text{ amp}$$

(b) $$I_{A\text{eff}} = 0.707 I_{dc}$$

Therefore $$I_{dc} = \frac{43.5}{0.707} = 61.5 \text{ amp}$$

(c) Since the voltage wave is sinusoidal, its maximum value is

$$E_m = \sqrt{2} \times 115 = 162.6 \text{ volts}$$

But the *average* d-c voltage required for the load *and* the arc drop is

$$E_{av} = E_{dc} + 15 = \frac{2}{\pi} E_m$$

Therefore

$$E_{dc} = \left(\frac{2}{\pi} E_m\right) - 15 = \left(\frac{2}{\pi} \times 162.6\right) - 15 = 88.5 \text{ volts}$$

(d) $$\text{kw output} = \frac{88.5 \times 61.5}{1{,}000} = 5.45$$

(e) Since the ratio of transformation between primary and each half of the secondary is $2{,}300/115 = 20:1$,

$$I_P = \frac{61.5}{20} = 3.075 \text{ amp}$$

Multianode Three-phase Mercury-arc Rectifiers. For comparatively large power applications it is customary to use multianode mercury-arc rectifiers that are served by three-phase sources of supply through suitable transformer equipment. The latter is often a three-phase core-type unit with the primary coils connected in Y or Δ and the secondary windings always in Y. As the following diagrams illustrate, the line terminals of such a three- or six-phase transformer secondary are connected, respectively, to the three anodes or six anodes of the mercury-arc rectifier, and the neutral point of the secondaries is wired to the cathode through the load. Moreover, the transformer is generally designed with an extremely high value of leakage reactance to permit the rectifier to deliver a load current that is reasonably steady; under such conditions of operation the average and effective values of the load current are approximately equal.

Assuming a *rectangular* wave form, the *average anode current* in a rectifier with q anodes will be

$$I_{A\text{av}} = \frac{I_{dc}}{q} \tag{104}$$

because the direct current in the load I_{dc} flows continuously and is made up of q equal parts, each of which is delivered by an anode that fires during $1/q$th part of a cycle. The *effective anode current*, being a measure of the heating value in a given resistor will, however, be

$$I_{A\text{eff}} = \frac{I_{\text{dc}}}{\sqrt{q}} \tag{105}$$

Remembering that an anode will fire only if its potential is *more positive* than the others, it should be understood from a study of Fig. 340 (for a three-anode rectifier) that anode 1 will deliver load current *only* when its

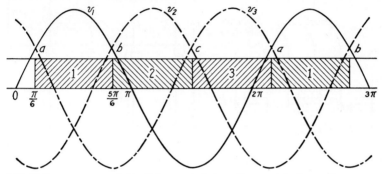

Fig. 340. Voltage waves and the firing periods in a three-anode three-phase mercury-arc rectifier.

voltage wave v_1 passes from a to b; similarly, anodes 2 and 3 will deliver currents *only* when the voltage waves v_2 and v_3 pass, respectively, from b to c and from c to a. Moreover, when an anode fires, one of the transformer secondary coils energizes a circuit consisting of the load and the arc within the rectifier during $1/q$th part of a cycle; this means, therefore, the effective value of the voltage for the part of the cycle indicated will have an average value of

$$V_{\text{dc}} = V_L + \text{arc drop} = \frac{q}{\pi} \sqrt{2}\, V \sin \frac{\pi}{q} \tag{106}$$

where V = effective voltage per transformer coil

Finally, the total kilovolt-ampere rating of the transformer that serves a q-anode rectifier and the load is equal to

$$\text{kva}_t = q \times \frac{VI_{A\text{eff}}}{1{,}000} = q \times \frac{VI_{\text{dc}}/\sqrt{q}}{1{,}000} = \frac{qVI_{\text{dc}}}{1{,}000} \tag{107}$$

because, as represented by Eq. (107), *each* of the q transformer coils has an output of $VI_{A\text{eff}}/1{,}000$ kva.

The Three-anode Rectifier. A diagram of a three-anode rectifier connected to a three-phase core-type transformer and a load is illustrated by Fig. 341a; although a starting anode is required as in Fig. 338, it was omitted for simplicity. Note particularly that the primary coils are connected in Δ, a common arrangement when a single transformer is used, and that the secondaries, as previously mentioned, are in Y. Thus, when an anode fires and a load current I_{dc} flows in *one* of the secondary windings, the *three* inductively coupled primary coils must carry currents whose sum is equal to I_{dc}/a, where a is the ratio of transformation of one primary coil with respect to one secondary coil, i.e., N_P/N_S. However,

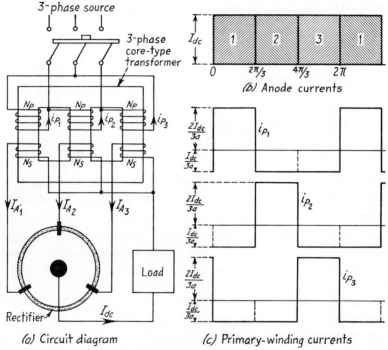

FIG. 341. Wiring and wave diagrams for a three-anode three-phase mercury-arc rectifier.

since the secondary winding that is carrying current is *directly* coupled to its corresponding primary winding (on the same core leg) and the latter is, in turn, in parallel with the other two series-connected primary windings (on the other two core legs), the total primary current I_{dc}/a will divide so that the current in the directly coupled primary will be $2I_{dc}/3a$ and the current in the other two primaries will be $I_{dc}/3a$. Another important point concerns the current directions in the primary windings. Assuming that a *normal* primary-current direction is *positive* when the anode of its corresponding directly coupled secondary is firing, it should be clear that the currents in other two series-connected primaries will *not* be in their normal directions and will therefore be negative.

The foregoing general analysis is illustrated by the circuit and wave diagrams of Fig. 341. For example, when anode 1 is firing for the period $2\pi/3$ radians and the secondary current in the *left* winding is I_{dc}, the corresponding primary-winding current i_{P_1} will be $+2I_{dc}/3a$ and the primary currents i_{P_2} and i_{P_3} in the *center* and *right* windings will be, simultaneously, $-I_{dc}/3a$. Next, when anode 2 is firing between $2\pi/3$ and $4\pi/3$ and the secondary current in the *center* winding is I_{dc}, the corresponding primary-winding current i_{P_2} will be $+2I_{dc}/3a$ and the primary currents i_{P_1} and i_{P_3} in the *left* and *right* windings will be, simultaneously, $-I_{dc}/3a$. Finally, when anode 3 is firing during the last third of the cycle, between $4\pi/3$ and 2π, and the secondary current in the *right* winding is I_{dc}, the corresponding primary-winding current i_{P_3} will be $+2I_{dc}/3a$, and the primary currents i_{P_1} and i_{P_2} in the *left* and *center* windings will be, simultaneously, $-I_{dc}/3a$. The wave diagrams of Fig. 341b therefore indicate that the primary-winding currents do not follow a simple uniform pattern in which positive and negative pulses vary similarly; each cycle of primary-winding current consists of a positive pulse for one-third of a cycle and a negative pulse for two-thirds of a cycle, with the magnitude of the positive current equal to twice the negative current.

EXAMPLE 13. The Δ-connected primary of a three-phase transformer is supplied with power from a 2,300-volt three-phase source, and the Y-connected secondary is connected to a three-anode mercury-arc rectifier, which delivers a d-c load of 100 kw at 220 volts. Assuming an arc drop of 15 volts, calculate: (a) the voltage across each transformer secondary; (b) the direct current in the load; (c) the effective anode (transformer secondary) current; (d) the effective current in each transformer primary; (e) the kva rating of the transformer.

Solution

(a) $$V_{dc} = (220 + 15) = \frac{3}{\pi} \sqrt{2}\, V \sin \frac{\pi}{3}$$

$$V = \frac{\pi \times 235}{3\sqrt{2} \times 0.866} = 200 \text{ volts}$$

(b) $$I_{dc} = \frac{100{,}000}{220} = 455 \text{ amp}$$

(c) $$I_{A\text{eff}} = \frac{455}{\sqrt{3}} = 263 \text{ amp}$$

(d) $$I_P = 263 \times \frac{200}{2{,}300} = 22.9 \text{ amp}$$

(e) $$\text{kva}_t = \frac{\sqrt{3} \times 200 \times 455}{1{,}000} = 157.5$$

532   ELECTRICAL MACHINES—ALTERNATING CURRENT

The Six-anode Rectifier. A schematic diagram of a six-anode rectifier connected to a load and the star-connected six-phase secondaries of a three-phase transformer is illustrated by Fig. 342a. Also, with the primaries in Δ, the transformer connections are similar to those shown in Fig. 320, where a three-phase source is changed to six-phase to serve a six-ring synchronous converter. The actual connections, exactly as they would be made for a three-phase core-type transformer and the six-anode rectifier with its load, are given in Fig. 343a; the latter diagram should be carefully compared with its schematic counterpart, Fig. 342a.

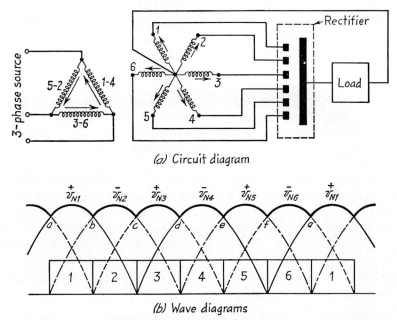

(a) Circuit diagram

(b) Wave diagrams

FIG. 342. Circuit and wave diagrams for a six-anode mercury-arc rectifier.

To understand how the circuit of Fig. 342a functions as the six anodes fire in succession, it is necessary to recognize the following conditions of operation:

1. The anode which fires will always be at a higher positive potential than the others.

2. The transformer secondaries that are connected to anodes 1 and 4 are directly coupled to, and are shown to occupy the same relative direction as, the primary labeled 1–4.

3. The transformer secondaries that are connected to anodes 3 and 6 are directly coupled to, and are shown to occupy the same relative (horizontal) direction as, the primary labeled 3–6.

4. The transformer secondaries that are connected to anodes 5 and 2

are directly coupled to, and are shown to occupy the same relative direction as, the primary labeled 5–2.

5. When anode 1 fires, transformer secondary N_1 is active, since its voltage v_{N_1} (Fig. 342b) is passing from a to b; moreover, the current in the directly coupled primary 1–4 will be in the direction shown and assumed to be *positive*.

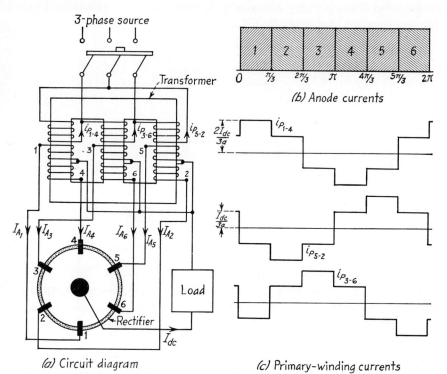

Fig. 343. Wiring and wave diagrams for a six-anode three-phase mercury-arc rectifier.

6. When anode 4 fires, transformer N_4 is active because, on the negative half of the cycle, terminal 4 is at a higher positive potential than terminal 1; this means that v_{N_4} (Fig. 342b), passing from d to e, is actually a *negative* voltage with respect to v_{N_1}, and that the current in the directly coupled primary 1–4 will be opposite to the direction shown, i.e., *negative*.

7. The statements made in (5) and (6) for windings 1–4 and its anodes 1 and 4 apply equally to windings 3–6 and 5–2 and their corresponding anodes 3 and 6 and 5 and 2; note particularly in Fig. 342b that v_{N_3} is indicated as positive in passing from c to d, v_{N_6} is labeled negative between f and a, v_{N_5} is shown positive for the firing period e to f, and v_{N_2} is represented as negative between b and c.

Referring next to Fig. 343b, the waveshapes of the primary currents are shown to correspond with the foregoing discussion as well as those made for the three-anode rectifier, Fig. 341. For example, when anode 1 fires, the primary current $i_{P_{1-4}}$ is *positive* and equal to $2I_{dc}/3a$, while the currents $i_{P_{5-2}}$ and $i_{P_{3-6}}$ are *negative* and equal to $I_{dc}/3a$; also, the primary currents $i_{P_{5-2}}$ and $i_{P_{3-6}}$ and their directions are similarly determined. The magnitudes of the primary currents with their directions, as anodes 1 to 6 fire in succession, are given in Table 12.

TABLE 12

Firing anode	Primary currents		
	$i_{P_{1-4}}$	$i_{P_{3-6}}$	$i_{P_{5-2}}$
1	$+2I_{dc}/3a$	$-I_{dc}/3a$	$-I_{dc}/3a$
2	$+I_{dc}/3a$	$+I_{dc}/3a$	$-2I_{dc}/3a$
3	$-I_{dc}/3a$	$+2I_{dc}/3a$	$-I_{dc}/3a$
4	$-2I_{dc}/3a$	$+I_{dc}/3a$	$+I_{dc}/3a$
5	$-I_{dc}/3a$	$-I_{dc}/3a$	$+2I_{dc}/3a$
6	$+I_{dc}/3a$	$-2I_{dc}/3a$	$+I_{dc}/3a$

EXAMPLE 14. A six-anode mercury-arc rectifier delivers a 4,000-kw load at 2,400 volts d-c. If the primaries of the three-phase transformer that serves the rectifier and load are connected in Δ and to a 66,000-volt three-phase source, calculate the following, assuming a 20-volt arc drop: (a) the direct current in the load; (b) the effective anode (transformer) current; (c) the voltage across each of the transformer secondary coils; (d) the kva rating of the transformer.

Solution

(a) $$I_{dc} = \frac{4,000}{2.4} = 1,667 \text{ amp}$$

(b) $$I_A = \frac{1,667}{\sqrt{6}} = 680 \text{ amp}$$

(c) $$V_{dc} = 2,400 + 20 = \frac{6}{\pi} \sqrt{2} \; V \sin \frac{\pi}{6}$$

$$V = \frac{2,420\pi}{6\sqrt{2} \times 0.5} = 1,790 \text{ volts}$$

(d) $$\text{kva}_t = \frac{\sqrt{6} \times 1,790 \times 1,667}{1,000} = 7,300$$

Ignitron Rectifiers. Since glass-enclosed rectifiers, Fig. 338, are generally fragile and dissipate heat losses with some difficulty, they are limited to the two-anode construction and for use in comparatively small-power single-phase circuits. For power requirements up to about 600 kw and 600 volts d-c current the three- and six-anode rectifiers, Figs. 341 and 343,

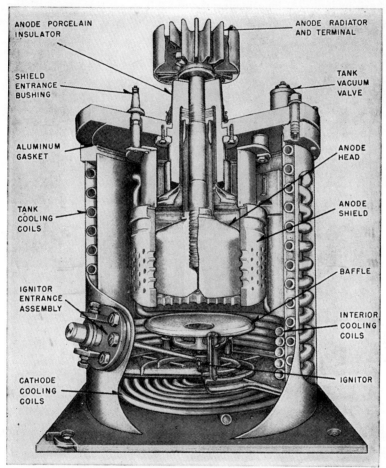

Fig. 344. Cutaway view of an *ignitron* rectifier unit. (*Westinghouse Electric Corp.*)

have been operated successfully from three-phase circuits. Moreover, for still higher ratings and voltages, rectifiers with six and 12 anodes have been constructed, but these are generally equipped with water jackets and cooling coils to carry off the heat of the arcs and with specially designed anode baffles and shields to overcome the possibility of *arc-back* or *backfire;* the latter results because hot spots tend to develop on the anodes, which, in turn, act as secondary sources of electrons. Then, when one such hot spot is at a negative potential with respect to another anode, breakdown occurs within the rectifier as a high anode-to-anode current, bypassing the cathode, flows to short-circuit the transformer; this abnormal condition causes a concentration of heat at the anodes, which are usually damaged. Other design improvements that tend to minimize arc-back include: (1) lengthening the arc (although this increases the arc drop and lowers the efficiency of the rectifier) and (2) providing each

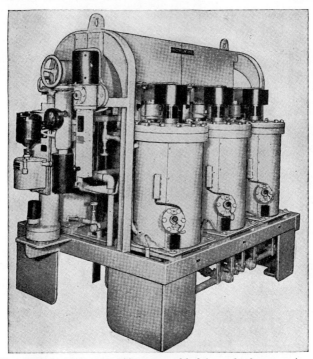

FIG. 345. Group of six *ignitron* rectifiers assembled for polyphase service. (*Westinghouse Electric Corp.*)

anode with a grid that functions, when excited, to start the anode current and thereafter prevents anode-to-anode current in the presence of hot spots.

Many of the difficulties experienced with multianode rectifiers have been largely overcome by the development and use of *single-anode tanks*, each with its own cathode; these are then combined in groups of three, six, or 12 and operated from a three-phase source through properly connected transformer and ignition equipment. Manufactured by the General Electric Company and the Westinghouse Electric Corporation under the trade name of *ignitron*, Fig. 344, the unit consists of a main anode of graphite, a mercury pool cathode, and a firing *ignitor*, all enclosed in a water-jacketed steel tank. Its unique operating feature is an ignition scheme, invented by Seplian and Ludwig, in which an arc is started at the cathode every time the anode fires and at the very instant the unit must function. The *ignitor*, as it is called, is a pointed rod of high-resistance refractory material, such as carborundum, which dips into the mercury pool. When a timed current impulse of about 20 to 40 amp is passed from the ignitor to the mercury, a spark occurs at the junction to form a small cathode spot; the latter then ionizes the mercury within a few microseconds and permits the anode to fire during an accurately measured period that is controlled by a phase-shifting circuit. Since the ignitron conducts only when it is firing, there is little danger from backfire.

Ignitrons are manufctured in many sizes and ratings. For small-power applications they are usually glass tubes with sealed-in components. The larger units are constructed with water-cooled tanks and, when assembled in groups for polyphase service, as illustrated in Fig. 345, have been used to deliver as much as 100,000 kw at voltages as high as 3,000. An extremely important advantage of the single-anode tank is that a failure of one unit in an assembly is readily corrected by replacing it with a spare; this is in contrast with a multianode rectifier whose entire capacity is lost when a defect develops.

Questions

1. What is a *converter?* Name three types of converter.
2. Under what operating conditions is it often necessary to use converter equipment? Give specific examples.
3. Why is it not possible to use transformers to convert from single-phase to two- or three-phase?
4. Describe the general construction of the synchronous converter.
5. At what speed does a synchronous converter operate?
6. Why are single-phase converters rarely used?
7. Upon what factors does the voltage between plus (+) and minus (−) brushes depend?
8. Explain why the d-c voltage between brushes in a single-phase converter is equal to $\sqrt{2}$ times the single-phase a-c voltage impressed across the slip rings.
9. How is the winding tapped in a three-phase converter?
10. How many taps must be made to the armature winding of a three-phase (*a*) four-pole converter? (*b*) six-pole converter?
11. Prove that the a-c voltage between slip rings in a three-phase converter is 0.612 times the d-c voltage between brushes.
12. How many taps must be made to the armature winding of a four-pole six-phase converter? How far apart are adjacent taps?
13. Prove that the d-c voltage in a six-phase converter is 2.83 times the a-c voltage between slip rings.
14. What advantages have six-phase converters compared with three-phase converters?

15. Explain carefully why the heating and output of a given converter changes as the number of phases increase.
16. How does the power factor at which a converter operates affect the heating and output of synchronous converters?
17. What important facts are indicated by a study of Table 11?
18. Describe the shunt-motor method of starting synchronous converters. What precautions must be taken when starting by this method?
19. Describe the auxiliary-motor method of starting synchronous converters. Why must the auxiliary induction motor have two poles fewer than the synchronous converter?
20. What two important functions are served by the squirrel cage when it is placed in the pole faces of a synchronous converter?
21. Describe carefully the exact procedure that must be followed in starting a converter when the squirrel cage in the pole faces is employed for this purpose.
22. In what direction does the armature turn, with respect to the revolving field, when a converter is started by the method described in Question 21?
23. Explain why the revolving field approaches zero speed as the armature approaches synchronous speed.
24. Upon what does the polarity at the d-c brushes depend when the converter finally reaches synchronous speed?
25. How is it possible to make a converter have a definite d-c brush polarity?
26. Referring to Fig. 320, explain how the synchronous converter is started on one-half of rated voltage and then connected to a rated-voltage source when the machine comes up to speed.
27. Why is it necessary to sectionalize the d-c field (see Fig. 321) when a synchronous converter is started by the method described in Question 21?
28. Why is it necessary to open the series-field diverter circuit when a converter is started by induction-motor action?
29. Give several reasons for the use of transformer equipment in connection with the operation of converters.
30. Indicate the four possible transformer connections for three-phase converters.
31. Why is it not possible to use a simple Y connection on the secondary side of the transformers when a three-phase converter must supply a three-wire d-c service?
32. Explain carefully how a zigzag transformer connection overcomes the difficulty described in Question 31.
33. What three connections may be used on the secondary side of a group of three transformers when changing from three- to six-phase?
34. Which of the connections in Question 33 must be used if the converter is to provide three-wire d-c service?
35. Why is the effect of armature reaction much less pronounced in synchronous converters than in equivalent d-c generators?
36. Explain why the commutators of synchronous converters are generally longer than those used on equivalent d-c generators.
37. Why is it desirable to have some degree of control over the d-c voltage of a converter?
38. What is the general effect of increasing the field excitation of a synchronous converter?
39. In what way is the d-c voltage controlled when the field excitation of a converter is changed? What per cent change is possible by this means?
40. What operating objections exist when the d-c voltage is controlled by field-excitation adjustment?

CONVERTERS AND RECTIFIERS

41. Explain how series reactance in the line wires may be used to control the d-c output voltage in a converter. What objection is there to this method?
42. Explain how the d-c voltage may be controlled by the use of a regulator.
43. Explain the principle of the synchronous booster in connection with d-c voltage control of converters.
44. Describe the construction of the synchronous booster converter, referring to Fig. 328.
45. Discuss the procedure that must be followed in connecting a synchronous converter in parallel with another such machine so that both may deliver power to a common load.
46. What precautions must be taken to assure stable operation when synchronous converters are connected in parallel?
47. Explain what would happen to two converters that are connected in parallel and feed a storage-battery load if the a-c power to one machine were suddenly interrupted.
48. What is the general practice to prevent a converter from "running away" when it is operated in parallel with others?
49. What methods may be used to alter the division of load between converters connected in parallel?
50. What is an *inverted converter*?
51. Does an inverted converter operate at synchronous speed? Explain.
52. Why is an inverted converter unstable when it delivers a lagging-power-factor load?
53. What is the general practice to prevent inverted converters from running away?
54. Explain how a separately excited inverter converter, with the exciter mounted on the converter shaft, tends to stablize its operation when it delivers a lagging power-factor load.
55. What is a *synchronous-synchronous frequency converter*? Under what conditions is it necessary to use such a machine?
56. What is an *induction frequency converter*?
57. Explain how it is possible to obtain a higher frequency than that of the source with an induction frequency converter.
58. Repeat Question 57 for a lower frequency.
59. What voltage is developed by an induction frequency converter when the frequency is increased? decreased?
60. Referring to Fig. 329, explain why the switch of the converter must be of the magnetic type and why it must be interconnected with the motor side of the starter for the driving motor.
61. Where are induction frequency converters generally used?
62. Describe the operation of a phase converter that changes single-phase to three-phase.
63. What practical application does the phase converter have?
64. Define a *rectifier*.
65. What useful function is performed by rectifiers?
66. List several general types of rectifier.
67. Distinguish between half-wave and full-wave rectification in circuits that are energized by a single-phase source.
68. Make sketches to illustrate full-wave rectification when: (a) two rectifiers are used; (b) four rectifiers are used to form a bridge circuit. List the advantages and disadvantages of each kind of connection.
69. Describe the construction of the *copper-oxide* (*dry-disk*) rectifier, and state the principle that governs its operation.

70. What determines the number of units that must be used and connected in series in each section of a full-wave copper-oxide rectifier?
71. Describe the construction of the *selenium* rectifier and state the principle that governs its operation.
72. What determines the area of the plates and the number of units that must be used in each section of full-wave selenium rectifier?
73. Describe the construction and the principle of operation of the *hot-cathode gaseous* rectifier. Distinguish between *anode* and *cathode* and show on a sketch the conventional direction of current flow in such a rectifier.
74. Make a complete wiring diagram showing how a Rectigon (or Tungar) is connected to a transformer and load for half-wave rectification.
75. Make a complete wiring diagram showing how two Rectigons (or Tungars) are connected to a transformer and load for full-wave rectification.
76. For what general kinds of service are Rectigons and Tungars employed?
77. What is meant by *tube drop*, why does it occur, and what is its approximate value in a Rectigon (or Tungar)?
78. Describe the general construction of a simple mercury-arc rectifier and indicate why its rating can be extended far beyond that possible with the hot-cathode gaseous type.
79. What special properties are possessed by mercury that makes it particularly suitable for use in an arc type of rectifier?
80. State the general principle that makes it possible to use two anodes in a single-envelope mercury-arc rectifier designed for full-wave rectification.
81. Make a sketch of a two-anode glass-enclosed mercury-arc rectifier and explain how it is started and how it operates to deliver a direct current to a load.
82. Explain how the *cathode spot* is formed at the surface of the mercury to sustain a continuous arc stream.
83. What is the importance of the comparatively high transformer leakage reactance in a mercury-arc rectifier system?
84. In a glass-enclosed mercury-arc rectifier, why is a large dome-shaped section provided in the envelope?
85. What is meant by *overlap* when referring to the operation of a two-anode mercury-arc rectifier?
86. Distinguish between the actual direct current in the anode and its effective value. How are they related in a two-anode mercury-arc rectifier?
87. What is the approximate range of arc drops in mercury-arc rectifiers? What determines whether its value is comparatively low or high?
88. Under what conditions is it desirable to employ multianode mercury-arc rectifiers that are energized by polyphase systems?
89. What connection is always used on the secondary side of the transformer (or transformers) in multianode rectifier systems?
90. In a multianode mercury-arc rectifier, what is the relation between the direct current in the load and (a) the average anode current? (b) the effective anode current?
91. During what part of a cycle does each anode fire in a multianode rectifier?
92. Make a wiring diagram showing a three-anode mercury-arc rectifier connected to a load and a three-phase transformer whose primaries are in Δ.
93. Repeat Question 92, but show three single-phase transformers properly connected instead of a three-phase transformer.
94. Make a sketch showing how the current varies in (a) one of the anodes of a three-anode rectifier; (b) each of the primaries of a Δ-connected three-phase transformer.

CONVERTERS AND RECTIFIERS 541

95. What advantages are possessed by a six-anode mercury-arc rectifier when compared with one having three anodes?
96. List the important operating conditions that prevail when a six-anode rectifier is delivering a load.
97. Make a wiring diagram showing a six-anode mercury-arc rectifier connected to a load and a three-phase transformer whose primaries are in Δ.
98. Draw diagrams illustrating the waveshapes in the anode circuits and the primary windings of the transformers for a six-anode mercury-arc rectifier.
99. What is meant by *arc-back*, and what operating condition is responsible for the phenomenon?
100. What constructional practices are employed in multianode mercury-arc rectifiers to prevent arc-back?
101. Describe the construction of the *ignitron*.
102. What important principle is responsible for the production of an arc at the surface of the mercury pool so that an ignitron can fire?
103. What important advantages are possessed by the ignitron as compared with multianode rectifiers?

Problems

1. It is desired to obtain 250 volts direct current from a single-phase converter. What a-c voltage must be used?
2. A single-phase converter has a rating of 2 kw and is operated from a 220-volt a-c source. Neglecting losses and assuming unity power factor, calculate: (a) the d-c output voltage and current; (b) the slip-ring current.
3. If the converter of Prob. 2 has an efficiency of 84 per cent and operates at a power factor of 0.92, calculate the slip-ring current.
4. A 7.5-kw single-phase inverted converter operates from a 240-volt d-c source. Assuming a full-load efficiency of 91 per cent and a load power factor of 0.83, calculate: (a) the slip-ring voltage; (b) the alternating-current output; (c) the direct-current input.
5. How many taps are brought out from the armature winding to the slip rings of: (a) a six-pole three-phase converter? (b) an eight-pole six-phase converter? As in Fig. 316, indicate by sketches how this is done.
6. A three-phase converter has a d-c rating of 120 kw and 600 volts. Assuming an efficiency of 94 per cent and a power factor of 0.95, calculate: (a) the direct-current output; (b) the slip-ring voltage; (c) the slip-ring current. (d) Also determine the kilovolt-ampere load on each of the transformers.
7. A 1,000-kw three-phase converter delivers full load at 600 volts. If the efficiency is 94 per cent and the power factor is 0.96, determine: (a) the direct-current output; (b) the a-c slip-ring current; (c) the kilovolt-ampere load delivered by each transformer.
8. A three-phase converter has a rating of 500 kw when operating at unity power factor. Neglecting any change in efficiency, what would be its rating when operating at a power factor of 0.80 (refer to Table 11, p. 493)?
9. A synchronous converter delivers 500-amp direct current. Assuming 100 per cent efficiency and unity power factor, determine the a-c slip-ring current if the converter has: (a) three rings; (b) six rings; (c) 12 rings.
10. A laboratory converter has the following rating: 10-kw six-phase, 120 volts d-c, 60 cycles, 1,800 rpm. Determine the following: (a) the number of poles in the machine; (b) the full-load direct-current output; (c) the a-c volts between adja-

cent slip rings; (d) the a-c slip-ring current, assuming an efficiency of 92 per cent and unity power factor.

11. What should be the kilovolt-ampere rating of each of the three transformers used in Prob. 10?

12. A 3,000-kw 750-volt d-c 60-cycle six-phase synchronous converter is fed from a three-phase 23,000-volt high-voltage line through a bank of three transformers that are connected in Δ on the high side and diametrical on the low side. (a) Make a winding diagram showing all connections for the problem (refer to Fig. 306). Assuming an efficiency and power factor of 96 per cent and 0.95 respectively, calculate: (b) the direct-current output; (c) the a-c voltage between adjacent slip rings; (d) the voltage of each transformer secondary coil; (e) the a-c slip-ring current; (f) the ratio of transformation of each of the three transformers.

13. A 4,000-kw 1,500-volt 12-phase converter operates at unity power factor and an efficiency of 96.5 per cent. Calculate: (a) the a-c voltage between adjacent slip rings; (b) the direct-current output; (c) the a-c slip-ring current.

14. A 250-kw six-phase 600-volt synchronous converter is operated as a d-c generator by being driven by a diesel engine. Assuming that normal temperature of the armature winding is not to be exceeded and that normal flux densities are to be maintained, calculate the permissible output current.

15. Referring to Table 11, p. 493, at what power factor should a six-phase converter operate to have the same relative output as a three-phase converter that operates at unity power factor?

16. Referring to Table 11, at what power factor should a 12-phase converter operate to have the same relative output as a three-phase converter that operates at unity power factor?

17. The three transformers that feed a three-phase converter are connected in Δ on the primary side and Y on the secondary side. If the slip-ring current is 471.5 amp, calculate: (a) the current in each transformer primary coil if the ratio of transformation is 21.7:1; (b) the line current on the primary side of the transformers.

18. If the slip-ring voltage in Prob. 17 is 367 volts, calculate: (a) the d-c voltage; (b) the voltage per transformer secondary coil; (c) the primary line voltage.

19. The d-c voltage of a three-phase synchronous booster converter is 600 volts when the field of the booster is unexcited. What maximum voltage must be generated in each phase of the booster winding if the voltage output is to be varied ± 10 per cent.

20. Solve Prob. 19 for a six-phase synchronous booster converter.

21. (a) What is the smallest number of poles that can be used in two synchronous machines if it is desired to convert from 60 to 50 cycles? (b) At what speed will the set operate?

22. A six-pole three-phase wound-rotor motor is to be used as an induction frequency converter. The stator is connected to a 60-cycle source. At what speed, and in what direction, must the rotor be driven if it is to develop: (a) 120 cycles? (b) 100 cycles? (c) 180 cycles? (d) 40 cycles?

23. If the voltage between slip rings in Prob. 22 is 120 at standstill, calculate the slip-ring voltages for the different frequencies.

24. A Tungar rectifier charges a storage battery at 5 amp and 6.4 volts. Assuming sinusoidal output variations for the half-wave rectifier and a tube drop of 10 volts, calculate the current and voltage ratings of the transformer secondary.

25. If the primary of the transformer in Prob. 24 is connected to a 118-volt source, compute the volt-ampere input, neglecting the exciting current.

CONVERTERS AND RECTIFIERS 543

26. An a-c ammeter (iron-vane type) registers 8.4 amp when placed in the load circuit of a half-wave Rectigon rectifier. What will be the deflection on a d-c ammeter (D'Arsonval type) when placed in the same circuit?

27. Two Tungar tubes are connected to a center-tapped transformer secondary for full-wave rectification and deliver a load current of 16 amp at 12.4 volts. (a) Draw a circuit diagram for the problem. Calculate: (b) the effective current in each half of the transformer secondary; (c) the voltage rating of the transformer secondary, assuming a tube drop of 12 volts.

28. A half-wave Rectigon rectifier delivers a d-c load of 10 amp at 20 volts. If the primary and secondary voltages of the transformer are, respectively, 115 and 62, calculate: (a) the tube drop; (b) the primary current, neglecting the exciting current.

29. Assuming a rectangular wave form, calculate the effective anode current in a two-anode mercury-arc rectifier for a d-c load of 50 amp.

30. A two-anode mercury-arc rectifier has an arc drop of 15 volts and delivers a d-c load at 120 volts. Calculate the maximum potential difference between cathode and anode.

31. Assuming rectangular wave forms, compute the effective anode currents for the following d-c loads in the given multianode mercury-arc rectifiers: (a) 75 kw, 230 volts, three anodes; (b) 1,500 kw, 1,200 volts, six anodes; (c) 5,000 kw, 2,400 volts, 12 anodes.

32. A two-anode mercury-arc rectifier delivers a load of 5.98 kw at 115 volts. If the primary of the transformer is connected to a 4,600-volt source, the tube drop is 15 volts, and a rectangular waveshape is assumed for the anode current, calculate: (a) the anode current; (b) the voltage rating of each of the transformer secondary windings; (c) the primary current, neglecting the exciting current; (d) the kilovolt-ampere rating of the transformer primary; (e) the kilovolt-ampere rating of the transformer secondary.

33. A three-anode mercury-arc rectifier supplies a 250-kw 230-volt load and is energized from a 4,600-volt three-phase source through a three-phase transformer whose primaries are in delta. Assuming an arc drop of 20 volts, calculate: (a) the d-c load current; (b) the voltage and kilovolt-ampere rating of each transformer secondary; (c) the current and kilovolt-ampere rating of the three-phase transformer.

34. A six-anode mercury-arc rectifier is served by a three-phase transformer having a rating of 5,000 kva and 13,800/930–1,860 volts. Assuming a 25-volt arc drop, calculate: (a) the d-c load voltage, current, and power; (b) the anode current.

APPENDIX 1

NATURAL SINES, COSINES, TANGENTS, AND COTANGENTS

0° to 2°   3° to 8°

	sin	cos	tan	cot			sin	cos	tan	cot	
0	.000000	1.00000	.00000	∞	90°	3°	.05234	.99863	.05241	19.081	87°
5'	1454	1.00000	145	687.55	55'	10'	524	847	533	18.075	50'
10'	2909	1.00000	291	343.77	50'	20'	.05814	831	.05824	17.169	40'
15'	4363	0.99999	436	229.18	45'	30'	.06105	813	.06116	16.350	30'
20'	5818	998	582	171.89	40'	40'	395	795	408	15.605	20'
25'	7272	997	727	137.51	35'	50'	685	776	700	14.924	10'
30'	.008727	996	.00873	114.59	30'	4°	.06976	.99756	.06993	14.301	86°
35'	.010181	995	.01018	98.218	25'	10'	.07266	736	.07285	13.727	50'
40'	1635	993	164	85.940	20'	20'	556	714	578	13.197	40'
45'	3090	991	309	76.390	15'	30'	.07846	692	.07870	12.706	30
50'	4544	989	455	68.750	10'	40'	.08136	668	.08163	12.251	20'
55'	5998	987	600	62.499	5'	50'	426	644	456	11.826	10'
1°	.017452	.99985	.01746	57.290	89°	5°	.08716	.99619	.08749	11.430	85°
5'	.01891	982	.01891	52.882	55'	10'	.09005	.99594	.09042	11.059	50'
10'	.02036	979	.02036	49.104	50'	20'	295	567	335	10.712	40'
15'	181	976	182	45.829	45'	30'	.0585	540	629	.385	30'
20'	327	973	328	42.964	40'	40'	.9874	511	.09923	10.078	20'
25'	472	969	473	40.436	35'	50'	.10164	482	.10216	9.7882	10'
30'	618	966	619	38.188	30'	6°	.10453	.99452	.10510	9.5144	84°
35'	763	962	764	36.178	25'	10'	.10742	421	.10805	.2553	50'
40'	.02908	958	.02910	34.368	20'	20'	.11031	390	.11099	9.0098	40'
45'	.03054	953	.03055	32.730	15'	30'	320	357	394	8.7769	30'
50'	199	949	201	31.242	10'	40'	609	324	688	.5555	20'
55'	345	944	346	29.882	5'	50'	.11898	290	.11983	.3450	10'
2°	.03490	.99939	.03492	28.636	88°	7°	.12187	.99255	.12278	8.1443	83°
5'	635	934	638	27.490	55'	10'	476	219	574	7.9530	50'
10'	781	929	783	26.432	50'	20'	.12764	182	.12869	.7704	40'
15'	.03926	923	.03929	25.452	45'	30'	.13053	144	.13165	.5958	30'
20'	.04071	917	.04075	24.542	40'	40'	341	106	461	.4287	20'
25'	217	911	220	23.695	35'	50'	629	067	.13758	.2687	10'
30'	362	905	366	22.904	30'	8°	.13917	.99027	.14054	7.1154	82°
35'	507	898	512	22.164	25'	10'	.14205	.98986	351	6.9682	50'
40'	653	892	658	21.470	20'	20'	493	944	648	.8269	40'
45'	798	885	803	20.819	15'	30'	.14781	902	.14945	.6912	30'
50'	.04943	878	949	20.206	10'	40'	.15069	858	.15243	.5606	20'
55'	.05088	870	.05095	19.627	5'	50'	356	814	540	.4348	10'
	cos	sin	cot	tan			cos	sin	cot	tan	

88° to 90°   82° to 87°

544

APPENDIX

Natural Sines, Cosines, Tangents, and Cotangents (*Continued*)

9° to 14° 15° to 20°

	sin	cos	tan	cot			sin	cos	tan	cot	
9°	.15643	.98769	.15838	6.3138	81°	15°	.25882	.96593	.26795	3.7321	75°
10′	.15931	723	.16137	.1970	50′	10′	.26163	517	.27107	.6891	50′
20′	.16218	676	435	6.0844	40′	20′	443	440	419	.6470	40′
30′	505	629	.16734	5.9758	30′	30′	.26724	363	.27732	.6059	30′
40′	.16792	580	.17033	.8708	20′	40′	.27004	285	.28046	.5656	20′
50′	.17078	531	333	.7694	10′	50′	284	206	360	.5261	10′
10°	.17365	.98481	.17633	5.6713	80°	16°	.27564	.96126	.28675	3.4874	74°
10′	651	430	.17933	.5764	50′	10′	.27843	.96046	.28990	.4475	50′
20′	.17937	378	.18233	.4845	40′	20′	.28123	.95964	.29305	.4124	40′
30′	.18224	325	534	.3955	30′	30′	402	882	621	.3759	30′
40′	509	272	.18835	.3093	20′	40′	680	799	.29938	.3402	20′
50′	.18795	218	.19136	.2257	10′	50′	.28959	715	.30255	.3052	10′
11°	.19081	.98163	.19438	5.1446	79°	17°	.29237	.95630	.30573	3.2709	73°
10′	366	107	.19740	5.0658	50′	10′	515	545	.30891	.2371	50′
20′	652	.98050	.20042	4.9894	40′	20′	.29793	459	.31210	.2041	40′
30′	.19937	.97992	345	.9152	30′	30′	.30071	372	530	.1716	30′
40′	.20222	934	648	.8430	20′	40′	348	284	.31850	.1397	20′
50′	507	875	.20952	.7729	10′	50′	625	195	.32171	.1084	10′
12°	.20791	.97815	.21256	4.7046	78°	18°	.30902	.95106	.32492	3.0777	72°
10′	.21076	754	560	.6382	50′	10′	.31178	.95015	.32814	.0475	50′
20′	360	692	.21864	.5736	40′	20′	454	.94924	.33136	3.0178	40′
30′	644	630	.22169	.5107	30′	30′	.31730	832	460	2.9887	30′
40′	.21928	566	475	.4494	20′	40′	.32006	740	.33783	.9600	20′
50′	.22212	502	.22781	.3897	10′	50′	282	646	.34108	.9319	10′
13°	.22495	.97437	.23087	4.3315	77°	19°	.32557	.94552	.34433	2.9042	71°
10′	.22778	371	393	.2747	50′	10′	.32832	457	.34758	.8770	50′
20′	.23062	304	.23700	.2193	40′	20′	.33106	361	.35085	.8502	40′
30′	345	237	.24008	.1653	30′	30′	381	264	412	.8239	30′
40′	627	169	316	.1126	20′	40′	655	167	.35740	.7980	20′
50′	.23910	100	624	.0611	10′	50′	.33929	.94068	.36068	.7725	10′
14°	.24192	.97030	.24933	4.0108	76°	20°	.34202	.93969	.36397	2.7475	70°
10′	474	.96959	.25242	3.9617	50′	10′	475	869	.36727	.7228	50′
20′	.24756	887	552	.9136	40′	20′	.34748	769	.37057	.6985	40′
30′	.25038	815	.25862	.8667	30′	30′	.35021	667	388	.6746	30′
20′	320	742	.26172	.8208	20′	40′	293	565	.37720	.6511	20′
50′	601	667	483	.7760	10′	50′	565	462	.38053	.6279	10′
	cos	sin	cot	tan			cos	sin	cot	tan	

76° to 81° 70° to 75°

ELECTRICAL MACHINES

Natural Sines, Cosines, Tangents, and Cotangents (*Continued*)

21° to 26°

	sin	cos	tan	cot	
21°	.35837	.93358	.38386	2.6051	69°
10'	.36108	253	.38721	.5826	50'
20'	379	148	.39055	.5605	40'
30'	650	.93042	391	.5386	30'
40'	.36921	.92935	.39727	.5172	20'
50'	.37191	827	.40065	.4960	10'
22°	.37461	.92718	.40403	2.4751	68°
10'	730	609	.50741	.4545	50'
20'	.37999	499	.41081	.4342	40'
30'	.38268	388	421	.4142	30'
40'	537	276	.41763	.3945	20'
50'	.38805	164	.42105	.3750	10'
23°	.39073	.92050	.42447	2.3559	67°
10'	341	.91936	.42791	.3369	50'
20'	608	822	.43136	.3183	40'
30'	.39875	706	481	.2998	30'
40'	.40141	590	.43828	.2817	20'
50'	408	472	.44175	.2637	10'
24°	.40674	.91355	.44523	2.2460	66°
10'	.40939	236	.44872	.2286	50'
20'	.41204	.91116	.45222	.2113	40'
30'	469	.90996	573	.1943	30'
40'	734	875	.45924	.1775	20'
50'	.41998	753	.46277	.1609	10'
25°	.42262	.90631	.46631	2.1445	65°
10'	525	507	.46985	.1283	50'
20'	.42788	383	.47341	.1123	40'
30'	.43051	259	.47698	2.0965	30'
40'	313	133	.48055	809	20'
50'	575	.90007	414	655	10'
26°	.43837	.89879	.48773	2.0503	64°
10'	.44098	752	.49134	353	50'
20'	359	623	495	204	40'
30'	620	493	.49858	2.0057	30'
40'	.44880	363	.50222	1.9912	20'
50'	.45140	232	587	768	10'
	cos	sin	cot	tan	

64° to 69°

27° to 33°

	sin	cos	tan	cot	
27°	.45399	.89101	.50953	1.9626	63°
10'	658	.88968	.51319	486	50'
20'	.45917	835	.51688	347	40'
30'	.46175	701	.52057	210	30'
40'	433	566	427	.9074	20'
50'	690	431	.52798	1.8940	10'
28°	.46947	.88295	.53171	1.8807	62°
10'	.47204	158	545	676	50'
20'	460	.88020	.53920	546	40'
30'	716	.87882	.54296	418	30'
40'	.47971	743	.54673	291	20'
50'	.48226	603	.55051	165	10'
29°	.48481	.87462	.55431	1.8040	61°
10'	735	321	.55812	1.7917	50'
20'	.48989	178	.56194	796	40'
30'	.49242	.87036	577	675	30'
40'	495	.86892	.56962	556	20'
50'	.49748	748	.57348	437	10'
30°	.50000	.86603	.57735	1.7321	60°
10'	252	457	.58124	205	50'
20'	503	310	513	.7090	40'
30'	.50754	163	.58905	1.6977	30'
40'	.51004	.86015	.59297	864	20'
50'	254	.85866	.59691	753	10'
31°	.51504	.85717	.60086	1.6643	59°
10'	.51763	567	483	534	50'
20'	.52002	416	.60881	426	40'
30'	250	264	.61280	319	30'
40'	498	.85112	.61681	212	20'
50'	745	.84959	.62083	107	10'
32°	.52992	.84805	.62487	1.6003	58°
10'	.53238	650	.62892	1.5900	50'
20'	484	495	.63299	798	40'
30'	730	339	.63707	697	30'
40'	.53975	182	.64117	597	20'
50'	.54220	.84025	528	497	10'
	cos	sin	cot	tan	

58° to 63°

APPENDIX

Natural Sines, Cosines, Tangents, and Cotangents (*Continued*)

33° to 38°   39° to 45°

	sin	cos	tan	cot			sin	cos	tan	cot	
33°	.54464	.83867	.64941	1.5399	57°	39°	.62932	.77715	.80978	1.2349	51°
10'	708	708	.65355	301	50'	10'	.63158	531	.81461	276	50'
20'	.54951	549	.65771	204	40'	20'	383	347	.81946	203	40'
30'	.55194	389	.66189	108	30'	30'	608	.77162	.82434	131	30'
40'	436	228	.66608	.5013	20'	40'	.63832	.76977	.82923	.2059	20'
50'	678	.83066	.67028	1.4919	10'	50'	.64056	791	.83415	1.1988	10'
34°	.55919	.82904	.67451	1.4826	56°	40°	.64279	.76604	.83910	1.1918	50°
10'	.56160	741	.67875	733	50'	10'	501	417	.84407	847	50'
20'	401	577	.68301	641	40'	20'	723	229	.84906	778	40'
30'	641	413	.68728	550	30'	30'	.64945	.76041	.85408	708	30'
40'	.56880	248	.69157	460	20'	40'	.65166	.75851	.85912	640	20'
50'	.57119	.82082	.69588	370	10'	50'	386	661	.86419	571	10'
35°	.57358	.81915	.70021	1.4281	55°	41°	.65606	.75471	.86929	1.1504	49°
10'	596	748	455	193	50'	10'	.65825	280	.87441	436	50'
20'	.57833	580	.70891	106	40'	20'	.66044	.75088	.87955	369	40'
30'	.58070	412	.71329	.4019	30'	30'	262	.74896	.88473	303	30'
40'	307	242	.71769	1.3934	20'	40'	480	703	.88992	237	20'
50'	543	.81072	.72211	848	10'	50'	697	509	.89515	171	10'
36°	.58779	.80902	.72654	1.3764	54°	42°	.66913	.74314	.90040	1.1106	48°
10'	.59014	730	.73100	680	50'	10'	.67129	.74120	.90569	1.1041	50'
20'	248	558	547	597	40'	20'	344	.73924	.91099	1.0977	40'
30'	482	386	.73996	514	30'	30'	559	728	.91633	913	30'
40'	716	212	.74447	432	20'	40'	773	531	.92170	850	20'
50'	.59949	.80038	.74900	351	10'	50'	.67987	333	.92709	786	10'
37°	.60182	.79864	.75355	1.3270	53°	43°	.68200	.73135	.93252	1.0724	47°
10'	414	688	.75812	190	50'	10'	412	.72937	.93707	661	50'
20'	645	512	.76272	111	40'	20'	624	737	.94345	599	40'
30'	.60876	335	.76733	.3032	30'	30'	.68835	537	.94896	538	30'
40'	.61107	.79158	.77196	1.2954	20'	40'	.69046	337	.95451	477	20'
50'	337	.78980	.77661	876	10'	50'	256	.72136	.96008	416	10'
38°	.61566	.78801	.78129	1.2799	52°	44°	.69466	.71934	.96569	1.0355	46°
10'	.61795	622	.78598	723	50'	10'	675	732	.97133	295	50'
20'	.62024	442	.79070	647	40'	20'	.69883	529	.97700	235	40'
30'	251	261	.79544	572	30'	30'	.70091	325	.98270	176	30'
40'	479	.78079	.80020	497	20'	40'	298	.71121	.98843	117	20'
50'	706	.77897	498	423	10'	50'	505	.70916	.99420	058	10'
						45°	.70711	.70711	1.00000	1.0000	45°
	cos	sin	cot	tan			cos	sin	cot	tan	

52° to 57°   45° to 51°

APPENDIX 2

LOGARITHMS OF NUMBERS

Number	0	1	2	3	4	5	6	7	8	9
10	0000	0043	0086	0128	0170	0212	0253	0294	0334	0374
11	0414	0453	0492	0531	0569	0607	0645	0682	0719	0755
12	0792	0828	0864	0899	0934	0969	1004	1038	1072	1106
13	1139	1173	1206	1239	1271	1303	1335	1367	1399	1430
14	1461	1492	1523	1553	1584	1614	1644	1673	1703	1732
15	1761	1790	1818	1847	1875	1903	1931	1959	1987	2014
16	2041	2068	2095	2122	2148	2175	2201	2227	2253	2279
17	2304	2330	2355	2380	2405	2430	2455	2480	2504	2529
18	2553	2577	2601	2625	2648	2672	2695	2718	2742	2765
19	2788	2810	2833	2856	2878	2900	2923	2945	2967	2989
20	3010	3032	3054	3075	3096	3118	3139	3160	3181	3201
21	3222	3243	3263	3284	3304	3324	3345	3365	3385	3404
22	3424	3444	3464	3483	3502	3522	3541	3560	3579	3598
23	3617	3636	3655	3674	3692	3711	3729	3747	3766	3784
24	3802	3820	3838	3856	3874	3892	3909	3927	3945	3962
25	3979	3997	4014	4031	4048	4065	4082	4099	4116	4133
26	4150	4166	4183	4200	4216	4232	4249	4265	4281	4298
27	4314	4330	4346	4362	4378	4393	4409	4425	4440	4456
28	4472	4487	4502	4518	4533	4548	4564	4579	4594	4609
29	4624	4639	4654	4669	4683	4698	4713	4728	4742	4757
30	4771	4786	4800	4814	4829	4843	4857	4871	4886	4900
31	4914	4928	4942	4955	4969	4983	4997	5011	5024	5038
32	5051	5065	5079	5092	5105	5119	5132	5145	5159	5172
33	5185	5198	5211	5224	5237	5250	5263	5276	5289	5302
34	5315	5328	5340	5353	5366	5378	5391	5403	5416	5428
35	5441	5453	5465	5478	5490	5502	5514	5527	5539	5551
36	5563	5575	5587	5599	5611	5623	5635	5647	5658	5670
37	5682	5694	5705	5717	5729	5740	5752	5763	5775	5786
38	5798	5809	5821	5832	5843	5855	5866	5877	5888	5899
39	5911	5922	5933	5944	5955	5966	5977	5988	5999	6010
40	6021	6031	6042	6053	6064	6075	6085	6096	6107	6117
41	6128	6138	6149	6160	6170	6180	6191	6201	6212	6222
42	6232	6243	6253	6263	6274	6284	6294	6304	6314	6325
43	6335	6345	6355	6365	6375	6385	6395	6405	6415	6425
44	6435	6444	6454	6464	6474	6484	6493	6503	6513	6522
45	6532	6542	6551	6561	6571	6580	6590	6599	6609	6618
46	6628	6637	6646	6656	6665	6675	6684	6693	6702	6712
47	6721	6730	6739	6749	6758	6767	6776	6785	6794	6803
48	6812	6821	6830	6839	6848	6857	6866	6875	6884	6893
49	6902	6911	6920	6928	6937	6946	6955	6964	6972	6981

APPENDIX

Logarithms of Numbers (*Continued*)

Number	0	1	2	3	4	5	6	7	8	9
50	6990	6998	7007	7016	7024	7033	7042	7050	7059	7067
51	7076	7084	7093	7101	7110	7118	7126	7135	7143	7152
52	7160	7168	7177	7185	7193	7202	7210	7218	7226	7235
53	7243	7251	7259	7267	7275	7284	7292	7300	7308	7316
54	7324	7332	7340	7348	7356	7364	7372	7380	7388	7396
55	7404	7412	7419	7427	7435	7443	7451	7459	7466	7474
56	7482	7490	7497	7505	7513	7520	7528	7536	7543	7551
57	7559	7566	7574	7582	7589	7597	7604	7612	7619	7627
58	7634	7642	7649	7657	7664	7672	7679	7686	7694	7701
59	7709	7716	7723	7731	7738	7745	7752	7760	7767	7774
60	7782	7789	7796	7803	7810	7818	7825	7832	7839	7846
61	7853	7860	7868	7875	7882	7889	7896	7903	7910	7917
62	7924	7931	7938	7945	7952	7959	7966	7973	7980	7987
63	7993	8000	8007	8014	8021	8028	8035	8041	8048	8055
64	8062	8069	8075	8082	8089	8096	8102	8109	8116	8122
65	8129	8136	8142	8149	8156	8162	8169	8176	8182	8189
66	8195	8202	8209	8215	8222	8228	8235	8241	8248	8254
67	8261	8267	8274	8280	8287	8293	8299	8306	8312	8319
68	8325	8331	8338	8344	8351	8357	8363	8370	8376	8382
69	8388	8395	8401	8407	8414	8420	8426	8432	8439	8445
70	8451	8457	8463	8470	8476	8482	8488	8494	8500	8506
71	8513	8519	8525	8531	8537	8543	8549	8555	8561	8567
72	8573	8579	8585	8591	8597	8603	8609	8615	8621	8627
73	8633	8639	8645	8651	8657	8663	8669	8675	8681	8686
74	8692	8698	8704	8710	8716	8722	8727	8733	8739	8745
75	8751	8756	8762	8768	8774	8779	8785	8791	8797	8802
76	8808	8814	8820	8825	8831	8837	8842	8848	8854	8859
77	8865	8871	8876	8882	8887	8893	8899	8904	8910	8915
78	8921	8927	8932	8938	8943	8949	8954	8960	8965	8971
79	8976	8982	8987	8993	8998	9004	9009	9015	9020	9025
80	9031	9036	9042	9047	9053	9058	9063	9069	9074	9079
81	9085	9090	9096	9101	9106	9112	9117	9122	9128	9133
82	9138	9143	9149	9154	9159	9165	9170	9175	9180	9186
83	9191	9196	9201	9206	9212	9217	9222	9227	9232	9238
84	9243	9248	9253	9258	9263	9269	9274	9279	9284	9289
85	9294	9299	9304	9309	9315	9320	9325	9330	9335	9340
86	9345	9350	9355	9360	9365	9370	9375	9380	9385	9390
87	9395	9400	9405	9410	9415	9420	9425	9430	9435	9440
88	9445	9450	9455	9460	9465	9469	9474	9479	9484	9489
89	9494	9499	9504	9509	9513	9518	9523	9528	9533	9538

LOGARITHMS OF NUMBERS (*Continued*)

Number	0	1	2	3	4	5	6	7	8	9
90	9542	9547	9552	9557	9562	9566	9571	9576	9581	9586
91	9590	9595	9600	9605	9609	9614	9619	9624	9628	9633
92	9638	9643	9647	9652	9657	9661	9666	9671	9675	9680
93	9685	9689	9694	9699	9703	9708	9713	9717	9722	9727
94	9731	9736	9741	9745	9750	9754	9759	9763	9768	9773
95	9777	9782	9786	9791	9795	9800	9805	9809	9814	9818
96	9823	9827	9832	9836	9841	9845	9850	9854	9859	9863
97	9868	9872	9877	9881	9886	9890	9894	9899	9903	9908
98	9912	9917	9921	9926	9930	9934	9939	9943	9948	9952
99	9956	9961	9965	9969	9974	9978	9983	9987	9991	9996

INDEX

A

Action, generator, 1, 23
  motor, 35
Adjustment, degree of compounding, 108, 109, 196
All-day efficiency, 139, 183
Alnico magnets, 40
Alternating-current generators (*see* Alternators)
Alternators, 10, 29, 225
  armature windings in, 236
  construction of, 225
  efficiency of, 257, 258
  frequency of, 229
  generated voltage in, 233
  hunting of, 263
  losses in, 257
  open-circuit test of, 255
  parallel operation of, 259
  phasor diagram for, 251
  regulation of, 243, 245, 252
  resistance test of, 254
  short-circuit test of, 255
  stator of, 232
  steam-driven, 226
  synchronizing of, 259
  voltage of, 242
  voltage drops in, 246
  windings in, 236
Ammeter, clamp-on, 311
Amortisseur, 475
Applications, dynamo, 113, 183, 196, 215
Arago, D. F., 346
Armature, voltage drops in, 246
Armature core, 49
Armature reactance, 244, 248, 405
Armature reaction, 199, 204
  in alternators, 244, 246
    voltage drop, 248

Armature reaction, in converters, 505
  in direct-current generators, 113
  in direct-current motors, 172
  pole laminations to counteract, 116
  in series motors, 405
Armature resistance, 187, 244, 246
Armature reversing, 175, 176
Armature windings, 1, 6, 11, 45, 51
  circulating currents in, 66
  coil span in, 55
  dead elements in, 76
  doubly reentrant, 58
  duplex-lap, 58
  frog-leg, 80
  lap, 53
  multielement coil, 71
  multiplex, 59, 68
  parallel paths in, 51, 59, 65
  simplex-lap, 53, 59, 62, 73
  simplex-wave, 51, 61, 65, 67, 76
  singly reentrant, 58
  types of, 53
  wave, 53
Automatic controllers, 394
Automatic starter, 149, 177
Autotransformers, 19, 305
  conducted power in, 307
  transformer power in, 307

B

Balance coil, 197
Bearings, 45
Booster, constant-current, 207
  series, 205, 507
  track return, 206
Braking, 135
  dynamic, 210, 393
  electrical, 207
  regenerative, 212

Brush rigging, 45, 52
Brush shifting, 172
Brushes, 13
Building up, conditions for, 98

C

Capacitors, 426
Centrifugal switch, 422, 423, 432
Characteristics, d-c generator, 94
Circuit, transformer, 293
Clamp-on ammeter, 311
Coil, balance, 197
  double-element, 72
  frog-leg, 56
  reactance, 198
  triple-element, 75
Coil span, 55
Commutating poles, 13, 47, 116, 172, 505
Commutation, 32, 39, 114, 121, 123, 172
Commutator, 11, 13, 49
  cross-connected, 80
  sparking of, 195
  undercutting of, 51
Commutator pitch, 58
Commutator segment, 49
Compensating windings, 119, 174
Compensator, 372
Compounding, degree of, 108, 109, 196
  generator, 107
Concatenation, 388
Condenser, synchronous, 466, 467
Connections, delta, 238, 352, 487
  equalizer, 78, 81
  star, 237, 352
Consequent-pole windings, 389
Construction, generator and motor, 45, 225
Control, speed (*see* Speed control)
  Ward Leonard, 166, 169
Controllers, automatic, 394
  compound-motor, 147
  drum, 214
  series-motor, 146
  shunt-motor, 147
Converters, 1, 481
  armature reaction in, 505
  commutating poles in, 505
  control of, 506
    induction regulator, 507
    series-reactance, 506

Converters, control of, series-resistance, 506
  conversion ratios, 491
  currents in, 492
  field sectionalizing of, 497
  frequency, 1, 5, 474, 481, 511
  heating of, 490, 493
  induction frequency, 511, 512
  inverted, 570
  parallel operation of, 509
  phase, 481, 514
  polyphase, 486
  ratings of, 490, 493
  relative heating of, 493
  relative output of, 493
  rotary, 1, 481
  series booster, 507
  single-phase, 483
  six-phase, 488, 501, 503
  starting of, 494
  synchronous, 481
    construction of, 482
    starting of, 494
      auxiliary-motor, 494
      induction-motor, 494
      reduced-voltage, 496
    transformer connections for, 497
    zigzag, 500
    transformers for, 497
      delta-star connections of, 502
      zigzag connections of, 500
  synchronous booster, 507
  synchronous-synchronous frequency, 511
  three-phase, 486, 498
  transformer connections for, 497
  transformers for, 497
    Δ-Y connections of, 502
    zigzag connections of, 500
  two-wire d-c, 498
  types of, 481
  uses of, 481
  voltage control of, 506
Copper-oxide rectifiers, 518
Core, armature, 49
Current, rotor, 360

D

Damping, synchronous-motor, 474
Delta connections, 238, 352, 487

INDEX

553

Diactor regulator, 104
Direct-current generator (*see* Generators, direct-current)
Distortion, flux, 115, 199
Distribution factor, 241
Diverter, series-field, 109
Diverter-pole generators, 200
Double-cage rotor, 370
Dynamic braking, 210, 393
Dynamos, 45
  applications of, 113, 183, 196, 215
  efficiency of, 183
  power loss in, 183
  rating of, 183
  special, 196
Dynamotors, 203

E

Eddy-current loss, 184, 257
Eddy currents, 297, 301
Efficiency, 139, 183
  all-day, 304
  alternator, 257
  conventional, 187, 192
  generator, 186
  importance of, 193
  induction-motor, 364
  motor, 190
  transformer, 301, 303
Electrical braking, 207
Electrolytic rectifiers, 516
Electromagnetic induction, 268
Equalizer, 128
Equalizer connections, 78, 81
Excitation, 41
Exciter, 1
  pilot, 228
  synchronous-motor, 452

F

Factor, distribution, 241
  pitch, 239
Faraday's law, 23
Field, resultant, 356
  revolving, 232, 354
  salient-pole, 6
  series, 47
  shunt, 47
Field flashing, 100
Field poles, 6, 45

Field reversing, 175
Field structures, 2
Fields, main, 40
Flashing, field, 100
Flux, armature reaction, 199
  leakage, 72, 286
  mutual, 285, 294
Flux distortion, 115, 199
Force, 37
Frequency, 30
  rotor, 359
Frequency converter, 1, 5, 474, 481, 511
Frog-leg coil, 56
Frog-leg windings, 69, 80
Full-voltage starting, 370

G

Gears, 407
Generated voltage, direction of, 27
Generator, action, 1, 23
Generator principles, 23
Generators, 1
  alternating-current, 10, 29, 225
    armature windings in, 236
    construction of, 45, 225
    efficiency of, 257, 258
    frequency of, 229
    generated voltage in, 233
    hunting of, 263
    losses in, 257
    open-circuit test of, 255
    parallel operation of, 258
    phasor diagram for, 251
    regulation of, 243, 245, 252
    resistance test of, 254
    short-circuit test of, 255
    stators, 232
    steam-driven, 226
    synchronizing of, 259
    voltage of, 242
    voltage drops in, 246
    waterwheel, 226
    windings in, 236
  direct-current, 24
    applications of, 183, 215
    armature reaction in, 113
    automobile, 9
    building up of, 98
    characteristics of, 94
      no-load, 95

554 ELECTRICAL MACHINES

Generators, direct-current, compensating windings for, 119, 174
  compound, 8, 94, 106, 198, 201
    cumulative, 107
    degree of compounding of, 108, 109, 196
    differential, 107
  construction of, 45
  diverter-pole, 200
  efficiency of, 186
  flat-compound, 108
  flux distribution in, 115
  interpoles for, 116
  load on, 134
  operation of, 134
  overcompound, 108
  parallel operation of, 124, 129
  principles of, 23
  rating of, 195
  regulation of, 103
  selection of, 196
  series, 94, 111, 170, 204
  shunt, 7, 94, 198
    building up of, 98
    parallel operation of, 125
    self-excited, 105
    separately excited, 7, 96
    test on, 96, 101
    voltage control of, 104
  speed of, 9
  testing of, 187
  third-brush, 198
  three-wire, 196
  train, 9
  types of, 7
  undercompound, 108
  voltage of, 134
  voltage characteristics of, 9
  voltage equation for, 26
  diverter-pole, 200
  steam-driven, 226
  third-brush, 198
  three-wire, 196
  train, 9
  waterwheel, 226

H

Harmonics, 239
Hunting, alternator, 263
  synchronous motor, 474

Hysteresis, 184
Hysteresis loss, 184, 257, 297, 301

I

Ignitron, 535
Impedance, equivalent, 291
  synchronous, 253
  transformer, 290
Induction, electromagnetic, 268
  mutual, 269
Induction frequency converter, 511, 512
Induction motors (*see* Motors, induction)
Induction-voltage regulator, 336
Inerteen, 430
Insulation, transformer, 274
Interpoles, 13, 47, 116, 172
Inverted converter, 570

L

Laminations, 49, 298
  squirrel-cage, 270
Lap windings, 53, 58, 351
  equalizer connections in, 78, 81
Leakage flux, 72, 286
Leakage reactance, 287
Lenz's law, 28, 288, 410
Load, 9
Loading, transformer, 283
Losses, 188
  brush, 185, 257
  copper, 185, 257
  dynamo, 186
  eddy-current, 184, 25⁻
  electrical, 184, 257
  exciter, 257
  friction, 184, 257, 297, 301
  hysteresis, 184, 257, 297, 301
  rotational, 184, 257
  stray-load, 257
  stray-power, 189
  ventilation, 257
  windage, 184, 257

M

Machines, 1
  alternating-current, 10, 29, 225
  compound, 41
  direct-current, 21

INDEX

Machines, direct-current, commutating poles for, 13
   main fields in, 40
   rotating electrical, 1
   series, 41
   shunt, 41
Magnetic neutral, 172
Magnetism, residual, 97
Magnetization curve, 97
Magnets, alnico, 40
   permanent, 40
Measurement, armature-resistance, 187
   interpole-winding resistance, 188
   rotational loss, 188
   series-field resistance, 188
   shunt-field resistance, 188
Mercury-arc rectifiers (*see* Rectifiers, mercury-arc)
Motor action, 35
Motor principles, 23
Motors, 1
   adjustable-speed, 136, 168
   alternating-current, braking of, 392
      starting of, 17
      types of, 14, 401
   applications of, 183, 215
   automatic controller for, 394
   automatic starters for, 149
   braking of, 135, 207, 392
   compound, 11, 141
      dynamic braking of, 212
      plugging of, 209
      starters for, 142, 149, 177
      torque of, 156
   constant-speed, 136, 163
   construction of, 45
   control of, 154, 168, 214
      series-parallel, 214
      Ward Leonard, 166
   controllers for, 146
   differential-compound, 164
   direct-current, armature current in, 172
      braking of, 207
      brush shifting in, 172
      characteristics of, 134
      classification of, 135
      commutation in, 39
      compensating windings for, 119, 172
      constant-speed, 163
      construction of, 45

Motors, direct-current, control of, 168
      dynamic braking of, 210
      efficiency of, 190
      plugging of, 208
      reversing of, 175
      speed characteristics of, 159
      speed control of, 165
      speed regulation of, 163
      starting of, 12, 140
      torque characteristics of, 154
      types of, 11
   efficiency of, 190, 364
   fractional-horsepower, 14, 141, **421**
   hysteresis, 441
   induction, 346
      applications of, 382
      automatic starter for, 371
      blocked-rotor test of, 366
      classes of, 382
      concatenation of, 388
      dynamic braking of, 393
      efficiency of, 364
      frequency method of control of, 391
      no-load test of, 364
      plugging of, 393, 395
      principle of, 346
      rotor of, 350
      rotor current in, 360
      rotor frequency of, 359
      rotor power of, 360
      rotor torque of, 362
      rotor voltage in, 359
      shaftless, 383
      speed control of, 380, 387, 388
      starting of, 368, 370
      starting current in, 382
      starting torque of, 364, 382
      stator of, 308, 381
      stator resistance test for, 364
      two-speed, 391
      winding of, 351, 381, 387
      wound-rotor, 380
         operating characteristics of, 386
         speed control of, 387
   interpole, 175
   loading of, 153
   parallel-connected, 135
   plugging of, 208, 393, 395
   polyphase, 346
      classification of, 15
   rating of, 195

Motors, reluctance, 440
    synchronous, 440
  repulsion, 436, 438
    principle of, 433
  repulsion induction, 412, 439
  repulsion-start, 431
  reversing of, 175
  Schrage, 392
  selection of, 135, 196, 217
  series, 11, 15, 141, 170, 401
    controllers for, 146
    large, 403
    small, 404
    torque of, 156
  series-parallel operation of, 214
  shaded-pole, 15, 408
  shaftless, 383
  shunt, 11, 141
    plugging of, 208
    starters for, 142, 149
    torque of, 156
  single-phase, 401
    classification of, 14
    hysteresis, 441
    reluctance, 440, 473
      synchronous, 440
    reluctance-start, 415
    repulsion, 402, 436, 438
    repulsion induction, 412, 439
    repulsion-start, 402, 431
      performance of, 437
    series, 403
    shaded-pole, 408
      characteristics of, 411
      principle of, 410
      speed control of, 413
      torque of, 414
    split-phase (*see* split-phase, *below*)
  small, 401
  speed characteristics of, 159
  speed control of, 154, 165, 380, 387, 388, 391
  speed regulation of, 163
  split-phase, 402, 407
    automatic starters for, 443
    auxiliary winding in, 417, 422
    capacitor, 414, 419
    capacitor-start, 419, 424, 427
    centrifugal switch for, 422
    concentric winding for, 419
    electromagnetic relay for, 423

Motors, split-phase, hoist control for, 427
    main winding in, 417, 422
    reversing of, 431
    single-value capacitor, 430
    standard, 419
    two-value capacitor, 428
    starter for, 443
  squirrel-cage, 15, 370
    double, 370, 382
    operating characteristics of, 380
  starters for, 142
    four-point, 145
    three-point, 143
  starting of (*see* Starting)
  starting torque of, 364, 381
  supersynchronous, 475
  synchronous, 4, 9, 15, 402, 440, 449, 482
    amortisseur for, 475
    applications of, 476
    automatic starter for, 456
    construction of, 450
    damping of, 474
    dual-purpose, 470
    exciters for, 452
    high-starting-torque, 458
    hunting of, 474
    induction-motor starting of, 453
    loading of, 460
    power-factor adjustment of, 464
    principle of operation of, 459
    simplex, 458
    speed of, 462
    starting of, 453
  synchronous-induction, 472
  synchronous reluctance, 440
  Telechron, 442
  universal, 401
    applications of, 407
    characteristics of, 404
    constant-speed governor for, 408
    geared, 407
    principle of, 402
    speed of, 407
  variable-speed, 136, 163
  wound-rotor, 379, 386

P

Phase converters, 481, 514
Pitch, coil, 74, 239

INDEX

Pitch, commutator, 58
  fractional, 239
Pitch factor, 239
Plugging, 208, 393, 395
Polarity, additive, 314
  subtractive, 314
  transformer, 313
Pole shoe, 45
Poles, commutating, 13, 47, 116, 172, 505
  field, 6
Power, developed, 139
  rotor, 360
Power loss, 183
Prime mover, 19, 135
Principle, generator, 23
  induction-motor, 346
  motor, 23
  repulsion, 433
  synchronous-motor, 459
Pyranol, 430

R

Rating, continuous-duty, 196
  dynamo, 183, 195
  intermittent-duty, 196
Ratio, current, 279
  induced-voltage, 279
  transformation, 279
  turn, 279
  voltage, 279
Reactance, armature, 244, 248, 405
  equivalent, 291
  induction-motor, 367
  leakage, 287
  synchronous, 253
  transformer, 290
Reactance coil, 198
Reactance voltage, 121, 243
Reaction, armature (*see* Armature reaction)
Rectifiers, 515
  arc-back in, 535
  backfire in, 535
  copper-oxide, 518
  electrolytic, 516
  full-wave, 516
  half-wave, 516
  hot-cathode gaseous, 521
  ignitron, 535
    polyphase connection of, 536

Rectifiers, mechanical, 32, 516
  mercury-arc, 524
    cathode spot in, 526
    multianode, 528
    operation of, 525
    single-anode, 524
    six-anode, 532
    three-anode, 529
    three-phase, 528
    two-anode, 524
    wave shape in, 527
  Rectigon, 516, 521
    ballast resistor in, 522
  selenium, 520
  silicon-type, 516
  three-phase connection of, 521
  tube drop in, 522
  Tungar, 516, 521
    ballast resistor in, 522
Rectigon, 516, 521
Regulation, alternator, 243, 245
  generator, 103
  speed, 163
  transformer, 286, 293, 296, 299
  voltage, 103
Regulation curves, 106, 244
Regulator, Diactor, 104
  induction voltage, 336
  Tirrell, 104
Relay, electromagnetic, 423
Residual magnetism, 97
Resistance, armature, 187, 244, 246
  equivalent, 291
  induction-motor, 367
  interpole-winding, 188
  series-field, 188
  shunt-field, 188
  transformer, 290
Reversing, armature, 175, 176
  field, 175, 176
  motor, 175
Revolving field, 232, 354
Rigging, brush, 45, 51
Right-hand rule, 28
Rotary converters, 1, 481
Rotation, direction of, 175
Rotor, 45, 350
  double-cage, 370
  skewed, 350
  squirrel-cage, 370
  wound, 16, 379

Rotor frequency, 359
Rotor torque, 362

S

Saturation curve, 98, 105, 255
Sectionalizing, field, 497
Segment, commutator, 49
Selection, dynamo, 196
Self-inductance, 272
Series-field diverter, 109
Shaftless motor, 383
Shunt generators (see Generators, shunt)
Skewing, rotor, 350, 432
Slip, 358
Sludging, oil, 274
Span, coil, 55
Speed, rotor, 358
   slip, 358
   synchronous, 16, 225, 253, 357
Speed control, 154, 387, 391
   direct-current motors, 165
   frequency converter, 392
   induction motors, 380, 387, 388
Speed regulation, 163
Spider, 49
Squirrel cage, 15, 370, 380, 382, 422
Star connection, 237, 352
Starters, automatic, 149, 177
   counter-emf, 150
   current-limit, 151
   four-point, 145
   motor, 142
   three-point, 143
   time-limit, 150, 177, 209, 212
Starting, autotransformer, 372
   compensator, 372
   direct-current motor, 140
   full-voltage, 370
   induction motor, 368, 370
   line-resistance, 374
   line-voltage, 371
   open-Δ, 373
   part-winding, 377, 380
   reduced-voltage, 372, 374
   synchronous motor, 453
   wound-rotor, 17, 375
   Y-Δ, 17, 375, 377
Starting torque, 159, 196, 364, 421
Stator, 3, 45, 232, 348
Stray-power loss, test for, 189

Switch, centrifugal, 422, 423, **432**
Synchronizing, 259
Synchronous capacitor, 466, 467
Synchronous condenser, 466, 467
Synchronous converters (see Converters, synchronous)
Synchronous impedance, 253
Synchronous motor (see Motors, synchronous)
Synchronous reactance, 253
Synchronous speed, 16, 225, 253, 357
Synchronous-synchronous frequency converter, 511
System, constant-current booster, 207
   negative feeder-booster, 206
   series booster, 205
   series-parallel, 214

T

Test, blocked-rotor, 366
   load, 101, 364
   no-load, 96, 364
   open-circuit, 255, 297
   resistance, 254
   short-circuit, 255, 293
   stator-resistance, 364
   stray-power loss, 189
Third-brush generators, 198
Thury system, 112
Tirrell regulator, 104
Torque, 35, 154
   d-c motor, 37
   rotor, 362
   starting, 159, 196, 364, 421
Transformation ratio, 279
Transformer circuit, 293
Transformer voltage, 268
Transformers, 18, 268
   cases for, 274
   constant-current, 334
   construction of, 269
   converter, 497
   copper losses in, 301
   core losses in, 271, 297, 301
   core type, 270
   current, 309
   current ratio in, 279
   Δ-Δ connections of, 322
   Δ-Y connections of, 322
   distribution, 281, 284

INDEX

Transformers, eddy-current loss in, 297, 301
  efficiency of, 301
    all-day, 304
    maximum, 303
  equation of, 275
  equivalent circuit of, 293
  hysteresis loss in, 297, 301
  impedance of, 290
  instrument, 309
  insulation of, 274
  laminating of, 298
  leakage flux in, 272
  load operation of, 318
  loading of, 283
  losses in, 297
  main, 328
  mutual inductance in, 272
  no-load operation of, 318
  open-circuit test of, 293
  open-delta connection of, 326
    power in, 327
  parallel operation of, 316
  polarity of, 313
    additive, 314
    subtractive, 314
  pole-type, 284
  potential, 309, 312
  power, 19, 281
  primary of, 269
  ratio of transformation of, 279
  reactance of, 290
  regulation of, 286, 293, 296, 299
  resistance of, 290
  Scott connection of, 328
  secondary of, 269
  self-inductance in, 272
  shell type, 271, 312, 332
  short-circuit test of, 293
  sludging of, 274
  static, 268
  step-down, 19, 279
  step-up, 19, 279
  T-T connections of, 328
  tapping of, 281
  teaser, 328
  three-phase, 331
  three-phase connections of, 321
  V-V connections of, 325
  voltage ratio in, 279
  voltages in, 275

Transformers, Y-Δ connections of, 323
  Y-Y connections of, 321
  zigzag connection of, 500
Tungar, 516, 521

V

Voltage, alternator, 242
  direction of, 27
  generated, 233, 359
  generation of, 137
  induced, 288
  reactance, 121, 243
  rotor, 359
Voltage drop, armature-reactance, 247
  armature-reaction, 248
  armature-resistance, 246
Voltage equation, 26

W

Ward Leonard control, 166, 169
Windings, armature, 1, 6, 11, 45, 51
  circulating currents in, 66
  coil span in, 55
  compensating, 119, 174
  concentric, 419
  consequent-pole, 389
  dead element in, 76
  doubly reentrant, 58
  duplex-lap, 58
  fractional-pitch, 239
  frog-leg, 69, 80
  full-pitch, 239
  half-coiled, 235
  induction-motor, 351
  interpole, 118
  lap, 53, 58, 351
  multielement, 72
  multiplex-lap, 59
  multiplex-wave, 68
  parallel paths in, 51, 59, 65
  simplex-lap, 53, 59, 73
  simplex-wave, 51, 53, 61, 65, 67, 73
  singly reentrant, 58
  stator, 351
  two-speed, 391
  types of, 53
  wave, 51, 53
  whole-coiled, 236, 351
Wound rotor, 16, 380

ANSWERS

Chapter 2

1. (a) 13.05 volts, 16 amp, 208.8 watts; (b) 5.8 volts, 36 amp, 208.8 watts.
2. 1 volt, 0.5 volt, 2 volts. **3.** 250 volts. **4.** 600 rpm. **5.** 230 volts. **6.** 115 volts.
7. (a) 270; (b) 6. **8.** 540, 12. **9.** (a) 225 volts; (b) 2.056×10^6 maxwells.
10. (a) 45 cps; (b) 50 cps; (c) 41.7 cps. **11.** Six poles. **12.** 500 rpm. **13.** 2.48 lb.
14. 0.93 lb-ft. **15.** (a) 146 lb; (b) 44.1 lb-ft. **16.** 67 lb-ft. **17.** 480 amp.
18. 570 amp.

Chapter 3

1. (a) $Y_s = 9$; (b) $Y_c = 14$; (c) $Y_s = 12$; (d) $Y_s = 16$; (e) $Y_s = 27$; (f) $Y_s = 19$.
2. (a), (b), and (c) = 2; (d) = 8; (e) = 10; (f) = 12. **3.** (a) 53.3 amp; (b) 175 amp.
4. (a) 8; (b) 18; (c) 16; (d) 30; (e) 24. **6.** (a) Double; (b) single; (c) triple; (d) single;
(e) double. **7.** (a) 90 amp; (b) 180 amp by each brush arm, 90 amp by each path.
8. (a) $Y_c = 37$ or 38; (b) $Y_c = 46$ or 47; (c) $Y_c = 76$; (d) $Y_c = 57$.
9. Proof of Prob. 8. **10.** (a) 4; (b) 6; (c) 2; (d) 4; (e) 6; (f) 8. **11.** 110 amp, 165 amp.
12. (a) No; (b) yes; (c) no; (d) yes. **13.** (a) $Y_c = 151$; (d) $Y_c = 91$.
14. Proof of Prob. 13. **15.** $Y_s = 4$, $Y_c = 16$, yes. **16.** (a) No; (b) yes; (c) yes;
(d) no. **17.** (a) 6; (b) 3; (c) 1; (d) 83.3 amp; (e) 166.7 amp. **18.** 240 volts.
19. 2. **20.** (a) 108; (b) 1 and 109. **21.** 36. **23.** (a) 20; (b) 16.
24. $Y_s = 12$, $Y_c = 95$.

Chapter 4

1. 1,800. **2.** 1,700 amp-turns. **4.** 12.5 per cent. **5.** 265 volts.
6. 130.5 volts, 89.5 volts. **7.** 242.8 volts. **8.** 250 volts, 2.08 ohms.
9. 50 ohms, 42 ohms, 68 ohms. **10.** 612.5 amp. **11.** 600 amp. **12.** 9,530 watts.
13. 187 and 3,009 amp-turns. **14.** 0.06 ohm. **15.** 1,037 amp-turns.
16. 133.8 volts. **17.** 28.75 kw. **18.** 625.4 volts, 317.7 kw. **19.** (a) 12,242 watts;
(b) 4,489 watts; (c) 194 watts; (d) 5,226 watts; (e) 4,800 watts. **20.** (a) 0.113 ohm;
(b) 2,300 amp-turns at no load, 2,525 amp-turns at full load. **21.** 8.
22. (a) 230 volts, 6.9 kw; (b) 16.9 kw.

Chapter 5

1. 218.8 volts. **2.** 27.6 amp. **3.** 12.5 volts. **4.** (a) 585 volts; (b) 667 amp.
5. 452 rpm. **6.** 13.1 per cent. **7.** (a) 522.1; (b) 7,730 watts, 10.36 hp. **8.** 10.4 hp.
9. 9.2 ohms. **10.** 3.07 ohms. **11.** 908 amp. **12.** 331 amp. **13.** 70.4 lb-ft.
14. 26 hp. **15.** (a) 1,685 rpm; (b) 6.83 per cent. **16.** 3.14 ohms. **17.** 1,435 rpm.
18. 113.8 amp. **19.** 888 rpm. **20.** 4.72 ohms. **21.** 295 lb-ft. **22.** (a) 2,250 rpm;
(b) 18.4 per cent. **23.** 18.8 hp. **24.** 56.7 hp. **25.** (a) 2,990 watts, 7.17 per cent;
(b) 1,770 watts, 530 watts; (c) 4.05 per cent, 1.21 per cent. **26.** 480 watts.

Chapter 6

1. 18,650 watts. **2.** 165 kw, 15 kw. **3.** 1,200 watts. **4.** 815 watts, 745 watts.
6. 1,565 watts, 85.6 per cent. **7.** 75 hp. **8.** (a) 87.2 per cent; (b) 14,100 watts;
(c) 5,800 watts, 8,300 watts. **9.** 83.8 per cent. **10.** (a) 61.9 amp; (b) 59.9 amp;
(c) 13.15 kw; (d) 85.2 per cent. **11.** 47.5 amp. **12.** 87.5 per cent. **13.** 83.6 per cent.
14. (a) 47.4 amp, 47 amp; (b) 11,750 watts; (c) 90.3 per cent. **16.** (a) 1.93 kw;
(b) 386 kwh; (c) $5.79. **17.** $19.80. **18.** 64 hp. **19.** (a) 125 kw; (b) 587 amp;
(c) 500 amp; (d) 87 amp toward the balance coil. **20.** $I_L = 543.5$ amp, $I_N = 0$.
21. $\text{kw}_{236} = 40$, $\text{kw}_{115+} = 80$, $\text{kw}_{115-} = 40$. **23.** 5.18 amp, 24.8 volts.
24. 1.19 ohms. **25.** 0.42 ohm.

Chapter 7

1. 60 cps. **2.** 800 rpm. **3.** 34 poles. **4.** 4.32 volts. **5.** 4.8 volts. **6.** 664 volts.
7. (a) 12; (b) 3. **8.** (a) 48; (b) 4. **9.** (a) $k_p = 0.978$; (b) $k_d = 0.955$.
10. (a) 1,270 volts per phase; (b) 2,200 volts between terminals. **11.** 1,805 volts.
12. 1,240 volts. **13.** 219 amp. **14.** 192.5 amp. **15.** 16.3. **16.** 9.77 per cent.
17. 12.8 per cent. **18.** $Z_S = 6.62$, $X_S = 6.6$. **19.** 49.6 per cent.
20. -14.3 per cent. **21.** 50.3 per cent. **22.** 1 per cent. **23.** 30 per cent.
24. 90.4 per cent. **25.** 29 kw. **26.** 13.1 kw. **27.** 86.4 per cent. **28.** (a) 917 amp;
(b) 231 amp; (c) 885 amp. **29.** 32.8°.

Chapter 8

1. (a) 7.2×10^5 maxwells; (b) 80,000 maxwells per sq in.; (c) 120 turns. **2.** 197.
3. (a) 6,600 volts, 440 volts; (b) 15:1. **4.** 2.085 amp, 41.7 amp.
5. (a) 2,100 turns, 70 turns; (b) 2.96×10^6 maxwells. **6.** 8 amp. **7.** (a) 4,420 volts;
(b) 19.4:1. **8.** 4,800/480 volts, $a = 10$; 4,800/240 volts, $a = 20$;
2,400/480 volts, $a = 5$; 2,400/240 volts, $a = 10$. **9.** 2,400–240/120 volts;
2,340–240/120 volts; 2,280–240/120 volts; 1,200–240/120 volts;
1,140–240/120 volts. **10.** 6,900, 30:1; 6,727, 29.2:1; 6,555, 28.5:1; 6,382, 27.7:1;
6,210, 27.1:1. **11.** 3.54 per cent. **12.** (a) 236 volts; (b) 9.75:1.
13. (a) $R_e = 0.80, 0.008$; (b) $X_e = 1.4, 0.014$; (c) $Z_e = 1.61, 0.0161$.
14. (a) $IR_e = 33, 3.3$; (b) $IX_e = 58, 5.8$. **15.** 1.5 per cent. **16.** 2.5 per cent.
17. -0.4 per cent. **18.** (a) 10.2, 31.6, 29.9; (b) 5.22 per cent. **19.** (a) 1.25, 4, 3.8;
(b) 2.82 per cent. **20.** (a) 123 watts; (b) 266 watts. **21.** (a) 62.5 kva; (b) 44 kw.
22. (a) 280 watts; (b) 303 watts; (c) 256 watts. **23.** (a) 515 watts; (b) 612 watts;
(c) 667 watts. **24.** $P_e = 48$ watts, $P_h = 112$ watts. **25.** (a) 3.78, 1.18, 3.6;
(b) 3.404 per cent. **26.** 97.9 per cent. **27.** 98.04 per cent. **28.** 97.9 per cent.
29. 98.05 per cent. **30.** (a) 2.62 per cent; (b) 96.46 per cent; (c) 20.3 kva;
(d) 97.02 per cent. **31.** 96.09 per cent. **32.** (a) 13.6 kw; (b) 18.4 kw.
33. (a) 10 amp; (b) 7.4 amp. **34.** $I_P = 50.25$ amp, $I_S = 4.55$ amp, 10 kva.
35. (a) Diagram; (b) 30 kva. **36.** 6, 12, 18, 24, 84, 90, 96, 108, and 114 volts.
37. 55.4 amp. **38.** (a) 80:5; (b) 16. **39.** 2,360 volts. **40.** (a) 127.6 volts;
(b) 104.4 volts. **41.** 15.4 amp. **42.** 11.58 kva, 8.42 kva. **43.** $\text{kva}_1 = 6.5$, $\text{kva}_2 = 11$.
44. $\text{kva}_1 = 81.5$, $\text{kva}_2 = 43.5$. **45.** 40,000 volts. **46.** 2,540 volts. **47.** (a) 4.97:1;
(b) 250 kva, 200 kw; (c) 942 amp; (d) 32.8 amp; (e) 942 amp; (f) 18.95 amp.
48. (a) 75 kva; (b) 43.5 kva. **49.** (a) 75.5 kva; (b) 34.7 kva. **50.** (a) $\text{kva}_t = 50$;
(b) $P_1 = 49.65$ kw, $P_2 = 19.65$ kw; (c) 150 kva. **51.** (a) 65.3 amp; (b) 188.5 amp;
(c) 75 kva. **52.** (a) 50.3 amp; (b) 201 amp; (c) 23.1 kva.
53. 0–55 volts; 0–110 volts; 165–275 volts; 110–330 volts.

ANSWERS 563

Chapter 9

1. 468 rpm. **2.** (a) 0.04; (b) 3.36 per cent. **3.** 12 poles. **4.** (a) 54 rpm; (b) 720 rpm.
5. (a) 299 volts, 60 cps; (b) 11.96 volts, 2.4 cps; (c) 500 volts, 100 cps. **6.** (a) 0.078;
(b) 7.83 per cent; (c) 0.856; (d) 158 lb-ft; (e) 89.7 per cent. **7.** (a) 82.4 amp;
(b) 15.4 amp. **8.** 3.87 ohms. **9.** (a) 22.2 volts; (b) 1,164 rpm, 1.8 cps.
10. (a) 275 watts; (b) 115 watts; (c) 3,260 watts; (d) 163 watts; (e) 3,097 watts;
(f) 4.15 hp; (g) 12.75 lb-ft; (h) 84.8 per cent; (i) 0.85. **11.** (a) 0.627 ohm;
(b) 0.327 ohm; (c) 0.7 ohm. **12.** 957 rpm. **13.** (a) 26.1 lb-ft; (b) 6.52 lb-ft.
14. 21.9 hp, 67.2 lb-ft. **15.** (a) 185 watts; (b) 111 watts; (c) 8,054 watts;
(d) 403 watts; (e) 9.95 hp; (f) 46 lb-ft; (g) 89.1 per cent; (h) 0.912. **16.** 15.6 lb-ft.
17. 200 volts. **18.** 31.3 amp. **19.** (a) $0.583T_{FL}$, $4.62I_{FL}$; (b) $2.67I_{FL}$.
22. 720 rpm, 514 rpm, 300 rpm. **23.** (a) Six and eight poles; (b) 12 and 16 poles.
25. 183 volts. **26.** 1,125 rpm. **27.** (a) 135 lb-ft; (b) 205 lb-ft.

Chapter 10

1. (a) and (b) = fractional; (c) and (d) = integral. **2.** 0.587. **3.** 311 watts.
4. Per cent regulation = 6.7, per cent slip = 6.67. **5.** (a) 0.754 lb-ft; (b) 143.5 oz-in.
6. 31 per cent. **7.** 40 per cent.
8. Use 353 turns per coil, with wire three gage numbers larger.
11. 5.2 lb-ft, 1,000 oz-in. **12.** (a) Six poles; (b) two poles; (c) six poles.
13. (a) 4 per cent; (b) 3.33 per cent; (c) 5.83 per cent. **14.** 185 µf. **15.** 225 rpm.
17. (a) 6; (b) 0.684. **19.** (a) 60 per cent; (b) 0.71; (c) 9.13 lb-in.
20. 60 per cent, 45.5 per cent.

Chapter 11

1. 600 rpm, 300 rpm, 720 rpm. **2.** 36 poles. **3.** 24 poles.
4. 10 and 12 poles, 600 rpm. **5.** (a) 41 kw; (b) 51.3 kva. **6.** 7.5 amp.
7. 274 lb-ft. **8.** 65 per cent tap, 1,430 volts. **9.** 51.8 per cent.
10. 438 amp, 0.032. **11.** 1,825 kva. **12.** 1,160 kva. **13.** (a) 0.891; (b) 5,610 kva.
14. 276 kva, 0.797 leading. **15.** (a) 2,980 kva; (b) 0.918; (c) 0.867 leading.
16. 0.637 leading, 706-kva input. **17.** 1,500 kva, 0.277 leading. **18.** (a) 2,660 kw;
(b) 2,800 kva; (c) 0.95. **19.** 985 kw. **20.** 3,000 hp.

Chapter 12

1. 177 volts. **2.** (a) 311 volts, 6.43 amp; (b) 9.1 amp. **3.** 11.78 amp. **4.** (a) 169.5
volts; (b) 53.3 amp; (c) 34.4 amp. **5.** (a) 9; (b) 24. **6.** (a) 200 amp; (b) 367 volts;
(c) 212 amp; (d) 44.8 kva. **7.** (a) 1,667 amp; (b) 1,745 amp; (c) 369 kva. **8.** 338 kw.
9. (a) 471.5 amp; (b) 236 amp; (c) 118 amp. **10.** (a) Four poles; (b) 83.3 amp;
(c) 42.4 volts; (d) 42.8 amp. **11.** 3.62 kva. **12.** (b) 4,000 amp; (c) 265 volts;
(d) 530 volts; (e) 2,070 amp; (f) 43.4:1. **13.** (a) 275 volts; (b) 2,670 amp; (c) 652 amp.
14. 241 amp. **15.** 0.865. **16.** 0.834. **17.** (a) 21.7 amp; (b) 37.6 amp.
18. (a) 600 volts; (b) 212 volts; (c) 4,600 volts. **19.** 21.2 volts. **20.** 21.2 volts.
21. (a) 12 poles and 10 poles; (b) 600 rpm. **22.** (a) 1,200 rpm; (b) 1,000 rpm;
(c) 2,400 rpm; (d) 400 rpm. For (a), (b), and (c) the direction of rotation should be
opposite to the revolving-field direction; for (d) it should be in the same direction.
23. (a) 240 volts; (b) 200 volts; (c) 360 volts; (d) 80 volts. **24.** 7.87 amp, 36.5 volts.
25. 287 va. **26.** 5.34 amp. **27.** (a) Drawing; (b) 12.55 amp; (c) 54.2 volts.
28. (a) 7.9 volts; (b) 8.5 amp. **29.** 35.3 amp. **30.** 289 volts. **31.** (a) 188 amp;
(b) 510 amp; (c) 600 amp. **32.** (a) 36.8 amp; (b) 144.3 volts; (c) 1.63 amp; (d) 7.5 kva;
(e) 10.62 kva. **33.** (a) 1,087 amp; (b) 213.5 volts, 134 kva; (c) 29.1 amp, 402 kva.
34. (a) 1,230 volts, 2,185 amp, 2,690 kw; (b) 896 amp.